BREAKDOWN IN HUMAN ADAPTATION TO 'STRESS' VOLUME I

BREAKDOWN IN HUMAN ADAPTATION TO 'STRESS'
Towards a multidisciplinary approach

VOLUME I

Part 1: Psychological and sociological parameters for studies of breakdown in human adaptation
J. Cullen and J. Siegrist (editors)

Part 2: Human performance and breakdown in adaptation
H.M. Wegmann (editor)

VOLUME II

Part 3: Psychoneuroimmunology and breakdown in adaptation: interactions within the central nervous system, the immune and endocrine systems
R.E. Ballieux (editor)

Part 4: Breakdown in human adaptation and gastrointestinal dysfunction: clinical, biochemical and psychobiological aspects
J.F. Fielding (editor)

Part 5: Acute effect of psychological stress on the cardiovascular system: Models and clinical assessment
A. L'Abbate (editor)

Compendium of papers presented in workshops sponsored by the Commission of the European Communities as advised by the Committee on Medical and Public Health Research. (Dublin, Ireland (Dec. 1982), Utrecht, the Netherlands (Dec. 1982), Köln, FRG (Jan. 1983), Bad Homburg, FRG (Feb. 1983), Dublin, Ireland (March 1983) and Pisa, Italy (April, 1983))

BREAKDOWN IN HUMAN ADAPTATION TO 'STRESS'
Towards a multidisciplinary approach

VOLUME I
Part 1: Psychological and sociological parameters for studies of breakdown in human adaptation
  J. Cullen and J. Siegrist (editors)
  Health Care and Psychosomatic Unit, Dublin, Ireland
  Dept. of Medical Sociology, University of Marburg, Marburg, Federal Republic of Germany

Part 2: Human performance and breakdown in adaptation
  H.M. Wegmann (editor)
  Institute of Aerospace Medicine, Cologne, Federal Republic of Germany

1984 **MARTINUS NIJHOFF PUBLISHERS**
a member of the KLUWER ACADEMIC PUBLISHERS GROUP
BOSTON / THE HAGUE / DORDRECHT / LANCASTER
for
THE COMMISSION OF THE EUROPEAN COMMUNITIES

Distributors

for the United States and Canada: Kluwer Boston, Inc., 190 Old Derby Street, Hingham, MA 02043, USA
for all other countries: Kluwer Academic Publishers Group, Distribution Center, P.O.Box 322, 3300 AH Dordrecht, The Netherlands

Library of Congress Cataloging in Publication Data

```
Main entry under title:

Breakdown in human adaptation to "stress."

     "Compendium of papers presented in workshops
sponsored by the Commission of the European Communities
as advised by the Committee on Medical and Public Health
Research.(Dublin, Ireland (Dec. 1982), Utrecht, the
Netherlands (Dec. 1982), Köln, FRG (Jan. 1983), Bad
Homberg, FRG (Feb. 1983), Dublin, Ireland (March 1983),
and Pisa, Italy (April 1983))"
     Vol. 2.edited by R.E. Ballieux, J.F. Fielding, and
A. L'Abbate.
     Contents: v. 1. pt. 1. Psychological and sociological
parameters for studies of breakdown in human adaptation /
J. Cullen and J. Siegrist, editors. pt. 2. Human perform-
ance and breakdown in adaptation / H.M. Wegmann, editor
-- v. 2. pt. 3. Psychoneuroimmunology and breakdown in
adaptation / R.E. Ballieux, editor -- [etc.]
     1. Medicine, Psychosomatic--Addresses, essays, lectures.
2. Adaptation (Physiology)--Addresses, essays, lectures.
3. Stress (Psychology)--Addresses, essays, lectures.
I. Cullen, John H.  II. Ballieux, R.E.  III. Commission
of the European Communities.  IV. Commission of the
European Communities. Committee on Medical and Public
Health Research. [DNLM: 1. Stress, Psychological--
Complications--Congresses. 2. Adaptation, Psychological
--Congresses. 3. Disease--Etiology--Congresses. WM 172
B828 1982-82]
RC49.B727  1983          616.08         83-19333
ISBN 0-89838-608-X (set)
```

ISBN 0-89838-608-X (set)
ISBN 0-89838-606-3 (v.I)
ISBN 0-89838-607-1 (v.II)
EUR 8943 I EN

Book information

Publication arranged by: Commission of the European Communities, Directorate-General Information Market and Innovation, Luxembourg

Copyright/legal notice

© 1984 by ECSC, EEC, EAEC, Brussels-Luxembourg.

All rights reserved. No part of this publication may be reproduced, stored in a retrieval system, or transmitted in any form or by any means, mechanical, photocopying, recording, or otherwise, without prior written permission of the publishers,
Martinus Nijhoff Publishers, 190 Old Derby Street, Hingham, MA 02043, USA.

Neither the Commission of the European Communities nor any person acting on behalf of the Commission is responsible for the use which might be made of the following information.

PRINTED IN THE NETHERLANDS

PREFACE

The widespread interest in "stressful" aspects of contemporary society which contribute to its burden of illness and diseases (e.g. gastro intestinal, cardiovascular) has led to a large number of statements and reports which relate the manifestations to a maladaptation of the individual. Furthermore, recent research suggests that under some conditions stress may have a more generalized effect of decreasing the body's ability to combat destructive forces and expose it to a variety of diseases.

Breakdown in adaptation occurs when an individual cannot cope with demands inherent in his environment. These may be due to an excessive mental or physical load, including factors of a social or psychological nature and task performance requirements ranging from those which are monotonous, simple and repetitive to complex, fast, decision-taking ones. Experience shows however that not all people placed under the same conditions suffer similarly, and it follows that to the social and psychological environment should be added a genetic factor influencing, through the brain, the responses of individuals.

It is clear that, besides human suffering, this "breakdown in adaptation" causes massive losses of revenue to industry and national health authorities. Thus a reduction in "stress", before "breakdown" occurs, or an improvement in coping with it would be very valuable.

An area of the Commission of the European Communities multiannual coordination programme in Medical and Public Health research considers "breakdown in human adaptation" occurring when facing the industrial life style in countries of the European Community or a too rapid change of the urban/industrial environment. The problem is of great economic importance from the points of view of industrial productivity, movement of workers and health care of all the population.

The principal aim of the programme is to determine a series of measurable parameters which could indicate, first <u>a-posteriori</u> but then <u>a-priori</u> the tendency or initiation of breakdown in adaptation and the form it may take if no counter measure was introduced. Such a programme, it is hoped, could contribute to a wider integration of European know-how, promote a better understanding of biological parameters underlying organic correlates of stress, and lead eventually to applied routine procedures. Harmonized and standardized measurements would give values for parameters from various laboratories in the European Community; these will be compared, integrated in a wider context and, results fed back at regular intervals to scientists taking part in the exercise for discussion and interpretation.

The broad outline of this programme calls for studies to:

1.- Investigate, simultaneously if possible, the sociological, psychological and endocrinological factors which may indicate "high risk groups" among the population,

2.- Develop reliable but relatively simple indicators which will allow the identification of persons likely to suffer a breakdown in adaptation,

3.- Establish the mechanisms by which certain factors lead to a higher risk in adaptive breakdown and the manifestations of performance, cardiovascular and gastro intestinal dysfunction,

4.- Use the information gathered to establish situations which will avoid or minimize breakdown in adaptation.

It is evident that to be successful, the combined expertise of researchers in biology, biochemistry, physiology, sociology, psychology, epidemiology and clinical sciences would be required. Through coordinated studies, these scientists could adapt and compare methods, follow results and test hypotheses since standardized or at least harmonized protocols would have been used by the various disciplines. This coordination will, it is hoped, allow the identification of real gaps and make specific recommendations as research develops to achieve the aims stated.

In this context one should mention studies to determine whether or not different categories or causes of adaptive breakdown are selectively correlated with pathogenic tendencies at the level of specific organs. The answer to this question would greatly facilitate the assessment of the relevancy of specific versus non specific organic markers of potential diseases.

Finally, because the programme would be directed towards one single but crucial goal with a well coordinated multinational and interdisciplinary approach, it is hoped to establish firm, productive and long lasting links between scientists, in Europe and elsewhere.

The publications in this series stem from a number of workshops held in 1982-1983 under the auspice of the Commission of the European Communities in the framework of the above mentioned programme, as advised by the Committee for Medical and Public Health Research.

These are not proceedings, but a compendium of recent overviews, in some cases examples of methods, and reports on specific topics, or approaches used or envisaged in the various disciplines where there could be fruitful collaboration in the overall study of breakdown in human adaptation to some stressful conditions. The purpose is to help identify important problems, indicate interdisciplinary research possibilities offered and through the references quoted make available sources of further information on methods and aspects outside the normal specific field of the scientist concerned.

The separation into five parts each dealing more specifically with one aspect is only done for the ease of the reader and the publication should be considered as a whole.

The Commission of the European Communities and its Committee for Medical and Public Health Research wish to thank all those who contributed through their participation in the elaboration of this programme and the organization of the meetings.

TABLE OF CONTENTS

VOLUME I

PART 1 – PSYCHOLOGICAL AND SOCIOLOGICAL PARAMETERS FOR STUDIES OF BREAKDOWN IN HUMAN ADAPTATION

I. GENERAL OVERVIEWS

Towards a taxonomy of methods: a general overview of psychological approaches in the study of breakdown of human adaptation
J. Cullen   3

Psychological field study techniques: overview and needs
D. Broadbent   38

Psychological field study techniques: a critical evaluation
Katherine R. Parkes   53

Sociological parameters in studies of breakdown: a selective overview
J. Siegrist, Karin Siegrist and I. Weber   61

Use of psychological indices in epidemiological studies: overview and needs
C.D. Jenkins   83

Stressful life events and illness: a review with special reference to a criticism of the life-event method
J. Aagaard   98

II. CONCEPTUAL APPROACHES

A lifetime prospective study of human adaptation and health
M.E.J. Wadsworth   122

Psychosocial and psychophysiological factors in the design and the evaluation of working conditions within health care systems
T. Cox   135

The relation of social to pathophysiological processes: evidence from epidemiological studies
M.G. Marmot   144

Unemployment and health: a review of methodology
S.C. Farrow   149

Ontogenetic development and breakdown in adaptation: a review on psychosocial factors contributing to the development of myocardial infarction, and a description of a research program
P. Falger, A. Appels and R. Lulofs   159

Physiological issues in establishing links between psychosocial factors and cardiovascular illness
T. Theorell   188

White collar occupation and coronary prone behaviour
J. Siegrist   198

III. METHODS

Psychological methods: an overview of clinical applications
A. Steptoe   208

Psychological factors in the breakdown of human adaptation: some methodological issues
C. Mackay   219

Monitoring signs of decrease in human adaptation: use of quantitative measures available in official statistics
T. Theorell   241

Inventory of stressful life-events (ILE)
J. Siegrist and K.H. Dittmann   251

The Norwegian female climacteric project (VOS)
Aslaug Mikkelsen and A. Holte   257

Questionnaire for organisational stress (VOS)
N. van Dijkhuizen   262

A scale for measuring the marital relationship among males
M. Waltz   267

PART 2 — HUMAN PERFORMANCE AND BREAKDOWN IN ADAPTATION

Human performance in transport operations: introductory remarks
H.M. Wegmann   275

I. AIR TRANSPORT

Air crew workload
S.R. Mohler and H.D. Nichamin — 279

Safety, individual performance and mental workload in air transport: Oedipus as Icarus
Patricia Shipley — 285

Stress management in air transport operations: beyond alcohol and drugs
F.H. Hawkins — 305

Reasons for eliminating the "age 60" regulation for airline pilots
S.R. Mohler — 322

Human factors education in European air transport operations
F.H. Hawkins — 329

II. ROAD TRANSPORT

Behaviour research in road traffic
J. Moraal and J.B.J. Riemersma — 363

Some theoretical considerations on accident research
M.L.I. Pokorny and D.H.J. Blom — 374

Accident of bus drivers – practical and methodological problems
D.H.J. Blom and M.L.I. Pokorny — 385

Effects of alcohol on driving performance: a critical look on the epidemiological, experimental and psychosocial approaches
E.A. Sand — 394

Investigations on the influence of continuous driving on the motion activity of vehicle drivers
M. Lemke — 409

III. SEA TRANSPORT

Human performance in seafaring
A. Low and H. Goethe — 422

Stress factors and countermeasures in navigation
G. Athanassenas — 449

Ship of the future: human problems and performance
H. Boehm 457

Accidents on board merchant ships
N. Rizzo and F. Amenta 469

Sleep data sampled from the crew of a merchant marine ship
P. Knauth, R. Condon, F. Klimmer, W.P. Colquhon, H. Hermann and J. Rutenfranz 484

IV; SPECIAL REVIEWS

Transport operators as responsible persons in stressful situations
P. Branton 494

Stress response as a function of age and sex
Joan Vernikos 509

Drugs and transport operations
A.N. Nicholson 522

Mechanical vibration in transport operations
L. Vogt 552

V. METHODS

Continuous electrophysiological recording
T. Akerstedt and L. Torsvall 567

Dimensions of flight crew performance decrements: methodological implications for field research
R.C. Graeber, H.C. Fouschee, J.K. Lauber 584

Methodology in workstress studies
G.C. Cesana and A. Grieco 606

VOLUME II

PART 3 — PSYCHONEUROIMMUNOLOGY AND BREAKDOWN IN ADAPTATION: INTERACTIONS WITHIN THE CENTRAL NERVOUS SYSTEM, THE IMMUNE AND ENDOCRINE SYSTEMS

Immunology for nonimmunologists: some guidelines for incipient psychoneuroimmunologists
N. Cohen 625

Neuroendocrine interactions with brain and behaviour: a model for psychoneuroimmunology ?
B. Bohus   638

Psychoneuroimmunology
R. Ader   653

Emotions, immunity and disease: an historical and philosophical perspective
G. F. Solomon   671

Immunoglobulins as stress markers ?
H. Ursin, R. Mykletun, E. Isaksen, R. Murison, R. Vaernes and O. Tønder   681

Problems of clinical interdisciplinary research - investigation into bronchial asthma as a paradigm
Margit von Kerekjarto   691

Factors involved in the classical conditioning of antibody responses in mice
R. Gorczynski, S. MacRae and M. Kennedy   704

The bone marrow, our autonomous morphostatic "brain"
W. Pierpaoli   713

Immune regulation of the hypothalamic - hypophysial - adrenal axis: a role for thymosins and lymphokines
N. HALL, J. McGillis, B. Spangelo, D. Healy and A. Goldstein   722

Stress and immune response: parameters and markers
R. Ballieux   732

PART 4 - BREAKDOWN IN HUMAN ADAPTATION AND GASTROINTESTINAL DYSFUNCTION: CLINICAL, BIOCHEMICAL AND PSYCHOBIOLOGICAL ASPECTS

The brain and the gut
F.P. Brooks   743

The role of psychiatric assessment in the management of functional bowel disease
D.H. Alpers   750

Application of psychological measures in epidemiological studies of gastrointestinal disease: a critical opinion
J. Fielding 761

Stress-related nicotine abuse and disorders of the gastrointestinal tract
J.R. Bennett 765

Use of quantitative methods for the study of psychological factors in ulcer patients
G.C. Lyketsos 782

Stress, the immune system and GI function
A.S. Peña 795

Clinical recognition of stress related gastrointestinal disorders in adults
J.F. Fielding 799

Stress and inflammatory bowel disease (IBD)
Gisela Huse-Kleinstoll, Th Küchler, A. Raedler, K.H. Schultz, ... 807

Upper GI bleeding lesions related to- or associated with- stress
M. Deltenre, A. Burette and M. de Reuck 819

PART 5 — ACUTE EFFECT OF PSYCHOLOGICAL STRESS ON CARDIOVASCULAR SYSTEM: MODELS AND CLINICAL ASSESSMENT

I. SYSTEMS INTERPLAY IN STRESS RESPONSE

Need for clinical models: physiopathological versus epidemiological study
L. Donato 833

Psychosocial stress: endocrine and brain interactions and their relevance for cardiovascular processes
B. Bohus and J.M. Koolhaas 843

Hormonal response to acute stress: focus on opioid peptides
R.E. Lang, K. Kraft, Th. Unger and D. Ganten 854

II. MYOCARDIAL INFARCTION

Clinical studies

Emotional stress and heart disease: clinical recognition and assessment
K. McIntyre and C. Jenkins — 861

Possibilities and limitations of longterm studies on the effect of psychological stress on cardiovascular function
A. Perski and T. Theorell — 884

Interaction between short- and long-term stress in cardiovascular disease
J. Siegrist — 892

Clinical clues of neuro-humoral interpretation of the genesis of coronary spasm
A. L'Abbate — 900

Provocative testing for coronary spasm
G. Specchia and S. de Servi — 916

Hemodynamic characterization of different mental stress tests
L. Tavazzi, G. Mazzuero, A. Giordano, A.M. Zotti, G. Berolotti — 923

Experimental studies

Thoracic autonomic nerves regulating the canine heart
J.A. Armour — 931

Nervous coronary constriction via α-adrenoreceptors: counteracted by metabolic regulation, by coronary β-adrenoreceptor stimulation or by flow dependent, endothelium-mediated dilation
E. Bassenge, J. Holtz, R. Busse and M. Giesler — 949

III. CARDIAC ARRHYTHMIAS

Clinical studies

Clinical clues to psychological and neuro-humoral mechanisms of arrhythmogenesis
D.C. Russell — 961

Clinical clues and experimental evidence of the neuro-humoral interpretation of cardiac arrhythmias
P. J. Schwartz, E. Vanoli and A. Zaza — 975

IV. ARTERIAL HYPERTENSION

Clinical studies

Blood pressure control during mental stress
J. Conway, N. Boon, J. vann Jones and P. Sleight   98

Somatic responses to acute stress and the relevance for the study of their mechanisms
A.W. von Eiff, H. Neus and W. Schulte   99

Neurohumoral factors involved in the pathogenesis of hypertension
J.L. Elghozi, L.C. Jacomini, M.A. Devynck, L.A. Kamal, J.F. Cloix, M.G. Pernollet, H. de The and P. Meyer   100

Experimental studies

Results of experimental studies favouring the hypothesis of the influence of stress on the genesis of hypertension
G. Mancia, A. Ramirez, G. Bertinieri, G. Parati, A. Zanchetti   101

Animal models for the assesment of stress on arterial blood pressure
D.T. Greenwood, P.W. Marshall and C.P. Allott   102

V. METHODS

Validation and quantification of mental stress tests, and their application to acute cardiovascular patients
A. Steptoe   103

Methods and limits for the detection of the response of coronary circulation to acute stress
A. Biaggini   104

LIST OF CONTRIBUTORS

J. Aagard,
Institute of psychiatric demography
Aarhus psychiatric hospital,
DK-8420 Riskov, Denmark

A.L.'Abbate,
Fisiologia Clinica,Istituto del
Consiglio Nazionale delle Ricerche
presso l'Università degli Studi di
Pisa,
Via Savi 8, Pisa, Italia.

R. Ader,
University of Rochester,
300 Crittenden Bvard.,
Rochester, NY 14642, USA.

T. Akerstedt,
Laboratory for clinical stress
research,
National Institute for psychosocial
factors and health,
10401 Stockholm, Sweden.

C.P. Allott,
Bioscience Department, Imperial
Chemical Industries, Plc,
Pharmaceuticals Division,
Alderley Park, Macclesfield,
Cheshire, U.K.

D. Alpers,
Division of gastroentrology,
Washington University school of
Medecine,
660 S. Euclid Ave.
St. Louis, MO 63110, USA.

F. Amenta,
Centro internazionale radiomedico,
CIRM,
via dell'Archittetura 41,
I-00144 ROMA, Italy

A. Appels,
State University of Limburg,
School of Medecine, Dept. of medical
psychology,
MAASTRICHT, The Netherlands

J. Armour,
Dalhousie University, Department of
physiology and biophysics,
Halifax, Nuova Scotia B3H 4H7,
Canada

G. Athanassenas,
State mental hospital,Psychiatric
department III, DAPHNI-ATHENS,Greece

R. Ballieux,
Department of clinical immunology,
University hospital,
Catharijnesingel 101,
3511 GV Utrecht, The Netherlands

E. Bassenge,
Albert Ludwigs Universität,
Herman Herder Str. 7,
D-7800 Freiburg 1, FRG.

J. Bennett,
Hull Royal infirmary,
Anlaby Road,
Kingston upon Hull HU3 2JZ, U.K.

G. Bertinieri,
Università di Milano,
Istituto di clinica medica IV,
via F.Sforza 35,
20122 MILANO, Italy

G. Bertolotti,
Centro di riabilitazione di Veruno,
VERUNO (NO), Italy

A. Biaggini,
Institute of clinical physiology-CNR
Via Savi 8,
I-56100 PISA, Italy

D. Blom,
Netherlands institute for preventive
health care / TNO,
P.O.B. 124,
2300 AC Leiden, The Netherlands

H. Boehm,
Bremen polytechnic, Department of
nautical sciences,
Werderstr. 73,
2800 Bremen, FRG.

B. Bohus,
Department of animal physiology,
University of Groningen,
P.O.B. 14,
9750 AA Haren, The Netherlands.

N. Boon,
Cardiac department, John Radcliffe
Hospital,
Headington, Oxford, U.K.

P. Branton,
22 Kings Gardens,
London NW6 4PU, U.K.

D. Broadbent, Medical Research Council,
University of Oxford, Oxford, U.K.

F. Brooks,
University of Pennsylvania hospital,
Department of medecine.GI section,
574A Maloney Bdg.,3400 Spruce St.
Philadelphia, Pa 19104, USA

A. Burette
Université Libre de Bruxelles,
Hopital Brugmann, G.I. departement
Bruxelles, Belgium

R. Busse
Albert Ludwigs Universität,
Herman Herder Str. 7,
D-7800 Freiburg 1, FRG.

G. Cesana,
Istituto di medicina del lavoro,
Clinica del lavoro L.Devoto dell'
università,
Via S. Barnaba 8,
20122 Milano, Italy

J. Cloix,
INSERM, U 7, CNRS LA 318,
Faculté de médecine, departement
de pharmacologie,
Necker-Enfants malades,
156 rue de Vaugirard,
75015 Paris, France

N. Cohen,
Agricultural university, Department
of experimental animal morphology and
cell biology,
Wageningen, The Netherlands

W. Colquhon,
Medical Research Council perceptual and cognitive performance unit,
Laboratory of experimental psychology, University of Sussex,
Brighton, Sussex BN1 9QG, U.K.

R. Condon,
Medical Research Council perceptual
and cognitive performance unit,
Laboratory of experimental
psychology,
University of Sussex,
Brighton, Sussex BN1 9QG, U.K.

J. Conway,
Cardiac department, John Radcliffe
hospital,
Headington, Oxford, U.K.

T. Cox,University of Nottingham,
Department of psychology - stress
research, Nottingham NG7 2RD, U.K.

J. Cullen,
Health care and psychosomatic unit,
Garden Hill, E.H.B. Box 41A,
Dublin 8, Ireland

M. Deltenre,
Université Libre de Bruxelles,
Hopital Brugmann, GI Department
Bruxelles, Belgium

M. Devynck,
Inserm U7, CNRS LA 318, Faculté
de médecine Necker-Enfants malades
156 Rue de Vaugirard,
75015 Paris, France

N. van Dijkhuizen,
SWO/DPKM, Ministry of defence,
P.O.B. 20702,
2500 ES Den Haag, The Netherlands

K. Dittmann,
Institute of medical sociology,
Faculty of medecine,
University of Marburg,
Marburg, FRG.

L. Donato,
CNR Institute of clinical physiology and institute of pathology,
University of Pisa,
Via Savi 8,
56100 Pisa, Italy

A. von Eiff,
Medizinische Universitätsklinik,
Venusberg, D-5300 Bonn 1, FRG.

J. Elghozi,
INSERM U7, CNRS LA 318, Faculté
de médecine Necker-Enfants malades
156 Rue de Vaugirard,
75015 Paris, France

P. Falger,
State University of Limburg school
of medecine, Department of medical
psychology,
Maastricht, The Netherlands

S. Farrow,
Department of epidemiology and
community medecine, Welsh National
School of medecine, Heath Park,
Cardiff, Wales, U.K.

J. Fielding,
Department of medecine and gastroentorology, The Charitable Infirmary, Jervis St.,Dublin 1, Ireland

H. Foushee,
Man-vehicle systems research division, NASA, Ames Research center,
Moffett Field, California 94035 USA

D. Ganten,
German Institute for high blood pressure research and Department of Pharmacology, University of Heidelberg, Im Neuenheimer Feld 366
D-6900 Heidelberg, FRG.

M. Giesler,
Albert Ludwigs Universität,
Herman Herder Str. 7,
D-7800 Freiburg, FRG.

A. Giordano,
Centro di riabilitazione di Veruno
Fondazione clinica del lavoro
Veruno (NO), Italy

H. Goethe,
Bernhard -Nocht-Institute for nautical and tropical diseases, Department of nautical medecine,
2000 Hamburg, FRG.

A. Goldstein,
Department of biochemistry,
The George Washington University
College of health sciences,
Washington DC 20037, USA.

R. Gorczynski,
Ontario cancer institute,
500 Sherbourne St.
Toronto, Ontario M4X 1K9, Canada

R. Graeber,
Man-vehicle systems research division, NASA, Ames research center,
Moffett Field, California 94035 USA.

D.T. Greenwood,
Bioscience Department, Imperial Chemical Industries Plc, Pharmaceutical division, Alderley Park,
Macclesfield, Cheshire, U.K.

A. Grieco,
Istituto di medicina del lavoro,
Clinica del lavoro L. Devoto dell'
Università, via S. Barnaba 8,
20122 Milano, Italy

N.Hall,
Department of biochemistry, The George Washington University, college of health sciences,Washington DC 20037, USA.

F. Hawkins,
Schiphol airport,
P.O.B. 75577,
1118 ZP Amsterdam, The Netherlands

D. Healy,
Pregnancy research branch, National Institute of child health and development,
Bethesda, Maryland 20205, USA.

H. Hermann,
ESSO A.G.
Kapstadring 2,
D-2000 Hamburg 60, FRG.

A. Holte,
Institute of behavioural sciences in medecine, University of Oslo,
P.O.B. 1111, Blindern, Oslo 3,
Norway

J. Holtz,
Albert Ludwigs Universität,
Herman Herder Str. 7
D-7800 Freiburg 1, FRG.

G. Huse-Kleinstoll,
Department of Medical Psychology
Medical Clinic, University Hospital
Hamburg-Eppendorf, FRG.

E. Isaksen,
Institute of physiological psychology, University of Bergen,
Bergen, Norway

L. Jacomini,
INSERM U7, CNRS LA 318, Faculté de médecine Necker-Enfants malades,
156 Rue de Vaugirard,
75015 Paris, France

C. Jenkins,
Department of preventive medecine and community health, University of Texas medical branch,
Galveston, Texas 77550, USA.

L. Kamal,
INSERM U7, CNRS LA 318, Faculté de médecine Necker-Enfants malades,
156 Rue de Vaugirard,
75015 Paris, France

M. Kennedy,
Ontario cancer institute,
500 Sherbourne St.,
Toronto, Ontario M4X 1K9, Canada

Margit von Kerekjarto,
Universitäts-Krankenhaus Eppendorf
Medizinische Klinik,
Martinistr. 52,
2000 Hamburg, FRG.

F. Klimmer,
Institut für Arbeitsphysiologie an
der Universität Dortmund,
Ardeystr. 67,
D-4600 Dortmund 1, FRG.

P. Knauth
Institut für Arbeitsphysiologie an
der Universität Dortmund,
Ardeystr. 67,
D-4600 Dortmund 1, FRG

J. Koolhaas,
Department of animal physiology,
State University of Groningen,
P.O.B. 14,
9750 AA Haren, The Netherlands

K. Kraft,
German Institute for high blood pressure research and Department of
Pharmacology, Universität of
Heidelberg,
Im Neuenheimer Feld 366,
D-6900 Heidelberg, FRG.

Th. Küchler,
Department of medical psychology,
University hospital,
Hamburg-Eppendorf, FRG

R.E. Lang,
German Institute for high blood pressure research and Department of
Pharmacology, University of
Heidelberg,
Im Neuenheimer Feld 366,
D-6900 Heidelberg, FRG.

J. Lauber,
Man-vehicle systems research division, NASA, Ames research center,
Moffett Field, California 94035 USA

M. Lemke,
Volkswagenwerk AG, Nutzfahrzeug-Entwicklung,
D-3180 Wolfsburg, FRG.

A. Low,
Bernard Nocht Institute for nautical
and tropical diseases, Department of
nautical medecine,
2000 Hamburg, FRG.

R. Lulofs,
State University of Limburg school
of medecine, Department of medical
psychology,
Maastricht, The Netherlands

G. Lyketsos,
Dromokaiton mental hospital,
Athens, Greece

S. MacRae
Ontario cancer Institute,
500 Sherbourne St.,
Toronto, Ontario M4X 1K9 Canada

C. Mackay,
Medical Division, Health and safety
executive,
25 Chapel St.,
London NW1 5DT, U.K.

G. Mancia,
Università di Milano, ospedale
Maggiore, Istituto di clinica
medica IV, via F. Sforza 35,
20122 Milano, Italy

M. Marmot,
London School of Hygiene and tropical medecine,
Keppel St.,
London WC1 7HT, U.K.

D.W. Marshall,
Bioscience Department, Imperial
Chemical Industries Plc, Pharmaceuticals Division, Alderley Park,
Macclesfield, Cheshire, U.K.

G. Mazzuero,
Centro di riabilitazione di Veruno,
Fondazione clinica del lavoro,
Veruno (NO), Italy

J. McGillis,
Department of biochemistry, The
George Washington University
college of health sciences,
Washington DC 20037, USA.

K. McIntyre,
Harvard University School of Public
health,
Boston, Massachusetts 02115, USA

P. Meyer,
INSERM U7, CNRS LA 318, Faculté de
Médecine Necker-Enfants malades,
156 Rue de Vaugirard,
75015 Paris, France

Aslaug Mikkelsen,
Institute of behavioural sciences in medecine, University of Oslo,
P.O.B. 1111, Blindern,
Oslo, Norway

S. Mohler,
Wright State University School of Medecine,
Dayton, Ohio 45401, USA.

J. Moraal,
Institute for perception TNO,
3769 ZG Soesterberg, The Netherlands

R. Murison,
Institute of physiological psychology, University of Bergen,
Bergen, Norway

R. Mykletun,
Rogoland Research Institute,
Stavanger, Norway

H. Neus,
Medizinische Universitätsklinik,
Venusberg, D-5300 Bonn 1, FRG.

H. Nichamin,
Wright State University School of Medecine, Dayton, Ohio 45401, USA.

A. Nicholson,
Royal Air Force Institute of Aviation Medecine,
Farnborough, Hampshire, U.K.

G. Parati,
Universitá di Milano, Istituto di clinica medica IV, via F. Sforza 35,
20122 Milano, Italy

Katherine Parkes,
Department of experimental psychology
University of Oxford, Oxford, U.K.

A. Pena,
University hospital, Department of gastroentorology,
Leiden, The Netherlands

M. Pernollet,
INSERM U7, CNRS LA 318, Faculté de Médecine Necker-Enfants malades,
156 Rue de Vaugirard,
75015 Paris, France

A. Perski,
National Institute for psychosocial Factors in Health, P.O.B. 60210,
10401 Stockholm, Sweden

W. Pierpaoli,
Institute for integrative biomedical research,
Ebmatingen, Switzerland

M. Pokorny,
Netherlands Institute for preventive health care/TNO,
P.O.B. 124,
2300 AC Leiden, The Netherlands

A. Raedler,
Department of Medical psychology
University hospital,
Hamburg-Eppendorf, FRG.

A. Ramirez,
Università di Milano, Istituto di clinica medica IV,
Via F. Sforza 35,
20122 Milano, Italy

M. de Reuck,
Université Libre de Bruxelles,
Hopital Brugmann, GI Department
Bruxelles, Belgium

J. Riemersma,
Institute for perception / TNO
3769 ZG Soesterberg, The Netherlands

N. Rizzo,
Centro Internazionale Radiomedico
C.I.R.M.,
via dell'Architettura 41,
00144 Roma, Italy

D. Russell,
Cardiovascular Research Unit,
University of Edinburgh
Edinburgh, U.K.

J. Rutenfranz,
Institut für Arbeitsphysiologie an der Universität Dortmund,
Ardeystr. 67,
4600 Dortmund1, FRG.

E. Sand,
Université Libre de Bruxelles,
Campus Erasme, Laboratoire d'épidémiologie et de Médecine sociale,
880 Route de Lennik,
1070 Bruxelles, Belgium

P. Schwartz,
Institute of cardiovascular research Giogio Sisini, University of Milano
via F.Sforza 35, 20122 Milano, Italy

W. Schulte,
Medizinische Universitätsklinik,
Venusberg, D-5300 Bonn, FRG.

K. Schulz,
Department of medical psychology,
University hospital
Hamburg-Eppendorf, FRG.

S. de Servi,
Division of Cardiology,
Policlic S. Matteo, University of
Pavia,
27100 Pavia, Italy

Patricia Shipley,
Department of occupational psychology, Birkbeck College,
University of London,
Malet Street,
London WC1 E7HX, U.K.

Karin Siegrist,
University of Marburg, Faculty of
Medecine, Department of Medical
Sociology,
Marburg, FRG.

J. Siegrist,
University of Marburg, Faculty of
Medecine, Department of Medical
Sociology,
Marburg, FRG.

P. Sleight,
Cardiac Department, John Radcliffe
Hospital,
Headington, Oxford, U.K.

G. Solomon,
Fresno County Department of Health,
P.O.B. 11867, Fresno, California
93775, USA.

B. Spangelo,
Department of Biochemistry, The
George Washington University,
College of Health Sciences,
Washington DC 20037, USA.

G. Specchia,
Division of Cardiology, Policlinic
S. Matteo, University of Pavia,
27100 Pavia, Italy.

A. Steptoe,
St. George's Hospital Medical
School, University of London, Department of psychology, Cranmer Terrace
London SW17 ORE, U.K.

L. Tavazzi,
Centro di riabilitazione di Veruno
Fondazione Clinica del Lavoro,
Veruno (NO), Italy

H. de The
INSERM U7, CNRS LA 318, Faculté de
Médecine Necker-Enfants malades
156 Rue de Vaugirard,
75015 Paris, France

T. Theorell,
National Institute for psychosocial factors in health,
P.O.B. 60210,
10401 Stockholm, Sweden.

O. Tønder,
The Gade Institute, Department of
Microbiology and Immunology,
University of Bergen,
Bergen, Norway

L. Torsvall,
Laboratory for Clinical Stress
Research, National Institute for
psychosocial factors and health,
1040 Stockholm, Sweden

Th. Urger,
German Institute for high blood
pressure Research and Department of
Pharmacology, University of Heidelberg, Im Neuenheimer Feld 366,
D-6900 Heidelberg, FRG.

U. Ursin,
Institute of physiological psychology, University of Bergen,
Bergen, Norway.

R. Vaernes,
Institute of physiological
psychology, University of Bergen,
Bergen, Norway

J. vann Jones
Cardiac Department, John
Radcliffe Hospital,
Headington, Oxford, U.K.

E. Vanoli,
Institute of cardiovascular
Research Giorgio Sisini, Università di Milano, via F. Sforza,
20122 Milano, Italy

Joan Vernikos,
Biomedical Research Division, NASA
Ames Research Center, Moffett Field,
California 94035, USA.

L. Vogt,
DFVLR, Institut für Flugmedizin,
P.O.B. 906058, 5000 Köln 90, FRG.

M. Wadsworth,
Medical Research Council, National
Survey of Health and Development,
University of Bristol, Department of
Community Health,
Whiteladies Road,
Bristol BS8 2PR, U.K.

M. Waltz,
Universität Oldenburg,
Westerstr. 2
Oldenburg, FRG.

I. Weber,
University of Marburg, Faculty of
Medecine, Department of Medical
Sociology,
Marburg, FRG.

H. Wegmann,
DFVLR, Institut für Flugmedizin,
P.O.B. 906058,
5000 Köln 90, FRG.

H. Weiner,
Neuropsychiatric Institute, UCLA,
760 Westwood Plaza,
Los Angeles, California 90024 USA.

A. Zanchetti,
Università di Milano, Istituto di
Clinica Medica IV,
via F. Sforza 35,
20122 Milano, Italy.

A. Zaza,
Institute of Cardiovascular Research
Giorgio Sisini, Università di Milano
via F. Sforza 35,
20122 Milano, Italy.

A. Zotti,
Centro di riabilitazione di Veruno,
Veruno (NO), Italy.

PART 1

Psychological and sociological parameters for studies of breakdown in human adaptation

edited by

J. Cullen and J. Siegrist

TOWARDS A TAXONOMY OF METHODS - A GENERAL OVERVIEW OF PSYCHOLOGICAL
APPROACHES IN THE STUDY OF BREAKDOWN OF HUMAN ADAPTATION

John Cullen
Health Care and Psychosomatic Unit, Garden Hill,
E.H.B. Box 41A, 1 James's Street, Dublin, 8 (Ireland)

Because many of the papers in this volume may contain terms and arguments which are unfamiliar or difficult for the non-behavioural scientist this paper will try to explain itself to other professionals and specialists across the disciplinary boundaries. In this way it is hoped that more scientists and clinicians will be encouraged to seek to know more about these methods and, better still, perhaps to begin to use them in studies of breakdown of adaptation. It may be regarded as an initial step towards the development of comprehensive kits of methods for use in those studies and concerted actions and collaborations which the Programme may encourage throughout member countries.

General Purpose of the Psychological Methods:

Over the past decade and a half an enormous increase in interest in the role of psychosocial factors in health and disease has occurred. Methodologies deriving from the behavioural sciences have explored, delineated and quantified psychological phenomena implicated in human adaptation processes in the course of a human life-span experience. Of necessity there has been a concerted drive to elaborate test methods which can quantify human behavioural or experiential factors so that these can be related to measures of physical health status, physiological indices or the epidemiologists measures of morbidity and mortality. And, of course, pari passu with these developments in the behavioural sciences there has also been a very substantial advance in the possibility of measurement on the somatic or physiological side of the psychobiological equation.

Equally important over the same recent period has been the development of computer-based data-processing together with powerful statistical techniques and their related soft-ware packages. This has presented for the first time in human history, possibilities for convenient and quick, but very complex, multifactorial studies (e.g., using ANOVA-analysis of variance; factor analysis; time-series analysis; etc.) and analyses. This makes possible the study of much larger and more comprehensive models of the adaptation process and its breakdown in disease. Many more variables can be included and more complex hypotheses can be tested. Because the psychobiological equation implies the measurement of larger numbers of variables it will be important for those entering into studies of this kind for the first time to recognise the need for statistical advice at an early stage in order to ensure the rich harvest of insights which can be gained with complex data.

With this kind of statistical power the psychological methods have been utilised in:

- The search for explanations of variance in mortality and morbidity patterns
- The search for "risk factors"
- The search for correlates of physiological change or presence of pathology
- The search for comprehensive pathophysiological mechanisms and explanatory models
- The search for a comprehensive or holistic basis for care delivery or therapeutic or preventive interventions

In the following pages some of the more prominent tests will be described together with the sorts of variables they measure and what these mean in any comprehensive explanatory model of breakdown of adaptation. As a first step towards a comprehensive taxonomy of methods some of those tests which most usually are used to test the following dimensions will be described:

1. Life events, stresses or changes
2. Occupational stressors
3. Coping strategies
4. Behaviour style
5. Personality factors
6. Changes in mood, alertness or engagement level in daily life
7. Mental health: depression and/or anxiety
8. Mental efficiency or dysfunction
9. General health status - "wellness v illness"

From this range a fairly comprehensive "kit" could be selected for use in a wide variety of studies. For most a reference or two will be given and this with the advisory potential of those listed as willing to participate in the Community Programme should enable newcomers to this kind of research to find a satisfactory approach for most applications.

In addition to the dimensions listed above the paper will briefly introduce the reader to some issues which seem important for the future development of laboratory or clinical types of psychophysiological measures or of biological markers or indices.

1. Life Events, Stresses or Changes

Life events inventories, schedules of recent experiences and other similar measures attempt to quantify the exposure of an individual to life changes and stresses. By inventory the cumulative exposure over a stated period may be added up and so an approach to a dosimetry of stressor stimulation can be made. These inventories e.g., the Social Readjustment Rating Scale (SRRS), (Holmes and Rahe, 1967); the Life Events Inventory (LEI), (Cochrane and Robertson, 1973); the Psychiatric Epidemiology Research Interview (PERI), (Dohrenwend, et al., 1978), rank order in terms of stressfulness a wide range of commonly occurring life change events that are part of the shared life-experience of mankind such as loss by bereavement, change in job or financial status, health change, change in marital status and so on, together with social problems or changes like shift of domicile or trouble with the law. These inventories have been cross-culturally elaborated and validated with relatively minor adjustments being required to conform to local cultural mores or traditions. They have also more recently been developed with forms of special relevance for different age-groups in the life-cycle e.g., the Adolescent Life Change Event scale, (Yeaworth, et al., 1980; Coddington, 1972); Dimensions of Life Experience, (Chiriboga and Dean, 1979); The Life Events Checklist, (Johnson and McCutcheon, 1980).

A criticism of these inventories has been that they only give a measure of historical exposure of an individual to events which are likely to be upsetting but do not give a measure of the level of upset involved. For example, the loss of a spouse or close relative may on occasion be more of a blessing than a disaster. Similarly, promotion in a job can impact differently on different persons. More complex test inventories have been developed which include subjective ratings of the impact of the events on the individual concerned, e.g., the Dimensions of Life Experiences, (Chirboga and Dean, 1979); the Review of Life Experience (Role), (Hurst, et al., 1978); the Life Experiences Survey (LES), (Johnson and Sarason (1979). Some of these issues have been discussed in the writings of George Brown in London (Brown and Harris, 1978) and of the

Dohrenwends in the U.S.A. (Dohrenwend and Dohrenwend, 1981) where the reader may wish to take up their study more intensively.

The tests have proved to be of continuing usefulness in all their forms especially for prospective cohorts, clinical history quantification and in the explanation of varience in morbidity or mortality patterns or as measures of the contribution of stressful life events to risk.

For future research on these tests much work needs to be done on the contribution of occupational physical stresses to lowering resistance to life-change stresses. Exposure to excessive noise, shift-work or industrial toxins may reduce resistance and coping ability. Exposure to multiple interacting events may have multiplicative rather than additive effects but we do not yet have studies which also measure physiological changes which can test this type of hypothesis. The temporal proximity of events may be similarly compounded in their effect and so may temporal order effects. Complexity of the events, their severity and individual highly specific vulnerability to some types of life-changes will have to be researched in the future. Even as the inventories stand at present, however, they could be used prospectively in "at risk" registers and in psychological post-mortems or post-morbidity onset studies in systematic ways by those who have access to clinical populations.

2. Occupational Stressors

In increasing volume over the past decade or so studies have been carried out which show significant effects on health of psychosocial factors arising in working life. The recognition of the importance of occupational stressors arises in this volume quite correctly because the Commission of the European Community has requested a special interest in those factors as part of the Programme of researches.

The life change event inventories, referred to in the preceding section of this paper, contains some items which record changes in working life. An early study by Theorell (1974) finds that occupational life changes are significantly associated with the incidence of myocardial infarction. Only 17 per cent of his controls compared with 41 per cent of the patients reported changes at work during the year prior to illness. Eight important changes at work were reported - *"Change to a different line of work; retirement from work; major change in work schedule; increased responsibility; decreased responsibility; trouble with boss; trouble with colleagues; unemployed for more than one month"* (Theorell, 1974, p. 107).

Many more comprehensive tests for work-related factors have been developed and extensively validated and applied. These questionnaires seek to identify the load of stress incurred in working life, the job demands, the social supports available, the way roles and the structure of the organisation together with the physical working environment impinge on the subject. Other physical work stressors such as toxins interact with psychosocial ones. In addition, differences in patterns of occupational stresses have been described for "white-collar" workers (Cooper and Marshall, 1978) and "blue-collar" workers (Poulton, 1978). Occupational grade has been shown to correlate significantly with coronary heart disease amongst British civil servants (Marmot, et al., 1978).

The following tests are amongst the more widely used for exploring exposure to occupational factors which may lead to breakdown in adaptation and health:

ISR/NIOSH - P-E Fit Scales (Person-Environment Fit Scales), Institute for Social Research, Michigan/National Institute for Occupational Safety and Health (Caplan, et al., 1975).
These scales have been developed for the measurement of a relatively broad range of conceptually distinct variables such as occupational psychosocial factors (e.g., workload, skill utilisation, role stressors, job security); job-related strain factors (e.g., job dissatisfaction, boredom), social supports from such sources as co-workers, supervisors, family, etc., and physical and mental health outcomes (e.g., depression, anxiety, gastrointestinal and cardiovascular complaints, etc.). Caplan, et al. (1975) have utilised these scales to examine occupational differences in job

demands and their possible relationships to health measures including both physiological and biochemical indicators.

Job Diagnostic Survey: JDS (Hackman and Oldham, 1976). The Job Diagnostic Survey is one of the most widely used measures in contemporary studies of people at work and may be used with staffs from a very wide range of grade levels. It provides a measure of four different dimensions of an individual's experience of work, viz: (i) "job dimensions" defined in terms of the degree to which the job involves skill variety, task identity, autonomy, and so on; (ii) "critical psychological states" including, e.g., experienced meaningfulness of the work; (iii) "affective reactions to the job" measures such personal affective reactions as general job satisfaction, and a variety of specific satisfactions, e.g., with pay, security, supervision, etc., and, finally (iv) "individual growth need strength" measures the strength of the respondent's desire to obtain "growth" satisfaction from his or her work. Recent studies have demonstrated systematic relationships between JDS measures and mental health outcomes, particularly among blue collar workers.

The Work Environment Scale (WES), Moos (1981). The WES measures an individual's perception of ten aspects of his or her work setting, grouped in terms of relationships dimensions, personal growth or goal orientation dimensions and system maintenance and system change dimensions. Developed in the U.S., these scales have been extensively used in a number of practical applications, and have been related to a variety of health and wellbeing indicators, including cardiovascular measures.

Job Demands and Job Decision Latitude Scale,(Karasek, 1979). Developed in the context of national survey data from Sweden and the United States, Karasek's (1979) model predicts that psychophysiological strain results from the interaction of job demands and job decision latitude. Job decision latitude refers to the working individual's potential personal control over his tasks and his workplace conduct during the working day. The job demands scale measures three categories of psychological stressors at work: those related to work load, unexpected tasks and personal conflict. These measures have been related to such important health and wellbeing indicators as exhaustion, depression, absenteeism, pill consumption, job dissatisfaction, psychophysiological stress reactions, and cardiovascular diseases.

Life and Work Tedium Questionnaire,(Kafry and Pines, 1980). Considerable interest has recently been expressed in the possible health and wellbeing consequences of intensive and protracted engagement in specific health-care provider roles. One specific consequence, popularly known as "burnout", has attracted considerable attention. A number of scales for the measurement of experienced "burnout" have been formulated. These include the "Life and Work Tedium Questionnaire" developed by Kafry and Pines (1980) and the Maslach

Burnout Inventory (MBI), developed by Maslach and Jackson (1981). Three subscales are included in the MBI measuring, respectively, emotional exhaustion, depersonalisation, and personal accomplishment. Psychometric analyses have shown that the scale has both high reliability and validity as a measure of burnout.

3. Coping Strategies

Ways of coping with psychosocial problems may be more or less efficient and effective. It is now well established that if coping is inadequate adaptation will breakdown and health will suffer. For example, long-term effects may occur in the bereaved if coping with the loss is disturbed or inadequate. Major life change events make heavy demands on the repertoire of coping skills and workplace stresses may make excessive and unremitting demands on the workers coping resources. Stressful demands may evoke quite different attempts to adapt or cope in different individuals. Some may confront the situation directly, others may habitually avoid this direct approach. One worker has dichotomized individual differences by referring to types as "blunters" or "sensitizers" (Miller, 1979).

Animal studies have addressed this issue also. Weiss (1972) has elegantly explored the role of good coping in offsetting stressful conditions. The availability of good coping opportunities through control over the situation or learning its regularities or predictability may indeed reduce breakdown, as evidenced by peptic ulceration, and enhance adaptation. Seligman (1975) in a series of experiments in animals and in man has demonstrated that if successful coping is denied or access to coping behaviour is not available then exposure to stressful demands leads to behaviour breakdown in a form he terms "learned helplessness". In this state which has some analogies with depressive illness coping attempts cease and adaptation breaks down.

There are several tests which address this dimension of coping style. One of the most prominent is Richard Lazarus': Ways of Coping Checklist (Folkman and Lazarus, 1980).

A facet of the psychology of coping which has been found to be a significant issue for adaptation is concerned with the habitual perceptions which individuals have about the "locus of control" in their lives and in their health. This seeks to identify whether an individual perceives the influences in his life to be amenable to control and influence by himself or whether they are perceived to be largely or entirely matters of fate or of external control by others or by external events.

The following are some widely used and validated locus of control tests:

Rotter's Internal-External Locus of Control Scale, (Rotter, 1966).

A detailed description of this scale is provided in a monograph by Rotter (1966). The Rotter Internal-External Locus of Control Scale is a 23-item forced choice questionnaire, and is scored in the "external" direction, that is, the higher the score the more "external" the individual. The scale thus differentiates between "internal" and "external" locus of control individuals. It has been used extensively in a very wide range of settings including psychophysiology, psychosomatics, clinical medicine and psychiatry. Early assessment devices for measuring locus of control, such as Rotter's (1966) I-E scale have undergone extensive refinements in recent years. An excellent account of these developments is presented by Lefcourt (1981).

Multidimensional Health Locus of Control Scales (MHLC) (Wallston, et al., 1978; Wallston and Wallston, 1981)
The MHLC scales measure three distinct dimensions: Internality (IHLC); Chance Externality (CHLC); and Powerful Others Externality (PHLC). Development of this scale is outlined in Wallston and Wallston (1981). These MHLC scales may be used in rather different ways, for example as either independent *or* dependent variables, as in the measurement of treatment outcomes, etc.

4. Behaviour Style

The most notable tests in this area are those which detect and quantify the habitual so-called Type A behaviour. Frequently detected facets of this behaviour style include: "competitiveness, time urgency, impatience, aggressiveness, etc.". Jenkins later in this volumee addresses many of the problems associated with this type of behaviour. The Type A behaviour pattern has mainly been studied in connection with cardiovascular disease where considerable evidence has accumulated with regard to its role as a risk factor for coronary heart disease. The recent report of the U.S. National Heart, Lung and Blood Institute Review Panel on Coronary-Prone Behaviour and Coronary Heart Disease (1981) gives a useful access to controversies and issues surrounding this topic. More recently the specificity of the behaviour pattern for cardiovascular disease alone is being challenged.

Perhaps the most widely used test for this behaviour pattern is the Jenkins Activity Scale (Jenkins, et al., 1979).

The aggressiveness component of the Type A behaviour pattern has more recently been gaining prominence as a very important contributor to the riskiness of the behaviour style. Tests have been developed to measure habitual levels of aggressivity and hostility in individuals e.g., State-Trait Anger Scale (STAS), (Spielberger, 1980).

A question with regard to these behaviour patterns has been whether they are inborn or whether they are learned behaviour patterns. It has also been suggested that the behaviour may rather be a response to certain kinds of workplace psychosocial factors and are therefore to be seen as an attempt at adaptive behaviour in unfavourable organisational environments. A start has been made into answering these questions in recent years with the development of tests for Type A behaviour in children, e.g., the MYTH: Matthews Youth Test for Health (Matthews and Angulo, 1980).

In addition to our need to know if certain environments, e.g., at work can induce habitual behaviour of the kind outlined here it is also important to know if the behaviour can be changed by education, counselling or other means. There is some evidence that it can. It is not unknown, however, whether having insight into one's Type A behaviour pattern reduces its risk potential.

There are other habitual behaviour patterns which have not been systematically studied or quantified, e.g., stimulus seeking; "slow-suicide", anti-health: smokers, addicts, etc.; fatalism or helplessness-prone behaviour and some others which seem problematic for maintaining adaptation and health. Finally, there are the behaviour patterns which arise from socially assigned habitual roles such as those assigned with gender - sex-role behaviour patterns. A test which has contributed to studies in this very important area is the BEM Sex-Role Inventory (BEM, 1974). This test reflects behaviour patterns which issue from perceptions of desirable or undesirable "feminine" or "masculine" behaviour patterns. With increasing diffusion of traditional rigid sex-role assignments in society and the workplace these behaviour patterns may be mismatched with new environmental demands and therefore give rise to stressful situations.

5. Personality Factors

These factors are conceived of as enduring personal _traits_ as opposed to more transient _states_ (e.g., of mood or temporary reality based anxiety and so on). There is a very considerable literature on tests which measure these enduring personality factors and on their correlations with physiological or health-related measures. These personality factors have been shown to help to explain some of the variance in biological responsivity or vulnerability.

The most widely used tests are:

The Cattell 16 Personality Factor Test (16PF).
(Cattell, et al., 1970).

AND

The Eysenck Personality Inventory (EPI).
(Eysenck and Eysenck, 1964).

These tests must be mentioned here because they provide necessary typological information about individual differences which is required for some types of health and adaptation studies. Nevertheless they are complex to administer and score and require experienced psychologist interpretation of data. For this reason they are not perhaps as readily available and transparent in transdisciplinary studies as the other tests.

6. Changes in Mood, Alertness or Engagement Level in Daily Life

These subjective states have been measured with a variety of tests and an extensive review and critique of them is provided elsewhere in this volume by Mackay. They have proved extremely useful and valid in tracking variations in mood over hours or days in, for example, studies of circadian rhythms or in relation to dose-response curves with psychotropic drugs and so on. There is an emerging research literature which is relating this type of tracking of subjective states to changes in physiological and biochemical parameters in the various kinds of biological rhythms. Examples of these tests are:

The Profile of Mood States (POMS),
(McNair, et al., 1981).

And

The Bond and Lader Scale
(Bond and Lader, 1974).

In addition, there is evidence that there may be individual differences in patterns of arousal and alertness throughout the diurnal cycle. "Morningness" and "Eveningness" may be partially innate or acquired and there is a test designed to access this dimension - the Morningness-Eveningness Scale (Horne and Ostberg, 1976). This test has been useful in exploring the question of fit between individual characteristic and environmental or occupational demands, for example in shift-work which is now known to make substantial demands on the adaptation mechanisms at both psychosocial and biological levels.

7. Mental Health: Depression and/or Anxiety

Depression has highly interesting biological facets which are not yet clearly understood. In one phase of a depressive illness cortisol the "stress hormone" par excellence, is markedly elevated for a prolonged period. Later it falls to low levels. Furthermore, depressive-like reaction states seem to be a component of the risk factors for a variety of illnesses including myocardial infarction. Anxiety, on the other hand, seems to be a fairly constant feature in the subjective response to environmental stress and has well known physiological arousal aspects.

There are now available extensively used and validated mental health tests which most commonly measure degrees of depression and anxiety at both sub-clinical and at full clinical levels of severity. The most commonly used of these tests are:

- The General Health Questionnaire (GHQ) in versions containing either 60, 30, 28 or 12 items. (Goldberg, 1979)
- The Crown-Crisp Experiential Index (CCEI) formerly the Middlesex Hospital Questionnaire (MHQ) (Crown and Crisp, 1979).
- The Beck Depression Scale (Beck, 1961, 1967)
- The Taylor Manifest Anxiety Scale (Taylor, 1953)

All of these tests are extremely useful especially in clinical studies and the possibility of using comparable measures of this kind in multi-national concerted action researches may begin to elucidate the apparently potent role of these major psychological aspects of breakdown in adaptation and the onset of illness.

8. Mental Efficiency or Dysfunction

This area has received extensive study in vigilance and other types of experiments under stressful conditions in psychological laboratories. However, measures of the occurrence of alterations in everyday mental functions such as cognitive and memory ones have been largely neglected. Broadbent, et al. (1982) with his Cognitive Failures Questionnaire (CFQ) has addressed this topic. The test seeks to identify lapses of memory function, attention, cognition, etc., which are involved in the course of the ongoing everyday adaptation to the environment. Research to date seems to suggest that habitual patterns exist in individuals which may be more permanent "trait" measures than environmentally determined fluctuations in "state".

Much further work remains to be done in this field. It will be important because it is clear that cognitive processes are at the core of human adaptation processes whether with regard to performance in challenging conditions or in the maintainance of health under exposure to the stressors arising from life-change events or occupational demands.

9. General Health Status - "Wellness" v. "Illness"

Breakdown of adaptation implies the emergence of general dysfunction in the body systems. The emergence of specific diseases or syndromes is not the only pathway by which the breakdown may manifest itself. Stress researchers have emphasised that many of the body system changes they observe, e.g., in the endocrine system or the cardiovascular system may be the precursors of disease. Along with this it is well known, now, that demands on health-care delivery systems and their soaring costs come as much from unspecific "unwellness" and symptoms as much as from fully fledged classical diseases or syndromes. For this reason measures of general "wellness" or general levels of "illness" can be extremely useful in identifying stressful situations where psychosocial demands may be beginning to overwhelm the coping and adaptive resources. Unspecific symptom patterns may, for example, emerge in stressful working conditions and, indeed carry over into after work and domestic life.

A number of tests have been developed which are designed to monitor the emergence of these dimensions. Examples of the more commonly used tests are:

- The Minnesota Multiphasic Inventory (MMPI) (Dahlstrom and Welsh, 1960)
- The Cornell Medical Index (Brodman, et al., 1956)
- The Psychosomatic Symptom Checklist (Gurin, et al., 1960).
- The General Health and Adjustment Questionnaire (GHAQ) (Tasto, et al., 1978).

In this kind of research it is often necessary to track patterns of wellness and illness over periods of time. Jenkins (1980) has devised a health diary - The Monthly Health Review - which is most useful for this purpose. It enables observation of patterns of symptoms and illnesses and how these relate to temporal patterns of change in environmental demand or psychosocial stress load.

10. Laboratory Measures and Clinical Psychophysiological Measures

Laboratory, field and clinical methodologies have been discussed by various participants in these workshops. It is not intended to review these here. Rather it may be useful to suggest some needs for further research on methodology in these areas. Firstly, in the psychological side of the psychobiological equation emphasis has most often been on performance measures, signal-detection errors and so on as indicators of breakdown in adaptation. Phenomenological observations of signs of subjective distress, agitation, anger and so on have not been systematized in universally accepted ways. There is much need for an expansion of research to evolve good methods for quantification of these types of variables.

On the biological side of the equation there has been a tendency to measure non-clinical parameters, e.g., skin conductance or heart-rate more on the basis of their availability, laboratory tradition or cost rather than on their clinical relevance for good pathophysiological models of disease. This present Programme of the European Community on breakdown of adaptation may bring workers from both sides of the psychobiological equation together so that these more urgent and relevant parameters may be measured.

In addition it seems necessary to reiterate a

need previously expressed by the author (Cullen, 1979) for the development of systematic dynamic profiles of biological responsivity to psychosocial and/or physical stressors which would measure:

1. Change to an inappropriate level in an inappropriate direction of thresholds for arousal or the activation of a biological system.
2. Shifts to faster or slower rise-times in the activation of a biological parameter.
3. Changes in the maximum or minimum levels of response.
4. Development of sustained plateaux of responses - their duration, time of onset and level.
5. Slower or faster decay-time (recovery, return to base-line) after activation of a biological system.

These sorts of dynamic profiles would help us to discern the "stress-prone" or those who are involved in the development of a process of breakdown in psychobiological adaptation. Most clinical biomedical assessments do not approach things in this way. Perhaps this Community Programme of research will facilitate a widening and a sharing of paradigms in this area. This would progress research into and knowledge of pathophysiological mechanisms and would also provide a context for the much neglected but very necessary research on ways to reverse the adaptation breakdown processes.

11. Towards a Taxonomy of methods and the criteria for Their Evaluation

This overview of the more prominent test procedures has been offered in the context of an attempt to explain briefly for the non-specialist the parameters they measure and how these parameters relate to each other. Sometimes the measures seem to be immediately relatable to other measures but this is not always so and the intervening variables have not yet been systematically studied. Some suggested needs for further research on these matters have been mentioned.

With regard to the criteria for validation of methods perhaps we may need to answer some or most of the following questions:

1. What factors do they purport to measure?

2. What factors do they actually measure?

3. How do they measure them?

4. Are the measures of independent variables in the same logical order or class as the dependent variables?

5. Can quantitative measures of the psychological variables be mapped in a meaningful way onto a physiological or pathophysiological model with

quantifiable parameters?

6. Are the biological measures or biomedical measures which we use as dependent variables, adequately criticised on grounds of relevance, salience, clarity, comprehensiveness and up-to-dateness?

7. How do the psychological measures relate to each other within tests and between tests?

8. Do we need another test?

9. Have we adequate theories, constructs or models of processes to support the tests?

10. Are these models as well elaborated for the psychological processes as for the biological processes and vice-versa?

11. Are the psychological measures applicable, useful and acceptable in the clinic, in field-work, in laboratory studies or in only one or some of these?

12. Can they be applied conveniently to large populations?

13. How?

14. Is their purpose, rationale and rationality likely to be transparent or opaque for non-specialists or other

professionals in trans-disciplinary or inter-disciplinary studies.

15. Do they work best in acute or chronic situations or both?

Perhaps, as the work proceeds in this Programme better "kits" of methods will become available and prove their usefulness in this most necessary study of breakdown in human adaptation and health.

BECK, A.T. (1967). Depression: Clinical, Experimental and Theoretical Aspects. New York: Harper and Row.

BECK, A.T.; WARD, C.H.; MENDELSON, M.; MOCK, J. and ERBAUGH, J. (1961). An inventory for measuring depression. Arch. Gen. Psychiat., 4, 561-571.

BEM, S.L. (1974). The measurement of psychological androgny. J. Consult. Clin. Psych., 42, 155-162.

BOND, A. and LADER, M. (1974). The use of analogue scales in rating subjective feelings. Brit. J. Med. Psychol., 47, 211-218.

BROADBENT, D.E.; COOPER, P.F.; FITZGERALD, P. and PARKES, K.R. (1982). The Cognitive Failures Questionaire (CFQ) and its correlates. Brit. J. Clin. Psychol., 21(1), 1-16.

BRODMAN, K.; ERDMANN, A.J. Jr., and WOLFF, H.G. (1956) Cornell Medical Index Health Questionnaire (Manual). New York, N.Y.: Cornell University Medical College.

BROWN, G.W. and HARRIS, T. (1978). Social origins of Depression. London: Tavistock.

CAPLAN, R.D.; COBB, S.; FRENCH, J.R.P.; VAN HARRISON, R. and PINNEAU, S.R. (1975). Job Demands and Worker Health (main effects and occupational differences). National Institute for Occupational Safety and Health Report No. NIOSH - 75-160. Cincinatti, Ohio, 45226.

CATTELL, R.B.; EBER, H.W., and TATSUOKA, M.M. (1970). Handbook for the 16 Personality Factor Questionnaire. Illinois: I.P.A.T.

CHIRIBOGA, D.A. and DEAN, H. (1978). Dimensions of stress: perspectives from a longitudinal study. J. Psychosom. Res. 22, 47-55.

COCHRANE, R. and ROBERTSON, A. (1973). The life events inventory: a measure of the relative severity of psychosocial stressors. J. Psychosom. Res. 17, 135-139.

CODDINGTON, D.R. (1972). The significance of life-change events as etiologic factors in diseases of children: A study of the normal population. J. Psychosom. Res., 16, 7-18.

EYSENCK, H.J. and EYSENCK, S.B.G. (1964). Manual of the Eysenck Personality Inventory. University of London Press Ltd. (available from NFER Publishing Co.).

FOLKMAN, S., and LAZARUS, R.S. (1980). 'An analysis of coping in a middle-aged community sample'. J. Hlth. Soc. Beh., 21, 219-239.

GOLDBERG, D. (1979). Manual of the General Health Questionnaire. NFER Publishing Co., Windsor, Berks.

GURIN, G.; VEROFF, J. and FELD, S. (1960). Americans View Their Mental Health. New York: Basic Books.

HACKMAN, J.R. and OLDHAM, G.R. (1976). Motivation through the design of work: Test of a theory. Organiz. Behav. Hum. Perform., 16, 250-279.

HOLMES, T.H. and RAHE, R.M. (1967). The social readjustment rating scale. J. Psychosom. Res., 11, 213-218.

HORNE, J.A. and OSTBERG, O. (1976). A self-assessment questionnaire to determine morningness-eveningness in human circadian rhythms. Int. J. Chronobiology, 4, 97-110.

HURST, M.W.; JENKINS, C.D. and ROSE, R.M. (1978). The assessment of life change stress: a comparative and methodological inquiry. Psychosom. Med., 40, 126-141.

JENKINS, C.D.; KROGER, B.E.; ROSE, R.M. and HURST, M. (1980) Use of a Monthly Health Review to Ascertain Illness and Injuries. Am. J. Pub. Hlth., 70(1), 82-84.

JENKINS, C.D.; ZYZANSKI, S.J. and ROSENMAN, R.H. (1979). Jenkins Activity Survey Manual. New York. The Psychological Corporation.

JOHNSON, J.H. and MCCUTCHEON, S. (1980). Assessing life stress in older children and adolescents: Preliminary findings with the life events checklist. In: I.G. Sarason and C.D. Spielberger (Eds.) Stress and Anxiety, Vol. 7, Hemisphere Publishing Corporation, London.

JOHNSON, J.H. and SARASON, I.G. (1979). Recent developments in research on life stress. In: V. Hamilton and D.M. Warburton (Eds.) Human Stress and Cognition: An Information Processing Approach. John Wiley & Sons, Chichester, Chapter 7.

KAFRY, D. and PINES, A. (1980). The experience of tedium in life and work. Human Relations, 33(7), 477-503.

COOPER, C.L. and MARSHALL, J. (1978). Sources of Managerial and white collar stress. In: Cooper, C.L. and Payne, R. (eds.) Stress At Work. Chichester: John Wiley and Sons.

CROWN, S. and CRISP, A.H. (1979). Manual of the Crown-Crisp Experiential Index. Hodder and Stoughton, London.

CULLEN, J. (1979). Coping and health - a clinician's perspective. In: Levine, S. and Ursin, H. (eds.) Coping and Health, pp 295-322. New York: Plenum Press.

DAHLSTROM, W.G. and WELSH, G.S. (1960). An MMPI handbook: A guide to use in clinical practice and research. Minneapolis: University of Minnesota Press.

DOHRENWEND, B.S. and DOHRENSEND, B.P. (Eds.) (1981). Stressful Life Events and their Contexts (Monographs in Psychosocial Epidemiology Series). Prodist: New York.

DOHRENWEND, B.S.; KRASNOFF, L.; ASKENASY, A.R. and DOHRENWEND, B.P. (1978). Exemplification of a method for scaling life events: the PERI life events scale. J. Hlth. Soc. Beh., 19, 205-229.

KARASEK, R.A. (1979). Job demands, job decision latitude, and mental strain: Implications for job redesign. Administrative Science Quarterly, 24, 285-308.

LEFCOURT, H.M. (1981). Research with the Locus of Control Construct. Volume I. Assessment Methods. New York: Academic Press.

MARMOT, M.G.; ROSE, G.; SHIPLEY, M. and HAMILTON, P.J.S (1978). Employment grade and coronary heart disease in British civil servants. J. Epidem. Community Health, 32, 244-252.

MASLACH, C. and JACKSON, S.E. (1981). The measurement of experienced burnout. Journal of Occupational Behaviour, 2, 99-113.

MATTHEWS, K.A. and ANGULO, J. (1980). Measurement of the Type A behaviour pattern in children: Assessment of children's competitiveness, impatience-anger, and aggression. Child Dev., 51: 466-475.

MCNAIR, D.M.; LORR, M. and DROPPLEMAN, L.F. (1981). Manual for the Profile of Mood States. Educational and Industrial Testing Service, San Diego, California, 92107.

MILLER, S.M. (1979). When is a little information a dangerous thing? Coping with stressful events by monitoring versus blunting. In: Levine S. and Ursin, H.: Coping and Health. New York: Plenum Press.

MOOS, R.H. (1981). Work Environment Scale Manual, Consulting Psychologists Press: Palo Alto, California.

NATIONAL HEART LUNG AND BLOOD INSTITUTE REVIEW PANEL (1981) Coronary-Prone Behaviour and Coronary Heart Disease. Circulation, 63, 1199-1215.

POULTON, E.C. (1978). Blue collar stressors. In: Cooper, C.L. and Payne, R. (eds.) Stress at Work. Chichester: John Wiley & Sons.

ROTTER, J.B. (1966). Generalised expectancies for internal versus external control of reinforcement Psychological Monographs, 80, (1, whole No. 609).

SELIGMAN, M.E.P. (1975). Helplessness. San Francisco: W.H. Freeman and Co.

SPIELBERGER, C.D. (1980). Preliminary manual for the State-Trait Anger Scale (STAS). Centre for Research in Community Psychology, University of South Florida.

TASTO, D.L.; COLLIGAN, M.J.; SKJEI, E.W. and POLLY, S.J. (1978). Health Consequences of Shift Work. National Institute for Occupational Safety and Health, Report No. NIOSH 78-154, Cincinatti, Ohio, 45226.

TAYLOR, J.A. (1953). A personality scale of manifest anxiety. J. Abn. Soc. Psychol., 48, 285-290.

THEORELL, T. (1974). Life events before and after the onset of a premature myocardial infarction. In: Dohrenwend, B.P. and Dohrenwend, B.S. (eds.): Stressful Life Events - Their Nature and Effects. New York: John Wiley & Sons.

WALLSTON, K.A. and WALLSTON, B.S. (1981). Health Locus of Control Scales. In: H.M. Lefcourt (ed.) Research with the Locus of Control Construct. Volume I. Assessment Methods, Chapter 6. New York: Academic Press.

WALLSTON, K.A.; WALLSTON, B.S. and DEVELLIS, R. (1978) Development of the Multidimensional Health Locus of Control (MHLC) Scales. Health Education Monographs, 6, 161-170.

WEISS, J.M. (1972). Influence of psychological variables on stress-induced pathology. In: Ciba Foundation Symposium No. 8. Physiology, Emotion and Psychosomatic Illness. Amsterdam: Elsevier Press (Excerpta Medica).

YEAWORTH, R.C.; YORK, J.; HUSSEY, M.; INGLE, M.E.; and GOODWIN, T. (1980). The development of an adolescent life change event scale. Adolescence Vol. XV, No. 57, 91-97.

PSYCHOLOGICAL FIELD STUDY TECHNIQUES - OVERVIEW AND NEEDS

D. Broadbent
Medical Research Council, University of Oxford, United Kingdom

The special requirements of field studies

The area to be considered here is the type of investigation which goes outside the laboratory, but examines only one particular situation rather than sampling an entire population as epidemiologists may do. Examples are the study of student nurses in two particular hospitals by Parkes (1982), of workers in an insurance company by Rissler and Elgerot (described in English by Frankenhaeuser, 1981), or of noise-exposed workers in a particular plant by Cohen et al. (1976). Compared with laboratory studies, this kind of investigation is bound to have larger sources of error; compared with studies of whole populations it may be partial. If one takes some specific problem, such as the effect of standing up at work rather than sitting down, measurements in the laboratory can be controlled for difference in food intake, in environmental temperature, and for variability in the work itself. A field study probably cannot; and yet will look only at the effect of standing in one particular context, perhaps for female workers, with an incentive payment scheme, and when transport to work is easy. Epidmiological studies may not have the control found in the laboratory, but they will have more generality. In the example they could include men, other payment systems, and different parts of a country.

However, the intermediate field study has some major advantages to go with its disadvantages. In studying breakdown of adaptation, there are some variables that cannot be simulated in the laboratory; such as the kind of motivation, or the effect of very long exposure. Thus, the three studies mentioned above could not have been done in the laboratory. Conversely, the large scale epidemiological study has several potential weaknesses. For instance, the quality of the data may be poor as the collection may lie in many hands and supervision may be difficult. More important, there may be unknown correlations between the variables in the study and other hidden ones in the population. In the whole economic system, paced workers are likely to live in different cities from unpaced ones, to have different payment systems, networks of social interaction, and so on; it is helpful to have a field study in which paced and unpaced are compared in one factory with a standard rate and method of pay and homogeneous conditions of life (Broadbent and Gath, 1981).

The major advantage of the field study from the point of view of this workshop is that it allows parallel use of several techniques. Within a single factory or community it may be possible to use not only questionnaire data or abstract morbidity statistics, but also physiological methods quite close to those of the laboratory, and even methods of measuring performance similar to those used in the laboratory. The extra dimension given to the research in this way should greatly assist in the interpretation of each kind of result. But it is unlikely that any one laboratory, or indeed country, can possess all the skills needed to exploit this advantage of field studies. Hence, we ought to look with special closeness at the opportunities for field studies within a concerted action.

In what follows, each kind of method is briefly assessed to indicate what is available now and what needs development. The survey is biased towards areas of research and of national effort that are close to the writer; this is itself one of the weaknesses that a concerted action should try to overcome. In a cross-disciplinary field involving may different journals as well as countries, it is particularly likely that each specialist requires help from others even in getting information.

Attention is concentrated upon occupational stress; but much of what is said will apply equally to studies of community organisation or general human ecology.

Questionnaires : Situational aspects

Ideally, any field study should look at human response to objectively defined features of the situation; people in noise greater than 95 dBC compared with those below, or those who have changed home address in the last six months compared with those that have not. But every situation has many aspects so that, for example, different persons working on an assembly line may use more or less gross muscular movements, may have more or less spare time before the work moves on, may have a mixture of different assemblies compared with a constant type, may have more or less serious consequences attached to error, and so on. It is therefore necessary to specify the situation from these points of view; the best way is to ask questions more or less systematically. There are now available questionnaires that cover a number of aspects of work from these points of view (McCormick et al., 1972; Hackman and Oldman, 1975; Banks et al., 1981) and if these are used investigators can be assured of comparability between their conditions and those of other studies.

Because of the very large number of detailed characteristics that a situation can have, it is particularly useful to have a few dimensions or general properties on which situations can be matched even though they differ in detail. For example, Karasek (1979, 1981) showed that a number of questions about jobs in the U.S. (Quinn and Shepherd, 1974) fell into two factors. One corresponds broadly to the amount of work that had to be done, and the other to the discretion or autonomy that the person has over the way the work is performed. Although the questions first emerged from nationwide surveys they can usefully be used on their own in field studies, and make a very short way of assessing situations on these two important characteristics (e.g., Parkes, 1982). It is clear that each dimension is related to important psychological effects. For example, Broadbent (1982) notes a number of instances in which demand seems to be related predominantly to anxiety and not to depression; whereas discretion behaves differently. However, it is not yet clear what the dimensions mean in terms of bodily change, disease or performance.

These dimensions alone are not enough to specify jobs; for example, the Job Diagnostic Survey has extra dimensions of feedback and importance, as well as distinguishing variety from responsibility, in predicting job satisfaction. A little further development is probably needed therefore in finding the best supplements to Karasek's dimensions, though a great deal has already been achieved.

One extra dimension that should be mentioned specifically is that of social support. This is to some extent separate from features of the job; it refers rather to the extent to which the person is helped by or can share troubles with other people. A number of studies have shown that close social links of this kind do not merely improve the state of the person; they actually act as a defence or interacting factor. Thus a person with good support may be equally satisfied with a situation whether it is demanding or not; but a person with low support may show large effects (La Rocco et al., 1980). Useful short questionnaires are provided by House (1981, p. 71).

More generally, the whole social environment can be assessed by questionnaire for qualities other than its supportiveness. The nurses on a ward in a hospital, or a team packing chocolates in a factory, may offer much support and yet be so organised that it is very unclear what the person should actually do. Equally, from knowing the degree of support or of clarity one may not know how much freedom or autonomy the group allows each person. Moos (1981) provides questionnaires that assess such qualities of the social situation.

In addition to the chronic features of the situation, the individual person may also be affected by traumatic events or changes in way of life; again there are a number of standard methods for assessing the severity of the events that have taken place. These range from the well-known questionnaire of Holmes and Rahe (1967) to the intensive interviewing of Brown and Harris (1978). As with the other areas we have been discussing, there is a trade-off between the detail obtained and the saving of time in the questioning which is of real importance because the field study depends on the goodwill of the people studied, and too long an enquiry may easily lose that goodwill. In the case of life events, the difficulty with a quick questionnaire is that some events may be caused by the individual rather than purely occurring spontaneously. If we find an association between getting divorced and subsequent illness, that may be due to the divorce being an early symptom of the illness rather than the illness being caused by the divorce. Hence the emphasis of critical authors upon rather careful determination of the exact circumstances of the life vents (Kasl, 1978).

Finally, the same caution should be given about all questionnaire methods of assessing the situation. If the questions are answered by the person whose health, physiology, or performance is also being assessed, then the answers may be coloured by the state of the person and so produce a false association. Ideally the situation should be as assessed by other people, not the one whose responses to the situation are being measured (Jenkins et al., 1975). If this is not possible, then the description of the situation by the person themselves is certainly interesting, but it must always be remembered that one is studying the relationship between, say, a partiular symptom and the perception of the situation, not the situation itself. To be concrete, when Broadbent (1982) shows that people in demanding jobs have higher anxiety, he is in fact showing that the anxious people describe their own jobs as demanding. Other people might or might not think the anxious people were over-worked!

Questionnaire : Individual characteristics

It is already well-established that the effect of a situation may depend on the nature of the individual exposed to it, and if possible therefore some measures should be taken to determine how the particular people being studied compare with other samples. There are perhaps three main characteristics that are important; though they may not be independent. The three are motivation, temperament, and attitude or cognition. Under the first heading, it has been reported that the effect of autonomy in the job depends on the strength of the person's need to realise their potential through work; those who treat the job more instrumentally, as merely a way of getting money, may be less dependent on having an interesting job (Blood and Hulin, 1967; Jackson et al., 1981). A rather similar finding is that reported by Broadbent (1982), that the high levels of anxiety found in assembly line workers appear only in those who report themselves as working for satisfaction; those working for money are if anything less anxious when paced! Unfortunately, although there are ways of assessing some individual motives there is perhaps less agreement on the major dimensions than there is on describing the external situation. They may in any case be socially influenced or determined, and thus need care in trans-European work. Well-known motives for which measures exist are the need for achievement (Atkinson and Raynor, 1974; the measures here are projective tests that are hard to apply, but questionnaires do exist: Argyle and Robinson, 1962). Perhaps we should include here Type A personality, although that might be regarded as temperamental.

As a definite example of temperamental factors one could take the dimension of introversion-extroversion. Although usually assessed by questionnaire (Eysenck and Eysenck, 1975), it is clear that it has biological correlates and may modify the response in conditions that affect the arousal system. Thus it has repeatedly been found to be an important factor in studies of circadian rythm and shift work; and it now seems well-established that the effects of caffeine are in appropriate circumstances in opposite directions in introverts and extroverts, both factors interacting with time of day (Revelle et al., 1980). Our own group has, in unpublished work, confirm a finding of Totman that introverts show more virus shedding after experimental exposure to rhinovirus than extroverts do. In this case the questionnaires are well studied, translated, and have been used in several countries.

The attitudinal or cognitive aspects are more recent developments. However, it seems that the belief in one's importance in determining events may be important in defining the person against the bad effects of failure (Cohen et al., 1976); a widely used measure is the questionnaire of

Rotter (1966). In similar fashion, it seems that nurses who report many failures of attention, memory, and similar cognitive functions are particularly vulnerable to subsequent stressful situations (Broadbent et al., 1982). These quentionnaires are in an intermediate state; they are reasonably well developed in their countries of origin, but would probably need further work to be used in other languages or societies.

Questionnaires : Mental health

If we now turn from specification of the situation to looking at the effects on the person, the outstanding change of the past decade has been availability of validated measures of mental health. At one extreme, the uncertainty of psychiatric classification has been greatly reduced by means of standardised means of examination such as the Present State Examination (Wing et al., 1974). These require interview by a well trained examiner, but if the time and the necessary interviewer are available they provide the best indicator of the person's state of mind, assessed in standard form. The practical difficulties are not trivial in large-scale studies however; and there is the problem that the examination covers all forms of metal illness, including psychosis. Questioning directed at some of these topics may be unacceptable in some situations, as suggesting that the respondent is 'mad' in the popular rather than the medical sense.

If this is so, or if time is short, then there are now sets of written questions which have been validated against the more thorough examinations, and give reasonable results. The General Health Questionnaire and the Crown-Crisp Experiential Inventory (formerly the Middlesex Hospital Questionnaire) are good examples, each of which has been validated against psychiatric assessment; and also has been employed successfully in industrial contexts (Goldberg, 1972; Crown and Crisp, 1966, 1979). The GHQ has perhaps been more widely used, but the MHQ had the original advantage of having separate scales of anxiety, depression, bodily symptoms, and obsessionality. As noted earlier, the effects of a situation on one of these may sometimes be different from those on other scales, and it is therefore desirable to discriminate rather than to have only a global score of 'ill-health'. For instance, in our unpublished work on prediction of results of experimental exposure to rhinovirus we have found that obsessional symptoms rather than other scales predict the amount of nasal secretion produced in the experimental cold. The GHQ also now has separate scales however, so if a long form of it is used separate scores can be obtained (Goldberg and Hillier, 1979). For other countries, translation and revalidation may be necessary; the measures work well in languages of the Indian sub-continent (Cochrane et al., 1977). That is not much help for European use, but suggests cross-cultural applicability.

Psychological and biological measures

Here there is a major distinction between the measures that can be obtained non-invasively and those that need some intrusion such as the taking of blood or urine. The latter are often more analytic scientifically but arouse more reluctance in the people being studied. Studies such as those of Frankenhaeuser and her collaborators show what can be achieved however. The highest priority measure has for some years been that of the catecholamines, which show very frequent changes in stressful situations. However, they do also change in situations that are pleasantly exciting or arousing, and it is therefore important that recent studies are also measuring cortisol (Frankenhaeuser and Johansson, 1982). In accordance with the views and evidence put forward by Henry and Stephens (1977), the catecholamines may indicate the existence of a challenge to the person, but the most serious personal consequences appear only when the challenge is inadequately met, and cortisol may be a better indicator of that state. There is still need for further evidence about relations between these measures and some of those mentioned earlier. Furthermore, it would be helpful to have more evidence relating these measures to incidence of subsequent disease. Methods of assay are well developed in certain laboratories, but it is fair to say that there is a certain amount of expertise involved that needs to be acquired by any group that wishes to undertake such analyses for itself. Often, it will be economic to co-operate with a laboratory that already has the skills.

As time goes by, it will doubtless be desirable to measure responses of the immune system, but at present there appear to be rather a large number of possibilities.

Turning to measures obtained non-invasively, many of laboratory interest may be difficult to collect under field conditions. For example, EEG measures still require equipment that is quite hard to transport and the measures themselves may be disturbed by electrical features of the environment. Rather dogmatically, the same difficulty may be put forward as a reason for giving low priority to electrodermal and to such cardiovascular measures as finger plethysmography. Temperature fluctuations and problems in securing good attachment of the recording equipment can be a nuisance in many practical situations.

A borderline measure is that of flicker fusion or two-flash threshold. Although the equipment is fairly portable, it may need a correct level of lighting in the place where the measure is being taken. It also needs an interruption of the ordinary activity, which is less true of some of the other measures. However, it has been successfully used in a number of European countries as a measure related to arousal (see, e.g., Weber et al., 1980). Interestingly, it has been rather neglected in the U.K., and this is probably an example of the need for information exchange.

From the point of view of field study, measures such as heart rate or blood pressure are particularly attractive. (Interpreting heart rate broadly, as the best score may nowadays be a complex transform of the succession of intervals, rather than just the average.) These measures can be reliably obtained with a minimum of inconvenience to the person concerned, and a number of robust pieces of equipment are commercially available. It is no longer necessary, for example, to have a doctor inflate a cuff manually and read a column of mercury. At the same time heart rate and blood pressure are responsive to various forms of stress (e.g., Obrist et al., 1978). They are therefore strong candidates for inclusion in field studies.

It would perhaps be desirable to have more data relating them to psychological measures on the one hand, and to prediction of disease on the other. In contrast to the biochemical measures, the main need for a laboratory wishing to include such measures would probably be advice not on the collection of the data, but on deriving the most interesting analyses from it once collected (Mulder, 1980).

Performance measures

Traditionally, the measurement of psychological function, rather than questionnaire response, has been a matter for the laboratory. Over the last twenty years a number of interesting findings have emerged, that seem to dissociate various environmental variables in terms of the kind of effects they produce. For example, the average speed of unpaced work seems to increase late in the day, and to decrease with barbiturates or alcohol; whereas the variability of rate of work (or performance on a paced task) seems rather to be affected by sleeplessness, noise, or chlorpromazine (Broadbent, 1971). Effects on reaction time from barbiturates are exaggerated by making the reaction stimulus hard to see, while the effects of amphetamine are rather exaggerated by increasing uncertainty about the time at which the reaction signal is going to arrive (Sanders, 1981). The effect of time of day on a task with a high memory load is different from that on a task with a low load, the best performance being early in the day rather than late (Monk et al., 1978). There are various theories about these effects, but at an empirical level the agreement between various laboratories is quite impressive.

The interesting point about the current situation is that such measures can now be taken in the field as well as the laboratory. Portable tests of simple functions such as serial reaction time have been available for a few years and are beginning to be used (Glenville and Wilkinson, 1978). The position is transformed, however, by the arrival of small cheap microcomputers, often fitting into a brief case, which can present a person at home or in the factory with complex situations and measure many aspects of their behaviour. For example, my own group has devised a test lasting ten minutes that measures sixteen aspects of a person's attention, such as their ability to ignore distractors outside a certain distance from fixation. These measures relate, for example, to time of day or to obsessionality. The various measures regarded as significant by theorists such as Sanders or Broadbent can now be obtained in field situations as well as in the laboratory. There is clearly a need to determine the relationship between such measures and the more familiar answers to questions or changes in physiological variables.

Statistical techniques

A point that deserves emphasis is the fact that field studies require fairly sophisticated statistics to make full use of the data collected. Laboratory experiments make use normally of the simple forms of analysis of variance appropriate for orthogonal factors, since it can usually be arranged that any factor is applied in an experiment independently of any other. In the field, the number of people in each of several classifications is most unlikely to be equal, and thus factors are correlated even though, in the population as a whole, they are not. For example, the ratio of repetitive to non-repetitive workers in a factory may be greater amongst women than amongst men. Analysis of the effects of repetition must therefore eliminate any correlates of sex. There are alternative ways of doing this, each of which makes assumptions that may or not be appropriate for the particular situation under study. Perhaps the most popular technique for large-scale epidemiological studies is to use multiple regression, but this has some weaknesses for the type of problem we are considering. For example, we have already seen that some of the relationships are interactive rather than additive; a low cognitive failure score, or a high level of social support, does not simply subtract from a person's score of symptoms. It actually alters the size of the effect of other variables such as job stress. One way introducing such effects into multiple regression is to include variables that are the product of two simple variables; but sometimes the effect may be some other function of the two. Another approach is to dichotomise variables and use various developed forms of analysis of variance; but this may lose information. In some cases furthermore one may extract important causal information from measuring a large number of variables and by estimating simultaneously the best single representation of subsets of the variables, and the relations between them (Jöreskog and Sörbom, 1981). It will not in general be the case for such techniques to be well-known, in their most advanced form, in psychological or medical research teams. Provision for spreading knowledge about them is probably needed.

Conclusions

From what has been said, certain needs appear that might be met within the framework of the concerted action.

a. For a number of questionnaire measures, there is a need for translation and for fresh validation in different national contexts.
b. For measures of motivation in particular, there is probably a need for development of new measures. These will require particular attention to cultural differences between countries; for example, is it the case that the satisfactions obtained from work are the same in Germany as in England ?
c. There is relatively little information about the physiological and biochemical correlates of some of the psychological measures, and this information will probably require joint action between different teams. Similarly, the extent to which one can make satisfactory prediction of disease from, e.g. raised catecholamine levels requires further evidence.
d. Provision is needed for teams without the necessary expertise to obtain assays of certain substances elsewhere; or to get advice on the correct use to make of measures they can probably obtain (e.g., heart

rate measures).

e. In particular, statistical expertise needs to be disseminated. Possible ways of doing this include not only the designation of centres than can provide expert advice, but probably also the holding of workshops at which experts in each new technique can communicate it. These could profitably be 'Bring Your Own Data' workshops!

f. Because of the problems of multi-discipline research, spread across several languages, an information exchange between people working in this area would be very welcome. This might for example take the form of circulating a list of abstracts of papers produced by each co-operating laboratory, around all the other collaborating teams.

g. Because field opportunities are relatively rare, it would make the best use of any that arise, if teams from other disciplines and countries were to be made aware of them and to collect data at the same time.

h. If on the other hand the actual data collection were to be done by one team only, then it would be helpful if a group of expert advisers in topics not the special field of the main team were to be called together at the research planning stage, to discuss the plan and advise on possible changes. These might conveniently be called Research Planning Workshops.

REFERENCES

Argyle M. and Robinson P. (1962) Two origins of achievement motivation. British Journal of Social and Clinical Psychology, 1, 107-120

Atkinson J.W. and Raynor J.O. (1974) Motivation and Achievement. Wiley

Banks M.H., Jackson P.R., Stafford E.M. and Warr P.B. (1981) An Empirical Evaluation of the Job Components Inventory. Memo No; 392, Social and Applied Psychology Unit of the Medical Research Council, Sheffield, England

Broadbent D.E. (1971) Decision and Stress. Academic.

Broadbent D.E. (1982) Some relations between clinical and occupational psychology. Keynote address: International Association of Applied Psychology, Edinburgh, Scotland.

Broadbent D.E. and Gath D. (1981) In Salvendy G. and Smith M.J. (Eds) Machine Pacing and Occupational Stress. Taylor & Francies: 243-252

Broadbent D.E., Cooper P.F., Fitzgerald P. and Parkes K.R. (1982) The Cognitive Failures Questionnaire (CFQ) and its Correlates. British Journal of Clinical Psychology, 21, 1-16

Blood M.R. and Hulin C.L.(1967) Alienation, environmental characteristics, and worker responses. Journal of Applied Psychology, 51, 284-290

Brown G.W. and Harris T. (1978) The Social Origins of Depression. Tavistock

Cochrane R., Hashmi F. and Stopes-Roe M. (1977) Measuring psychological disturbance in Asian immigrants to Britain. Social Science and Medicine, 11, 157-164

Cohen A. (1976) The influence of a company hearing conservation programme on extra-auditory problems in workers. Journal of Safety Research, 8, 146-162

Cohen S., Rothbart M. and Phillips S. (1976) Locus of control and the gegenerality of learned helplessness in humans. Journal of Personality and Social Psychology, 34, 1049-1056

Crown S., Crisp A.H. (1966) A short clinical diagnostic self-rating scale for psycho-neurotic patients. British Journal of Psychiatry, 112, 917-923

Crown S. and Crisp A.H. (1979) Manual of the Crown-Crisp Experiential Index. London, Hodder and Stoughton

Eysenck H.J. and Eysenck S.B.G. (1975) Manual of the Eysenck Personality Questionnaire. London, Hodder and Stoughton

Frankenhaeuser M. (1981) In Gardell B. and Johansson G. (Eds) Working Life Wiley, pp. 213-234

Frankenhaeuser M. and Johansson G. (1982) Stress and Work : psychobiological and psychosocial aspects. Address to International Congress of Applied Psychology, Edinburgh, England

Glenville M. and Wilkinson R.T. (1979) Portable devices for measuring performance in the field. Ergonomics, 22, 927-934

Goldberg D.P. (1972) The detection of psychiatric illness by questionnaire. London, Oxford University Press

Goldberg D.P. and Hillier V.F. (1979) A scaled version of the General Health Questionnaire. Psychological Medicine, 9, 139-145

Hackman J.R. and Oldham G.R. (1975) Development of the Job Diagnostic Survey. Journal of Applied Psychology, 60, 159-170

Henry J.P. and Stephens P.M. (1977) Stress, Health and the Social Environment. Spinger Verlag

Holmes T.H. and Rahe (1967) The social readjustment scale. Journal of Psychosomatic Research, 11, 213-218

House J.S. (1981) Work Stress and Social Support. Addison-Wesley, p. 71

Jackson P.R., Paul L.J. and Wall T.D. (1981) Individual differences as moderator of reactions to job characteristics. Journal of Occupational Psychology, 54, 1-8

Jenkins G.D., Nadler D.A., Lawler E.E. and Cammann (1975) Standardized observations: an approach to measuring the nature of jobs. Journal of Applied Psychology, 60, 171-181

Jöreskog K.G. and Sörbom D. (1981) Analysis of Linear Structural Relationships by the Method of Likelihood (LISREL). User's Guide. University of Uppsala, Sweden

Karasek R. (1979) Job demands, job decision latitude and mental strain. Implications for job redesign. Administrative Science Quarterly, 24, 285-306

Karasek R.A. (1980) Job socialisation and job strain. The implications of two related psychosocial mechanisms for job design. In Gardell B. and Johansson G. (Eds) Working Life. London, Wiley, pp. 75-94

Kasl S.V. (1978) In Cooper C.L. and Payne R. (Eds). Stress at Work. MIT Press, pp. 3-48

LaRocco J.M., House J.S. and French J.R.P. (1980). Social support, occupational stress, and health. Journal of Health and Social Behaviour, 21, 202-218

McCormick E.J., Jeanneret P.R. and McLean R.C. (1972) A study of job characteristics and job dimensions as based on the Position Analysis Questionnaire (PAQ). Journal of Applied Psychology, 56, 347-367

Monk T.H., Knauth P., Folkard S. and Rutenfranz J. (1978) Memory based performance measures in studies of shift work. Ergonomics, 21, 819-826

Moos R.M. (1981) Work Environment Scale Manual. Palo Alto, California, Consulting Psychologists Press

Mulder G. (1980) The Heart of Mental Effort., Groningen, Doctoral Thesis

Obrist P.A., Gaebelin C.J., Teller E.S., Langer A.W., Grignolo A., Light K.C. and McCubbin J.A. (1978) The relationship among heart rate, carotid dP/dt and blood pressure in humans as a function of the type of stress. Psychophysiology, 15, 102-115

Parkes K.R. (1982) Occupational stress among student nurses: a natural experiment. Journal of Applied Psychology, 67 (6), 784-796

Quinn R.P. and Shepard L.J. (1974) The 1972-73 Quality of Employment Survey. IRS, University of Michigan

Revelle W., Humphreys M.S., Simon L. and Gilliland K. (1980) The interactive effect of personality, time of day, and caffeine. A test of the arousal model. Journal of Experimental Psychology: General, 109, 1-31

Rotter J.B. (1966) Generalised expectations for internal versus external control of the environment. Psychological Monograph, 80, Whole, No.609

Sanders A.F. (1981) In Salvendy G. and Smith M.J. (Eds) Machine Pacing and Occupational Stress. Taylor and Francis, pp. 57-64

Weber A., Fussler C., O'Hanlon J., Gierer R. and Grandjean E. (1980) Psychophysiological effects of repetitive tasks. Ergonomics, 23, 1033-1046

Wing J.K., Cooper J.E. and Sartorius N. (1974) Measurement and Classification of Psychiatric Symptoms. Cambridge, Cambridge University Press

PSYCHOLOGICAL FIELD STUDY TECHNIQUES - A CRITICAL EVALUATION

K. R. Parkes,

Department of Experimental Psychology, University of Oxford, United Kingdom

This paper discusses two of the most widely used types of psychological measures employed in field studies of occupational adaptation: measures of task characteristics and the work environment more generally, and measures of mental health and well-being. Broadbent's paper in this volume points out the advantages of using measures of this type, and instances some of those presently available. However, we should hesitate to recommend existing instruments to colleagues in other disciplines without critical evaluation of their suitability and limitations. The main focus of this paper is therefore to review some of the psychometric and methodological issues involved in the use of such questionnaires to study breakdown of adaptation in work settings.

Measures of task characteristics and the work environment.

The Job Diagnostic Survey (JDS) (Hackman & Oldham, 1975) is the most familiar of the instruments intended to assess perceptions of task characteristics, and its use has been reported in numerous studies. However, it has also been the subject of considerable criticism, most recently in two extensive review articles. Discussing issues involved in the measurement of perceived task characteristics, Aldag et al. (1981) raise serious doubts as to the psychometric adequacy of the JDS, commenting on the unacceptable levels of discriminant and substantive validity, the uncertain dimensionality, and the lack of test/re-test validity data. The authors conclude that 'sole reliance on such indices can no longer be justified'. From a wider perspective, Roberts and Glick (1981) criticise the lack of clarity in the underlying theoretical formulation; the failure of most empirical research to test appropriately the relationships predicted by the model; the lack of dimensional stability; and the weakness of the rationale on which the computation of the overall score from the component subscales is based. Both reviews emphasise the need for multi-method, multi-trait approaches to the measurement of job characteristics, and discuss methodological difficulties inherent in research which relies on self-report by job incumbents as the sole source of information about the characteristics of tasks. Other authors have demonstrated the biasing effects on JDS scores of different frames of reference and job attitudes (O'Reilly et al., 1980), and of different types of social and informational cues in experimental studies of perceived task characteristics (O'Reilly & Caldwell, 1979).

Much less information has been accumulated about the Work Environment Scale (WES) (Moos, 1981), which is intended to assess ten different dimensions of the social climate in work settings. The issues of discriminant validity, dimensionality, and substantive validity raised above are largely unexplored in relation to this questionnaire, and it is clear that further psychometric evaluation and validation of the WES are much needed. In particular, examination of the inter-scale correlations presented by Moos (1981) suggests that there is considerable overlap between the different subscales. Whilst this may imply that favourable conditions tend to co-exist, it is more probable that the subscales fail to discriminate clearly the characteristics they are intended to assess. Factor analyses carried out on WES responses from student nurses (Parkes, unpublished data) strongly suggest that the data would be better described with relatively few dimensions. Broadbent's paper also discusses the two dimensions of job demand and job discretion, identified by Karasek (1979) as the critical characteristics of work situations, and notes in addition the importance of social support, and the measures developed by House (1981). Our analyses suggest that job demand, job discretion and social support are essentially orthogonal, and this conceptualisation (which is not unlike that of Payne, 1979) provides one way of describing the work environment in relatively few dimensions, for which reliable measures are currently available.

The uncertain dimensionality of both the JDS and the WES scales raises the general issue of how many dimensions are necessary to provide an adequate description of task characteristics, and of social climate in work settings, and the more specific question, addressed by Burt (1976), of how many dimensions are required to account for the covariation between work characteristics and the affective/behavioural responses of job incumbents. This issue has not been resolved, and determination of the minimum number of dimensions which adequately define the critical characteristics of the work situation in this context should be regarded as a major current research requirement. In relation to proposed European collaboration, a further important question arises as to whether the dimensions are stable across cultures, and across different types of organisations.

Measures of affective states

The problems posed by the need for short and readily-administered measures of mental health and well-being also require some further discussion. The General Health Questionnaire (GHQ) (Goldberg, 1972, 1978) and the Middlesex Hospital Questionnaire (MHQ) (Crown & Crisp, 1979), mentioned in Broadbent's paper, have been widely employed in occupational and community settings, but most of the validation data were obtained from clinical populations and general practice patients.

The GHQ was originally developed as a means of identifying potential clinical 'cases', and Goldberg (1978) reviews studies which show it has acceptable validity for this purpose, although more dubious results have been reported from recent work (Benjamin et al., 1982; Tarnolpolsky et al., 1979). In addition, Goldberg (1978) provides evidence for the validity of the GHQ as a continuous quantitative measure of psycho-neurotic disturbance in general practice patients. Whilst a scaled version of the GHQ has been derived by factor analytic techniques, the evidence for the validity of the subscales produced is not strong (Goldberg & Hillier, 1979).

The MHQ differs from the GHQ in that it was devised as a self-report instrument intended to discriminate between different neurotic syndromes in clinical groups (Crown & Crisp, 1966). Validation of the subscales has predominantly been concerned with demonstrating the correspondence between diagnostic categories and appropriate subscale scores, and with showing that the score profiles of neurotic patients are higher than those of normal subjects (Crown & Crisp, 1979; Crisp et al., 1978). However, the differential meaning of MHQ subscale scores in normal subjects is by no means clear. Two studies (Wing et al., 1978; Parkes, unpublished data) in which the Present State Examination (PSE) (Wing et al., 1974) psychiatric screening interview has been used with non-clinical groups, suggest that among normal subjects non-specific neurotic symptoms are much more frequent than symptoms characteristic of specific neurotic disorders. Both sets of data indicate that the majority of normal subjects do not experience any specific neurotic symptoms severe enough to be rated; although only a minority of subjects (15-20%) are entirely free of non-specific symptoms, such as tension, worry and irritability. Normal subjects' responses to the MHQ must therefore relate predominantly to their experience of non-specific symptoms, and consequently are unlikely to discriminate clearly between questionnaire items relating to different types of neurotic disorder.

Consistent with this, examination of the correlations between neurotic syndromes assessed by the PSE, and subscale scores from a modified form of the MHQ developed for occupational settings and used in our study, failed to demonstrate subscale validity. Whilst the total symptom score on the MHQ was significantly related to total neurotic symptoms rated on the PSE ($r=0.56**$), providing evidence of the overall validity of the MHQ as a measure of neurotic distress, the pattern of syndrome correlations with the subscales of anxiety and depression was inconsistent, and did not provide convincing evidence of the discriminant validity of these subscales. In general, the PSE ratings (irrespective of which syndrome they represented) showed the most significant relationships with the somatic symptom subscale, suggesting that subjects tended to endorse somatic symptoms on the MHQ to indicate their experience of affective distress, regardless of its nature as assessed by the trained interviewers.

The findings outlined above are consistent with evidence from a number of studies, reviewed by Parkes (1982), which indicate that self-report symptom checklists do not provide an effective means of discriminating different affective disorders. All such checklists tend to show large general factor loadings, and high inter-scale correlations (in some instances as high as 0.75), with the discrimination of anxiety and depression giving rise to the most serious problems. Hoffmann and Overall (1978) question whether the Hopkins Symptom Checklist, widely used in the U.S.A., provides a valid basis for discriminating different aspects of psychopathology, or whether it actually represents no more than a global index of distress, and this doubt applies equally to other symptom checklists. This question clearly requires further investigation if subscale scores from the GHQ and MHQ are to be used to discriminate different affective responses in studies of occupational adaptation, as in the material presented by Broadbent (1982). Whilst this would undoubtably be desirable in principle, and whilst Broadbent's data suggests that the MHQ anxiety and depression subscales may be assessing different (although related) affective responses, there is at present little information as to how these scales should be interpreted.

A further point which should also be considered in relation to symptom checklists, is the lack of a validated and widely-used self-report measure of irritability, in spite of the fact that this was one of the most frequently rated symptoms in the PSE data. Snaith et al. (1978) described a self-report scale which included measures of inward-directed and outward-directed irritability, but included only limited validation data. Further development of such measures would be of value, as an addition to the more familiar measures of anxiety and depression.

Relationships between work characteristics and affective responses

Since questionnaires which assess the work situation and questionnaires which assess mental health and other aspects of affective well-being, such as work satisfaction, are frequently administered together (or a single questionnaire which combines both sets of variables is used), it is important to consider not only the adequacy of each type of measure in isolation, but also the additional problems which may arise when relationships between them are examined. Not only do cross-sectional self-report data preclude causal inference, as Broadbent's paper points out, but interpretation of such data is confounded by other methodological problems: (i) potential spurious inflation of relationships, for instance, by consistent self-report tendencies, or by extrinsic factors such as salary levels, significantly related to both job characteristics and measures of well-being; (ii) common method variance, resulting from use of self-report questionnaires with similarly phrased questions and response formats for both independent and dependent variables; (iii) domain overlap, in which the content areas of the items assessing job characteristics are conceptually similar to those assessing affective states, thereby potentially creating in-built correlations between these measures. Aldag et al. (1981) point out that as a consequence of these methodological deficiencies the findings of almost all studies using the JDS are questionable. However, the problems are by no means confined to this particular approach, and the great majority of studies relating task characteristics to employee responses are open to one or more of these criticisms.

Some indication of the extent to which correlations between self-report measures are artificially inflated is evident from studies in which information about the work situation is obtained not from the job incumbent, but from supervisers or peers. These independent ratings tend to show much lower correlations with self-report measures of the incumbent's affective state than do the corresponding ratings obtained from the incumbent (for example, Kiggundu, 1980). Similarly, when objective information about employee responses, such as sickness/absenteeism, productivity or performance are analysed in relation to self-report measures of the perceived task characteristics, the correlations tend to be much lower than those between self-report measures and sometimes not in accordance with predictions.

Conclusions

These findings do not imply that subjective information is unimportant or that existing questionnaires for the assessment of perceived task characteristics and mental health variables should be abandoned entirely. However, they do suggest that the data should be interpreted with caution, and that reliance on self-report questionnaires as the only sources of information about both the work situation and the responses of employees should be avoided. In relation to a proposed 'package' of techniques for collaborative projects, some general suggestions for widening the scope of the information obtained and increasing its usefulness in the study of breakdown in adaptation are outlined below:

(i) Parallel forms of assessment should be available in addition to self-report versions, so that assessment of task characteristics can be carried out not only by the job incumbent but also by other individuals informed about the job concerned, for instance, supervisers and peers. The use of observers trained to make standardised observations should also be considered. Jenkins et al. (1975) showed that aspects of jobs such as work meaningfulness, variety, and autonomy, more usually assessed by self-report, could be successfully rated by observers. The multi-method multi-trait approach they adopted, in which interview and observation techniques were used to assess each of several dimensions of the work situation, illustrates how information about job characteristics from more than one source can be systematically structured and analysed. A similar approach has recently been adopted by Johnson et al. (1982) in relation to the Job Descriptive Index, a measure of job satisfaction. For the purposes of proposed European collaboration, it would also be desirable for parallel forms in different languages to be developed, although cross-cultural translation poses considerable problems, as noted by Bhagat and McQuaid (1982).

(ii) More attention should be given to alternative methods of assessing jobs, particularly those which emphasise objective aspects of the work situation (see, for instance, Elias et al., 1982; Globerson & Crossman, 1976; McCormick, 1976; Schwab & Cummings, 1976), and which avoid the preconceptions relating to task dimensionality and the nature of relevant task characteristics inherent in the JDS. The possibility of using quantitative data routinely collected for management purposes, such as production records, staff numbers, and hours worked, might also be considered. Such information can provide an objective measure of workload over a particular time period, and may be particularly useful in longitudinal research.

(iii) Particular types of research work may impose demands for other forms of assessment instruments. For instance, in longitudinal prospective studies involving assessment of work conditions, assessment of the individual's expectations at the time of entry to employment may be desirable. Thus, for example, discrepancies between expectations and actual perceptions of the work situation have been found to predict early discharge from the Navy (Hoiberg & Berry, 1978). In other studies, such as those based on a person-environment fit model of occupational adaptation, a form of assessment instrument which allows subjects to record perceptions of their ideal work situation may be needed, so that congruency between perceived and desired job attributes can be examined. The WES includes forms for both expectations and ideal perceptions, and Moos (1981) provides normative data.

(iv) If symptom checklists such as the GHQ or MHQ are to be used to assess levels of psychiatric distress, they should be validated against screening interview ratings on a representative sub-group of the population prior to the main data collection (Tarnolpolsky et al., 1979). This validation should be specifically directed towards whatever types of overall measures or subscale scores are to be used in the subsequent research.

(v) More objective types of information, which do not rely on self-report (such as sickness and absenteeism, performance, superviser's reports, and medical records, psychophysiological measures) may provide valuable indices of occupational adaptation.

(vi) The use of multiple indices to assess single constructs, such as social support, work demand or well-being, is particularly valuable, since this allows forms of multivariate analysis in which measurement error can be separated from errors of prediction (Joreskog & Sorbom, 1981).

References

Aldag, R.J., Barr, S.H. & Brief, A.P. (1981) Measurement of perceived task characteristics. Psychological Bulletin, 90, 415-431.

Benjamin, S. Decalmer, P. & Haran, D. (1982) Community screening for mental illness: a validity study of the General Health Questionnaire. British Journal of Psychiatry, 140, 174-180.

Bhagat, R.S. & McQuaid, S.J. (1982) Role of subjective culture in organizations: a review and directions for future research. Monograph. Journal of Applied Psychology, 67, 653-685.

Broadbent, D. E. (1982) Some relations between clinical and occupational psychology. Keynote address: International Association of Applied Psychology. Edinburgh, England.

Burt, R.S. (1976) Interpretational confounding of unobserved variables in structural equation models. Sociological Methods and Research, 5, 3-52.

Crisp, A.H., Gaynor Jones, A. & Slater, P. (1978) The Middlesex Hospital Questionnaire: a validity study. British Journal of Medical Psychology, 51, 269-280.

Crown, S. & Crisp, A.H. (1966) A short clinical diagnostic self-rating scale for psycho-neurotic patients. British Journal of Psychiatry, 112, 917-923.

Crown, S. & Crisp, A.H. (1979) Manual of the Crown-Crisp Experiential Index. London: Hodder and Stoughton.

Elias, H.J., Gottschalk, B. & Staehle, W.H. (1982) Arbeitsstrukturierung auf der grundlage der dualen arbeitssituationsanalyse. Zeitschrift fur Arbeitswissenschaft, 36, (8 NF), 1-8.

Globerson, S. & Crossman, R.F.W. (1976) Nonrepetitive time: an objective index of job variety. Organizational Behaviour and Human Performance, 17, 231-240.

Goldberg, D.P. (1972) The detection of psychiatric illness by questionnaire. London: Oxford University Press.

Goldberg, D.P. (1978) Manual of the General Health Questionnaire. Windsor: NFER Publishing Company.

Goldberg, D.P. & Hillier, V.F. (1979) A scaled version of the General Health Questionnaire. Psychological Medicine, 9, 139-145.

Hackman, J.R. & Oldham, G.R. (1975) Development of the Job Diagnostic Survey. Journal of Applied Psychology, 60, 159-170.

Hoffmann, N.G. & Overall, P.B. (1978) Factor structure of the SCL-90 in a psychiatric population. Journal of Consulting and Clinical Psychology, 46, 6, 1187-91.

Hoiberg, A. & Berry, N.H. (1978) Expectations and perceptions of navy life. Organizational Behaviour and Human Performance, 21, 130-145.

House, J.S. (1981) Work Stress and Social Support. Cambridge, Mass.: Addison-Wesley.

Jenkins, G.D., Nadler, D.A., Lawler, E. E. & Cammann, C. (1975) Standardized observations: an approach to measuring the nature of jobs. Journal of Applied Psychology, 60, 171-181.

Johnson, S.M., Smith, P.C. & Tucker, S.M. (1982) Response format of the Job Descriptive Index: assessment of reliability and validity by the multitrait-multimethod matrix. Journal of Applied Psychology, 67, 4, 500-505.

Joreskog, K.G. & Sorbom, D. (1981). Lisrel (Version V): analysis of linear structural relationships by the method of maximum likelihood. Chicago, Illinois: International Educational Services.

Karasek, R. (1979) Job demands, job decision latitude and mental strain: implications for job redesign. Administrative Science Quarterly, 24, 285-308.

Kiggundu, M.N. (1980) An empirical test of the theory of job design using multiple job ratings. Human Relations, 33, 339-351.

McCormick, E.J. (1976) Job and task analysis. In M.D. Dunnette (Ed.), Handbook of Industrial and Organizational Psychology. Chicago: Rand McNally.

Moos, R.H. (1981) Manual of the Work Environment Scale. California: Consulting Psychologists Press, Inc.

O'Reilly, C.A. & Caldwell, D.F. (1979) Informational influence as a determinant of perceived task characteristics and job satisfaction. Journal of Applied Psychology, 64, 2, 157-165.

O'Reilly, C.A., Parlette, G.N. & Bloom, J. (1980) Perceptual measures of task characteristics: the biasing effects of differing frames of reference and job attitudes. Academy of Management Journal, 23, 118-131.

Parkes, K. (1982) Field dependence and the factor structure of the General Health Questionnaire in normal subjects. British Journal of Psychiatry, 140, 392-400.

Payne, R. L. (1979) Demands, supports, constraints and psychological health. In: Mackay, C.J. & Cox, T. (Eds.) Response to stress: occupational aspects. Guildford, England: IPC Science and Technology Press.

Roberts, K. H. & Glick, W. (1981) The job characteristics approach to task design: a critical review. Journal of Applied Psychology, 66, 2, 193-217.

Schwab, D.P. & Cummings, L.L. (1976) A theoretical analysis of the impact of task scope on employee performance. Academy of Management Review, 1, 23-35.

Snaith, R.P., Constanopoulos, A.A., Jardine, M.Y. & McGuffin, P. (1978) A clinical scale for the self-assessment of irritability. British Journal of Psychiatry, 132, 164-171.

Tarnopolsky, A., Hand, D.J., McLean, E.K., Roberts, H. & Wiggins, R.D. (1979) Validity and uses of a screening questionnaire (GHQ) in the community. British Journal of Psychiatry, 134, 508-15.

Wing, J.K., Cooper, J.E., & Sartorius, N. (1974) Measurement and classification of psychiatric symptoms. Cambridge: Cambridge University Press.

Wing, J.K., Mann, S.A., Leff, J.P. & Nixon, J.M. (1978) The concept of a 'case' in psychiatric population surveys. Psychological Medicine, 8, 203-217.

SOCIOLOGICAL PARAMETERS IN STUDIES OF BREAKDOWN:
A SELECTIVE OVERVIEW

Johannés Siegrist,Karin Siegrist,Ingbert Weber

Department of Medical Sociology
Faculty of Medicine,University of Marburg,FRG

ABSTRACT

   This paper gives a critical overview over two main areas of sociological contributions to epidemiologic and clinical studies of human disease:parameters related to work and parameters related to social networks and social support.Due to limited space,this paper concentrates on one disease condition:ischemic heart disease(IHD).In a further section,links between social and biological processes are discussed,and a conceptual sociological approach is developed which relates critical experiences of active distress to impaired longterm control of social status. This approach is elaborated to some extent in order to illustrate possible avenues of future research which strengthen cross-fertilization between medical and social sciences in the framework of a concerted action on studies of breakdown in human adaptation.

INTRODUCTION

   It is now widely accepted that epidemiologic information provides medicine with a third source of evidence besides experimental basic sciences as well as clinical sciences. The scope of epidemiology is broadened to the extent that chronic disease has a major impact on health. Socio-economic and sociocultural characteristics become increasingly important,and this is one of the starting points of transdisciplinary cooperation between medical and social sciences. It has been stated again and again that the most telling weakness of traditional epidemiological research in the role of social factors in the etiology of disease is the oversimplification of sociological parameters(McQueen et al.1982).Social variables are often treated as conceptually unitary or at best used as an index based on inadequate components. Differentials in morbidity and mortality have been documented,for example,by several indicators of social inequality such as level of education,income,occupational status,residential status,or subjective judgements of social position. Although some of the results obtained with these conventional sociological parameters are quite impressive,indicators themselves are not accurate measurements of real social life settings which are far more complex.Only recently sociology has developed multi-dimensional concepts of social inequality which identify homogeneous socio-economic strata by means of causal models(Blau 1977,Bertram 1981). Socio-structural configurations of inequality seem to cluster around theoretically meaningful concepts such as degree of control and autonomy in the work setting(Kohn 1980)or degree of social resources under conditions of insecurity(see below). A second criticism has been raised with regard to static rather than dynamic nature of conventional sociological parameters which have been largely derived from one-point measurement. Again,only recently sociology has moved towards career approaches and towards time-series analysis.

This paper is far from providing a comprehensive overview over recent conceptual and methodological developments in the field.It concentrates on two main areas of sociological contributions to epidemiologic and clinical studies of human disease:on parameters and respective results related to work and on parameters and respective results related to social networks and social support.Both areas describe chronic potentials for stressful experiences. The area of subacute recent life changes as an additional important source of stressful experiences is documented by the paper of Aagard(in this volume).Other areas of sociological and psychological investigations in the field of stress research are documented as well in this volume showing the selective nature of uni-disciplinary approaches and the need for comprehensive interdisciplinary approaches. This holds true for every disease category,not only for IHD which is given special emphasis in this paper.

WORK-RELATED STRESSORS

Every student of work-related cardiovascular epidemiology is faced with an obviously paradoxical finding: on the one hand type A coronary prone behavior has been linked to the incidence of IHD as an independent risk factor (Dembroski et al., 1978). It has also been demonstrated that type A is prevalent among higher economic or educational groups (Rosenman et al., 1975). On the other hand, the incidence of IHD is especially high among lower socio-economic groups, even if one controls for somatic risk factors. Rose and Marmot (1981) for example, demonstrate a ratio of 3.6 between highest and lowest occupational groups in a large sample of civil servants, analyzing coronary mortality. Finnish data show an increased mortality from IHD in unskilled workers (Koskenvuo et al., 1980).
The risk of sudden cardiac death after first myocardial infarction has been linked to low educational level in a very careful study which revealed that after controlling for type of documented arrhythmia, sudden death is over three times as high in the low education group as compared to the better educated one. (Weinblatt et al., 1978). How can we explain these discrepant findings? To some extent, it may be possible to solve the problem by detection of measurement bias or by re-analysis of sampling procedures. However, as differences are marked, we think that distinct types of experiences of work stress and related conditions of everyday life may prevail in either socio-economic group. In higher socio-economic strata, especially in white-collar occupations, several protective mechanisms such as a good social support, a relatively healthy life style or high degree of social security may lead to the fact, that coronary-prone behavior pattern becomes the main or one of the main sources of socio-emotional stress experience which is thought to **precipitate** development and onset of IHD, whereas among lower class working populations, especially among blue-collar workers, heavy socio-environmental burden such as workload, including mental and physical stres-

sors, increased social instability and insufficient social support in the face of crisis are the types of stressors commonly experienced. Examples of class-related workload associated with increased IHD risk are given in this section, and a critical review suggests the introduction of theoretically meaningful sub-classifications and dimensions of stress experience.

Class-related working conditions are commonly analyzed in terms of (a) physical working conditions, (b) psychomental workload, and (c) reward structure of the work place. Study designs until now have been rather simple relating working conditions as independent variables to neuroendocrine parameters (under experimental conditions), to changes in standard risk factors (under experimental or field conditions) or to morbidity/ mortality from IHD (in field studies). Onle few studies integrate several types of working conditions or provide a basis for integrating working and extrawork living conditions and no study is known which relates all three kinds of depending variables in a longitudinal design.

However, several pieces of evidence exist which may be linked together in more comprehensive studies in the near future. Some instructive examples of such pieces of evidence are shortly presented.

Among risky physical working conditions, exposure to cardiotoxic substances, heavy lifting and noise have been linked to increased risk for IHD. All those physical stressors are more common in lower socio-economic strata, especially in unskilled and skilled blue-collar workers. The following three cardiotoxic substances have been shown to have a negative impact on the cardiovascular system: carbonmonoxide, carbondisulphid, and nitroglicerine-nitroglycol (Bolm-Audorf et al., 1983). Results relating physical activity to IHD are ambiguous, as it is well known that high physical activity has protective effects on the cardiovascular system. However, some types of unusual and stressful physical movements, such as heavy lifting, seem to affect blood pressure and thus in the long run impair the cardiovascular system (Theorell et al., 1977).

The best studied physical stressor with relation to IHD is
noise. Several studies have shown dose-related effects of
noise on endocrine secretion such as epinephrine or norepinephrine, on free-fatty acids as well as on cholesterol level
(Bolm-Audrof et al., 1983). Longitudinal studies have linked
time of exposure to high level of noise to incidence of high
blood pressure and hypertension, as well as to a prevalence
of IHD (Capellini et al., 1974). Of course, simultaneous presence of several work stressors seems to be especially typical
for blue-collar work settings. In a large retrospective case-
control study on subjects with early myocardial infarction
(Siegrist 1983) it has been shown that 60% of industrial workers with rate-fixing experienced stressful levels of noise
as an aggravating condition. Shiftwork and noise were related
as well. Simultaneous stressful presence of noise and time
pressure (C=.29; $p<0.001$), of noise and interruptions (C=.35;
$p<0.001$), of noise and heavy physical work (C=.35; $p<0.001$)
have been documented in AMI-patients compared to healthy
controls. The same holds true for physical stressors such as
heat/cold (C=.51; $p<0.001$) and toxic substances (C=.40; $p<0.001$). Similar results can be obtained from nation-wide
records on blue-collar workers undergoing medical rehabilita-
tion after first acute myocardial infarction. Occupations significantly overrepresented as compared to respective male working population in the same age group are metal workers, saw-
yers and wood working mechanists, precision instrument makers,
unskilled workers, miners and furnacemen (Bolm-Audorf et al.,
1983).

Well documented conditions of work stress have been elaborated
in numerous studies on psychomental workload. Overwork or
work overload has been the most prominent variable. In his
large studies on Bell Telephone employees, Hinkle et al.(1968)
showed that individuals who worked full time and attended evening colleges during the same time period ran a higher risk of
dying from AMI. In a prospective study on construction workers
in Sweden it was shown that frequency of reported psychosocial
job strain was elevated among those who suffered subsequent

AMI, especially among those who previously had exhibited no signs of heart disease (Theorell et al., 1977b). Concrete workers over the age of 5o with psychosocial work strain had an especially high risk of AMI. It is assumed that the older concrete workers suffered from chronic workload due to the fact that norms for group piece work as a basis for wages were increasingly difficult to fulfill with advanced age (see also Orth-Gomer et al., 1983).

A series of impressive studies relating changes in standard risk factors to changes in workload is now available. Most of them look particularly at white-collar occupations. Friedman et al. (1958) in a pioneering study observed that the blood cholesterol level of bank accountants with moderate chronic workload increased when they worked under pressure of a deadline to close their books. Bank accountants going through periods of high workload have also been shown to respond with more vulnerability (heart rate acceleration) when exposed to a standardized physical test than otherwise (Friedman et al., 1958). Kornitzer et al. (1979) studied two bank firms in Belgium, one semi-nationalized and one private, and showed a higher incidence of hypertension and of AMI in the private firm. Scores of subjective workload were found to be significantly elevated in employees of the private bank. Similar results are available as to air traffic controllers: Cobb and Rose (1973) who compared employees on particularly demanding workplaces (working at airports with high amount of traffic) to those on less demanding workplaces found a 60% higher rate of hypertension incidence in the group with quantitative overload.

Conditions of blue-collar worksettings such as repetitive work, pace system, time urgency, irregular work time have been studied experimentally as well as under field conditions (e.g. Cooper et al., 1978; Weber, 1983 as overviews). The innovative work of Frankenhaeuser et al. (1976) has introduced a two-dimensional model of stress experience specifically relevant for blue-collar working conditions. They

showed in several experimental studies that combinations of
high demands and low control were especially predictive for
endocrine changes as well as for concomitant feelings of effort
and distress. Karasek et al. (1982) have put forth the hypothesis that high demands in combination with inadequate stimulation and low margin for decision making may be an important
risk constellation for IHD. So called "strain jobs" entail many demands, often a rush for time or role conflicts,
and at the same time a poor margin for decision making. Data
from two large populations in the US and Sweden until now confirm that prevalence of AMI was highest in occupational groups
doing strain work. In the Swedish study, for example, for the
combination "hectic work pace" and "lack of control over work
pace" the relative risk for AMI under age 55 is 2.o (Alfredsson et al., 1982).

This two-dimensional theoretical approach to the study of
psychomental workload in blue-collar working conditions is of
special importance with regard to recent developments in the
field of psycho-neuroendrocrinologic stress research to be
outlined below.

Other stressful work conditions which are peculiar to lower
socio-economic groups are not outlined here as no systematic
attempt of analysis is made in this paper (see Orth-Comer et
al., 1983, Cooper et al., 1978 and Weber, 1983 for overview).

With regard to poor reward as an essential dimension of stress
experience at the work place, insufficient wages and earnings,
lack of socio-emotional support, self esteem and feed-back at
the work place and poor career opportunities have been the
best studied indicators until now. In the well-known Framingham study, recent results have shown that cerebrovascular disease incidence in a ten-year-follow-up has been significantly
higher in men who said to have poor chances to get a satisfying income. This result remained significant after controlling for age, blood pressure, cholesterol, and cigarette smoking (Eaker et al., 1983). Liljefors and Rahe (1970) showed in
a twin study that dissatisfaction with the own income and edu-

cational level was the most important psychosocial discriminating factor. Low social support has been shown to be a very important intervening variable in relations between high demands, low control and experiences of work stress (e.g. Cox et al., 1981, see below). A prospective study on 1o.ooo male employees of the Israel civil service showed that age-standardized incidence rates of AMI were significantly higher in subjects who at the beginning reported heavy conflicts with superiors at their working places (Medalie et al., 1973).
Poor career opportunities as a risk condition of IHD has been reported in the work of Hinkle et al. (1968), and downward-mobility has been shown to be related to an elevated risk of subsequent IHD in several retrospective and consecutive studies (e.g. Weber, 1983). In a large retrospective case-control study of patients with early myocardial infarction, for example, we could show, that about 22% of AMI subjects as compared to about 9% in the control group underwent forced downward mobility (Weber, 1983).
Taken together, these results show that crude indicators of social inequality such as socio-economic position can be subdivided into meaningful dimensions of stress experience which per se or in combination can be linked to risk factors, morbidity or mortality from IHD. These stressful working conditions interfere with social biographies of individuals in several ways, often weakening socio-emotional support, increasing the potential for marital discord and the vulnerability in the presence of critical life-events. This issue is elaborated in the next section.

SOCIAL SUPPORT

Social support is not a clear concept. It is defined as "harmonious social relations" (Sterling et al., 1981), "affectively positive interactions with others" (Henderson, 1977). It has something to do with attachment behavior and social bonds 'and is produced predominantly in the primary group. Other levels of social integration, however, are perhaps equally rele-

vant: membership in different kinds of cohesive groups, size
and quality of social networks, and integration on the level
of society at large. To whatever definition of support one adheres, there is some evidence for an unequal distribution of
this resource in the different classes of society: if we take
stability of the family, size of network and diversity of relations or closeness of marital bonds the lower class seems
to be disadvantaged (Berkman, 1977; Core, 1978). On the other
hand, social bonds seem to be of utmost importance in the history of our species: following a socio-biological argument,
there has been preferential selection for supportive competences as the survival of the group depended on mutually supportive behaviors. A basic assumption underlying most research
on support as a factor of relevance to health and disease is
that good social support protects individuals from the negative impact of environmental stressors. As to the mechanisms
that are at work and as to the specificity of the processes
linking society, behavior, emotions, physiology and disease
outcome there is no consensus as yet.

Parallel to the concept of stress the concept of social support is supposed to be meaningful at the following levels of
argumentation: (a) at the level of society it means strong integration via cultural patterns; (b) at the level of interactive behavior between individuals it means cognitive, emotional and instrumental assistance and positive feed back; (c)
at the level of emotion it means a feeling of security, lowered anxiety and aggression, better chances of relaxation; (d)
physiologically, social support should lead to a smaller increase of endocrine secretion of stress hormones, e.g. epinephrine, norepinephrine and cortisol; (e) as to disease processes, lack of social support should result in an increased vulnerability.

Most of the studies on social support and IHD make an implicit
or explicit use of a model that links all five levels although
the strength of these links is not very high and their quality
is rarely discussed. Another critical point is that best evidence for associations between behavior, emotions, physiology

and disease outcomes comes particularly from animal studies
(Henry et al., 1977). The problem to draw conclusions as to
humans is obvious: behaviors and emotions may have different
meanings for humans, as they are interwoven with individual
histories of experiences and cognitions. Finally a series of
methodological shortcomings in the field of social support
must be mentioned, such as weak measures of the criterion variable, problems of validity and reliability of support measures as well as lack of conceptual clarification of dimensions
of support, poor comparability of designs and operational measurements, and inconsistent results between comparable studies.

Recently efforts have been made to define social support more
clearly as a concept in theoretical contexts, as a variable in
empirical designs, and as a factor in statistical models to be
tested.
Considerations that seem to be helpful on the conceptual level converge as to certain desiderata of future research.
First we should get better knowledge about the time structure
of supportive relationships, the way in which they are embedded into the biography, the ways in which they interact with
other factors over time (Pearlin, 1981; Siegrist 1983). It
might be promising to relate such reflections on the dynamics
of support to reflections on identity and identity formation,
as processes of identity formation can be described in terms
of what is going on in supportive networks (Siegrist, 1983).
Second, we need a more clear-cut definition of particularly
good support distinctive from social control, and we need
better knowledge of what makes the quality of supportive interactions (House, 1981; Siegrist, 1983). Third, we should work
on the distinction between the general climate of support and
the problem-related or situation-related supportive interaction (House, 1981; Siegrist, 1983).
Also more studies are needed like the one by Greif and Frese
(1983), dealing with questions such as: Is social support predominantly a subjective phenomenon? Is it a generalized feeling like well-being or something more specific? Is support
experienced at the work place related to measures of inter-

actional characteristics of the work environment? As every
student in the field knows there has been a long controversy
on buffering versus direct effects of social support. Some of
these problems seem to have been resolved by Kessler (1982)
who presents a long discussion with detailed information on
the possibilities of distinguishing between provoking agents
and vulnerability factors, or, more generally, on estimating
and interpreting modifyer effects in linear models, including
suggestions as to the assessment of reciprocal causation.

On the basis of these reflections available empirical evidence
seems only moderately convincing. More conceptually and methodologically sophisticated studies are needed in order to give
sociologically meaningful information. However, some pieces of
evidence are present right now. For example, there is evidence
for an inverse relationship between cohesiveness of society or
its subsystems and risk for IHD or high blood pressure (Cassel, 1974; Marmot et al., 1976). Results can be interpreted in
terms of buffering mechanisms: good support weakens the impact
of stressors on emotional and physiological reactions. But, at
an aggregate level it seems difficult to disentangle overall
supportiveness of a society from overall stressfulness of a
society. In addition, several components of complex processes
of social change are not necessarily working into the same direction (for example: complexity of a social system, diversity and ambiguity of role demands, frames of integration into
cohesive groups, availability of strong and of weak ties). To
put it in a more general way: in measuring the impact of social support we ought to keep separate the macrosocial environment and its demands from institutionalized frames of coping behavior, where processes of identity formation and of
development of interactive competences take place.

Transferred to the level of behavior interaction frames can be
defined such as to sum up individual chances of obtaining
supports. A closer look at contexts or frames of support
shows that several indicators of poor interactive resources
are in fact associated with a higher risk of IHD. F.c marital
status, family size and network of friends have been reported

as important characteristics: subjects living alone, being divorced or widowed run a higher risk of IHD (especially men), subjects without close friends, wothout social contacts outside work and family and without sufficient size of network are at increased cardiovascular risk (Berkman et al., 1979)and as Berkman (1977) showed, those interactive resources are usually less developed in the lower than in the middle class. In lower socio-economic groups threats to socio-emotional bonds due to increased social instability and higher rates of critical negative life events have been reported in several studies (e.g. Dohrenwend et al., 1974; Brown et al., 1979). The work of Brown et al. on depressive women is of special interest in this context. It shows a much higher vulnerability in blue-collar as compared to middle class women, the working class women being more often without opportunity of getting a job and without close confident. More recently Turner (1981) demonstrated that for Canadian lower class women there was no correlation between lack of social support and depression, when stress level was low or medium, but when stress level was high, lack of support and depression turned out to be closely related. Stress levels did not have that importance for middle class women. It seems that in terms of lack of socio-emotional support, distinctive risk constellations have to be delineated, for example cumulative effects of chronic interpersonal difficulties and amount and subjective impact of negative life changes (Siegrist et al., 1980), or relations between high workload and low degree of support as to problems of overload (House, 1981; Siegrist et al., 1980). These latter findings have been reported in studies on increased IHD risk. Processes of unfavorable social comparison should be mentioned as well: recent results have shown a significantly higher risk of AMI in men whose wives had higher educational degrees (Haynes et al., 1980). Lack of socio-emotional support, increased threats to social stability, impaired networks and poor social coping resources are thought to prevail among lower socio-economic groups. Again, multidimensional analysis of stressful experiences in terms of highly demanding situations and at the same

time low degree of opportunities of control may be fruitful.
If it holds true that stressful effects of social inequality
are produced by cumulative effects of heavy socio-environmental burden at the work place and low social support in the presence of interpersonal conflicts and negative life events,
these different areas of stress experience should result in
a common pathway of neuroendocrine changes which eventually
affect the development of IHD by means of direct relationships. How can we conceive of such a common psycho-neuroendocrine pathway? Without neglecting important indirect links between social situations, modifications of risk factors by behavioral means (such as smoking, faulty diet, physical inactivity) and IHD, it is these direct links which provide a fruitful avenue for transdisciplinary research between social and
medical sciences. The next section tries to outline an approach
to such an common psycho-neuroendocrine pathway.

RELATION OF SOCIAL TO BIOLOGICAL PROCESSES

As mentioned at the beginning, this section develops a specific and highly tentative argument on direct relations between
certain social conditions or risk situations characterized by
high demand and low control, neuroendocrine reactions and pathophysiological outcomes relevant to the development of IHD.
It goes unsaid that several links within this argumentation
are empirically still unproven, and that the main purpose is
to provide a conceptual approach to transdisciplinary research
in this special field. However, the final section will give
some empirical evidence too along these lines.
1. It is possible to relate some of the most important precursors of IHD - essential hypertension, atherosclerosis, myocardial necrosis, spasms and ventricular premature beat at least
partially, in a functional way to sustained neurohormonal imbalance.
2. Neurohormonal imbalance may be produced, among other processes, by synergistic activation of the sympathetic-adrenomedullary system (with enhanced release of catecholamines) and

of the pituitary adrenal-cortical system (with enhanced release of ACTH and corticosteroids).

3. These synergistic effects are highly probable in situations which elicit two distinct behavioral patterns during a given time period or in short time intervals: the "fight-and-flight-reaction" and the "conservation-withdrawal-reaction" (Henry et al., 1977).

4. These evolutionary old patterns of coping with socio-environmental stressors are associated, in mammals and also in humans, morphologically and functionally, with two hormonal systems and their somewhat separate morphological substrate: the "fight-andflight-reaction" being related to the amygdaloid complex in the limbic system of the brain and subsequently to the sympathetic-adrenal-medullary system, and the "conservation-withdrawal-reaction" being related to the hippocampal complex of the limbic system and subsequently to the pituitary adrenal-cortical axis (Henry et al., 1977).

5. The defense response, and the conservation-withdrawal response respectively, deal with experiences or anticipations of threat to socio-emotional bonds and affiliation and with experiences or anticipations of threat to maintenance of physical or social status. It is their relation to perceived loss of basic rewards and self-esteem which allows them to play an improtant role in the complex interactions between external world, higher nervous activity and somatic regulations. Frequency, duration and intensity of these reactions may trigger pathological developments in the cardiovascular system via enhanced biosynthesis and release of related neurohormones (Henry et al., 1977).

6. Stressors which threaten socio-emotional bonds and/or maintenance of social status can be responded to by active or passive coping. During active coping, the individual has the feeling that it is necessary and possible to fight against threats (predominance of the defense response), whereas passive coping can be characterized as giving-up reaction after experience of powerlessness and/or helplessness (predominance of the conservation-withdrawal reaction). Active coping without

success, continuous efforts without reward, intense threat to one's ambition to control a relevant situation, and exorbitant or overwhelming demands upon one's adaptive capacities - these seem to be classes of critical experience which create feelings of irritation, anger, frustration and dissatisfaction. It is probable that during these experiences both stress axes are activated, i.e. the sympathetic- adreno-medullary and the pituitary-adrenal-cortical system. We propose to label these classes of critical experiences "active distress".

7. "Active distress" has been shown to be elicited in experimental situations of task performance where degree of controllability was low (Frankenhaeuser et al., 1976; Frankenhaeuser et al., 1982). This may hold true for several characteristics of work settings as well as for critical life circumstances. Finally, subjects with coronary-prone behavior may react to experiences of active distress with special intensity, as they are obviously vulnerable to threats to personal control over demanding situations.

An empirical test of these thoughts includes, as far as cardiovascular pathology is concerned, the following steps:
- demonstration of sympathetic-adrenal-medullary and pituitary-adrenal-cortical activation in response to active distress in experimental as well as in real life situations;
- demonstration of clinical evidence and relevance of neurohormonal imbalance due to synergistic action of the two stress axes, i.e. demonstration of their impact on early and/or decisive stages in the development of precursors of ischemic heart disease;
- demonstration of epidemiologic links between situations and/ or dispositions which mainly elicit active distress and the disease outcome postulated by theory.

Besides quoted experimental evidence of the first step, a large bulk of information is now available on relations between catecholamines, ACTH, cortisol (as well as sex steroid hormones and several brain peptides) and pathophysiology of the cardiovascular system as to the second step (e.g.McQueen et al., 1982; Siegrist, 1983; Karasek et al., 1982; Sterling

et al., 1981; Henry et al., 1977). This contribution gives
some preliminary information on the third step, linking refined sociological parameters to experiences of active distress
and to subsequent disease onset.

CONCEPTUAL DEVELOPMENTS

It has been pointed out that sociology suffers from lack of
longitudinal studies, and that social risk situations have
been conceptually derived from one-point measurement. Social
inequylity as measured by socio-economic position, specific
job characteristics, status inconsistency, lack of social
support etc. is considered as a chronic condition of stress
experiences which, due to chronicity itself, is thought to be
predictive of disease onset. However, such static rather than
dynamic concepts are of limited success. Researchers in the
field become increasingly aware of the need of studying careers or time patterns of stress experiences rather than stable characteristics. This is the point where we propose to
adopt a career perspective to the two-dimensional model of
stress developed by Frankenhaeuser et al.(1976; 1982) and by
Karasek et al (1982). As outlined earlier in this paper, this
model defines those job characteristics as specifically stressful which induce simultaneously high demand and low situaional control. Adopting a career perspective to this model
means to analyze conditions of low control and high demand not
only at the level of job characteristics but also, and especially so, at the level of occupational careers. The following conditions are interpreted as situations of impaired longterm control despite continuous efforts and activities of involved individuals:
- impact of the business cycle on individual job conditions
(during periods of rapid economic growth as well as during
periods of "recession" (Brenner et al.,1982));
- lack of opportunities for promotion in the presence of individual achievement motivation;
- increased demands on one's work due to cut down of personal;

- forced downward mobility or forced occupational changes (due to technological changes (Weber, 1983));
- forced overtime work due to low income;
- insecurity of employment while sustained efforts are made to maintain achieved position.

These and similar conditions are thought to stimulate recurrent efforts without giving reward in a longterm perspective.
Even worse, these conditions threaten achieved social status and potentials of self-esteem, emotional reward and sense of socially mirrored identity. Of course, analogous conditions can be found in extra-job aspects of individual biographies (which are heavily affected by the former) such as threats to affiliative bonds or disruption of status continuity between parents and children. The social roles in work and family seem to be at the core of reproduction of basic social rewards and of feelings of self-esteem. Environmental changes that threaten identity as to these social roles in the presence of sustained efforts to maintain identity thus are powerful triggers of experiences of active distress. These experiences are longlasting and difficult to cope with because they disrupt taken-for-granted routines of everyday life and even threaten essential goals of life.

This conceptual approach is demonstrated in the following figure 1 : threat to social status and presence of high need to control environmental demands critically lower the threshold for adaptive coping with experiences of active distress caused by daily social stressors. Exposure to chronic and/or subacute social stressors thus becomes critical if in addition impaired longterm control over social status as well as high need for control are present in an individual. In terms of a career approach, marked signs of non-adaptive coping in the presence of these conditions are pivots which initiate breakdown at a behavioral, emotional and physiological level.
Analyses of occupational and extra-job careers in the light of these thoughts should be able to move beyond static descriptions of persons at risk. We conclude from this argument that combined information on conditions of status control, psycholo-

logical pattern(need for control)and social stressors,as well as on indicators of non-adaptive coping is needed in order to predict subjects at risk of cardiovascular breakdown.

Figure 1: Impaired longterm control over status and non-adaptive coping with active distress:a conceptual scheme

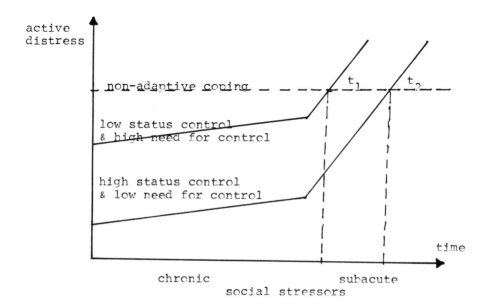

An appropriate test of this hypothesis calls for prospective time-series analysis combining sociological,psychological, endocrinological and clinical information.Prediction of premature(t_1)breakdown calls for developing and assessing valid markers of signs of non-adaptive coping at behavioral,emotional and physiological levels. Psychology until now has concentrated on the conservation-withdrawal syndrom and reactions of helplessness,hopelessness and depression as well as vital exhaustion(eg Henry et al 1977,see also the contribution of Falger et al.in this volume). However,in terms of the concept outlined above,sustained anger and hopelessness,effort and distress are thought to be the critical promoters of emotional breakdown with cardiovascular impact.

Preliminary results from an ongoing prospective study on 416 initially IHD free male blue-collar industrial workers (age 25-55) show that two indicators of non-adaptive coping with long- or short-term social stressors are simultaneously present in only 6 subjects (1,4%) of the study population. These indicators are "sustained anger and hopelessness during the last year" and "prolonged sleep disturbances unrelated to bodily pain or noise". The only two manifestations of new coronary events within one year occurred in this subgroup. Observed incidence in this subgroup, thus, is significantly higher than expected incidence in the study population ($p < 0.0001$). It is important to note that most of these subjects with marked signs of non-adaptive coping suffered from excessive workload as a chronic condition, irregular work schedules or job insecurity and that mean frequency of experienced serious negative life events during the last two years was 2.3 per person. Non-adaptive coping such as emotional upset and severe sleep disturbances emerges from social contexts which generate long-term as well as short-term experiences of active distress of high intensity. Cumulation and reinforcement of these experiences in the framework of impaired status control may critically trigger cardiovascular breakdown(Siegrist et al., 1983). Further research is needed to substantiate this approach as a step towards an integration of biological and sociological concepts (Henry, 1982) and towards a cross-fertilization between medical and social sciences. The concerted action on studies of breakdown in human adaptation provides a promising framework for such an approach.

REFERENCES

Alfredson, L., Karasek, R., Theorell, T. 1982.
  Myocardial infarction risk and psychosocial work environment: an analysis of the male Swedish working force.
  Soc. Sci. Med. 16: 463-468.
Berkman, L. 1977.
  Social networks, host resistance and mortality. a follow up study of Alameda County residents. Diss., University of California, Berkeley.

Berkman, L.F., Syme, S.L. 1979.
Social networks, host resistance and mortality: a nine-year follow-up study of Alameda County residents. Amer. J.Epidemiol. 109: 186-204.
Bertram, H. 1981.
Sozialstruktur und Sozialisation. Luchterhand Verlag, Darmstadt.
Blau, P.M. 1977.
Inequality and heterogenity. New York.
Bolm-Audorf, U., Siegrist, J. 1983.
Occupational morbidity data in myocardial infarction: a case-reference study in West-Germany. J.Occup.Med.25: 367-371.
Brenner, H., Mooney, H.A. 1982.
Economic changes and sex-specific cardiovascular mortality in Britain. Soc. Sci. Med. 16:431-442.
Brown, C.W., Harris, T.O. 1979.
Social origins of depression. Tavistock, London.
Capellini, A., Maroni, M. 1974.
Indagine clinica sull'ipertensione arteriosa e la malattia coronarica e loro eventuali rapporti con l'ambiente di lavoro in operai di un'industria chimica. Medicina del lavoro 65:297.
Cassel, J. 1974.
Hypertension and cardiovascular disease in migrants: a potential source of clues. Int.J.Epidemiol. 3:204-205.
Cobb, S. Rose,R.M. 1973.
Hypertension, peptic ulcer and diabetes in air traffic controllers. J.American Med. Assoc. 224:489.
Cooper, C.L., Payne, R. (eds.) 1978.
Stress at work. Wiley and Sons, New York.
Cox, T., McKay,C.,J. 1981.
A transactional approach to occupational stress. In: Corlett, N., Richardson, J. (Editors), Stress, work, design and producitvity. John Wiley, Chichester.
Dembroski, T.M, Weiss, S.M.,Shields, J.L., Haynes, S.G., Feinleib, M. (Editors) 1978.
Coronary prone behavior. Springer Verlag, Berlin et al.
Dohrenwend, B.S., Dohrenwend, B.P. (Editors) 1974.
Stressful life events: their nature and effects. Wiley& Sons, New York.
Eaker, E.D., Feinleib, M., Wolf, P. 1983.
Psychosocial factors and the 10-years incidence of CVA in the Framingham heart study. News Letter of the Int. Soc. Cardiol., Council of Epidemiology,January.
Frankenhaeuser, M., Gardell,B. 1976.
Underload and overload in working life: outline of a multidisciplinary approach. J.hum.Stress 35.
Frankenhaeuser, M., Lundberg, O. 1982.
Psychoneuroendocrine aspects of effort and distress as modified by personal control. In: Bachmann,W., Udris, I. (Editors), Mental load and stress in activity - European approaches. VEB, Berlin, GDR.
Friedman, M., Rosenman, R.H., Carol, V. 1958.
Changes in serum cholesterol and blood clotting time in

men subjected to cyclic variation of occupational stress. Circulation 17:852.

Gore, S. 1978.
The effect of social support in moderating health consequences of unemployment. J.hlth Soc.Behav. 19:157-165.

Greif, S., Frese, M. 1983.
Social support - real or illusionary? A validation-study of a measure of social support. (unpublished manuscript).

Haynes, S.G., Feinleib, M., Kannel, W.B. 1980.
Psychosocial factors and CHD-incidence in Framingham - results from an 8-year follow-up study. Amer.J.Epidemiol. 108:229.

Henderson, S. 1977.
The social network, support and neurosis. The function of attachment in adult life. Psychiatry 131: 185-191.

Henry, J.P. 1982.
The relation of social to biological processes in disease. Soc.Sci. Med. 16:369-380.

Henry, J.P., Stephens, P. 1977.
Stress, health and the social environment. Springer Verlag New York.

Hinkle, L.E., Whitney, L.H., Lehman,E.W., Dunn, J., Benjamin, B., Kink, R., Plakum, A., Flekinger, B. 1968.
Occupational education and coronary heart disease. Science 161:238.

House, J.S. 1981.
Work stress and social support. Eddison & Wesley, Reading.

Karasek, R.A., Schwartz, J, Theorell, T.1982.
Job characteristics, occupation and coronary heart disease. Final report, Columbia University, New York (unpublished).

Kessler, R.C. 1982.
Methodological issues in the study of psychosocial stress: measurement, design and analysis. Paper presented at the 10th World Congress of Sociology, Mexico City.

Kohn, M.L. 1980.
Social stratification and parental values. Bethesda. (manuscript).

Kornitzer, M.D., Dramaix, M., Cheyssen, H. 1979.
Incidence of ischemic heart disease in two Belgian cohorts followed during 10 years. European J. Cardiol. 9:455.

Koskenvuo, M., Kaprio, J., Kesäneimi, A., Sarna, S. 1980.
Differences in mortality from ischemic heart disease by marital status and social class. J.Chron.Dis. 33:95-116.

Liljefors, I., Rahe, R.H. 1970.
An identical twin study of psychosocial factors in CHD in Sweden. Psychosom.Med. 33: 523.

Marmot, M.G., Syme, S.L. 1976.
Acculturation and coronary heart disease in Japanese Americans. Amer.J. Epidemiol. 104:225-247.

McQueen, D., Siegrist, J.1982.
Social factors in the etiology of chronic disease: an overview. Soc. Sci. Med. 16: 353-367.

Medalie, J.H., Kahn, M.H., Neufeld, H.E., Riss,E., Goldbourt, U. 1973.

5-years myocardial infarction incidence. J.Chron.Dis. 26: 329.

Orth-Gomer, K., Perski, A., Theorell,T. 1983.
Psychosocial factors and cardiovascular disease - a review of the current state of our knowledge. Stockholm. (unpublished manuscript).

Pearlin, L.I., Liebermann, M.A., Menaghan, E.G., Mullan, J.T. 1981.
The stress process. J.Hlth Soc.Behav. 22:337-365.

Rose, G. Marmot, M. 1981.
Social class and coronary heart disease. Brit.Heart J. 45: 13-19.

Rosenman, R.A., Brand, R.J., Jenkins, D., Friedman, M., Strauss, R., Wurm, M. 1975.
Coronary heart disease in the Western Collaborative Group Study : final follow up experience of 8 1/2 years. J. Amer. Med. Assoc. 233:872-877.

Siegrist, J., Dittmann, K., Rittner, K.,Weber, I. 1980.
Soziale BElastungen und Herzinfarkt. Enke Verlag, Stuttgart.

Siegrist, J. 1983.
Psycho-social coronary risk constellations in the work setting. In: Benson, H., Gentry, W., de Wolff, C. (Editors), Behavioral Medicine: Work, stress and health. Amsterdam (in press).

Siegrist, J., Dittmann, K.H., Matschinger, H., Siegrist, K., Weber, I., Brockmeier, R. 1983.
Severe socio-emotional distress and risk of myocardial infarction - first results of a prospective study on bluecollar workers. (submitted to CVD Epidemiologic News Letter).

Siegrist, K. 1983.
Soziale Unterstützung: Das Soziale, wo es hilft. Marburg, (unpublished manuscript).

Sterling, P., Eyer, J. 1981.
Biological basis of stressrelated mortality. Soc. Sci.Med. 15:3-42.

Theorell, T., Floderus-Myrhed, B. 1977.
"Workload" and risk of myocardial infarction - a prospective psycho-social analysis . Intern. J. Epidemiol. 6: 17-21.

Theorell, T., Olsson, A., Engholm, G. 1977.
Concrete work and myocardial infarction. Scandinavian J. of work, environment and health 3:144.

Turner, J. R. 1981.
Social support as a contingency in psychological wellbeing. J.Hlth Soc. Behav. 22:357-367.

Weber, I. 1983.
Berufstätigkeit, psychosoziale Belastungen und koronares Risiko. (unpublished Ph.D.-Dissertation, University of Marburg).

Weinblatt, E., Ruberman, W., Goldberg, J.D., Frank, C.W., Shapiro, S., Chadhary, B.S. 1978.
Relation of education to sudden death after myocardial infarction. New Engl.J. Med. 299:60-65.

USE OF PSYCHOLOGICAL INDICES IN EPIDEMIOLOGICAL STUDIES OVERVIEW AND NEEDS

C. David Jenkins
Department of Preventive Medicine and Community Health
University of Texas Medical Branch
Galveston, Texas 77550 - U.S.A.

ABSTRACT
    This paper gives an overview of the methodological stages in the development of quantitative measures of psychosocial and behavioral factors which can be used in large scale population studies. The state of knowledge regarding the involvement of psychological and social factors in the breakdown of human adaptation is probably closer to being an art than a science at this point in time. The "road map" suggested in this paper may offer guidance in assisting behavioral scientists to develop the specific quantitative methods which are necessary for more definitive study of psychosocial factors in human adaptation.

INTRODUCTION
    The past two decades have witnessed a spreading realization among leading scholars in public health and medicine internationally that psychological, behavioral and social factors have a critical impact on the natural history of disease as well as on the delivery and acceptance of health care services. A clear statement of the realization of the importance of behavioral factors is expressed in the opening paragraphs of a recent publication by the Institute of Medicine, of the National Academy of Sciences in the United States (Hamburg et al., 1982):

> "The heaviest burdens of illness in the United States today are related to aspects of individual behavior, especially long-term patterns of behavior often refered to as "lifestyle". As much as 50% of mortality from the ten leading causes of death in the United States can be traced to lifestyle. Known behavioral risk factors include cigarette smoking, excessive consumption of alcoholic beverages, use of illicit drugs, certain dietary habits, reckless driving, non-adherence to effective medication regimens, and maladaptive responses to social pressures."

This realization places the behavioral sciences alongside of anatomy, biochemistry, cell biology, pharmacology, and others, as basic sciences to the fields of medicine and public health. The social and behavioral sciences are thus challenged to create the conceptual models, the constructs, and the methodology for elucidating the role of psychosocial processes in the promotion of health, the etiology of illness, the recovery process, and the utilization of health services.

The primary indicator of the location of a field of knowledge along the continuum between art and science is the degree to which its major concepts can be operationalized and measured. In the early stages of studying a scientific question the methods used must necessarily be intuitive and exploratory. The purpose at this stage is to obtain a feeling for the factors and processes which are involved, to try to put these into words, and to make some predictions or generate some hypotheses about the phenomenon under study. The next stage of the process is to develop measures of key variables which are reliable and valid and to conduct a study on a large enough group of individuals to provide an acceptably powerful test of the hypotheses. Clinical experience and unstructured clinical studies are particularly well suited to hypothesis generation, and more structured, large-scale studies - often epidemiological - are particularly well suited for hypothesis testing. There are of course many exceptions because both clinical studies and epidemiologic studies can generate and test hypotheses.

The science of psychology over the past 90 years has been concerned with the study both of overt behavior and subjective phenomena such as knowledge, beliefs, emotions, attitudes and motives. Psychometrics is concerned with the development of mathematical methods for the measurement of psychosocial and behavioral phenomena. Psychometrics entered its first period of rapid development in the 1920's and 1930's. The availability of high speed computers introduced a new period of rapid development of psychometric methodologies since about 1960. Psychometric technology has been under-utilized to date in health and medical research and practice, but the field is ready to give a strong affirmative answer to the question of whether it is possible to develop and use in epidemiologic studies reliable and valid indicators of wide ranges of psychosocial and behavioral characteristics and processes.

E. L. Thorndike, a leading figure in the field of psychometrics is quoted as having stated in 1914: "Whatever exists, exists in some amount, and can be measured." Given this challenge, and reviewing the current usual practices in the use of psychosocial variables in health studies, it becomes apparent that among their several roles in health research, scientists are called upon to achieve the potential of measuring the key concepts in our emerging sciences. This requires the following operational steps:

1) clarification of concepts and their statement in operational terms;
2) quantification of the concepts through development of measurement methods which are reliable and valid;
3) standardization of these measures on a variety of reference populations;
4) creation in the attitudes of scientists in other fields of a commitment to invest the time and resources necessary for development of rigorous measures of psychosocial variables;
5) development of an appreciation of the need for careful cross-cultural translation of psychosocial instruments, and not merely verbal translation without regard for cultural equivalence;
6) interdisciplinary development of hypotheses and explanatory models and selection of psychosocial variables based on this approach, rather than including variables primarily because they are popular or easy to measure.

A large number of psychological and sociological indicators have undergone the sequence of research development outlined above. This is particularly true for measures of anxiety, depression, neuroticism, introversion-extraversion, social status, and a variety of attitude and morale variables used by industrial psychologists. Systematic research into factors involved in the breakdown of human adaptation is relatively new, however, and many of the most promising concepts dealing with this issue have not yet been subjected to the sequence of psychometric development which is required to move this research area along the road from art toward science.

With so much developmental work needing to be done, on so many variables, in so many centers, is there some common plan we can offer which could help researchers develop rigorous quantitative methods of measuring new variables more efficiently in terms of time, effort and resources? I believe there is a road map by which we can move from clinical art to quantitative science as regards our use of psychosocial and behavioral variables in large-scale studies of human adaptation and its breakdown.

The road map shows our journey to have seven stages. I want to illustrate each of these stages using the series of studies by which the theory and measurement of the coronary-prone Type A behavior pattern was developed over the last twenty years (Rosenman et al., 1964). This is not the only example of the use of this scientific road map, but it is the one with which I am most familiar. The purpose of the discussion that follows is not to illustrate research on the Type A pattern, but rather to demonstrate a planned sequence of research which any team can follow in moving from any behavioral science concept to a quantitative epidemiological methodology.

THE ROAD MAP

1. "Soak yourself" in the situation.
2. Make explicit the bases for clinical judgement.
3. Develop an initial assessment procedure.
4. Evaluate and improve reliability
   . inter-rater (or alternate forms)
   . stability over time
   . internal consistency
5. Evaluate validity
   . content validity
   . concurrent validity
   . predictive validity
   . construct validity
6. Analyse _why_ the relationships exist:
   . biological, psychological, social, cultural linkages
7. Refine the assessment procedure to make it as powerful, as automated, and as cost-effective as possible.

1. "Soak yourself" in the situation

The first part of the journey is to "soak yourself in the situation". Persons seeking to develop a measurement method, or merely wanting to understand a new concept fully, cannot do so merely by reading about that concept or receiving a quick briefing. The same principle holds true for any biomedical, sociological, or behavioral issue. One must develop a deeply intuitive grasp of the problem and of the "local language" used to discuss the problem. In my first contacts with Dr. s.. Meyer Friedman, and Ray Rosenman, the cardiologists in San Francisco who originated the concept of the coronary-prone Type A behavior pattern, I was struck with their use of a private terminology. They used ordinary English words but they conveyed special meanings: such words as motorizing, polyhedral thinking, the lateral smile, stylistics... It was necessary to soak oneself in the language and lore of this brilliant, clinically insightful research group in order to see the Type A phenomena as they saw it and thereby obtain an authentic grasp of the concept which they had formulated. In addition, it was necessary to get an intuitive sense of the perceptions which these clinicians could not put into words. To do this it was necessary to seek out examples of Type A and Type B behavior, to share in patient interviews with the clinicians where possible, and where this was not possible, to have the clinicians do imitations of the nonverbal components of the behaviors and styles which they believed to be important.

In this early stage of trying to absorb and understand, it is important for the consultant or visiting scientist not to seek to impose his or her own concepts or frame of reference prematurely, because this will distort the true dimensions of the concepts under study. Dr. s Friedman and Rosenman realizing that they were cardiologists dealing with behavioral concepts, invited several behavioral scientists to help them translate their observations into "scientific terms". One consultant, a Freudian psychiatrist, listened carefully to their description of the Type A coronary-prone behavior pattern, and after a few minutes of careful reflection, concluded that the pattern did not really exist, as described, because it did not fit into psychoanalytic theory.

Another consultant, a specialist in psychological testing, administered an array of psychological tests to samples of Type A and Type B men, but found no significant differences between the groups on the scales. The psychologist could easily distinguish the Type A from the Type B men by the way they entered the room and engaged in the testing activities, but the scores did not separate the groups. This consultant felt frustrated and unable to help.

To avoid the limitations exhibited by these two consultants, it is important to enter the new conceptual situation with the purposes of listening and not lecturing, learning and not correcting, in short, to "soak in the situation", holding one's own theories, methods and old frames of reference in abeyance until new ways of sensing and perceiving, labeling and organizing are learned.

"Soaking oneself in the situation" involves looking for very different kinds of examples of the concept, in this case the Type A behavior pattern. It involves taking notes of many kinds of observations while one is with the clinician or content expert, and then sorting them out, discussing common themes, exceptions, and paradoxes. Then one has to "test the limits", that is to look for the dividing point between Type A and Type B. It soon became clear, in this particular instance, that Type A was a multidimensional concept and that there would therefore be separate boundaries to the concept on each of its dimensions. A valuable approach to obtaining greater focus for the concept is to establish all the things it is not. In the present example, it was important to show that Type A was not the same as stress, and that time urgency was not the same as merely doing things rapidly.

2. Make explicit the bases for clinical judgement

The second stage of the journey involves translating the understanding gained in Step One above into operational statements. It is important that these statements be cast in observable terms. For example, it is less useful to say that restlessness is a mark of Type A than it is to say that rhythmic, impatient movements of the hands or feet (such as drumming the fingers) are a sign of Type A.

Collecting and cataloging the signs and symptoms of the concept under study is an important part of this second stage. After our first period of intensive work with Dr.s Friedman and Rosenman, we collected all our notes about the signs and symptoms of Type A, the characteristics looked for and asked about, typical underlying attitudes and values and ways to elicit them, and found we had some 37 indicators of behavior type which fell into 8 categories. These were then presented to the content experts for review, correction and modification. The corrected list then formed the beginning for a systematic approach to measuring the Type A pattern. The 8 categories of characteristics were style of speech, style of thought processes, values, response to specific questions, style of interpersonal relations, facial expressions, gestures and movements, and rhythm of breathing.

3. Develop an initial assessment procedure

The third stage of the research process involves constructing a "first draft" of a procedure for measurement. A variety of approaches are possible: clinical interviews, performance tests (including experimental challenge situations), structured precoded interviews, and self-administered questionnaires. The selected method of assessment needs to be pilot-tested on groups of persons whose status is known with regard to the variable being studied. This begins the process of test construction. In Type A research this process was followed for the structured interview of Friedman and Rosenman, the performance tests and rating scales of Dr. Rayman Bortner, and for the Jenkins Activity Survey, a self-administered computer-scored questionnaire for the Type A behavior pattern.

4. Evaluate and improve reliability

The fourth stage of the research sequence is to evaluate the reliability of this initial assessment technique. Three types of reliability are commonly considered, but not all of them may be relevant to a given assessment procedure. The three types of reliability are:
a) Inter-rater reliability. This asks the question of how consistently different judges rate the same object. The equivalence of parallel forms of the same test is also an example of inter-rater reliability. Unles there is a high rate of agreement on "who is a Type A", it would be fruitless to pursue more advanced research questions.
b) Stability over time. How consistently is the same object rated on different occasions? A different degree of "test-retest reliability" will be expected depending on the nature of the concept being evaluated. One would expect greater consistency over time for the measurement of enduring characteristics such as intelligence, extraversion, or the Type A behavior pattern, whereas the measurement of naturally fluctuating

states, such as fatigue or nervous tension, would be expected not to have so high a level of consistency on consecutive assessments, particularly if these assessments were done after long intervals of time. If a method purporting to assess a stable characteristic takes on very different values over repeated measures, one has reason to suspect something wrong with the measurement method.

c) Internal consistency. This asks the question whether the scale or measure is tapping a homogeneous construct. High internal consistency is a valuable property of a scale if the characteristic measured is postulated to be unidimensional. However, if a construct is known to be multidimensional, like the Type A behavior pattern, too high an internal consistency coefficient for a scale might lead one to suspect that it may be measuring only a single part of the total construct. More commonly, the problem is that a scale has too low an internal consistency coefficient. This finding suggests the concept has been diffusely defined or that many items in the scale are only peripherally relevant to its central idea.

Unless an assessment approach has a high degree of reliability, measurement of the desired concept becomes so diluted with random and systematic error that a clear rating of which individuals and groups possess more or less of the characteristic cannot be determined, and the true nature of the relationship of the variable under study with other known variables remains poorly estimated.

Reliability alone is not enough to guarantee that a measure is useful. Some scales, tests and clinical methods have excellent technical qualities which give them a high reliability, as measured by all three approaches above, and yet these methods predict nothing and explain nothing. What they lack is validity.

5. Evaluate and improve validity

The fifth stage in developing an effective measurement technique for psychosocial or epidemiological variables is to evaluate - and if possible to improve - validity. Psychometricians have identified five different approaches to assessing validity. A landmark paper generally credited for expanding the understanding of validity is that by Cronbach and Meehl (1955).

Since that time other writers have expanded on this notion. This paper will just touch on the main points.

The five approaches to assessing validity, ranging from the simplest to the most sophisticated, are as follows:

a) "Face" validity. This is at once the most naive and the most commonly used criterion of validity for psychosocial measures used in health studies. This approach merely asks whether there is in the mind of the scientists conducting the study an a priori or commonsense similarity between the measurement technique and the condition they are trying to assess. Thus, a series of questions asking for evaluations of the attributes of physicians would have face validity as a measure of attitudes toward this profession. This simple acceptance of face validity does not consider the possibilities that the answers of many persons to these questions may be influenced by attitudes toward authority figures in general, general tendencies to agree with statements irrespective of whether they are couched in positive or negative terms, or desires of respondents to give socially acceptable answers. In addition, this approach does not take into account the possibility that some persons may hold negative views toward physicians but be unwilling to express them in response to direct questions. Thus, while face validity is the easiest to achieve, it may not provide information which truly reflects underlying attitudes or predicts future behaviors. "Face validity" is the level at which most measures start, but there is a great risk of major error if one uses a scale which merely appears valid, unless a higher level of validity can be demonstrated.

b) Content validity. This approach to validity can be achieved when the full range of the concept under study is known and can be systematically sampled. For example, a biology test based on systematic sampling of statements made in a biology textbook would have content validity. It is easier to develop content-valid measure for areas of knowledge than it is for attitudes, although the latter could be accomplished if one could collect an exhaustive list of attitudinal statements about a particular class of objects and then sample systematically from these.

c) Concurrent validity. This approach requires that an accepted criterion of validity be available for each subject under study, and that the items of the proposed test instrument administered at the same time as the criterion was obtained be statistically capable of identifying persons in the same way they are labelled by the criterion. An example from medicine would be the use of blood glucose levels as a test for the presence of diabetes mellitus. In this case, the criterion of validity would be the clinical signs and symptoms of diabetes (not including blood glucose), and the evidence for concurrent validity would be the degree to which blood glucose levels could be used to accurately identify clinical diabetics. In our initial studies to develop a self-administered test for the Type A behavior pattern, we used the clinical interview of Friedman and Rosenman to assess the degree of concurrent validity of individual test items. We then developed a scale combining valid items through use of discriminant function equations, and checked the concurrent validity of the scale score by its association with the criterion of the cardiologists clinical interview rating. The next step was to determine whether scale scores discriminated between coronary heart disease patients and healthy persons. This also tested concurrent validity but with a more stringent, biomedical criterion.

d) Predictive validity. This approach also requires a criterion, but differs from concurrent validity in that the criterion measure is obtained at a later point in time than the data comprising the test procedure. For this type of validity the prediction is not only statistical but is future oriented. Biological risk factors for coronary heart disease were established by the method of predictive validity. Similarly, the Activity Survey Type A Scale achieved predictive validity only when it had been shown that among men free of coronary disease, those with high Type A scores had a higher risk of developing new CHD over the next four years than those with low Type A scores. Although both concurrent and predictive validity depend upon the availability of an accepted criterion, predictive validity is considered the stronger type of evidence and is more likely to be reflective of a truly causal association than is concurrent validity alone. The latter is more subject to retrospective bias, confounding of predictor with outcome, and selective entry and exclusion from the sample.

e) Construct validity. This approach to validity is more complex and theoretical, and it is based on the accumulated evidence from a number of studies rather than a single definitive demonstration. Construct validity asks such questions as: "What is the overall meaning of the measured variable? How does it fit into the established framework of knowledge? If our measured variable really assesses what we claim for it, with what other variables should it correlate highly? With what other variables will it correlate not at all?" When these data are collected, do the answers fit our predictions? If they do not, we have to change our beliefs about what the measured variable really reflects. The pursuit of construct validity gives the researchers much better understanding of what a particular variable really means and whether the test for it measures what is claimed for it. Construct validity is the approach to use when there is no readily available, theoretically satisfactory criterion of validity. Some examples of construct validity studies involving the Type A behavior pattern are the following: Early in the development of the Type A pattern, Friedman and Rosenman argued that Type A people were not neurotic and neurotics were not Type A. Other psychologists felt that Type A behavior was one expression of stress and psychological problems. Studies correlating measures of Type A with previously validated scales of psychological adjustment, such as the Sixteen Personality Factor Inventory or the Minnesota Multiphasic Personality Inventory showed Type A to have very low (from 0.00 to 0.25) but usually positive correlations with scales reflective of neuroticism and psychopathology. Later work has generally supported these findings, although occasionally higher correlations have been observed. The conclusion thus was that Type A and neuroticism are not mutually exclusive, but rather they are uncorrelated. The findings in a sense, refuted both sets of clinical predictions.

Another hypothesis was that if Type A people were really hard-driving and achievement oriented there would be more of them proportionally in higher prestige job levels than in lower occupational levels. Calculating the mean Type A scores of persons at different levels of an occupational prestige scale showed a highly significant tendency for more Type A persons to be situated in higher level jobs. The findings had sufficient strength to be supportive of the original hypothesis, but inasmuch as there were persons of both types at all job levels, and the relation of Type A to coronary disease was stronger rather than weaker when occupational level was statistically controlled, the analysis added clear evidence that Type A was not merely a surrogate measure for socioeconomic status.

Brozek and colleagues in Minnesota (1966) found that the Active Scale of the Thurstone Temperament Schedule was predictive of future coronary disease, and other researchers showed that measures of Type A were highly correlated with this particular scale. This provided added construct validity for the Type A measure by supporting the inference that the excessive activation, which is a common element both to measures of Type A and the Thurstone Active Scale, was in fact a risk factor for coronary disease in different populations.

6. Analyse why the relationships exist

The sixth stage in development of a psychosocial or behavioral measure for use in epidemiologic studies of health and disease is to analyse why the relationship holds. In the case of health studies, one would search for psychological, psychophysiological and biological linkages between the variable shown to be a valid predictor and the health outcome which one desires to predict and ultimately to control. In the case of the work with the Type A behavior pattern, we asked whether Type A might have only an indirect association with CHD risk, such as perhaps through a strong relationship to an already established risk factor. Exhaustive studies by a number of research teams have shown that Type A is not secondary to more commonly and previously recognized risk factors but that it makes an independent contribution to prediction of future CHD. Similarly, in the psychological realm, stratifying by other psychosocial risk factors for CHD, or covarying them out statistically has not diminished the predictive strength of Type A but shown it to be independent of the other psychological variables against which it has been tested to date.

Having ruled out the more obvious possibilities for secondary associations, researchers turned to more direct studies of explanatory biological mechanisms which might link Type A behavior to CHD risk. These studies involve examination of physiological, hormonal and neurohumoral differences between Type A and Type B persons. More recently, it has seemed that the primary differences between Type A and Type B persons are not in resting levels of these physiological and endocrine parameters, but rather in the reactivity of these systems to challenge circumstances.

Several research groups are now searching for differences between Type A and Type B persons in a host of cardiovascular and psychophysiological processes. A great many promising initial findings have been published (Dembroski et al., 1978), but there is still a great distance to go to understand the total picture. This direction of research could not take place until reliable, valid and convenient methods for measuring the important behavioral variable had been developed.

7. Refine the assessment procedure to make it as powerful, as automated and as cost-effective as possible

The steps we have outlined for the development of psychosocial measures, if pursued successfully, have brought us to the position of having a clearly defined and operationalized concept and one or more measures of it which are dependable and which really do quantify the concept as defined. The continuing work to develop construct validity and analyse why the measure relates to outcomes as it does has placed our newly measured variable in the context of other related knowledge and provided insights into the dynamics of the mechanisms which it reflects. Now at the final stage of the sequence, the time has come to take another look at the studied variable and our measurement techniques. On the basis of what we have now learned, is it possible to make further refinements in the assessment procedure? Or perhaps develop an entirely new, more powerful and more convenient method? This kind of evolution and progress is constantly taking place in biochemistry, endocrinology, cell biology and other basic sciences. It also takes place in behavioral measurement.

It is important in this context to strike a balance between constant flux of measurement methods, which leads to confusion and non-comparability of research results, and excessive rigidity, where a method of inquiry or assessment is used to the exclusion of newer more effective procedures. Perhaps the ideal course is for researchers to use well established measures of a variable simultaneously with new measures of that same variable so that comparability and a smooth transition can be achieved.

It has been our experience with psychosocial and behavior variables that easily administered and automated methods are the ideal toward which to strive. This is of course with the understanding that the quick, automated method is not seriously less reliable and less valid than the more difficult and expensive procedure.

In research into the Type A behavior pattern the first well developed procedure was the structured interview of Friedman and Rosenman. This required carefully trained interviewers and about 20 to 25 minutes of time from each subject. The logistics of the interviewing task were such that it took more than a year to complete the intake process in an epidemiological study involving 3,500 men. In contrast, the self-administered computer scored questionnaire developed to measure the Type A pattern can be administered to thousands of subjects in epidemiologic surveys in a few days. The further refinement of questionnaire - printing it on mark-sense forms - eliminates the time-consuming and expensive step of key-punching answers onto a computer card or into a terminal. The mark-sense form is directly machine-readable, the answers and their identifying case numbers are read onto magnetic tape cassettes and scored directly by computer.

Increased efficiency and automation must be the last stages, and not the first stages, in test development. Premature abbreviation of scales and automation can lead to loss of validity and reliability, and what is even more serious, the degree of loss may never be fully realized. It is clearly best that the developmental process be fully followed at the beginning of the research measurement process so that a clinically adequate optimally powerful version of the measurement method is available to be used as a criterion of validity against which to test abbreviated procedures.

CONCLUSIONS

This is a valuable topic for the European Community to be addressing in the 1980's. It is a critical issue for the health and welfare of communities and nations. The sciences of psychometrics and behavioral measurement have well established methodologies which can be applied to the task of developing specific measures of those concepts and variables which theory and clinical understanding identify as salient to issues of human adaptation. This will be an exciting decade as new work is done in the quantification of the dependent variable, human adaptation, which has biological, psychological, sociological and cultural manifestations. This will also be a time for developing more focused and more powerful measures of those independent variables which tend to break down or to build up the level of human adaptation.

REFERENCES

Brozek, J., Keys, A., Blackburn, H., Personality differences between potential coronary and noncoronary subjects. Ann NY Acad Sci 134: 1057-1064, 1966.

Chun, K., Cobb, S., French, J. R. P. Jr., Measures for psychological assessment. A guide to 3,000 original sources and their applications. Ann Arbor: University Michigan, Institute for Social Research, 1975.

Cronbach, L. J., Meehl, P. E., Construct validity in psychological tests. Psychol. Bulletin 52:281-302, 1955.

Dembroski, T. M., Weiss, S. M., Shields, J. L., Haynes, S. G., Feinleib, M. (Eds.), Coronary-prone behavior. New York: Springer, 1978.

Hamburg, D. A., Elliott, G. R., Parron, D. L. (Eds.), Health and behavior: Frontiers of research in the biobehavioral sciences. Institute of Medicine, National Academy of Sciences, Washington D. C., National Academy Press, 1982.

Jenkins, C. D., Zyzanski, S. J., Rosenman, R. M., The Jenkins Activity Survey, Form C. New York: Psychological Corp., 1979.

Rosenman, R. H., Friedman, M., Straus, R., Wurm, M., Kositchek, R., Hahn, W., and Werthessen, N. T., A predictive study of coronary heart disease: The Western Collaborative Group Study. J. Amer. Med. Assn. 189: 15-22, 1964.

STRESSFUL LIFE EVENTS AND ILLNESS: A REVIEW WITH SPECIAL REFERENCE TO A CRITICISM OF THE LIFE EVENT METHOD.

Jørgen Aagaard
Institute of Psychiatric Demography,
Aarhus Psychiatric Hospital,
DK-8240 Risskov,
Denmark.

INTRODUCTION

During the last decades a considerable number of quantitatively based studies on relationships between one and other indices of psycho-social stress and illness have been published. However, in many instances it remains unclear how these associations are to be interpreted, partly owing to methodological problems and limitations in the research strategies applied (Gunderson & Rahe, 1974, Dohrenwend & Dohrenwend, 1974, 1981, Rabkin & Struening, 1974 and Rahe & Arthur, 1978).

The purpose of this review is to point out some results and methodological issues based on the literature and experience from own investigations on psycho-social stress and various diseases in childhood, and hypertension and duodenal ulcer in adult (Aagaard, 1979, Aagaard et al., 1980, Aagaard et al., 1981a, Aagaard et al., 1981b, Aagaard, 1982, Aagaard & Kristensen, 1982, Aagaard et al., 1983 and Aagaard et al., in press).

In principle, at least two different approaches exist in analysing associations between psycho-social stress and illness, one mainly emphasizing personality traits including intrapsychic conflicts, coping etc., the other mainly stressful life events.

In a traditional psychiatric/psycho-analytical concept, which has been the framework in analysing the psychosomatic disorders, e.g. hypertension and duodenal ulcer, the main emphasis has been laid on personality traits. In 1913 Jaspers delimited the reactive psychosis as a disease entity and set up the etiological criteria. His viewpoint has had some impact on clinical psychiatry (e.g. Wimmer, 1916 and Færgeman, 1945). For him life events are seen as provoking factors of some psychoses, but they are assessed partly on individually based reaction patterns, partly on predisposing factors involving an unspecific lowering of resistence. Within the psychiatric crisis-theory (Cullberg, 1976) stressful life events are also considered provoking, but here their impact are assessed in a psycho-dynamic concept, and the person's psycho-social level of development and the accessibility of social support. Lazarus (1966) has pointed out the importance of personality traits in the comprehension of the effects of stressful life events; for instance persons with low self-esteem may experience and misunderstand many situations as stressful, and react with helplessness with a higher risk of illness, and persons characterized by pronounced defence-mechanisms or repressions may have less ability to cope adequately with a certain stressful situation. Lazarus has underlined the individual aspect i.e. what is stressful to one person need not be stressful to another person, and has underlined that the most important modifier in the effect of stressful life events is the person's coping strategy. This approach is in accordance with Hinkle (1958, 1974), who has suggested that the psychologically or socially conditioned limitations in coping with difficulties are more important than stressful life events for risk of illness.

In the different investigations on stressful life events the dependent variables are 1) onset of certain physical or psychiatric disorders, 2) the course of certain diseases, 3) various psychiatric symptoms, 4) susceptibility to unspecific illness, and 5) utilization of health services. In some instances a distinction has been made between illness and illness behaviour, in other instances between a general susceptibility to illness and onset or course of a certain disease. The concept of a general susceptibility or the unspecific approach is mainly based on Selye's theories (1956), whereas the specific approach is based on new psychosomatic concepts (e.g. Lipowski et al., 1977). The last mentioned approach involves an analysis of the relative effect of biological, social and psychological factors including stressful life events on a certain disease.

In the present investigation the following terminology has been applied:

By determinants respectively predictors is meant statistical risk indicators computed from cross-sectional respectively prospective investigations.

By psycho-social stress is meant an accumulation of unfavourable social factors and/or life events which may be potentially stressful.

By social factors is meant relatively stable aspects as level of education, income, social status, residence etc. Unfavourable social factors refer to a relative lack i.e. low level of education, low income etc.

By life events is meant changes in circumstances of life such as marriage, divorce, unemployment etc., which may be stressful.

By psychological distress is meant an accumulation of non-specific psychological symptoms. In present investigation a standardized inventory according to Kühl & Martini (1981) has been applied.

The life event questions applied in the present studies were a modified version of Holmes & Rahe's (1967) and Coddington's (1972) inventories. The introduction of the life event/LCU (Life-Change-Unit) method is considered as crucial. In this method a standardized selection of life events named Schedule of Recent Experiences (SRE) was used. These events were weighted by a professional panel in a concept of "social readjustment". By addition of these a priori determined weights on those life events which a person has experienced during a limited period, the variable named LCU was computed. This variable is comprehended as an operationalization of the psycho-social stress of shorter duration. Since the introduction, the life event method has been modified and improved several times, partly on the background of the criticism in which also Rahe (e.g. Rahe & Arthur, 1978) has been actively involved. The modifications concern 1) the selection of events, 2) the weighting procedure and 3) addition of various intervening variables between life events and disease.

In the following part of this paper some results from empirical investigations will be presented, as a background for a discussion of the relevance and applicability of the method.

1. STRESSFUL LIFE EVENTS AND ILLNESS. RESULTS FROM SELECTED EMPIRICAL INVESTIGATIONS.

Children, stressful life events and illness have been object for several investigations. However, already before parturition stressful life events may play a role for the children's future illnesses. Pregnant women with an accumulation of stressful life events are more likely to have complications at parturition (Gorsuch & Key, 1974 and Nuckolls et al., 1972).

In a prospective study (Egeland et al., 1980) a number of sociopsychological data were obtained during pregnancy. The health nurse assessed that about 10 per cent of the infant received inadequate care at one year of age. Stressful life events in pregnancy were found to be statistically predictive for inadequate care, but was also found to be of importance to the mother's mental health status, especially anxiety and insecurity.

Coddington's (1972) investigation concerns the frequency of life events in a population sample of children. This study has formed the reference group for many later investigations comparing life events reported by parents to sick children, and "our findings confirm the hypothesis that children in any patient population experience more significant life events preceding an illness than is to be expected in a healthy population" (Heisel et al., 1973).

The excess rate in recording of preceding stressful life events is recognized in several patient-control studies of children admitted to hospital (Kashani et al., 1981). A stress hypothesis has been confirmed for several physical diseases e.g. pyloric stenosis (Dodge, 1972), poisonings (Sibert, 1975), rheumatoid arthritis (Heisel, 1972) and cancer (Jacobs & Charles, 1980), however, a stress hypothesis is more evident as an explanation of children with a mental disorder or socio-pediatric patients. Epidemiological investigations have shown that parents to maladaptive school children (Sandler & Block, 1979), or to children with behavioural disorders (Gersten et al., 1974) more often report stressful life events than other parents. However, it must be emphasized that unfavourable social conditions of the family or a history of a mental disorder of the child give more explanation (statistically) for behavioural disorders than stressful life events (Gersten et al., 1977).

Own investigations within this area concern an analysis of the social history including preceding stressful life events for children (aged 1-14 years) who during one year were admitted to a pediatric department, as well as children who happened to consult a general practitioner. These two groups were follow-up examined with a registration of 1) admission, respectively readmission to hospital, 2) consultations at the general practitioner and 3) the parents' judgement of the child's health one year after the index hospitalization respectively consultation. Within the cross-sectional studies a multivariate analysis of determinants for diagnosis was performed, and within the prospective studies a multivariate analysis of the most important predictors for the children's later health was performed.

Determinants. An accumulation of unfavourable social factors was associated with psychosomatic and mental disorders in the material from hospital as well as the material from general practice. Unfavourable social factors were not associated with respiratory tract infections among the children in hospital. In the material from general practice no social explanation was found of e.g. otitis media chronica/recidivans. Various stressful life events were frequently reported before the index contact. In the material from hospital these events varied between diagnostic groups, thus a pronounced accumulation was found to psychosomatic and mental disorders, and some accumulation to respiratory tract infections in preschool children. The additional information (statistical explanation) on life events was pronounced.

Predictors. Admission respectively readmission to hospital was mainly explained by chronic or recurring diseases. Indices for psycho-social stress did not appear to be of considerable importance; however, the predictive effects of items concerning conflicts with peers, only-child, change to another school, mother beginning to work, were noticeable. Children with a psychosomatic or mental disorder at the index hospitalization were on the whole not readmitted. Poor health in the child judged by the parents had a very complex prediction pattern mainly concerning indices for more long-term psycho-social stress. In the material from hospital it was found that indices for chronic physical illness, psychosomatic or mental disorders in school children and respiratory tract infection in preschool children were predictive for poor health. Furthermore, poor health in mother was a pronounced predictor variable. Among the preschool children in the material from general practice solitary supporter was the most pronounced social predictor, however, among the school children both parents with full-time job was the most pronounced. Many consultations were partly predicted by illness, partly by indices for more short-term psycho-social stress. The most pronounced predictor items were: birth of brother or sister, change to another school, one of the parents' increased absence from home due to work, broken home, and addition of a third adult family member.

Young adults, mainly at university (Jacobs et al., 1969) or in military services (Cline & Chosey, 1972 and Rahe, 1974) have participated in several prospective investigations on stressful life events and illness. In the studies carried out in the U.S. Navy consultations to the soldiers' sick-quarters were the dependent variable. A high correlation was found between degree of previous psycho-social stress and number of consultations. Among young athletes life events have been reported to increase the risk of sport injuries (Padilla et al., 1976, Coddington & Troxel, 1980).

Among adults, many investigators have reported significant associations between stressful life events and a great number of severe physical or mental disorders.

Of severe physical diseases may be mentioned that stressful life events have been reported of etiological importance for e.g. tuberculosis (Hawkins et al., 1957), certain types of cancer (Solomon, 1969 and Schmale & Iker, 1971), hypertension and duodenal ulcer (Weiner, 1977 and Goldberg et al., 1980), coronary heart disease and myocardial infarction (Rahe et al., 1973, Theorell et al., 1975, Jenkins, 1976, Theorell & Floderus-Myrhed, 1977 and Kringlen, 1982). In a prospective, epidemiological investigation on middle-aged men Theorell reported that certain life events concerning increased work load were predictive of myocardial infarction. Jenkins (1976) has stated that work load and prolonged conflict situations are of importance as predictors of coronary heart disease.

Own investigations on adults concern patients suffering from hypertension, and patients about to have elective surgery for duodenal ulcer.

The material of patients suffering from hypertension consisted of all patients with essential hypertension in a medical out-patient clinic. The patients were followed prospectively during 2 years at 6-month intervals, where a standardized clinical programme was performed. Data concerning social and psycho-social factors including stressful life events were obtained at the start of the observation period. A multivariate predictor-analysis was performed on groups of patients with specified exacerbation in the clinical variables. Many of the patients with hypertension were bothered by symptoms due to disease, and many reported worries of different kinds and stressful life events. The disease symptoms were relatively independent of the severity of the disease and drug treatment groups. The statistical information of selected life events was similar to that of the clinical and social background variables in explaining aggravation of the disease. Severe exacerbation of the disease (e.g. morbid events) was mainly explained by higher WHO group and higher age, but information concerning symptoms due to the disease and retirement was also of importance. Worsening among patients with mild to moderate severe hypertension was predicted of items concerning personal problems, worries of different kinds and conflicts at home.

The duodenal ulcer material consisted of all patients with indication for elective surgical treatment for duodenal ulcer admitted to two surgical departments during a period of one year. Data concerning social, psychological, and psycho-social factors including stressful life events were obtained through a preoperative interview. At one year follow-up a blind clinical evaluation was performed and information concerning the patients' assessment of outcome was obtained. A multivariate predictor-analysis was performed. Most patient benefitted by treatment. Several improvements were noticed within the patients' work and leisure time activity, but the excess rate of symptoms indicating psychological distress remained unchanged during the one year follow-up. An operation with pyloroplasty was the most pronounced predictor for dumping, and unmarried individuals the most pronounced for a poor Visick grade or postoperative dyspepsia. The type of operation was without importance for the patients' assessment of outcome, but items such as long ulcer history and postoperative complications partly predicted those patients who at one year follow-up stated no improvement due to the operation. A preoperative accumulation of non-specific psychological symptoms predicted the clinicians' assessment. Indices for psycho-social stress or individual life events did not appear to be of any importance in prediction of outcome.

Mental disorders and stressful life events have been extensively examined several times, especially schizophrenia, various depressions, attempted suicide and minor psychiatric disorders (Dzegede et al., 1981, Day, 1981 and Finlay-Jones, 1981).

Jacobs & Myers (1976) found that schizophrenic patients admitted to hospital did not report more life events than a group of normal controls, however, patients with a depression reported twice as many life events as the schizophrenics (Jacobs et al., 1974).

In Brown & Birley's (1968) study of schizophrenics, patients were selected with a well-defined onset, and within 3 months before admission to hospital. By comparing this subgroup of schizophrenics with a control group was found that the schizophrenic patients reported twice as many stressful life events, most pronouncedly in the last three weeks prior to onset of the psychosis. No association was found between the character of the life events and the character of the schizophrenic psychosis. It was concluded that stressful life events may precipitate the time of onset of the psychosis.

Paykel et al. (1969) compared stressful life events reported by patients with depression, and a control group. The patients had in the preceding period experienced three times as many life events especially concerning increased conflicts at home, and illness or death within their family. With a similar technique Paykel et al. (1975) have interviewed patients admitted for attempted suicide. These patients had experienced four times as many life events as the control group, especially concerning increased conflicts with spouse, severe disease in the family and own severe physical disease.

Brown (1974) has pointed out that in instances where onset of a depression is precipitated by a life event this is often experienced as threatening. Brown (1981) has suggested that the link between severe stressful life events and onset of a physical disease may be mediated by an intervening psychiatric disturbance.

Several investigations on stressful life events, psycho-physiological symptoms and various psychiatric symptoms, have been published (Myers et al., 1972, Uhlenhuth & Paykel, 1973, Markush & Favero, 1974 and Dzegede et al., 1981). A consistent finding is that an accumulation of stressful life events is correlated strongly to an excess of psychiatric mainly neurotic symptoms. Finally it must be mentioned that in the earlier described prospective, epidemiological study of Theorell et al. (1975) which mainly concerned incidence of severe physical diseases, stressful life events were predictive to onset of neurosis.

Older people and the importance of life events have only to a lesser extent been investigated (Blazer, 1980). Severe stressful life events such as death of spouse (Madison & Viola, 1968, Parkers & Birthwell, 1971, and Jacobs & Østfeld,1977) or admission to a nursing home (Liberman, 1961 and Rowland, 1977) are reported associated with an excess of morbidity and mortality during the year after the event.

2. DISCUSSION OF THE LIFE EVENT METHOD.

In a great number of investigations associations have been found between stressful life events or indices for short-term psycho-social and many different physical and mental disorders among children as well as adults. It must be emphasized that these associations are not tantamount to the fact that stressful life events contribute to a pronounced explanation (i.e. statistical information). In Rahe's investigations in the U.S. Navy, life events yielded an explanation (statistically) to around 10% of the later reported spells of illness. In Brown's investigations only some of the patients reported preceding severe, threatening stressful life events, however, this was more frequent than in a comparison group.

It is essential to be attentive to the fact that stressful life events only refer to one dimension in a socio-medical disease model. Other dimensions involve social and psychological factors and biomedical, genetic variables with supposed relation to the disease.

Within the various investigations many different techniques and designs have been used. Some investigations, in particular those published before the 1970's, are from a methodological point of view criticizable. In a review from 1978 Rahe & Arthur wrote: "Initial conception of the life changes and illness work were necessarily simple and straightforward, but as the evidence for the validity of the general concept has mounted, it has also become necessary to think in terms of the complexity of the social, psychological and physiological variables involved. The brain, whatever its other higher integrative functions is now seen as the major endocrine organ as well. It is in the brain that psycho-social events, impinging as they do upon the perceptual systems, are transduced into physiological events which activate the neuroendocrine axis and other systems, such as autonomic control mechanisms. We believe that further studies of an epidemiological character which correlate life change and illness will be redundant. Instead, the enormously difficult task awaits us of filling in the crucial steps in an all-encompassing model which takes into account not only environmental variables but the sociological, psychological and physiological characteristics of the individual. We see a brigth future for this field of investigation. It is truly a field in which the efforts of sociologists, psychologists, psychiatrists, physiologists, endocrinologists, and pathologists will be necessary to provide a complete picture of the precise interrelationships between the vicissitudes of life and sickness". Thus, Rahe and associates have come to a research concept, which more correspond to the complexity of the problems, than those investigations only considering statistical associations between life events and one or another expression for illness. However, it must be emphasized that many partly unsolved methodological problems are associated with these complex models.

The following parts of this paper concern 1) the selection of life events, 2) the principles of weighting the events, 3) some of the psycho-physiological mechanisms involved in mediating the effect of the events, and 4) some problems related to design and supplementary variables.

2.1 The selection of the events.

Since the early study of Holmes & Rahe (1967) several similar life event inventories have been published (e.g. Paykel et al., 1971, Rahe, 1978, Coddington, 1972, Monoghan et al., 1979 and Dohrenwend et al., 1978).

Holmes & Rahe's 43 items life event inventory was based on a registration of "life events empirically observed to cluster at time of disease onset". This principle for selection gives rise to the presumption that some of the life events are reflecting consequences of illness rather than events of causal importance. For further information, see e.g. Brown (1974) or Hudgens (1974). Brown's interview-technique involves a careful determination of the time relation between an event and onset of a disease, and an assessment of the connection between the event and the disease.

In Holmes & Rahe's inventory is listed desirable as well as undesirable events. This approach is in accordance with the theory that the change itself, independently of the desirability of the event, constitutes the psycho-social stress (Levi, 1972). Several empirical investigations have, however, reported that only life events experienced as undersirable have an important impact on illness (e.g. Dohrenwend, 1973, Paykel, 1974, Vinokur & Selzer, 1975, Brown & Harris, 1978 and Grant et al., 1981). Some investigators have suggested that merely the more qualitative aspects of life events appear to be of significance. Among these aspects may be mentioned 1) the individual's experiences and appraisals of loss (Gersten et al., 1974 and Paykel, 1974), 2) the individual's responsibility for occurrence of an event (Paykel, 1974), and 3) which of the individual's spheres (family, friend, work) is mainly involved (Myers et al., 1972). These results necessitate some of the life events to have a qualitative dimension. As an example can be mentioned that a life event question as: "Change in number of arguments with spouse", may be expressed as increased conflicts with spouse in connection with the individual's worries due to the change. Many of the life event questions are relatively exact and well-defined, but for these questions which are vague or unclear a more specific formulation is necessary. Some of the life events listed are directly influenced by illness, and some of the events referring to the individual's social function and relation to others may be under the influence of illness. If these aspects are not taken into account, the associations discovered can hardly be interpreted as associations of etiological importance. The items referring to marriage, birth, death etc. are relatively independent of culture; this is opposite to many of the other items. Many concern work and money, and some are due to the cultural differences difficult to translate, e.g. the question: "Change in church activities". The more severe life events are of relevance. However, for many of the less severe events it appears somewhat accidental that exactly these items have been selected. It is a central question how much the selected events represent the disease precipitating events we are exposed to. No investigation has been found focusing on this problem, but as a general trend may be emphasized that later modifications of Holmes & Rahe's original inventory have added many more events.

Finally it should be mentioned that stressful life events beyond association to illness show a pronounced correlation to a number of social factors. Some epidemiological investigations have shown (e.g. Masuda & Holmes, 1978 and Goldberg & Comstock, 1980) that life events vary with 1) age, 2) level of education, 3) social class, 4) marital status and

5) race. Furthermore considerable intercorrelations have been found between life events. As many diseases, in particular some chronic somatic diseases and most mental disorders, are correlated with social factors (Andrews et al., 1978) it is evident that this must be taken into account. However, which social factors are included will depend on the diseases and the hypothesis. It must be stressed that an association between stressful life events and illness may be entirely misleading if the evaluation does not take into account the intercorrelations between social factors and stressful life events. It is only recently that multivariate statistical techniques have been applied in life event research (Cleary, 1981), but they have been applied throughout the present investigation.

It would be desirable through epidemiological investigations in some European countries to obtain information concerning the frequency of potentially stressful life events including intercorrelations to social factors, and to have the more methodological aspects of the inventory evaluated e.g. the test-re-test reliability (Horowitz et al., 1977).

2.2 The weighting of the events.

In Holmes & Rahe's inventory a panel of healthy individuals assessed the relative magnitude of adjustment. "Social readjustment includes the amount and duration in one's accustomed pattern of life resulting from various life events. As defined social readjustment measures the intensity and length of time necessary to accomodate to a life event, regardless of the desirability of this event". Holmes & Rahe's procedure of weighting is characterized by the fact that healthy persons from a specific theory assess a list of hypothetical events, i.e. a consensus a priori.

Many other principles of weighting are available. The panel may be constituted by healthy individuals with defined age, social group etc., or patients with well-defined diseases. The events may be hypothetical or earlier self-experienced. The principle of weighting may be the magnitude of social readjustment, strain, emotional disturbances or threat. Rahe (1978) and Theorell (1974, 1976) both have applied a priori determinated panel weights as well as the individual's own weighting. Paykel's (1971) principle in the weighting procedure is "how upsetting the event would be". In some investigations (e.g. Dohrenwend, 1973 and Gersten et al., 1974) different weights have been applied to undesirable and neutral events. In Brown's (1974) investigation a consensus a posteriori is applied concerning "the threat an event would pose to the average person". Furthermore, in some evaluations only a simple weighting i.e. application of equal relative weights to all events has been used (Gersten et al., 1974, Rahe, 1974 and Cleary, 1981). In some investigations a principle with statistically determined a posteriori weights has been applied (Ross & Mirowsky, 1979 and Shroat, 1981). In the present investigations a posteriori weights have been calculated through multivariate techniques as appropriate scoring and logistic regression (Ipsen & Feigl, 1966 and Bishop et al., 1975).

The different modifications in the weighting procedure are mainly due to insufficient statistical information in a traditionel evaluation. Theorell (1974) found a better statistical association of stressful life events to some severe physical diseases using the patients' own appraisals instead of the a priori derived panel weights.

Many investigations have been undertaken comparing dissimilarities in weighting life events in different countries, cultures and subgroups. In some investigations a high rank order correlation of the relative weights has been found across cultures (Rahe, 1969). However, many factors influence on the weighting. It has been reported that 1) young adults weight higher than older adults, 2) women higher than men, 3) level of education seems of importance, and 4) earlier self-experience events result in a higher score (Masuda & Holmes, 1978). Furthermore, Myers et al. (1972) reported that social class to a greater extent is associated with the weighting than to number of life events experienced.

Thus, the original principle of weighting is considered to be a passed historical phase in the development of quantitative methods. The principles to be applied in the future depend on the specific investigation, its design and hypothesis. In some instances the principle with subjective weights of self-experienced events will be preferred; in other instances the principle with a posteriori derived weights will be preferred, and in some instances relative weights are not needed at all. But in all instances it is necessary to take into account other variables of supposed impact e.g. social, biomedical, psychological, with the inference that the associations discovered can be reasonable and interpretable.

2.3 Psycho-physiological mechanisms

Consideration must be taken to the indicator of illness applied. The indicators concern e.g. 1) self-reporting of symptoms, 2) utilization of health services, 3) diagnoses of patients admitted to hospital, 4) diagnoses derived from a population based health survey.

As considerable differences exist between illness and illness behaviour (Mechanic, 1968), and the diagnostic tradition varies considerably, the many partly contradictory results published are not surprising. In the instances where the indicator refers to illness behaviour the concept mainly ought to be socio-psychological; however, when the indicator is a well-defined diagnosis some assumption on the psycho-physiological mechanisms may be taken into account.

In Kagan & Levi's model (1974) stressful life events are considered as the provoking factors, and the effect on the psycho-physiological reactions is supposed mediated through a psycho-biological programme which is assumed to be mainly genetic conditioned. The psycho-physiological reactions may result in illness. The model incorporates a number of intervening variables, which may modify i.e. promote or inhibit this effect. The model involves a so-called cybernetic system with possibility for feedback. The intervening variables may e.g. refer to social vulnerability, social network and support, the coping mechanisms, and biomedical/clinical risk factors.

Holmes & Rahe's original concept of an explanation of the impact of life events on illness is not further specified, however, presumably in accordance with Selye's theories (1956). These concern the physiological reactions - the general adaptation syndrome (GAS) - to mainly physical strain. In Selye's theory it has been emphasized that the GAS-reaction in itself does not result in illness, but if the reaction is prolonged or repeated frequently, and if a vulnerability in an organ exists, these reactions may result in illness.

Stressful life events cause a neurohormonal and physiological activation. This is particularly conspicuos on the haemodynamic reactions. But stressful life events involve other effects such as 1) function of thrombocyts and fibrinogen, 2) levels of lipids and cholesterol and 3) function of the immunological system. These effects are mainly mediated through the catecholamines and the corticosteroids (Weiner, 1977 and Rahe et al., 1974).

To explain why life events increase the susceptibility to unspecified illness, the psycho-physiological mechanisms only yield a modest and unspecific contribution. The assumption is that the individual becomes less capable to control patogenetic processes. Another matter is when certain diseases are considered (e.g. duodenal ulcer, hypertension, coronary heart disease). Here the concept is essential and has yielded a contribution to the comprehension of the etiology and pathogenesis. But why repeated activation of the physiological and neuro-hormonal mechanisms in some persons result in a specific disease, and in other persons does not have this effect is not satisfactory explained, unless Selye's assumption on a specific organ vulnerability is taken into account.

A biologically conditioned susceptibility to the effect of stressful life events can undoubtedly in some instances give an explanation to a certain disease. It seems more likely, however, that a susceptibility can be conditioned by biological as well as psychological and social factors.

During recent years a great activity within the biological stress-research has taken place. Several biological markers have been identified both for acute and for more prolonged strain. It is a characteristic trend that general theories as e.g. Selye's are partly abandoned, and that the main field of interest has moved from the responses in PNS to biochemical changes in CNS. Furthermore, the interest has changed from an attempt to find general mechanisms to an attempt to elucidate the individually based reactions. In a newly published review Weiner (1981) wrote:

"For many years it was believed that stressful life events acted upon the minds of persons and incited an emotional response which in some mysterious manner was translated into bodily changes that culminated in disease. In recent years it has, however, become apparent that the psychological response to experiences is a complex and individual matter. Different persons perceive and appraise an experience in personal ways. The meaning of the event to each person differs: for one it is trivial, while for another it may be portentous. The meaning of the event may also be processed differently by each person and may arouse qualitatively and quantitatively different emotional responses. Some persons then cope quite effectively with the experience or seek information that allows them to cope with it. Other persons fail to cope, cannot adapt, or do not seek and use information. Failure to adapt to life experience is believed to be the context in which many diseases begin".

The complexity of the psycho-physiological mechanisms has resulted in new strategies within the biological stress-research. Results from these efforts may later on infer considerable changes in the socio-psychological stress-research. This process has been in progress during recent years (e.g. Rahe & Arthur, 1978 and Lazarus, 1978).

2.4 Design and supplementary variables.

Several methodological problems are connected with the selection as well as the weighting of the life events. Some of these problems are, however, possible to solve. Furthermore, many methodological problems are connected with 1) the dependent variable i.e. the indicator for illness, 2) the design of the investigation, and 3) the selection of supplementary variables, i.e. various intervening variables between stressful life events and illness.

Attention should be drawn to the fact that having symptoms, consulting a physician, and getting a diagnosis, are quite different phenomena. Several people are suffering from mental or physical symptoms, but only some will consult a physician. Among the different symptoms and complaints reported to the physician only a few refer to a well-defined disease. Indeed, in many cases the problem is to exclude a physical explanation of symptoms or complaints. It is essential in a given study to define the indicator of illness in an exact way. This aspect has been somewhat neglected in several of the earlier reported investigations.

Vaillant et al. (1970), e.g., have suggested that stressful life events rather contribute to a comprehension of illness behaviour than to illness. It might be assumed that one way to cope with a stressful situation might be to enter a sick role (Mechanic, 1968). However, some prospective investigations have documented associations between stressful life events and increased risk for well-defined diseases. But the association to non-specific psychological symptoms indicating distress is also essential, as excess of non-specific psychological symptoms is a predictor variable to utilization of health services (Harrington, 1978).

The design of the investigations has most often been either prospective with onset or course of a certain disease as the dependent variable, or patient-control studies. Many methodological problems are connected with the patient-control studies in comparison of information obtained from the patients and from the controls. The patient, and the physician as well, are seeking an explanation of the actual disease. Furthermore, some patients attempt to deny obvious associations. In Brown's interview-technique it has been attempted to eliminate some of these bias.

It is suggested that some of the reported statistical associations from patient-control studies more refer to bias than to real associations. The fact that nearly all published patient-control investigations show significant associations between life events and illness must give basis to the idea that bias is an important explanation (Sackett, 1980).

In the investigations applying stressfull life events first published, no intervening variables were taken into account, but gradually various studies have incorporated supplementary variables, particularly on 1) coping, 2) social support, 3) unfavourable social factors, and 4) biomedical risk factors.

Coping, i.e. the Ego's conscious behaviour during stressful life events and/or psycho-physiological symptoms, is presumed to modify strongly the effect of stressful life events. Andrews et al. (1978) found that an inadequate coping had a similar effect (statistically) for mental disorders as an accumulation of stressful life events. Especially Lazarus (1966, 1974) has described in detail the different coping strategies and their importance.

The degree of social support available to the individual is of several investigators considered as the type of variable which most pronouncedly can modify the impact of stressful life events (e.g. Cobb, 1976, Jacobs & Myer, 1976, Kaplan et al., 1977, Lin et al., 1979 and Wilcox, 1981).

The fact that unfavourable social factors act as intervening variables of considerable importance appear from several investigations including the present. A rather consistent finding is that the effect of stressful life events for mental disorders acts on the basis of unfavourable social factors (Dohrenwend, 1973). Furthermore, social factors are intercorrelated with a number of other variables of importance to several diseases, e.g. smoking, drinking and eating habits.

Biomedical risk factors refer to the fact that if in relation to a certain disease there are biomedical variables of importance to onset or course, these variables must be taken into account in the analysis.

3. CONCLUSION.

Holmes & Rahe's method is considered as a pioneer work. The method has been criticized and modified several times as a consequence of the fact that a great number of research groups intensively and for many years have been working to elucidate some of these very complex associations. However, the many methodological problems must not obscure the fact that it is evident that stressful life events or indices for short-term psycho-social stress play a considerable role as a contributory cause to many physical and mental disorders.

Stressful life events have been regarded as one of the components in a multifactorial disease model. The effect on illness will, however, also depend on a large number of biological, social and psychological variables. The criticism presented must implicate that future quantitatively based investigations on stressful life events and illness also involve a large number of other variables. Further, in most instances it is necessary to apply multivariate statistical techniques in the evaluation.

A central question is, however, whether the limits have been reached for what quantitatively based investigations - however refined they might be - may yield to a further comprehension of the effect of stressful life events on illness.

Within the areas where association between stressful life events and illness is already evident it is hardly necessary to obtain further quantitatively based documentation. Within these areas it is rather a more qualitative approach which is required. This could either be in a concept according to Lazarus (1978) with more dynamic, process-oriented models, or in the clinical psychiatric concept applied in comprehension of the reactive psychosis.

Among those diseases, where a clinically based assumption considers psycho-social factors of importance, but where the documentation is lacking or contradictory, there is undoubtedly a need for further analyses of the relative effect of life events and other risk factors (biological, social, psychological). But the methodological issues discussed in this review must be taken into consideration in an attempt to apply improved research strategies within this very complex field.

REFERENCES

Andrews, G., Tennant, C., Hewson, D. and Schonell, M., 1978. The relation of social factors to physical and psychiatric illness. Am. J. Epidemiol., 108: 27-35.
Bishop, Y.M.M., Fienberg, S.E. and Holland, P.W., 1975. Discrete Multivariate Analysis. Theory and Practice. MIT Press, London.
Blazer, D., 1980. Life events, mental health functioning and the use of health care services by the elderly. Am. J. Public. Health, 70: 1174-1179.
Brown, G.W., 1974. Meaning, measurement, and stress of life events. In: B.S. Dohrenwend & B.P. Dohrenwend (editors), Stressful Life Events. Their Nature and Effects. Wiley, New York, pp. 217-243.
Brown, G.W., 1981. Life events, psychiatric disorder and physical illness. J. Psychosom. Res., 25: 461-473.
Brown, G.W. and Birley, J.I.T., 1968. Crises and life changes and the onset of schizophrenia. J. Health Soc. Behav., 9: 203-214.
Brown, G.W. and Harris, T., 1978. The Social Origins of Depression: A Study of Psychiatric Disorder in Women. Tavistock, Free Press, London.
Cleary, P.J., 1981. Problems of internal consistency and scaling in life event schedules. J. Psychosom. Res., 25: 309-320.
Cline, P.W. and Chosey, J.J., 1972. A prospective study of life changes and subsequent health changes. Arch. Gen. Psychiatry, 27: 51-53.
Cobb, S., 1976. Social support as a moderator of life stress. Psychosom. Med., 38: 300-314.
Coddington, R.D., 1972. The significance of life events as etiologic factors in the diseases of children. I. J. Psychosom. Res., 16: 7-18.
Coddington, R.D., 1972. The significance of life events as etiologic factors in the diseases of children. II. J. Psychosom. Res., 16: 205-213.
Coddington, R.D. and Troxel, J.R., 1980. The effect of emotional factors on football injury rates. A pilot study. J. Human Stress, 6: 3-5.
Cullberg, J., 1976. Krise og udvikling. Reitzel, København.
Day, R., 1981. Life events and schizophrenia: the "triggering" hypothesis. Acta Psychiatr. Scand., 64: 97-122.
De Faire, V. and Theorell, T., 1976. Life changes and myocardial infarction. Scand. J. Soc. Med., 4: 115-122.
Dodge, J.A., 1972. Psychosomatic aspects of infantile stenosis. J. Psychosom. Res., 16: 1-5.
Dohrenwend, B.S., 1973. Social status and stressful life events. J. Pers. Soc. Psychol., 28: 225-235.
Dohrenwend, B.S., 1973. Life events as stressor. A methodological inquiry. J. Health Soc. Behav., 14: 167-175.
Dohrenwend, B.S. (editor), 1981. Stressful Life Events and their Contexts. Neale Watson Academic Publications, New York.
Dohrenwend, B.S. and Dohrenwend, B.P. (editors), 1974. Stressful Life Events. Their Nature and Effects. Wiley, New York.
Dohrenwend, B.S., Krasnoff, L., Askenazy, A.R. and Dohrenwend, B.P., 1978. Exemplification of a method for scaling life events: the Peri life events scale. J. Health Soc. Behav., 19: 205-229.
Dzegede, S.A., Pike, S.W. and Hackworth, J.R., 1981. The relationship between health-related stressful life events and anxiety. An analysis of a Florida metropolitan community. Community Ment. Health J., 17: 294-305.

Egeland B., Breitenbucher, M. and Rosenberg, D., 1980. Prospective study of the significance of life stress in the etiology of child abuse. J. Consult. Clin. Psychol., 48: 195-205.
Finlay-Jones, R., 1981. Showing that life events are a cause of depression - a review. Austr. NZ. J. Psychiatry, 15: 229-238.
Færgeman, P.M., 1945. De psykogene psykoser (Disp.). Munksgaard, København.
Gersten, J.C., Langner, T.S., Eisenberg, J.G. and Orzek, L., 1974. Child behaviour and life events: undesirable change or change per se? In: B.S. Dohrenwend & B.P. Dohrenwend (editors), Stressful Life Events. Their Nature and Effects. Wiley, New York, pp. 159-170.
Gersten, J.C., Langner, T.S., Eisenberg, J.G. and Simcka-Fugan, U., 1977. An evaluation of the etiologic role of stressful life change events in psychological disorders. J. Health Soc. Behav, 18: 228-244.
Goldberg, E.L. and Comstock, G.W., 1980. Epidemiology of life events: Frequency in general populations. Am. J. Epidemiol., III: 736-753.
Goldberg, E.L., Comstock, G.W. and Graves, C.G., 1980. Psychosocial factors and blood pressure. Psychol. Med., 10: 243-255.
Gorsuch, R.L. and Key, M.K., 1974. Abnormalities of pregnancy as a function of anxiety and life stress. Psychosom. Med., 36: 352-372.
Grant, I., Sweetwood, H.L., Yager, J. and Gerst, M., 1981. Quality of life events in relation to psychiatric symptoms. Arch. Gen. Psychiatry, 38: 335-339.
Gunderson, E.K.E. and Rahe, R.H. (editors), 1974. Life Stress and Illness. Thomas, Springfield.
Harrington, R.L., 1978. Mental health component of multiphasic health testing services. In: M.F. Collen (editor), Multiphasic Health Testing Services. Wiley, New York, pp. 413-436.
Hawkins, N.F., Davies, R. and Holmes, T.H., 1957. Evidence of psychosocial factors in the development of pulmonary tuberculosis. Am. Rev. Respir. Dis., 75: 768-780.
Heisel, J.S., 1972. Life changes as etiologic factors in juvenile rheumatoid arthritis. J. Psychosom. Res., 16: 411-420.
Heisel, J.S., Ream, S., Raitz, R., Rappaport, M. and Coddington, R.D., 1973. The significance of life events as contributing factors in the diseases of children. J. Pediatr., 83: 119-123.
Hinkle, L.E., 1974. The effect of exposure to culture change, social change and changes in interpersonal relationships on health. In: B.S. Dohrenwend & B.P. Dohrenwend (editors), Stressful Life Events. Their Nature and Effects. Wiley, New York, pp. 9-44.
Hinkle, L.E., Jr., Christenson, W.N., Kane, F.D., Ostfeld, A.M., Thetford, W.N. and Wolff, H.G., 1958. An investigation of the relation between life experience, personality characteristics, and general susceptibility to illness. Psychosom. Med., 26: 278-295.
Holmes, T.H. and Rahe, R.H., 1967. The social readjustment rating scale. J. Psychosom. Res., 11: 213-218.
Horowitz, M.J., Schaefer, C., Hiroto, D., Wilner, N. and Levin, B., 1977. Life event questionnaires for measuring presumptive stress. Psychosom. Med., 39: 413-431.
Hudgens, R.W., 1974. Personal catastrophe and depression. A consideration of the subject with respect to medically ill adolescents, and a requiem for retrospective life events studies. In: B.S. Dohrenwend & B.P. Dohrenwend (editors), Stressful Life Events. Their Nature and Effects. Wiley, New York, pp. 119-134.

Ipsen, J. and Feigl, P., 1966. Appropriate scores for clinical and public health variables. Am. J. Public Health, 56: 1287-1295.
Jacobs, M.A., Spilken, A. and Norman, N., 1969. Relationship of life change, maladaptive aggression, and upper respiratory infections in male college students. Psychosom. Med., 31: 31-44.
Jacobs, S. and Myers, J., 1976. Recent life events and acute schizophrenic psychosis. A controlled study. J. Nerv. Ment. Dis., 162: 75-87.
Jacobs, S. and Ostfeld, A.M., 1977. An epidemiological review of the mortality of bereavement. Psychosom. Med., 39: 344-357.
Jacobs, S., Prusoff, B. and Paykel, E., 1974. Recent life events in schizophrenia and depression. Psychol. Med., 4: 444-453.
Jacobs, T.J. and Charles, E., 1980. Life events and the occurrence of cancer in children. Psychosom. Med., 42: 11-24.
Jaspers, K., 1913. Allgemeine Psychopathologia. Springer, Berlin.
Jenkins, C.D., 1976. Recent evidence supporting psychologic and social risk factors for coronary disease. N. Engl. J. Med., 294: 1033-1038.
Kagan, A.R. and Levi, L., 1974. Health and environment. Psychosocial stimuli. A review. Soc. Sci. Med., 8: 225-241.
Kaplan, B.H., Cassel, J.C. and Gore, S., 1977. Social support and health. Med. Care, 15: 47-58.
Kashani, J.H., Hodges, K.K., Simonds, J.F. and Hildebrand, E., 1981. Life events and hospitalization in children: A comparison with a general population. Brit. J. Psychiatry, 139: 221-225.
Kringlen, E., 1982. Coronary heart disease. Social Causation? Psychiatr. Soc. Sci., 2: 5-22.
Kühl, P.H. and Martini, S., 1981. Psykisk sårbare. Socialforskningsinstituttet. Publ. nr. 102. Teknisk Forlag, København.
Lazarus, R.S., 1966. Psychological Stress and the Coping Process. McGraw-Hill, New York.
Lazarus, R.S., 1978. A strategy for research on psychological and social factors in hypertension. J. Human Stress, 4: 35-40.
Lazarus, R.S., Averill, J.R. and Opton, E.M., 1974. The psychology of coping: Issues of research and assessment. In: G.V. Coelho, D.A. Hamburg and J.E. Adams (editors), Coping and Adaptation. Basic Books, New York, pp. 249-315.
Levi, L., 1972. Stress and distress in response to psychosocial stimuli. Acta Med. Scand., 528/Suppl.
Liberman, M.A., 1961. Relationship of mortality rates to entrance to a home for the aged. Geriatrics, 16: 515-519.
Lin, N., Simeone, R.S., Ensel, W.M. and Kuo, W., 1979. Social support, stressful life events, and illness: A model and an empirical test. J. Health Soc. Behav., 20: 108-119.
Lipowski, Z.J., Lipsitt, D.R. and Whybrow, P.C. (editors), 1977. Psychosomatic Medicine. Current Trends and Clinical Applications. Oxford University Press, New York.
Maddison, D. and Viola, A., 1968. The health of widows in the year following bereavement. J. Psychosom. Res., 12: 297-306.
Markush, R.H. and Favero, R.V., 1974. Epidemiologic assessment of stressful life events, depressed mood, and psychophysiological symptoms. A preliminary report. In: B.S. Dohrenwend & B.P. Dohrenwend (editors), Stressful Life Events. Their Nature and Effects. Wiley, New York, pp. 171-190.
Masuda, M. and Holmes, T.H., 1978. Life events: Perceptions and frequencies. Psychosom. Med., 40: 236-261.
Mechanic, D., 1968. Medical Sociology. Free Press, New York.

Monoghan, J.H., Robinson, J.O. and Dodge, J.A., 1979. The children's life events inventory. J. Psychosom. Res., 23: 63-75.
Myers, J.K., Lindenthal, J.S., Pepper, M.P. and Ostrander, D.R., 1972. Life events and mental status. A longitudinal study. J. Health Soc. Behav., 13: 398-406.
Nuckolls, K.B., Cassel, J. and Kaplan, B.H., 1972. Psychosocial assets, life crisis, and the prognosis of pregnancy. Am. J. Epidemiol., 95: 431-441.
Padilla, E.R., Rohsenow, J.D. and Bergman, A.B., 1976. Predicting accident frequency in children. Pediatrics, 58: 223-226.
Parkers, C.M. and Birthwell, S., 1971. Bereavement. Proc. R. Soc. Med., 64: 1-14.
Paykel, E.S., 1974. Life stress and psychiatric disorder. Applications and the clinical approach. In: B.S. Dohrenwend & B.P. Dohrenwend (editors), Stressful Life Events. Their Nature and Effects. Wiley, New York, pp. 135-150.
Paykel, E.S., Myers, J.K., Dienelt, M.N., Klerman, G.N., Lindenthal, J.J. and Pepper, M.P., 1969. Life events and depression. A controlled study. Arch. Gen. Psychiatry, 21: 753-760.
Paykel, E.S., Prusoff, B.A. and Myers, J.K., 1975. Suicide attempts and recent life events. A controlled comparison. Arch. Gen. Psychiatry, 32: 327-333.
Paykel, E.S., Prusoff, B.A. and Uhlenhuth, E.H., 1971. Scaling of life events. Arch. Gen. Psychiatry, 25: 340-347.
Rabkin, J.G. and Struening, E.L., 1976. Life events, stress, and illness. Science, 194: 1013-1020.
Rahe, R.H., 1969. Multi-cultural correlations of life change scaling. J. Psychosom. Res., 13: 191-195.
Rahe, R.H., 1974. Life change and subsequent illness reports. In: E.K.E. Gunderson & R.H. Rahe (editors), Life Stress and Illness. Thomas, Springfield, pp. 58-78.
Rahe, R.H., 1978. Life change clarification. Psychosom. Med., 40: 95-97.
Rahe, R.H. and Arthur, R.J., 1978. Life changes and illness studies. Past history and future directions. J. Human Stress, 4: 3-15.
Rahe, R.H., Bennett, L. and Romo, M., 1973. Subjects' recent life changes and coronary heart disease in Finland. Am. J. Psychiatry, 130: 1223-1226.
Rahe, R.H., Rubin, R.T. and Arthur, R.J., 1974. The three investigators study: Serum uric acid, cholesterol, and cortisol variability during stresses of everyday life. Psychosom. Med., 36: 258-268.
Ross, C.E. and Mirowsky, S., 1979. A comparison of life-event-weighting schemes: change, undesirability, and effect-proportional indices. J. Health Soc. Behav., 20: 166-177.
Rowland, K.F., 1977. Environmental events predicting death for the elderly. Psychol. Bull., 84: 349-372.
Sackett, D.L., 1979. Bias in analytic research. J. Chronic Dis., 32: 51-63.
Sandler, I.N. and Block, M., 1979. Life stress and maladaptation of children. Am. J. Community Psychol., 7: 425-440.
Schmale, A.H., Jr. and Iker, H.P., 1971. Hopelessness as a predictor of cervical cancer. Soc. Sci. Med., 5: 95-100.
Selye, H., 1956. The Stress of Life. McGraw-Hill, New York.
Shroat, P.E., 1981. Scaling of stressful life events. In: B.S. Dohrenwend & B.P. Dohrenwend (editors), Stressful Life Events and their Contexts. Neale Watson Academic Publications, New York, pp. 29-47.

Sibert, R., 1975. Stress in families of children who have ingested poisons. Br. Med. J., III: 87-89.
Solomon, G.F., 1969. Emotions, stress, the central nervous system, and immunity. Ann. NY. Acad. Sci., 164: 335-343.
Theorell, T., 1974. Life events before and after onset of a premature myocardial infarction. In: B.S. Dohrenwend & B.P. Dohrenwend (editors), Stressful Life Events: Their Nature and Effects. Wiley, New York, pp. 101-117.
Theorell, T. and Flodérus-Myrhed, B., 1977. "Workload" and risk of myocardial infarction - A prospective psychosocial analysis. Int. J. Epidemiol., 6: 17-21.
Theorell, T., Lind, E. and Flodérus-Myrhed, B., 1975. The relationship of disturbing life changes and emotions to the early development of myocardial infarction and other serious illnesses. Int. J. Epidemiol., 4: 281-293.
Uhlenhuth, E.H. and Paykel, E.S., 1973. Symptom intensity and life events. Arch. Gen. Psychiatry, 28: 473-477.
Vaillant, G.E., Shapiro, L.N. and Schmith, P.P., 1970. Psychological motives for medical hospitalization. JAMA, 214: 1661-1665.
Vinokur, A. and Selzer, M.L., 1975. Desirable versus undesirable life events. Their relationship to stress and mental distress. J. Per. Soc. Psychol., 32: 329-337.
Weiner, H. (editor), 1977. Psychobiology and Human Disease. Elsevier, New York.
Weiner, H., Hofer, M.A. and Stunkard, A.J. (editors), 1981. Brain, Behaviour, and Bodily Disease. Research Publications: Association for Research in Nervous and Mental Disease vol. 59. Raven Press, New York.
Wilcox, B.L., 1981. Social support, life stress, and psychological adjustment. A test of the buffering hypothesis. Am. J. Community Psychol., 9: 371-386.
Wimmer, A., 1916. Psykogene sindssygdomsformer. Sct. Hans Hospital. Jubilæumsskrift. Gad, København.
Aagaard, J., 1979. Social background and life events of children admitted to a paediatric department. Acta Paediatr. Scand., 68: 531-539.
Aagaard, J., 1982. Social factors and life events as predictors for children's health. A one-year prospective study after discharge from hospital. Scand. J. Soc. Med., 10: 87-93.
Aagaard, J., Amdrup, E., Aminoff, C., Andersen, D. and Sørensen, F.H., 1981a. A clinical and socio-medical investigation of patients five years after surgical treatment for duodenal ulcer. I. Behavioural consequences and psychological symptoms. Scand. J. Gastroenterol., 16: 361-367.
Aagaard, J., Amdrup, E., Aminoff, C., Andersen, D. and Sørensen, F.H., 1981b. A clinical and socio-medical investigation of patients five years after surgical treatment for duodenal ulcer. II. Association of social and psychological factors with surgical outcome. Scand. J. Gastroenterol., 16: 369-375.
Aagaard, J., Amdrup, E., Aminoff, C. and Sørensen, F.H., In press. A predictor analysis of clinical assessment of outcome among patients operated on for duodenal ulcer. A one-year prospective study. Scand. J. Gastroenterol.
Aagaard, J., Husfeldt, P. and Husfeldt, V., 1980. Sociale faktorer og socialbegivenheder for børn, der konsulterer den praktiserende

læge. I. Metode, materiale og børnenes opvækstmiljø. Ugeskr. Laeger, 142: 845-850.

Aagaard, J., Husfeldt, P. and Husfeldt, V., 1983. Social factors and life events as predictors for children's health. A one-year prospective study within a general practice. Acta Paediatr. Scand., 72: 275-281.

Aagaard, J. and Kristensen, B.Ø., 1982. Social factors and life events as predictors for disease progression in essential hypertension. As two-year prospective study. Psychiat. Soc. Sci., 2: 85-95.

A LIFETIME PROSPECTIVE STUDY OF HUMAN ADAPTATION AND HEALTH

M E J Wadsworth
Medical Research Council, National Survey of Health and Development
University of Bristol, Department of Community Health
Canynge Hall, Whiteladies Road, Bristol, BS8 2PR, U.K.

ABSTRACT

Analysis of data on childhood, adolescent and early adult mental and physical adaptation to childhood emotional stress is described, using information from a national birth cohort study of over five thousand males and females. The cohort has been investigated at frequent intervals from birth, so far to age 36 years. Current work on the mental and physical health of this cohort is outlined, with particular reference to present cardiovascular and respiratory health and their associations with lifetime patterns of health and experience of stressful circumstances and events.

INTRODUCTION

In a valuable review of the epidemiology of life stress Susser (1981) wrote: "when we turn to the individual level ... and to the question of how the impact of the social structure is translated into pathology, much that is said must rest on assertions of authority and faith". Although it would have been better to have said "social processes and experiences" rather than "social structure", Susser's remarks draw attention to the relative dearth of data in the study of the epidemiology of life stress.

This may be seen both in the relative lack of studies that begin from a suspected source of stress and that then go on to look for outcomes in the same individuals, and also in the lack of studies that investigate stress over long periods of time in the life of the individual, in ways other than those involving long-term recollection.

Much of the work on stress and its associations with illness assumes implicitly that serious effects are the result of long-term stress rather than of the single stressful experience. Carefully executed retrospective studies, such as that of Brown and Harris (1978) show that the natural history of health damaging stress may extend over a considerable time scale. In a similar fashion many psychological hypotheses about adult behaviour and coping suggest associations with early life and childhood experience. One way of checking some aspects of such hypotheses is the use of a prospective longitudinal design, and this paper describes work on stress in a British longitudinal prospective study. First, completed work is summarised and then work in progress is described.

METHOD

The study concerned is the British Medical Research Council's National Survey of Health and Development. This study began in 1946 as an investigation of birth and birth circumstances of all the children born in England, Wales and Scotland in one week in March of that year. Eighteen months later a one-third sample of all single, legitimate births to wives of non-manual and agricultural workers, and one in four of similar births to wives of manual workers was visited at home, and this was the first of a series of contacts that have been maintained until today, when sample members are aged 36 years.

Contact is maintained with 85% of those still living in Great Britain, approximately 3,500 individuals at the latest contact.

The subjects have been studied during childhood and adolescence at intervals of not more than two years, and they are now contacted about every five years. Information has been collected on a very wide range of topics, including medical, social, psychological and educational data. For each person in the study there exists a life history unencumbered by problems of long-term recollection. Findings have been published on many aspects of the medical and social data and the study is described by Atkins et al (1981).

A second generation study of the first-born offspring of cohort members is also being carried out. It aims to study in much greater detail the childhood family circumstances already shown in this cohort study to be of considerable influence in adolescence and early childhood, and to make comparisons of childhood between the generations, as well as looking for generation change and carry-over (Wadsworth, 1981a).

FINDINGS ON ONE LONG-TERM SOURCE OF VULNERABILITY TO LATER STRESSFUL CIRCUMSTANCES

In order to take full advantage of this lifetime prospective data a supposed source of vulnerability from the earliest time of life was selected, namely the loss of a parent through death, divorce or separation during the child's first five years of life. This experience at this particular age is commonly believed to be associated with an increased risk of behaviour disturbance in later life. It is also popularly supposed to be a time during which foundations are laid for the infant's future moral and intellectual potential. Hypotheses are largely based on the concept of these early years as a time when the child becomes aware of his or her emotional bond with and dependence on parents, and discovers the nature of this emotional attachment. Failure of such attachment at this time is thought to be associated with an increased risk of permanent emotional damage.

Childhood susceptibility to emotional disturbance and stress is well documented in terms of its immediate and short-term effect (for example Bowlby, 1975, Rutter, 1972 and Douglas, 1973). Possible long-term effects are much more difficult to determine and they have usually been sought in descriptions of childhood recollected in adult life (for example Brown and Harris, 1978), or in animal studies (reviewed by Rutter, 1972), or case studies (see, for example, Clarke and Clarke, 1976). But these methods of obtaining information about long-term effects are subject to certain limitations. There is good evidence that recollection over a long period of time is open not only to memory error but also to distortion, and that distortion is particularly likely in those who have experienced childhood emotional disturbance (Yarrow et al, 1970). On the other hand the animal studies are an important source of hypotheses, but the extent of usefulness of generalising from them to human circumstances is not known. And although case and clinic-based studies of children provide rich and detailed material, the true population dimensions of their findings remain unknown. The longitudinal study is therefore of value because it provides prospectively collected information over many years, and thus substantially avoids the problems of memory error and distortion. A longitudinal study which, like the National Survey, is also a population sample, has the additional advantage of showing the extent of the problem in the population at large, but it is subject to limitations of data collection, in that very detailed information cannot be obtained. A national longitudinal study therefore fulfills the need for an indicator of the population extent of associations found between childhood emotional disruption and adult life.

The National Survey of Health and Development can uniquely test some aspects of these hypotheses by comparing children who lost a parent or parents through death, divorce or separation by the time they were five years old with those who, until this age, lived with both parents. Information for comparison is taken from various times later in their lives, and differences are sought serially in behaviour and illness experience. The findings are summarised below in age order.

LATER ILLNESS AND BEHAVIOUR DISTURBANCE

a) Ages six to eight years

Relatively soon after the experience of parental death, divorce or separation it was not surprising to find that a statistically significant excess of these children had experienced sleep disturbance, nightmares and nocturnal enuresis (Douglas, 1973) when compared with other children.

b) Ages nine to eleven years

During their school years all the children were examined by school doctors on four occasions. Data collected when the children were aged eleven years enabled us to test the suggestion of Davies and Maliphant (1974) that "stress in early childhood may contribute to ... atypical autonomic activity", since at this examination doctors took pulse rates before carrying out a physical examination. For boys the rates were statistically significantly different and lower if the break was caused by parental divorce or separation, and for girls the significance occurred if the break was as a result of parental death (Wadsworth, 1976).

At the age of ten years teachers were asked to rate survey children's attitude to work on a four-point scale. Whatever the family's social class, children who had experienced parental divorce, death or separation before they were aged five years were, in comparison with all other children, significantly more likely to be rated as average, poor or lazy workers.

c) Ages twelve to twenty-six years

One kind of behaviour commonly thought to be associated with emotional disruption through loss of a parent in early life is delinquency. Criminal records of all English and Welsh school children were examined, and children who had experienced parental divorce, separation or death before they were aged five years were significantly more likely to have a criminal record. The nature of their crime was distinctly different from that of others; they were significantly more likely to be violent or sexual offenders (Wadsworth, 1979). Even after allowances were made, in a stepwise multiple regression analysis, for the effects of such intervening variables as social class, education and changes in home circumstances as a result of parental separation or death, this association remained statistically significant.

If this long-term apparent effect is really so marked we should expect to see its manifestations in ways other than delinquency. Perhaps some children expressed this effect in various other ways such as illness and other kinds of behaviour. Illnesses selected for investigation as possible outcome variables were psoriasis, migraine, asthma and epilepsy, however treated, and all hospitalised cases of stomach and doudenal ulcers, adult colitis and psychiatric affective disorders. Behaviour studied was conviction for a crime and, in women, illegitimate pregnancy. Table 1 shows the extent of differences between experience of these illnesses and kinds of behaviour in the whole study population when grouped according to family disruption. There were no differences in the experiences of epilepsy, migraine, asthma or psoriasis, and they have therefore been omitted. In order to test whether family circumstances, which might have had mitigating or exacerbating effects on the development of such problems, have influenced these findings, the possible biasing effects of social class, change in social class, birth order and family size were taken into account in a discriminant function analysis designed to test the predictive power of data on family life disturbance in infancy. Table 2 shows that even after allowances were made for these circumstances, differences in prevalence of illnesses were still considerable.

DISCUSSION OF THESE FINDINGS

This study provides confirmation of the hypothesis that early life experience of parental death, divorce or separation may be a source of vulnerability to certain behaviour and illness problems in later life. These findings raise three sets of important points for further study.

Table 1. Percentages of men and women in each family disruption group who experienced these illnesses or kinds of behaviour.

Illness & behaviour	Age at which family disruption occurred			No family disruption
	0-4 years	5-15 years	16-26 years	
delinquency*	28.6%	16.1%	16.0%	14.1%
stomach ulcers or colitis**	1.6%	0.7%	0.4%	0.4%
psychiatric illness**	6.3%	2.7%	9.8%	1.7%
% of each group experiencing one or more of the above	36.5%	19.5%	17.2%	16.2%
males - total in each group	126	149	238	1,683
delinquency*	7.9%	2.6%	1.5%	1.6%
psychiatric illness**	3.4%	3.3%	4.0%	1.5%
illegitimate birth(s)	6.0%	4.6%	5.9%	3.9%
% of each group experiencing one or more of the above	17.3%	10.5%	11.4%	7.0%
females - total in each group	116	152	273	1,494

* by 21 years
** by 26 years

Table 2. Prevalence of two kinds of illness (amongst all who were not delinquent) grouped by the discriminant analysis using data on infant family disruption, social class, birth order and family size.

	Stomach/peptic/duodenal ulcers by age 26 years	Psychiatric disorder by age 26 years
discriminated as stressed (total = 711)	25.3 per 1,000	23.9 per 1,000
discriminated as not stressed (total = 915)	7.7 per 1,000	14.2 per 1,000

a) How is vulnerability maintained?

In this study there is evidence for two kinds of factors that may maintain vulnerability. Vulnerability may be maintained, as the psychological hypotheses suggest, as a result of internalisation of the effects of emotional damage caused at this sensitive time in development. It is possible that the findings on pulse rate differences are an indication of internalisation. In this case the processes involved may be truly biopsychosocial. But it is also possible that, at least to some extent, vulnerability may be maintained or even enhanced in effect by the stigmatising nature of the experience of parental loss at this time. Evidence for this is to be found in the significantly lower assessments of performance that teachers made of these children. If stigma plays an important part, then the process of maintenance of vulnerability may primarily involve self-esteem.

b) Timing of outcomes of vulnerability

From this study two important aspects can be determined in the timing of when the individual will have an increased likelihood of a stressful outcome associated with the kind of vulnerability investigated here. As we already know (for example, Brown and Harris, 1978), vulnerability may be triggered into a stressful outcome by the experience of a particularly disturbing event, although specificity and sensitivity of prediction from an event such as the death of a loved one may not be very good in a random population. But to a certain extent some kinds of vulnerability peak times may be discerned. We should expect peaks at certain times in the life cycle, for example in late adolescence and at the time of a first job, that is the time of adult identity formation. We should also expect peaks at the community's peak times of marriage and child bearing, of separation and divorce, and in the years immediately before and after retirement.

If, as the findings from this study suggest, the socially stigmatising nature of some vulnerability factors is part of the mechanism that maintains vulnerability over long periods of time, we should also expect that, as time changes, vulnerability will change as sources of stigmatisation change. For example, as parental divorce and separation become more common it is anticipated that its stigmatising effect on children will be reduced, and opportunities to study such changes exist now within countries, and certainly between the European countries. Social expectations of effects of certain life experiences undoubtedly play an important role in the determination of outcomes (Wadsworth, 1981b).

c) The form of outcomes of vulnerability

No doubt the form that outcomes of vulnerability may take is partly determined by genetic predisposition, but unfortunately this study has no data to examine this aspect of vulnerability.

However, there will also be two sorts of cultural effects on the form of outcomes. First is the effect of the group in society within which the individual has been brought up, and of the group within which the individual lives. These effects may be seen in the style of response that the individual adopts. For example, certain kinds of criminal behaviour are more likely in some groups than in others. But the individual's stage in the life cycle will also have an effect on the form of outcome, as well as on its timing. We know, for example, that there are peak ages for crime, divorce and separation, and for alcohol dependence and addiction, and it is suggested that the form of outcome is also to an extent determined by the behaviour culturally appropriate to the individual's age.

CURRENT WORK ON THE COHORT

a) Recent data collection

Current work on this lifetime prospective study continues the search for factors indicating earliest vulnerability to mental and physical illness, and to stress. There is undoubtedly a need, particularly in cardiovascular research, for investigations in which host and environmental variables are considered together (Review Body, 1981). The opportunity to do this in a study of a population for whom whole life history data are available on a wide range of topics seems likely to complement and to add to existing findings, for example those of the Framingham study (Haynes et al, 1978a and b) and the work of Siegrist and Halhuber (1981) and of Siegrist et al (1982). The Framingham investigators drew attention to existing scepticism about the role of physiologic and social factors in the genesis of illness, and noted "the presence of few prospective studies on the subject".

When members of this prospective study were aged 36 years they were visited at home by research nurses, specially trained for this investigation, who collected data on health histories, exercise, eating, drinking and smoking habits, occupation, home, family and marital circumstances. They also used the Present State Examination (Wing et al, 1974) to ascertain the subject's current mental health, in terms of affective state, and they measured blood pressure, respiratory function (forced expiratory volume), height, weight and girth.

Using these data various indicators of current mental and physical health are being drawn up to enable each member of the cohort to be located on the distribution, within the cohort, of each health variable. Then the predictive value of contemporary data and the information from early life will be tested for their power to predict the health indicator variables, thus giving the opportunity to ascertain, in multivariate analyses, the strongest sources of discrimination of the least and the most healthy. Information on smoking, diet, exercise and vulnerability to ill health as a result of maladaptation, collected both contemporaneously and in earlier life, will be tested for discriminatory power, but only the information on maladaptation is discussed in this paper.

b) Data on maladaptation as predictors of adult health

Central to this investigation of vulnerability to ill health and its origins, is the concept of identity, formulated here under the headings cultural and extra-familial factors, familial factors and occupational factors, as shown in figure 1. Under each of these headings are data from which a ranking of identity will be constructed in terms of the degree of individual affiliation with each of these three sets of factors.

Figure 1. Factors used in the investigation of vulnerability to
ill health.

Cultural and extra-familial factors
Religious belief
Contacts with friends
Pastimes and clubs
Political affiliation
Cultural float or movement away from family of origin

Familial factors
Contacts with the family of origin
Recollections of family of origin
Social mobility
Personal marital history
Marital history of parents
Fertility and plans for future fertility

Occupational factors
Work histories
Self-assessments of stress
Self-assessment of life chances in work terms
Expectations of teachers and parents of occupational chances
Income

For our investigation of cultural and extra-familial factors we shall make use of the data on social class, religion and political affiliation in families of origin and at various times in the survey member's life, in order to investigate what might be called cultural float. That is to say, we shall be able to examine the degree and constancy of movement away from parental views, and compare the management of life crises and style of life amongst those who move very little from parents, with those who move away, and then return, and with those who move away and stay away (Wadsworth and Freeman, 1983).

The life history data on occupational factors will allow us to match actual occupational achievement with earlier expectations of achievement made by the subjects themselves and by others; and we shall be able to compare these with our projections from earlier life data of likely position at 36 years. We shall also be able to continue the work of Cherry (1978 and 1983) in examining the degree of self-selection in work stress, and the role played by personal and home life stress, in determining health at age 36 years.

The familial factors will supply life event data that have been collected over the whole lifetime of our subjects, which will allow us to examine temporal effects, as well as the possibility, raised earlier in this paper, of critical periods when some kinds of experience increase long-term vulnerability. These data, together with teachers' assessments and the Bortner questionnaire (Adler, 1977) will enable us to study the genesis - if there is one - of type-A behaviour, and to study the role of experience in its development.

FUTURE WORK

At a future contact we hope to be able to construct biological indicators of rate of change of fitness, particularly cardiovascular fitness, in order to test the associations of these data with degree of change in cardiovascular health. We hope to be able to include, at least, a measure of urinary catecholamine level. However, there remain problems in a large population study of laboratory handling of blood and urine samples; and as we have found in population scale measures of blood pressure, suitably reliable, compact, light-weight equipment for making measurements in homes is still being developed, and this needs to be encouraged.

It is thus intended that the study will continue to be a testing ground for hypotheses on lifetime sources of vulnerability to stress, and for the further exploration of a wide range of medical and social outcomes of vulnerability and stress experience at various ages.

REFERENCES

Adler, R.H., 1977. Personlichkeitszuge (typ A) bei Patienten Claudicudio Intermitins ergab die Bortner-Tests. Schweizerische Wochenschr. 107: 1833.

Atkins, E., Cherry, N.M., Douglas, J.W.B., Kiernan, K.E. and Wadsworth, M.E.J., 1981. The 1946 British birth survey. In: S.A. Mednick and A.E. Baert (Editors), Prospective Longitudinal Research. Oxford University Press.

Bowlby, J., 1975. Attachment and Loss. Pelican Books, London.

Brown, G.W. and Harris, T., 1978. Social Origins of Depression. Tavistock Press, London.

Cherry, N.M., 1978. Stress, anxiety and work. J. Occup. Psychol., 51: 259-270.

Cherry, N.M., 1983. Reports of nervous strain, anxiety and symptoms in a cohort of 32 year old men at work in Britain. In preparation.

Clarke, A.M. and Clarke, A.D.B., 1976. Early Experience: Myth and Evidence. Open Books, London.

Davies, J.G.V. and Maliphant, R., 1974. Refractory behaviour in school and avoidance learning. J. Child Psychol. and Psychiat., 15: 23-31.

Douglas, J.W.B., 1964. The Home and the School. MacGibbon and Kee, London.

Douglas, J.W.B., 1973. Early disturbing events and later enuresis. In: I. Kolvin, R.C. MacKeith and S.R. Meadows (Editors), Bladder Control and Enuresis. Spastics International Medical Publishers, London.

Haynes, S.G., Levine, S., Scotch, N., Feinleib, M. and Kannel, W.B., 1978a. The relationship of psychosocial factors to coronary heart disease in the Framingham study - I. Amer. J. Epidem., 107: 362-383.

Haynes, S.G., Feinleib, M., Levine, S., Scotch, N. and Kannel, W.B., 1978b. The relationship of psychosocial factors to coronary heart disease in the Framingham study - II. Amer. J. Epidem., 107: 384-402.

Rutter, M., 1972. Maternal Deprivation Re-Assessed. Penguin Books, London.

Review Body, 1981. Coronary-prone behaviour and coronary heart disease: a critical review. Circulation, 63: 1199-1215.

Siegrist, J. and Halhuber, M.J., 1981. Myocardial Infarction and Psychosocial Risks. Springer-Verlag, Berlin.

Siegrist, J., Dittman, K., Rittner, K. and Weber, I., 1982. The social context of active distress in patients with early myocardial infarction. Soc. Sci. and Med., 17.

Susser, M.W., 1981. The epidemiology of life stress. Psychol. Med., 11: 1-8.

Wadsworth, M.E.J., 1976. Delinquency, pulse rates and early emotional deprivation. Brit. J. Criminol., 16: 245-256.

Wadsworth, M.E.J., 1979. Roots of Delinquency. Martin Robertson, Oxford and Barnes and Noble, New York.

Wadsworth, M.E.J., 1981a. Social class and generation differences in pre-school education. Brit. J. Sociol., 32: No. 4, 560-582.

Wadsworth, M.E.J., 1981b. Social change and the interpretation of research. Criminology, 19: 53-76.

Wadsworth, M.E.J. and Freeman, S.R., 1983. Generation differences in beliefs: a cohort study of stability and change in religious beliefs. Brit. J. Sociol. In press.

Wing, J.K., Cooper, J.E. and Sartorius, N., 1974. The Measurement and Classification of Psychiatric Symptoms. Cambridge University Press.

Yarrow, M.R., Campbell, J.R. and Burton, R.V., 1970. Recollections of Childhood: a study of the Retrospective Method. Monographs of the Society for Research in Child Development. Serial No. 138, Vol. 35.

PSYCHOSOCIAL AND PSYCHOPHYSIOLOGICAL FACTORS
IN THE DESIGN AND THE EVALUATION OF WORKING
CONDITIONS WITHIN HEALTH CARE SYSTEMS

Tom Cox

Stress Research
Department of Psychology
University of Nottingham
Nottingham NG7 2RD
United Kingdom

ABSTRACT

This paper describes the role of psychosocial and psychophysiological processes in the maintenance of health and the aetiology of ill-health. It attempts to develop an additional approach to the evaluation of health care systems based on the recognised importance of these processes for well-being, and on the role that they play in determining the impact of health-care systems on the well-being of the care providers. It goes someway in describing the development, validation and application of a possible assessment procedure. Although such a package is discussed in the context of health-care systems, it could equally be applied to other occupational situations, including those developing around the use of computer-technology.

1. MODELS OF HEALTH

Over the last century, effective models and strategies have evolved for dealing with health related-problems and for providing health care. Until relatively recently these have largely been concerned with acute and infectious diseases, and injury. However, with the development of public health measures and advances in immunisation, pharmacotherapy and surgical techniques, afflictions such as smallpox and typhoid are now adequately controlled, and much of the effect of physical trauma can be retrieved. At least this is so for the developing countries. The developing world is still in the process of establishing such controls. Thus for the former, health priorities have necessarily shifted towards the problems posed by chronic disorders. Unfortunately, in this area, the traditional models and approaches are not so effective, largely because they tend to assume relatively simple and specific cause-effect relationships. Somewhat by contrast, current knowledge of chronic disorders is framed in terms of multifactorial aetiologies and complex prognoses, both involving a great diversity of risk and buffer factors. At the individual level the degree of vulnerability appears to interact with exposure to challenging or precipitating agents in the possible presence of buffers. This interaction can alter the probability of disorders occuring, and, if they occur, can modify their severity and duration. Interestingly, many of these differently conceived factors or agents are essentially psychosocial in nature, appearing to express their influence through the person's behaviour or pattern of neuro-endocrine activity. Good examples of such an approach to chronic disorder are provided by current research on coronary heart disease (see Glass, 1977), and on cancers (see Cox and Mackay, 1982).

While traditionally attention has focussed on coronary heart disease, there is now a growing interest in the mechanisms involved in the aetiology and development of cancers. Unfortunately, much of the psychological research in this area has been methodologically inadequate. Many studies have been based on retrospective investigations of the life styles and psychological states of cancer patients, compared with those of matched groups of non-cancer patients, largely ignoring the impact of "having cancer". Much of the data which is reliable is based on animal experiments, and its immediate relevance to human experience and illness can be questioned. However, what may be tentatively suggested is that certain psychological and social experiences do significantly alter endocrine activity, particularly that reflecting adrenal function. This in turn may interfere with one of several processes, for example, immune system activity or the favourability of the cellular environment, and allow the accelerated development of certain cancers. These 'stressful' experiences may be more associated with particular behavioural styles than others, and coping may increase exposure to carcinogens, e.g. through smoking, or perhaps, diet.

Whether the disorders and injuries of interest are acute or chronic, the descriptive models adopted must thus allow for the influence of two pathological processes: the first relating to the direct effect of the affliction, and the second relating to the person's overall experience and response to their condition, and to "stress" in general. The person's experience of "stress" may give rise to significant changes in attitudes, behaviour and neuro-endocrine state, all of which can influence prognosis (Cox, 1978). The effects of ill-health and "stress" are therefore intimately and interactively related, and both encompass changes in psychological state and behaviour as well as physiological function (e.g. Cox et al, 1982a). Both can be reflections of a "breakdown in adaptation", and the changes they induce offer possible avenues for the measurement of such breakdown.

2. STRATEGIES OF HEALTH CARE

There are several different ways of dealing with the effects of breakdown in adaptation; most can be described in terms of responsibility being assumed and action taken either by the individual, or by an organisation or society. The provision of health care systems is one example of the adoption of responsibility by society and of action initiated at that level.

Current attempts to provide cost-effective and patient effective health care have related to three areas of concern. First, there is a need to maintain a healthy population, and to take action to prevent illness and injury. Second, should these occur, there is a need to provide effective diagnosis and treatment, and to improve the health of those declared ill or injured. Third, but not least, there is a need to rehabilitate those that have been treated and to re-introduce them into a normal pattern of living. Overall our concern has been to protect and improve the general level of well-being as a means of ensuring an acceptable quality to life. Given the previous comments on the nature of health problems, health care must partly focus on the role of psychosocial factors and psychophysiological mechanisms; indeed on the "role" of "stress".

3. DESIGN AND EVALUATION OF HEALTH CARE

The importance of psychosocial and psychophysiological factors in health has obvious implications not only for the design and practice of health care but also for the evaluation of health care systems. Indeed, recognition of their roles begins to provide a model for the general design and evaluation of work and working conditions. It is obvious, for example, that the design of such systems should reflect a consideration of the factors which promote and protect as well as threaten good health. The determination of these factors, and the assessment of their consequences and effects, would also provide one adequate basis for the evaluation of levels of health. Needless to say, such an approach would be best constructed from a well validated theoretical position, such as that provided by "interactional" or "systems" approaches to models of stress (see Cox, 1978, Cox and Mackay, 1981).

Health care systems might be evaluated in two different but related ways. Traditionally they have been considered in terms of the patient care they provide, and the cost of that care. Evaluation has thus tended to be economic or client-centred. However, health care systems may also be assessed in terms of their impact on the care providers, the health care personnel. Interestingly, the very psychosocial and psychophysiological processes which influence the development of illness in the patient population may also affect the occupational well-being and job performance of those that care for them. Impaired well-being among staff within an organisation has been associated with low job satisfaction and suboptimal performance, and may be related to high sickness-absence, poor time keeping and high labour turnover. Such effects within any health-care system would undoubtedly reduce the quality of and effectiveness of patient care, or increase the cost of good patient care.

4. STRATEGY FOR DESIGN AND EVALUATION

Two particular sets of instruments or procedures are required to contribute to the sensible design of health care systems and to allow for the present form of evaluation. First, there needs to be a means of identifying the sources of stress existing within any system. Second, there is a need to be able to assess the effects of such stressors on general well-being, on job satisfaction and on job performance. Traditionally approaches to questions of performance and well-being at work have involved the assessment of organisational and ergonomic factors in job design, with little emphasis placed on the psychosocial or psychophysiological status of the individual. Somewhat by contrast, the "transactional" approach to stress (Cox, 1978; Cox and Mackay, 1981) provides a "systems" - based description of occupational problems and their effects, with a deliberate emphasis on these latter factors. Such an approach might provide the necessary theoretical framework for the development of the two sets of instruments.

4a. Theoretical Framework: Transactional Model of Occupational Stress

It has been suggested that "stress" is experienced when the person is not adequately matched to his or her job or work environment (Cox and Mackay, 1981). The important dimensions in this mismatch may be considered, inter alia, in terms of:

1. Demands

Demands are requirements or requests for action (or adjustment), whether essentially behavioural or cognitive. They usually involve some degree of decision-making and the exercise of skill. They may be imposed by the external environment, say as a function of work or the work-home interface, or may be internal reflecting the person's needs: material, social and psychological. There may be several important dimensions describing external (job) demands, for example, "pleasantness/unpleasantness" and "ease/difficulty" (Cox, 1980; Cox and Mackay, 1979). Demands usually involve a time base, and may be amplified by an acute sense of time urgency (type A behaviour: Zyzanski and Jenkins, 1970). Failure to adequately fulfil demands may be experienced as threatening.

The absolute level of demand would not appear to be the important factor in determining the experience of stress. More important is any discrepancy that exists between the level of demand and the person's ability (resources) to meet that demand. The size of this discrepancy appears to be an important determining factor in the stress process. However, the relationship between the discrepancy and the intensity of the stress experience may be curvilinear (Cox, 1978). Within reasonable limits stress can thus arise through <u>overload</u> (demand>personal resources) or through <u>underload</u> (demand<personal resources). The person's ability to cope with such an imbalance may be constrained and supported in different ways.

2. Constraints

Constraints operate as restrictions or limitations on free action or thought, reflecting a loss or lack of discretion and control. These may be imposed externally. For example, constraints may be imposed by the requirements of specific jobs, by the rules of the organisation; they may be role-related or reflect the beliefs and values of the individual.

3. Support

Support can be made available in different ways, most essentially through "social interaction": through advice and information, through practical help, or by providing emotional support or declaring empathy. It is possible that women need and are more sensitive to social support than men (Viney, 1980; Cox et al, 1983). Together the person's individual and socio-economic resources and the social support available to them represent the positive buffer factors operating in the "stress process".

4b. <u>Studies Within the Transactional Framework: Repetitive Work</u>

The study of industrial repetitive work practices, carried out at Nottingham, has been set within the broad framework of the transactional model (see Cox et al, 1979; Cox, 1980; Cox et al, 1982). Studies both in the laboratory and on the factory floor have highlighted the demands associated with repetitive work:

<u>task-inherent</u>: repetition of a simple motor act, constraints on behaviour (etc).

task-contextual: restricted level of stimulation, constraints on behaviour (etc)

organisational incentive pay schemes, lack of control (etc)

These have been described in more detail elsewhere (Cox et al, 1982). Although the detail of this scheme is obviously situation dependent, the broad categories or levels of demand may apply more generally.

The evidence gathered from simulation and experimental studies has demonstrated that the immediate effects of initial exposure to repetitive work can be detected in terms of changes in work performance, in mood and in activity in certain physiological systems. Results from recent industrial surveys have supported these findings, and highlighted the importance of social support (e.g. marital status), age and objective job demands (repetition, rotation and levels of attentional demand) in determining job perceptions, mood and well-being (Cox et al, 1983). These ideas and data support the formulation of a general hypothesis relevant to design and evaluation issues.

4c. Formulation of a General Hypothesis

The following hypothesis may thus be suggested: demands which are not matched to personal resources and which are associated with high constraint and a lack of support in coping are experienced as "stressful". This experience may, in turn, be reflected in sub-optimal performance, impaired mood, poor well-being and overall job dissatisfaction. It may also be associated with more general changes in behaviour, and with altered patterns of neuro-endocrine activity. Together these may be of significance for pathogenesis. Consistent with this hypothesis it has been variously shown that the combination of high constraint (low discretion) with high demand (workload) gives rise to the experience of stress and predicts various "stress" outcomes (Payne, 1979, 1981, 1982; Karasek, 1979).

5. MEASUREMENT

In the Nottingham studies, the identification of sources of stress within different occupational groups has followed a three stage process using a psychometric paradigm.

Initially there is the collection of data on the nature of possible stressors; this has often been by interview or open-ended questionnaire, or by observation, expert advice, or through workshop discussion. Second, a possible list of stressor items has been associated in a structured format with "intensity" or "frequency" scales (to assess experience), and then subjected to psychometric examination. This has often used factor-analytical techniques. Finally, the model so developed, its various items and dimensions, have been reflected back to the "client" population, and their acceptability and face validity tested.

The status of these "stressor" models and psychometric instruments then requires validation against measures reflecting the psychosocial dimensions which underpin them, and measures of their psychophysiological and health outcomes.

5a. Possible Evaluation Instruments and Procedures

In the author's studies on repetitive work, it has been necessary to seek parameters for the measurement of demands, and constraints and of the support factors (e.g. Karasek, 1979; House, 1980; Turner, 1981; Pearlin et al, 1981) together with methods of assessing workers general perceptions and descriptions of their work (Cox and Mackay, 1979; Cox, 1980). Effects on outcomes such as performance and self-reported mood (Mackay et al, 1978), well-being (Cox et al, 1983) and physiological status (S. Cox et al, 1981/2) have also been considered. These instruments and procedures now form the basis of an evaluation procedure, which will be applied to the study of the effects of working conditions on health care personnel, and also in studies on transport workers, and screen-based work.

The measurement of specific sources and levels of dissatisfaction in particular occupational groups is obviously situation dependent, and requires the development of special psychometric instruments appropriate to each new situation studied. The development of such an instrument is by necessity closely linked to the initial identification of occupational problems.

The development and validation of an integrated assessment procedure must involve controlled reliability and validation studies; together with the collection of normative data. Such a package of measures would allow the critical relationships within the proposed theoretical framework to be tested. Furthermore, the application of such a package could eventually involve:

(a) its use in the form of a job design (or re-design) checklist to produce "constructive anxiety" concerning the critical psychosocial and psychophysiological factors involved in the work under study,

(b) its use in describing (and evaluating) psychosocial and psychophysiological factors in existing work systems (with reference being made to a progressively expanding normative data base) and its

(c) its subsequent use in the development of job improvement programmes, and its use in evaluating (validating) job improvement and redesign programmes, being applied before and after any changes.

5b. Cross-cultural validation

The development of general interest in Europe and North America in this type of approach to stressful working conditions and health-related issues, has naturally led to initiatives in many different countries. This, in turn, has raised the interesting question of the cross cultural or international comparison and validation of the different initiatives. Such validation could occur in two possible, but not equally effective ways. First, the various instruments and procedures in use within one culture could be translated and parallel forms prepared for use in another culture. The feasibility of this measure-based validation would vary greatly with the nature of the cultures involved. Differences in social and linguistic function as well as differences in cognitive structure may render such a strategy unsuccessful. The second, and more likely, approach is to attempt validation at a conceptual level, examining the model which generated or was generated by the overall approach. This model could then be used to identify important variables and critical relationships, with local measures being developed and validated within each culture. These could then be used to examine the structure of the model in a way appropriate for and acceptable to each culture.

6. SUMMARY

From the ideas and data briefly reviewed here, it has been possible to suggest an additional approach to the design and evaluation of job design and working conditions. This focusses on their impact on the job occupant, and emphasises psychosocial and psychophysiological factors. It contrasts with more traditional forms of evaluation. Such a strategy might be applied to many different occupational groups; here it is discussed in relation to health care systems and personnel. In that setting it can be used to supplement existing economic and client-centred approaches to design and evaluation.

In applying the present strategy, fundamental questions concerning stress theory may be raised and examined, and the present proposals may have theoretical as well as practical value.

7. ACKNOWLEDGEMENTS

The author wishes to acknowledge the support of the (British) Medical Research Council during the preparation of this paper. The views expressed here are his own.

8. REFERENCES

Cox, S., Thirlaway, M., Cox, T., 1981. A dental roll technique for the measurement of salivary activity. Behaviour Research Methods and Instrumentation, 13, 40-42.

Cox, S., Cox, T., Thirlaway, M. and Mackay, C.J., 1982. Effects of simulated repetitive work on urine catecholamine excretion. Ergonomics, 25, 12; 1129-1141.

Cox, T., 1978. Stress, Macmillans, London.

Cox, T., 1980. Repetitive Work. In: C.L. Cooper and R. Payne (eds) Current Concerns in Occupational Stress, Wiley, Chichester.

Cox, T. and Mackay, C.J., 1979. The impact of repetitive work. In: Sell, R. and Shipley, P. (eds) Satisfaction in Job Design, Taylor and Francis, London.

Cox, T. and Mackay, C.J., 1981. A transactional approach to occupational stress. In: N. Corlett and J. Richardson (eds) Stress, Work, Design and Productivity. Wiley, Chichester.

Cox, T. and Mackay, C.J., 1982. Psychosocial factors and psycho-physiological mechanisms in the aetiology and development of cancers. Social Science and Medicine, 16, 381-396.

Cox, T., Thirlaway, M. and Cox, S., 1982a. Repetitive work, well-being and arousal. In: H. Ursin and R. Murison (eds) Biological and Psychological Bases of Psychosomatic Disease, Pergamon, Oxford.

Cox, T., Thirlaway, M., Cox, S. and Gotts, G., 1983. The nature and measurement of well-being. Journal of Psychosomatic Research, (in press).

Glass, D., 1977. Stress, behaviour patterns and coronary heart disease. American Scientist, 65, 177-187.

Karasek, R., 1979. Job demands, job decision latitude and mental strain: implications for job design. Administrative Science Quarterly, 24, 285-308.

Mackay, C.J., Cox, T., Burrows, G.C. and Lazzerini, A.J., 1979. An inventory for the measurement of arousal and stress through self-report. British Journal of Social and Clinical Psychology, 17, 283-284.

Payne, R., 1981. Stress in task-focused groups. Small Group Behaviour, 12, 253-268.

Payne, R. and Fletcher, B., 1982. Job demands, supports and constraints as predictors of psychological strain among school-teachers. Journal of Vocational Behaviour (in press).

Payne, R., 1979. Demands, supports, constraints and psychological health. In: C.J. Mackay and T. Cox (eds) Response to Stress: Occupational Aspects, IPC, Guildford.

Pearlin, L.I., Menaghan, E.G., Liberman, M.A. and Mullan, J.T., 1981. The stress process. Journal of Health and Social Behaviour, 22, 337-356.

Turner, J.T., 1981. Social support as a contingency in psychological wellbeing. Journal of Health and Social Behaviour, 22, 357-367.

Viney, L. 1980. Transitions. Cassell, Melbourne, Australia.

Zyzanski, S.J. and Jenkins, C.D., 1970. Basic dimensions within the coronary-prone behaviour pattern. Journal of Chronic Diseases, 22, 781-795.

THE RELATION OF SOCIAL TO PATHOPHYSIOLOGICAL PROCESSES:
EVIDENCE FROM EPIDEMIOLOGICAL STUDIES

M.G. Marmot
London School of Hygiene and Tropical Medicine
Keppel Street, London, WC1E 7HT

ABSTRACT

Epidemiological strategies have proved appropriate for studying aetiology of chronic diseases. Sound application of epidemiological techniques allows the testing of hypotheses. Such techniques have been applied only to a limited extent in studying psycho-social factors. This is, in part, owing to the difficulty of adapting both concepts and measures for study in different social and cultural groups. Studies of Type-A behaviour, occupational stress and social supports provide examples where this problem must be tackled.

The demands of large population studies mean that relatively simple measures are used in epidemiological studies. Therefore results from these studies must be taken together with other smaller, but more in depth, investigations.

EPIDEMIOLOGICAL APPROACH TO CAUSATION

Who "discovered" that smoking causes lung cancer? Was it a clinician who noticed that his last 3 cases of lung cancer had all been heavy smokers and said "I wonder if...?". Was it the molecular biologist, studying the action of constituents of tobacco smoke on cells? Was it the experimenter who taught beagle dogs to smoke? Was it the epidemiologist who, in a variety of ways, linked smoking with increased risk of lung cancer?

There is not a single answer. The elaboration of a hypothesis is not, by itself, a discovery of causation - evidence is required to support it; nor is the astute clinical observation - it requires the systematic estimation of lung cancer incidence in exposed and 'unexposed'; nor is the demonstration of a plausible biological mechanism - it tells us what can happen, not what does happen; nor the animal experiment - does it apply to humans? There were many hypotheses about the cause of lung cancer, cigarette smoking among them. The contribution of the epidemiologists was to take one of these hypotheses and test it. The epidemiologists showed smoking to be the likeliest cause: by showing that smokers were at increased risk, that the increased risk was

unlikely to be due to factors other than smoking, that there was a
dose-response relationship, that the distribution of lung cancer in
the population followed the distribution of smoking, that trends over
time in lung cancer were correlated with trends in cigarette smoking,
that ex-smokers had a lower risk than continuing smokers, and so on.

What possible insights does this hold for the study of social
and psychological factors? Epidemiological study must form one of
the strategies for demonstrating causation. But where psycho-social
factors have been included in epidemiological studies, not all the
strategies available have been utilised fully.

For example, the association between type A behaviour and coronary
heart disease (CHD) is among the strongest of any psycho-social factor.
The most compelling evidence for this association comes from the
Western Collaborative Group Study - a study of predominantly middle
class white American men (Rosenman et al 1976). Our confidence in
the Type-A:CHD association is increased by its demonstration in women
(Haynes et al 1980a), and in Europe as well as in the U.S.A. (French-
Belgian Collaborative Group 1982). What has not been shown is that
Type-A behaviour follows the distribution in the population of CHD.
Or to put it another way, very little work has been done on the extent
to which variations in the frequency of Type-A behaviour may explain
variations in CHD over time, geographically and between subgroups of
the populations. Could an increase in Type A behaviour have accounted
for the 20th century epidemic of CHD? and a decrease in Type-A
contributed to CHD decline in several countries? We have no information.
To what extent could differences in Type A behaviour account for
international differences in CHD occurrence? We have little information.

Such information as we have, within populations, is not
encouraging. In our Whitehall study of British civil servants, for
example, men in the lower grades of employment had more than three
times the risk of CHD death (Marmot et al 1978), but fewer were Type-A's
than men in the upper grades (Marmot 1982). In Framingham, working
women have less heart disease than men but equal prevalence of Type-A
behaviour (Haynes et al 1980b).

These conflicting data suggest either that differences in CHD
between subgroups of the population are not due to differences in
Type-A behaviour, or that the measurement (and perhaps the concept)

of Type-A behaviour is biassed with respect to culture and/or class. At the very least, one can say that subjecting the Type-A-CHD association to rigorous testing using epidemiological methods points out the strengths and weaknesses of current theories and methods. It points the way to further research needs.

RESEARCH IN DIFFERENT SOCIAL AND CULTURAL GROUPS

This brings us to a problem in pursuing an epidemiological strategy for testing causation. There are variations between social and cultural groups in the meaning of psycho-social concepts and in the methods of measurement. Type-A behaviour may to some extent be a reaction to environmental stressors, the style of which is influenced by the culture. For example, a high grade civil servant may react to perceived threat by striving harder and become classified as exhibiting Type-A behaviour. The working-class counterpart in a more limited job with less flexibility and control may react to stressors in a different way. Differences between them in Type-A behaviour may tell us little about differences between them in external stressors or internal reactions to them.

If we move away from characteristics of individuals, such as Type-A behaviour, to characteristics of the social environment, such as potentially stressful working conditions, the same problems must be faced. Do the occupational stressors identified in one country/culture, and the methods used to identify them, apply equally in a different culture? The answer will come partly from theory and knowledge of the cultures concerned, but also from empirical research.

An example from our own studies illustrates the problems of cross-cultural research. Japan is remarkable among industrialised countries for its low rate of CHD. Japanese migrants to the USA and their descendents have a CHD rate intermediate between the rate in Japan and the high rate in the USA (Syme et al 1975; Marmot et al 1975). We were interested to explore the hypothesis that Japanese culture is characterised by a high level of social supports which protect against the type of stress which increases CHD risk (Matsumoto 1970). We found that, among the Japanese in California, various measures of acculturation were associated with CHD prevalence (Marmot et al 1976) i.e. the more Westernized the Japanese, the higher

the CHD prevalence. This relationship was independent of diet or
other cardiac risk factors. Our interpretation was that Japanese in
California who were brought up in a Japanese way and who remain
affiliated to their ethnic group are more likely to be recipients
of social support than more Americanised Japanese. But we had no
direct measure of social supports. It is possible that other features
of Japanese culture may have been responsible for the lower CHD
prevalence among the traditionally Japanese men.

To test this hypothesis further, elsewhere, using the same
measures, is impossible because our measures were specific to Japanese
culture. One must therefore rely on studies that have used the same
concept but developed measures applicable to the culture under study.

One such study comes from the Alameda County Human Population
Laboratory (Berkman and Syme 1979). They classified people according
to their degree of participation in social networks, using as indicators:
marital status, contact with relatives and friends, membership of a
religious group, and membership of other organisations. Over 9 year of
follow-up the all-cause mortality risk in those with fewest 'connections'
was more than double that of those with most 'connections'- independent of
smoking, alcohol consumption, obesity and physical activity.

Using similar measures to these, we found in our study of civil
servants that the lower grade men, those at highest risk of CHD, had
less participation in social networks   (Marmot 1982).

A VARIETY OF STUDY DESIGNS

The unsatisfying feature of these studies, including our own, is the
gap between our theories and the measures we employ to test them. A
theory states that people with a high degree of social support will be
under less stress and therefore have lower disease risk. Yet we
count number of social contacts rather than measure social and emotional
support, and we measure disease outcome rather than stress. Part of the
reason is that we do not have measures that are applicable to the large-
scale population studies needed to show relationships to disease.

Studies in this area require a variety of study designs: in depth
sociological studies that attempt to characterise individuals more
accurately with respect to stress; psycho-physiological studies that
show changes in blood pressure, catecholamine output, cortisol etc in
response to stress; studies that link these variables to subsequent

disease risk; intervention studies at the social level that attempt to change exposure to stressors, and at the personal level that attempt to modify their effects.

Epidemiology requires relatively simple measures applicable to field surveys, and more work is needed in adopting techniques for population studies. Disease occurs in populations and epidemiology has a crucial role to play in testing hypotheses about causation.

REFERENCES

Berkman, L.F. and Syme, S.L., 1979. Social networks, host resistance, and mortality: a nine-year follow-up study of Alameda County residents. Amer. J. Epidemiol, 109:186-204.

French-Belgian Collaborative Group, 1982. Ischaemic heart disease and psychological patterns. In: E. Denolin (Editor), Psychological problems before and after myocardial infarction. Adv. Cardiol., 29, Karger, Basel, pp 25-31.

Haynes, S.G., Feinleib, M., and Kannel, W.B., 1980a. The relationship of psychosocial factors to coronary heart disease in the Framingham study. Amer. J. Epidemiol., III: 37-58.

Kagan, A., Harris, B.R., Winkelstein, W., Johnson, K.G., Kato, H., Syme, S.L., Rhoads, G.G., Gay, M.L., Nichaman, M.Z., Hamilton, H.B., Tillotson, J., 1974. Epidemiologic studies of coronary heart disease and stroke in Japanese men living in Japan, Hawaii and California - demographic, physical, dietary and biochemical characteristics. J. Chron. Dis., 27: 345-364.

Marmot, M.G., Syme, S.L., Kagan, A., Kato, H., Cohen, M.B., Belsky, J., 1975. Epidemiologic studies of coronary heart disease and stroke in Japanese men living in Japan, Hawaii and California: prevalence of coronary and hypertensive heart disease and associated risk factors. Amer. J. Epidemiol., 102: 514-525.

Marmot, M.G., Syme, S.L., 1976. Acculturation and coronary heart disease in Japanese Americans, Amer. J. Epidemiol., 104: 225-247.

Marmot, M.G., Rose, G., Shipley, M., Hamilton, P.J.S., 1978. Employment grade and coronary heart disease in British civil servants. J. Epidemiol. and Community Health., 32: 244-249.

Marmot, M.G., 1982. Socio-economic and cultural factors in ischaemic heart disease. In: Denolin, H. (Editor), Psychological problems before and after myocardial infarction. Adv. Cardiol., 29, Karger, Basel, 68-76.

Matsumoto YS, 1970. Social stress and coronary heart disease in Japan: a hypothesis. The Milbank Mem Fund Quart.,48: 9-36.

Rosenman, R.H., Brand, R.J., Sholtz, R.I., Friedman, M., 1976. Multivariate prediction of coronary heart disease during 8.5 year follow-up in the Western Collaborative Group Study. Amer. J. Cardiol.,37: 903-910.

Syme, S.L., Marmot, M.G., Kagan, A., Kato, H., Rhoads, G.G., 1975. Epidemiologic studies of coronary heart disease and stroke in Japanese men living in Japan, Hawaii, and California: introduction. Amer. J. Epidemiol., 102: 477-480.

UNEMPLOYMENT AND HEALTH: A REVIEW OF METHODOLOGY

S.C. Farrow
Department of Epidemiology
and Community Medicine
Welsh National School of Medicine
Heath Park
CARDIFF, Wales, U.K.

ABSTRACT

Unemployment presents a difficult problem for the researcher because of the so called 'healthy worker effect' and because of the inter-relationship between unemployment and other important social factors including social class, income and housing. This paper will review the mothods that have been employed in the literature on unemployment and health. This will include cohort studies of individuals and their families both unemployed and employed. It will include studies of factory closures and will examine large populations using both a cross sectional and time series approach. The central question is: does unemployment contribute to the development of desease.

DESCRIPTIVE AND COHORT STUDIES

Much of the earlier writing suffered from being largely descriptive and uncontrolled. In this thirties there were a number of major contributions (Lazarsfeld et al., 1933, Bakke, 1933, Pilgrim Trust 1938, Eisenberg et al., 1938). More recent studies from the Tavistock Institute (Hill 1977) and from the social and Applied Psychology Unit at Sheffield (Warr 1981) have concentrated on psychological well being and discussed in varying details the psychological stages that many unemployed go through. These stages are similar to those described elsewhere in the life events literature of major loss, for example from the death of a spouse. This concept of loss of employment has been described by Jahoda et al. (1980) in some detail. She emphasises that the benefits of employment go beyond the economic, that is simply earning a living, and she describes five latent or secondary consequences of employment. It imposes a time structure on the working day, it compels contacts and shared experiences outsige the nuclear family. It demonstrates goals and purposes beyond the scope of the individual. It imposes status. It enforces activity.

These descriptive studies have added much to our understanding of the relationship between unemployment and health.

This review will consider the DHSS Cohort Study, the National Training Survey and the British Regional Heart Study and finally the OPCS Longitudinal Study.

The DHSS Cohort Study selected 2300 men from 86 Unemployment Benefit Offices throughout Great Britain who registered in the autumn of 1978.

When one compares the use of health services by the cohort sample with the 1978 General Household Survey, a significantly higher proportion had visited a doctor during the previous 2 weeks and had been an in-patient during the previous three months.

Fagin obtained a sample of 20 families from this cohort where the man had been in continuous employment throughout 1977 and by the time of the first interview had been unemployed for at least 16 weeks (Fagin et al. 1981). Those men who depended on their job principally to earn a living seemed less psychologically affected than those who obtained other benefits. Fagin described the importance of these other benefits in the following terms: good relationship with peers at work, a strong identification with the skill involved in doing the job, a strong attachment to the institutional nature of the work, opportunities to escape from marital disharmony, support in maintaining an authoritarian role in the familiy.

In a number of families the man went through the predictable pattern of psychological responses described earlier. In others the changes were more profound and resulted in clinical depression and suicidal thoughts. These men were often treated with mild tranquilisers and antidepressants. Physical symptoms included asthmatic attacks, skin lesions, backache and headache. There was an increased use of tobacco and alcohol.

Some wives were similarly affected although for some there was an improvement in health and status particularly if the family now relied on her income. Following the onset of unemployment men with previous histories of poor health suffered relapses as did men with previous disabilities. This was marked where the man had overcome a particular disability and managed to work successfully and where this disability had not contributed to his unemployment. Once unemployed the disability took on new meaning.

Fagin also described changes in the children which were in agreement with much that has been written before. Children unter 12 were markedly affected with increasing sickness, worsening of behaviour at home and at school, truancy and a deterioration in school work. There was also an increase in proneness to accidents. Children over about 12 on the other hand seemed to accept a greater degree of responsibility. They would forego some of the usual teenage pastimes and concentrate on trying to take on a more adult role. There was a sort of premature maturity.

The National Training Survey interviewed 50,000 individuals in 1975-1976 about their lifetime labour market experiences. This provides information on unemployment spells and sickness spells that lasted for more than 3 months. What was looked at was sickness certification and no attempt was made to relate this o actual morbidity or mortality. The Survey was intended to investigate the effects of training particularly on future earnings and was not an ideal source for modelling the determinants of the incidence of sickness. Despite the deficiencies, however, this data set has been used by Metcalf and Nickell to discriminate between the effect of unempoyment on sickness and the effect of sickness on the unemployed. Sickness was said to be responsible for increasing unempoyment by 21 % whereas unemployment was said to be responsible for increasing sickness by 77 % (Metcalf et al. 1979).

The British Regional Heart Study involved over 7000 men

aged 40-59 randomly selected from the age-sex registers of representative general practices in 24 British towns. This study was not designed to look specifically at the effect of unemployment on health but to explain the substantial geographic variation in cardiovascular disease in Great Britain by evaluating the role of environmental, socioeconomic and personal risk factors. Employment status was recorded and an attempt was made to separate the unemployed (ill) where ill health was thought to have been responsible for unemployment from the unemployed (not ill). The results showed that there were significant differences between the unemployed (ill) and the employed regarding chronic bronchitis, obstructive lung disease and ischaemic heart disease. With regard to unemployed (not ill) there was a significant increase in ischaemic heart disease. Neither unemployed group showed an increased in hypertension (Cook et el. 1982).

The OPCS Longitudinal Study was based on a one percent sample of the population of England and Wales from the Census of 1971 linked to information on births, deaths and cancer registration from 1971-1975. Analysis of males 15-64 by employment status showed high standardised mortality ratios amongst those unemployed who were seeking work. This analysis separated those off work sick and those permanently sick from those unemployed seeking work. The standardised mortality ratios were: all causes, employed 86, unemployed seeking work 130; for circulatory diseases, employed 88, unemployed seeking work 115 and for accidents and violence, employed 93 and unemployed seeking work 222. In seeking an explanation of this excess the Report pointed to social class differences, the possibility that health did contribute to unemployment and the effect of unemployment. The excess was present in all specific disease categories and was approximately 50 % higher. In the case of accidents and violence it was two and a half times as high. This included suicide where unemployment may have contributed to the excess of deaths (Fox et al., 1982).

Although these findings were preliminary it would appear

that the OPCS Longitudinal Survey provides a sample which can be returned to and hopefully future reports will give the results of longer term follow up.

The Cohort Study is an extremely powerful method but many of its strengths can be lost by poor definition, poor selection of variables, small numbers and inadequate follow-up. Of the studies mentioned perhaps the OPCS Study comes closest to the ideal. The question of morbidity, however, was not addressed.

The larger the sample size the more problematic the interviewing of individuals and their families. The strength of the interview is the insight it gives into the problem; the opportunity to develop hypotheses. Fagin's study raised new questions but was not large enough to provide answers.

CASE STUDIES, CROSS-SECTIONAL, TIME SERIES STUDIES

A number of studies have been based on the closures of factories, in some instances having a profound effect on whole communities. These include the Norbest Sardine Factory, in Norway (Westin et al., 1977), The Nordhavn Shipyard in Denmark (Iverson et al., 1981), the Ryhope Colliery in the North of England (1970), the Steel Factory at St. Etienne in France (Ziegler, 1979) and several car factories in Michigan U.S.A. (Cobb et al., 1977).

To summarise the findings of the first four studies, there was an increase in emotional problems, anorexia, weight loss, in doctor consultation and in medication. Over the long term anxiety and depression increased as did cardiovaskular and respiratory problems.

The Michigan Study (Cobb et al., 1977) found increases in new peptic ulcers, and in the risk of ischaemic heart disease amongst those made redundant and a worsening of peptic ulcers amongst wives. In addition there were three suicides amongst those made redundant which although not statistically significant may point to the need to repeat these studies with a larger group. Amongst the physiological changes were increases in uric acid and cholesterol. The uric acid levels reached peaks during the pre-redundancy phase whereas the cholesterol levels showed gradual increase. In both cases these levels were higher in men who were unsupported by families. In addition blood pressure was raised immediately before and after redundancy.

The first problem with these studies was sample size. They were generally of the order of 100-200 employees and except in the Michigan Study without a control group. The second problem was the length of follow up which was usually 1 to 2 years. The third problem was the nature of the survey. Few were designed to answer the question - does unemployment affect health? Most obtained self reported changes in health. There was little independent observation and physical examination including blood pressure measurement was carried out only in the Nordhavn and Michigan studies.

Factory closures should provide an ideal cohort for long term follow up. In the past it would appear that the researchers have not made themselves available at the right time to take advantage of the opportunities.

The researchers of the 1930s knew perfectly well that high unemployment areas were also areas of low income, poor housing and poor diet, and that it would be extremely difficult to isolate the independent effect of unemployment.

One of the classic studies was carried out by Hans Singer (1937) and published as an interim paper for the Pilgrim Trust Unemployment Enquiry. He worked out for each of the county boroughs of England and Wales the difference in mortality rates for various causes of death over the period 1928 to 1933 and compared these with the difference in the level of unemployment over the same period. This method of differencing sought to eliminate the effects of the other variables and assumed that they remained substantially constant over the time period in question. Singer quoted significant correlations between unemployment and maternal and infant mortality, and between unemployment and and specific causes of death including diarrhoea and

and enteritis, diphtheria, scarlet fever and tuberculosis.
Stern in a well argued critique of the statistical method
reworked Singer's original numbers and concluded that only
in the case of maternal mortality was the correlation significant (Stern, 1981).

Morris & Titmuss followed Singer's technique of differencing mortality in an attempt to control for other socio-economic variables besides unemployment (Morris et al., 1944). They collected data from 83 County Boroughs in England and Wales on rheumatic heart disease mortality and unemployment rates. In addition they attempted to estimate lags between changes in unemployment and changes in mortality.

The first to attempt cross national comparisons was Fraser although his work is not widely known (Fraser, 1973). He studied data from 25 countries for the years 1955, 1960 and 1965.

The work of Brenner (1973, 1979), on the other hand has been widely discussed. He developed a model with aggregate National data over perhaps 50 years including data on mortality and a number of indices representing economic activity. These studies have consistently shown both in the United States and Europe that depression in a national economy is associated with changes in mortality after a variable lag period 1 to 2 years. Some doubt has been raised concerning the specificity of Brenner's equations and whether the omission of certain varibles might have produced bias (Gravelle et al., 1981).

Cross sectional studies or time series studies have basic methodolical weakness because of the difficulties of control and choice of vaiables. They are more often seen as a starting point for an 'association' rather than the ultimate in epidemiological methodology.

If we are interested in causality these studies are not likely to be of great value.

There are many academic 'sidetracks' to discuss including the choice of analytical method for such multi-variable data. Current preference seems to be for log linear models, but it is important not to get lost in the secundary issues.

Whatever the arguments about Brenner's methods he provides an intellectual framework for understanding the possible mechanism by which unemployment may affect health. This is seen in terms of both economic hardship and stress. This would account for increases in death rates amongst the unemployed themselves as well as the young and elderly who might be affected by the economic loss of the wage earner and the stress produced by the loss of employment. Brenner further hypothesises that the threat of loss is an important producer of stress amongst those still at work and may be important in increasing for example, ischaemic heart disease particularly amongst those in the 55-65 age group.

CONCLUSION

To return to the central question 'Does unemployment affect health'? An association between mortality and unemployment has been established using cross sectional and time series models and by longitudinal studies. The early attempts by Singer, and Morris and Titmuss used a method of 'differencing' in order to reduce the influence of other intercorrelated variables. Brenner's work and that of the OPCS have added further weight to the association. Nevertheless, none of the studies have established wheter unemployment is causally related to increasing mortality. On morbidity the closure studies and the National Training Survey have pointed to a number of increases in illness and sickness spells. There is some evidence that this has resulted in increased utilisation of health services (DHSS Cohort) and increases in medication (Nordhavn).

Despite a large number of different studies employing different strategies this brief review of the literature demonstrates the limit of our knowledge both as to the central question of causation and the possible size of the effect.

An examination of these studies and strategies indicates areas of research that might be useful in the future. Only limited gain can be expected from large scale cross sectional or time series analyses. The best opportunity comes from carefully designed cohort studies over many years. Major closures provide such opportunities but as yet they have been studied in only limited numbers and with an inadequate period of follow up. Longitudinal studies such as the National Training Survey provide a large sample of individuals that can be returned to periodically. It is important that significant health questions and objective measurements are incorporated into the study design of similar longitudinal surveys so that the question of whether and to what extent unemployment affects health can be answered.

REFERENCES

Bakke, E.W. 1933.
    The unemployed man. London.
Brenner, M.H. 1973.
    Foetal, infant and maternal mortality during periods of economic instability. International Journal of Health Services 3.2: 145.
Brenner, M.H. 1979.
    Mortality and the national economy. A review, and. the experience of England and Wales 1936-1976. Lancet 2: 568.
Cobb S., Kasl, S.V. 1977.
    The consequences of job loss. N.I.O.S.H. Research Report US. D. HEW. No. 77-224.
Cook, D.G., Cummins, R. O., Bartley, M.H. and Shaper, A.S. 1982.
    The health of unemployed middle aged men in Great Britain. Lancet 1: 1290.
Department of Employment and Productivity.
    Ryhope: A pit closure: a study in redeployment HMSO 1970.
Eisenberg, P., Lazarsfeld P.F. 1938.
    The psychological effect of unemployment. Psychological Bulletin 35: 358.
Fagin, L., Little, M. 1981.
    Unemployment and health in families. DHSS.
Fox, A.J., Goldblatt, P.O. 1981.
    Longitudinal Study. 1971-1975. O.P.C.S. H.M.S.O. Ls No. 1.
Fraser, R.D. 1973.
    An international study of general systems of financing health care. International Journal of Health Services 3.3.
Gravelle, H.S.E., Hutchinson, G.,Stern, J. 1981.
    Mortality and unemployment a cautionary note. C.L.E. Dis-

cussion Paper 95.
Hill, J. 1977.
    The social and psychological impact of unemployment: a pilot study. Tavistock Institute of Human Relations. 2T: 74.
Iverson, L., Klausen, H. 1981.
    Lukningen af Nordhavns-Vaerflet. Institut for Social Medicin. Kobenhavns Universitet. 13.
Jahoda, M., Rush, H. 1980.
    Work, employment and unemployment. Science Policy Research Unit, University of Sussex.12.
Lazarsfeld, P.F., Jahoda, M., Zeisl, H. 1933^2.
    Die Arbeitslosen von Marienthal. Psychologische Monographien. 5: 123. Hirzl, Leipzig.
Metcalf, D. Nickell, S.J. 1979.
    Notes on the incidence of unemployment and sickness spells of 3 months or more 1965-1975. C.L.E. Working Paper. 72.
Morris, J.N., Titmuss, R.M. 1944.
    Health and social change 1. The recent history of rheumatic heart disease. The Medical Officer. 2:69.
The Pilgrim Trust. 1938.
    Men without work. Macmillan. London.
Ramsden, S., Smee, C. September 1981.
    The health of unemployed men. Employment Gazette. London.
Singer, H. 1937.
    Unemployment and Health. Pigrim Trust Unemployment Enquiry Interim Report. 4.
Stern, J. 1981.
    Unemployment and its impact on morbidity and mortality. Centre for Labour Economics L.S.E. Discussion Paper. 93.
Warr, P.B. 1981.
    Some studies of psychological well being and unemployment. Social and Applied Psychology Unit, Memo. Sheffield. 43.
Westin, S., Norum, D. 1977.
    Nar sardinfabrikken Nedlegges. Institutt for Hygiene og Sozialmedisin, University of Bergen.
Ziegler, H. 1979.
    Santé et licenciement collectif: la santé des ouvries d' une gross enterprise de métallurgie avant, pendant et après un lincenciement collectif. Prévenir, 1: 1.

ONTOGENETIC DEVELOPMENT AND 'BREAKDOWN IN ADAPTATION'

A Review on Psychosocial Factors Contributing to the Development of Myocardial Infarction, and a Description of a Research Programme.

P. Falger, A. Appels, R. Lulofs *

State University of Limburg School of Medicine
Département of Medical Psychology

Maastricht/The Netherlands

SUMMARY: In this paper, we review the role of the Type A coronary-prone behavior pattern, stressful life changes, and manifestations of vital exhaustion and depression in the development of myocardial infarction from a dynamic ontogenetic psychological perspective. It is argued, that although some of these psychosocial factors have contributed to different extents to the general understanding of the pathogenesis of myocardial infarction, in future studies more attention should be paid to the specific historical and social contexts in which Type A behavior, stressful life changes, and vital exhaustion and depression are embedded, in order to gain more psychological insight into the long-term developmental structures that may lead to the onset of coronary heart disease and myocardial infarction.

A research programme is described, in which the complex interactions between these psychosocial factors are investigated in different phases of the life course of myocardial infarction patients and healthy subjects, by means of epidemiological studies, biographical investigations, and psychophysiological experiments. Also two instruments that were developed to this purpose, the Maastricht Questionnaire on vital exhaustion and depression, and an interview schedule assessing the exposure to and coping with life changes over the life course, are discussed, together with some results.

* This paper should be regarded the result of a collective effort. Although the first author is primarily responsible for writing this paper, all authors have contributed equally to the research presented here, and to the theoretical and methodological considerations that have made these studies possible.
Dr. Paul Falger is a gerontologist, concerned with developmental structures and psychosocial factors in the biographies of myocardial infarction patients. Prof. dr. Ad Appels is a personality psychologist, and the author of the Maastricht Questionnaire on vital exhaustion and depression prior to myocardial infarction, and of the Dutch version of the Jenkins Activity Survey. He serves as the coordinator of the research programme. Dr. Rutger Lulofs is a psychophysiologist involved in research on psychophysiological and biochemical characteristics of the Type A behavior pattern, and of vital exhaustion and depression.

Part of the research programme is supported by grants from the Dutch Heart Foundation.

1. INTRODUCTION

In this paper, we consider 'breakdown in human adaptation' as the outcome of protracted psychosocial and biological processes that may cover extensive phases of the human life-span. However much this may appear to be a truism, yet this statement poses serious questions with respect to the appropriate methodology to be applied, or about the feasibility of the theoretical models to be constructed, in order to be able to conduct psychosocial research from this vantage point. It is proposed here that a life-span developmental orientation, both in methodology and theory construction, may solve some of those problems. The remainder of this paper, then, will be devoted to some considerations with respect to our investigations on 'breakdown in adaptation' that are presently being conducted from a life-span developmental perspective. These studies investigate the multilevel interactions of some psychosocial, behavioral and psychophysiological parameters in the development of myocardial infarction (MI) in adulthood.

First, we will review some properties of our particular life-span developmental theoretical orientation, then we will consider established empirical evidence about the factors mentioned above that may fit into this psychological orientation. This paper will be concluded with a description of the instruments that are employed in our studies, and with some results.

2. ONTOGENETIC DEVELOPMENT AND 'BREAKDOWN IN ADAPTATION'

About a decade ago, life-span developmental psychology became established as a comprehensive theoretical and methodological orientation in studying human behavioral development, although the original notions, then rather isolated, date back to the 1930's (Bühler, 1968, originally published 1933). This orientation is concerned primarily with '..the description and explication of ontogenetic (age-related) behavioral change from birth to death' (Baltes & Goulet, 1970), and hence with examining '..intraindividual change in behavior across the life-span and with inter-individual differences (and similarities) in intraindividual change' (Baltes et al., 1977). The adjective 'life-span developmental' has become generally accepted as denoting the potential range of problems admissible for investigation. However, it would have been conceptually more accurate if emphasis would have been put instead on the manifold processes involved, in succintly speaking of 'ontogenetic' psychology (Schroots, 1982).

Therefore, both terms will be used interchangeably throughout this paper.

Conceiving of human development as a series of protracted processes that will evolve on distinct conceptual levels (biological, psychic, social) and that simultaneously feature synchronal and diachronal characteristics (Schroots, 1982), emerged only recently in the history of scientific thinking on development. As long as the Aristotelian notion of continuity as the fundamental principle guiding development dominated Western philosphical thought, little incentives existed that would challenge this epistemology (Mendelsohn, 1980). That is, in defining development as a fundamentally continuous, stable process that evolves according to mechanical and linear principles as Aristotle and the mainstream in philosophy up to the end of the nineteenth century did, development becomes, in principle, fully predictable. As a consequence, this mechanistic notion of (human) development is less amenable to speculative thought. '..Mechanistic explanations, in contrast to those (organismic ones) that rely on a coherence of many complex forms, turn the mind to search for a series of little steps. Discontinuities (in development) pose serious intellectual challenges to mechanistic explanations, because they do not yield easily to descriptions of a series of connected, small causes' (Kagan, 1980).

In opting for an organismic explanation of ontogenetic development, as we will do in this paper, we ascribe to an active organism model of man. '..A primary characteristic of this model is its representation of the organism as the source of acts, rather than as the collection of acts initiated by external force (as in the mechanistic model). Inquiry is directed toward the discovery of principles of organization (or, in our case, toward the forces that may threaten it), toward the explanation of the nature and relations of parts and wholes, structures and functions, rather than toward the derivation of these from elementary processes. In this model, change itself is accepted as given. This means that change is not itself explainable by efficient, material cause. This position (that is being increasingly underscored in current biocybernetic explanation, Bateson, 1972) results in a denial of the complete predictability of man's behavior. Furthermore, the nature of change is qualitative as well as quantitave. At each new level of organization, new system properties emerge which are irreducible to lower levels, and which therefore are qualitatively different from them' (Reese & Overton, 1970).

Thus, changes in psychosocial, behavioral, and psychophysiological parameters that may be indicative of 'breakdown in adaptation' are the basic referents of our life-span developmental interests here. '..These changes will reflect basic qualitative changes conceptualized as changes in levels of organization or stages. The basic outline of (our) analysis (then) is the description of structures at a given period, the ordering of such periods, the discovery of rules of transition from one period to another, and the determination of the experiential conditions that facilitate or inhibit structural change' (Reese & Overton, 1970).

We now turn to a concise discussion of the evidence on the role of psychosocial, behavioral, and psychophysiological factors in the development of MI in adulthood, and to the particular manner in which those factors may fit into our ontogenetic orientation. This evidence pertains to the Type A coronary-prone behavior pattern, to stressful life changes, and to manifestations of vital exhaustion and depression.

3. TYPE A CORONARY-PRONE BEHAVIOR IN THE DEVELOPMENT OF MI

At the end of the 1950's, the American cardiologists Friedman and Rosenman described a specific, overt behavioral disposition (Type A) that was associated with the prevalence of coronary heart disease (CHD) and MI in middle-aged men and women (Friedman & Rosenman, 1959; Rosenman & Friedman, 1961). The Type A coronary-prone behavior pattern has been defined as: '..An action-emotion complex that can be observed in any person who is aggressively involved in a chronic, incessant struggle to achieve more and more in less and less time, and if required to do so, against the opposing efforts of other things or other persons, (..) stemming from a fundamental and irretrievable sense of insecurity about the intrinsic value of the personality involved' (Friedman & Rosenman, 1974).

In two longitudinal investigations in large American populations, the Western Collaborative Group Study (WCGS), and the Framingham Heart Study (FHS), it was demonstrated that initially healthy, middle-aged subjects exhibiting the Type A behavior pattern suffered MI (at least) twice as often as the subjects who were not characterized by this peculiar behavior disposition (Type B), over follow-up periods of eight years (Rosenman et al., 1975; Brand et al., 1976; Haynes et al., 1978a). The Type A coronary-prone behavior pattern thus proved to constitute an independent psychological risk factor in the development of CHD and MI (Brand, 1978).

In the wake of the WCGS, and later of the FHS, a multitude of epidemiological studies was conducted. As a rule, these have corroborated the initial, prospectively established positive association between Type A behavior and CHD, and particularly so MI, also in socioeconomic contexts that were quite different from the original Southern-Californian middle class cultural setting of the early 1960's (Rosenman & Chesney, 1980). The components of Type A behavior that appear to be most consistently related to the later onset of CHD are competitive drive, and impatience (Rosenman & Chesney, 1980), and, probably most important, hostility and anger (Williams et al., 1980).

At the same time, psychologists with various backgrounds began to explore different psychological dimensions that may underly Type A behavioral characteristics (Matthews, 1982). The latter studies, for the most part small-scale laboratory experiments, were conducted with (much) younger and considerably smaller populations than the original middle-aged subjects. Also, Type A behavior assessment procedures were rather different.

In the WCGS, the Structured Interview (SI) method contains some 25 questions in which subjects are asked about their characteristic manners of responding to every-day situations that should elicit impatience, hostility and competitiveness from Type A individuals (Rosenman, 1978). During the SI, speech and psychomotor characteristics are also assessed. All aspects contribute to the final classification into Type A or Type B. The other instruments used for Type A assessment are both self-report measures. The Jenkins Activity Survey (JAS), also developed and employed in the WCGS, consists of some 50 items similar to those represented in the SI (Jenkins, 1978). The Framingham Type A scale (FTA) contains 10 items, assessing individual competitive drive, sense of time urgency, and perception of job pressure (Haynes et al., 1978b). All three Type A measures, however disparate these may seem, appear to be reliable. However, they do not assess the same aspects of the broad construct of Pattern A behavior (Matthews, 1982).

The explorations on the psychological characteristics of the Type A coronary-prone behavior pattern have focused to a large extent on the dimension of controllability (Glass, 1977). A series of experimental studies (with college-aged subjects) on various aspects of controllability suggests that Type A behavior may reflect a specific inclination to cope with stressful characteristics of the environment. This susceptible psychological disposition appears to be appreciably correlated with physiological mechanisms that facilitate sympathetic nervous system activity, and with specific neurohormonal processes, notably the discharge of catecholamines (in particular noradrenaline), that are known to contribute to the development of CHD. In this manner, the dimension of controllability may be central to both the psychological and the pathophysiological components of Type A behavior (For a more extensive discussion cfr. Matthews, 1982). Another possibly important psychological dimension of Pattern A behavior, that is based on the speech characteristics aspect of the SI, includes degree of self-involvement (Matthews, 1982). Finally, it should be noted that the scarce but intriguing experiments on the development of Type A behavior in childhood indicate that the Type A behavior pattern is probably induced through specific mother-child interactions (Matthews, 1978; Matthews & Angulo, 1980). Ever escalating parental standards of performance, in particular with respect to achievement striving when mediated through high parental expectations and aspirations, frequent approval and disapproval, a competitive and involved attitude, and authoritative discipline techniques probably play a decisive role in the development of Type A behavior in childhood. Some evidence suggests that a slight genetic component may also be involved in this development (Bortner et al., 1970; Matthews & Krantz, 1976).

Comment: When we now consider the above evidence from our life-span developmental perspective, in order to construe a coherent picture with respect to the factors impinging on the developmental course of the Type A behavior pattern, several questions remain to be answered. The first major issue deals with the specific historical context that might have been contingent to the emergence of this behavioral disposition, the second one (as part of the former) with the environmental characteristics that are hypothesized to elicit this behavior pattern.

It could be argued that the emergence of Type A behavior in men is primarily the result of the specific psychological experiences of particular cohorts in the general population (Baltes et al., 1978). That is, when the WCGS-sample was studied at the end of the 1950's, its male participants were, at inception, between 39 and 59 years of age. This means that a major part of this population was born and raised in the 1920's and 1930's, before and during the Great Depression, followed by the possibly equally traumatic experience of WW II, and subsequent adjustment to the most massive long-cycle economic upswing in the history of the U.S.economy (Eyer, 1980). If the above notion of controllability as a core dimension of Type A behavior holds true, it could be conjectured that the unprecedented exposure to prolonged environmental stressors like the socially disruptive economical conditions of the 1930's might have facilitated the emergence of Type A-like behavior as a potentially adequate response pattern with respect to these hardships. Also, the forced readjustment in both men and women to the exigencies of a national economy gauged for warfare (the antipode of the stalled economic activities during the Great Depression) at the beginning of the 1940's, that was prolongued in the long-lasting economic growth period in the two decades after WW II, the latter offering ample opportunities for competitiveness, might subsequently have reinforced, and further developed, the Type A behavior pattern. The results from a longitudinal study on the relationships between social history and individual life history in personality development could be said to corroborate the first part of our reasoning (Elder, 1979). This study explored the effects of Depression hardships upon family patterns and adolescent personality in two Southern-Californian cohorts born at the beginning and the end of the 1920's. It was found that the boys from the latter cohort, who were living in a home environment characterized by progressing economic deprivation and increasing parental discord, experienced the ambiguity of their daily environment, caused by a loss in social roles of the father, the strongest. This process resulted in feelings of prolonged (social) insecurity, of hostility, and in turning away from emotional bonds. (These findings thus bear a striking resemblance to the major dimensions of the definition of Type A behavior, as cited at the beginning of this section). In girls, on the other hand, these disruptive conditions enhanced the emotional, supportive ties between mother and daughter.

The second issue is concerned with contemporary environmental characteristics that may bring about Type A behavior. It could be argued that modern, large industrial organizations are optimally suited to elicit and further perpetuate Pattern A, because of the bureaucratic complexity and ambiguity of the organizational structures involved, and of the rather abstract nature of the actual work to be performed (in managerial positions). These organizational aspects result in a high degree of interdependence among employees. Indeed, Type A managers tend to describe their jobs as having more responsibility, longer working hours, and heavier work loads than Type B subjects would (Chesney & Rosenman, 1980). Yet, despite of these perceived job pressures, Type A subjects do not seem in general to report more job dissatisfaction or manifestations of distress. This observation most probably holds true for Type A behavior in a relatively stable, predictable corporational environment. However, it could be argued that in organizations undergoing thorough change, be it either because of expansion or due to current dwindling, severe job stress might ensue, because a Type A individual could perceive such potentially destabilizing changes as a threat to his sense of control over the work environment. Among the circumstances that might trigger such occupational distress in susceptible Type A persons could be repeated structural reorganizations that bear upon the content of the work to be performed, or blocked opportunities for further promotions, that could induce status inconsistency. Adequate research in the vein of contemporary orientations in organizational psychology, like socio-technical systems analysis, could perhaps provide us with more insights into the potentially pathogenic interactions between work environments and coronary-prone behavior. (It might be noted in passing, that the current notion about 'workaholics', who could be said to carry the job-involvement, competitive drive, and time urgency aspects of Type A behavior to their extremes, could appear as adequate and rational to its proponents in science, business and politics, regarding the present, peculiar environmental demands, as when Type A behavior emerged some decades ago. Despite of the seemingly odd behavioral consequences of working days of 16 hours or more that are reported, in particular sleeping for but very few hours and curtailed eating, these persons were apparently in optimal physical condition (Machlowitz, 1981)).

This discussion leads us now to a concise review of some literature that is concerned, in part, with a different operationalization of 'work stress', that is the role of stressful life changes as factors that may contribute to 'breakdown in adaptation'.

4. STRESSFUL LIFE CHANGES IN THE DEVELOPMENT OF MI

Most of the studies relating the occurence of stressful life changes (SLC) to MI have been performed in Sweden and Finland, by Theorell et al. and by Rahe et al., respectively. Since we recently have published a review on the main Theorell studies (Falger, 1983a), we will summarize those findings here, supplemented when necessary by the Finnish investigations.

Except for two longitudinal surveys, all studies employed a retrospective case-control design. Subjects were middle-aged men from Stockholm, mostly from upper occupational strata, who survived MI, and healthy controls, usually matched for age. The number of participants in both groups varied from 27 to 104. The instruments employed invariably consisted of a modified version of the Schedule of Recent Experiences from the Social Readjustment Rating Scale, a self-report measure by Holmes & Rahe (1967). Alternatively, a CHD-behavior questionnaire, a sociological history survey, and several measurements of work load, all self-constructed by the investigators, were included. The time intervals that were studied as a rule extended over a period of two years before MI. In most instances, a significant build-up in the amount of SLC prior to MI could be demonstrated, whereas controls did not report such alterations. Most SLC that were encountered occured in association with conflicts at work, or with respect to coping with increased responsibilities. Job satisfaction in MI-patients appeared to be significantly less than in controls. Also, MI-cases evaluated these SLC as psychologically more upsetting, or as requiring more efforts at readjusting, than controls would.

The results from the somewhat larger retrospective studies in Finland, that also employed a modified Schedule of Recent Experiences, in general corroborated those findings (Rahe et al., 1973; Rahe & Romo, 1974; Rahe et al., 1974). However, it should be noted here that MI-survivors (male as well as female) were compared with subjects of both genders who died immediately or shortly after MI, instead of with healthy controls. Also, the apparent tendency of a considerable build-up in SLC in the half year before MI, that was demonstrated in the lives of male patients, with a somewhat steeper gradient in non-survivors, appeared to be absent in the biographies of women.

In the first prospective study over a four months interval, exposure to SLC in 21 MI-patients who had resumed working after MI was compared with changes in catecholamines output, and in other stress-related biochemical parameters (Theorell et al., 1972). Measurements about those parameters and on exposure to SLC were repeated about every two weeks. The SLC that were reported most often concerned, once again, conflict at work, or changes in circumstances work was to be performed. These SLC were significantly correlated with elevated catecholamines- and uric acid outputs. (In two later studies, continuous psychophysiological measurements were applied while MI-patients were being interviewed about emotional characteristics of formerly experienced SLC (Theorell et al., 1977; Theorell, 1980). It could be demonstrated that blood pressure and heart rate were heightened, compared with previously established base-line levels when peculiar emotional aspects were touched upon. In some instances, even ventricular arrhythmias could be provoked, that were reproduced accordingly when later on in the course of interviewing the eliciting SLC were discussed again).

The last prospective investigation on a sample of some 6,500 middle-aged building construction workers followed-up for more than one and a half years, measured non-fatal MI and cardiac death as endpoints of prior exposure to SLC, and of other manifestations of psychological distress during that interval. Also, the incidence rates with respect to some major physical complaints unrelated to cardiovascular functioning, and blood pressure ratings (recorded at the end of the follow-up period) were included. It was found that the 51 cases with first MI that could be identified, reported a significantly higher mean score on the work load index, consisting of the combined SLC and distress indices, than controls did at the beginning of the survey. The single items that discriminated best between cases and controls were increased, or too much responsibilities at work, extensive periods of overwork, and prior episodes of unemployment, the latter items thus suggesting a rather irregular course of work. Another important finding in this respect was related to the occupational level of the majority of MI-cases. In the subgroup of concrete workers, the building-construction workers laboring hardest under often harsh working conditions, the incidence of MI was about twice as high as in the other occupations combined. In addition, it was demonstrated that surviving MI-patients had reported neurotic complaints more often than controls. Also, in that part of the control group in which subjects scored highest on prior exposure to SLC, and to psychological distress, significantly elevated mean blood pressure readings were obtained, when compared with the other controls.

Since these results underscore the importance of work-related stressors in the development of MI rather unequivocally, it is unfortunate that none of the established Type A assessment procedures was included in any of these studies. This could possibly have clarified further the complex interrelationships between Type A behavior, specific job demands, and ensuing (dis)satisfactions, as referred to in the previous section, both in higher occupational strata, as in the retrospective studies, and in blue collar workers.

Comment: When we return to our ontogenetic orientation in order to elaborate further our ideas on protracted psychosocial factors contributing to 'breakdown in adaptation', again several questions remain to be answered. One kind is concerned with the possible build-up of stressful episodes over the entire human life course, another with the seemingly different psychological appraisals regarding SLC in MI-patients and controls.

All studies cited so far have investigated the prevalence of SLC in the last two years, or in even shorter periods of time, before MI. However, it would be difficult to conjecture any valid reason why earlier segments of an individual's life course would not contain comparable episodes of SLC clusterings, that could contribute to an accumulation of SLC with relevant bearings upon the later development of MI (Hultsch & Plemons, 1979; Brim & Ryff, 1980). Of course, the crucial question whether people might be able to remember distant events in their lives, and the psychological and social impact thereof in any reliable manner, at first sight could pose serious practical restrictions (Jenkins et al., 1979). However, it has been demonstrated that any individual will subject the cognitive representation of his personal life history to continuing interpretations as a function of the variously evolving, socially determined contexts that he is living in, and that he will grant particular emphasis to the transformations determined both by expected and eruptive SLC (Cohler, 1982). This argument is corroborated by various psychiatric studies on personality development and mental health in middle-adulthood (Levinson, 1977; Vaillant, 1977; Gould, 1978), although these investigations may lack the methodological rigour of well-performed epidemiological studies in other domains. Therefore, a more complete appraisal of stressful episodes in the life course of MI-patients, and of the particular manners in which they may have been coping with those experiences, might contribute to our understanding of the complexity of the psychosocial and cognitive processes involved. This could lead us to a methodology that is optimally suited to represent crucial objective and subjective characteristics of biographical structures. Some aspects of psychological and social coping with SLC in MI-patients that might contribute an important part to such methodology have recently been elaborated.

As we already noted above, the MI-patients in the Theorell studies evaluated the single SLC that they had encountered as either psychologically more upsetting, or as requiring more efforts at readjustment (Lundberg et al., 1975; Lundberg & Theorell, 1976). This also held true for non-experienced SLC that MI-patients had to rate accordingly, as compared with healthy controls. In other studies in which cognitive appraisals of SLC in MI-patients were included, similar results have been found. Thus, rather chronic psychological problems in adjusting to life stress were reported as the major characteristic of the last two years prior to MI in an American sample of middle-aged heart patients (Thiel et al., 1973). In an Australian study, equally retrospective like the former, different scales representing the ideosyncratic interpretation of the emotional impact of SLC discriminated between patients with first MI, and subjects with severe angina pectoris, who were matched for age (Byrne, 1980; Byrne & Whyte, 1980). The MI-patients scored significantly higher on dimensions of upsettingness, depressivity, and helplessness than the other patient group did. Once again, it is rather unfortunate that no measurements of Type A behavior were included in any of these studies, since the latter of the reported emotional characteristics in particular could be conceived to be closely linked to the controllability dimension of Pattern A (Glass & Carver, 1980). On the other hand, one should be cautious with regard to over-generalizing such findings, since the ratings referred to above as a rule were calculated with respect to single SLC, and subsequently were added. Also, different subgroups of the general population experience different frequencies of total SLC over given intervals, depending on demographic composition, socioeconomic status, and the appropriateness of the SLC inventory involved. Such relationships, if not adjusted for, could lead to coincidental associations between SLC and health-related outcomes (Goldberg & Comstock, 1980). Beyond these thoughtful observations, the scarce literature on the possible relationships between experienced SLC and developmental stages suggests that in late adulthood, and in beginning old age as well, particular loss-related SLC associated with role changes are perceived as less stressful than in younger age groups (Muhlenkamp et al., 1975; Sands & Parker, 1979; Brim & Ryff, 1980 ; Silver & Wortman,

1980). These results point to the intriguing sociological notion of the human life course as a succession of social role transitions, that are superimposed upon the biological timetable (Neugarten & Datan, 1973; Elder, 1975; Neugarten & Hagestad, 1976). Those transitions and the SLC that are in general associated with these could either occur 'on time', that is in accordance with the social timetable contingent in a specific cohort from the general population, or 'off time'. The effects of the latter SLC could constitute major stressors if these will disrupt the expected sequences and rhythms of life. It could be conjectured, now, that the particular psychological adjustment problems in MI-patients, that we have referred to above, may stem from the circumstance that the SLC they had to cope with, were occurring unanticipated or 'off time', as compared with controls from the same cohorts.

Besides those remarks it should be noted that the social networks in which individuals are participating, and the social support these could engender, might moderate the ideosyncratic perceptions of and coping with SLC considerably (Cobb, 1976; Dean & Lin, 1977). In an analysis of the relationship of specific social network characteristics to CHD, Berkman (1982) concludes that network configurations in the home and work environments are not only likely to influence the incidence of major SLC, but also that many events may seriously disrupt or otherwise change the supportive characteristics of existing network relationships. It could be conceived, now, that a build-up in SLC over the last years, or months, before MI might put such considerable strain upon the 'buffering' capabilities of the networks in which the prospective MI-patient is partaking, that the perception of actual support will diminuish, or even diasappear. If the majority of SLC would be located in the work environment while the supportive quality of the corresponding network would be questionable, potential 'buffering' capabilities would be restricted to the home environment. If this particular support system, that as such is capable of engendering serious psychosomatic strains (Grolnick, 1972), would be characterized by chronic relational frictions that have gradually developed during marriage, then the individual concerned would have but few instances left to turn to for social support. Such initial faltering in social network structures may thus anticipate the subsequent individual 'breakdown in adaptation'. Such 'breakdown' may be foreshadowd by feelings of exhaustion and depression, as could be concluded from some of the studies cited above. It is to these manifestations that we now turn our attention.

5. VITAL EXHAUSTION AND DEPRESSION IN THE DEVELOPMENT OF MI

After considering the influence that Type A behavior may exert over several decades of the life-span of MI-patients, followed by scrutinizing the triggering effects of SLC in the last period before impending 'breakdown', we will now examine some psychological phenomena that may be marking final 'breakdown in adaptation'.

It appears, then, that some 65% of the patients who suffer MI or sudden cardiac death have visited a physician in the months prior to their infarction or untimely death. However, in general the latter was not able to recognize unequivocal threatening clinical symptoms. What peculiar psychosomatic symptoms did those patients experience that urged them to consult a physician ? Several investigators have interviewed patients, or their relatives, asking them what phenomena had been experienced in the last months before MI or sudden death. In all instances it was observed that not only some probably cardiac symptoms but also feeling like tiredness, and general malaise, were reported by substantial numbers of the investigated middle-aged patients of both genders.

An investigation by Kuller et al. (1972) demonstrated that more than half of 208 sudden death cases had suffered from fatigue/weakness, 28% had reported sleep disturbances/difficulties with sleeping in the last weeks prior to death, whereas breathlessness had been experienced by 42%, and chest pain by about one third of the cases. In a retrospective study by Alonzo et al. (1975) it was found that in 112 hospitalized patients surviving MI, 38% reported feelings of fatigue/weakness, 16% experiences of general malaise, and 14% of emotional changes, whereas chest pain was reported by two thirds, and breathlessness by one third of the sample. For 88 subjects who suffered sudden death before hospitalization, feelings of fatigue/weakness were reported with respect to about half of those cases, emotional changes in 20%, and general malaise in 17%. In an investigation by Rissanen et al. (1978) similar experiences of fatigue/weakness were ascribed to 32% of 118 subjects suffering untimely cardiac death, while cardiac symptoms, including angina pectoris and breathlessness, were ascribed to about one third of the sample. In a retrospective study that employed the Depression Scale of the Minnesota Multiphasic Personality Inventory in relation to clinical parameters, it was established that ventricular arrhythmias in 50 middle-aged MI-patients, as well as in 50 healthy 'high-risk' controls, were positively associated with enhanced MMPI Depression scores (Orth-Gomer et al., 1980).

Yet, in spite of this and some other empirical evidence in support of specific psychological precursors of MI and sudden cardiac death, valid psychometric instruments that will reliably measure those prodromal phenomena of exhaustion and depression have only rarely been constructed. One such survey was developed for this purpose by Appels and has already been used in a series of studies. After reviewing those studies, we will finally discuss some current investigations in which the previous life history of subjects who have suffered MI may clarify the biographical structures that may lead to 'breakdown in adaptation', and studies in which the underlying psychophysiological mechanisms leading to 'breakdown in adaptation' are explored.

A. *Development of the MAASTRICHT QUESTIONNAIRE (MQ)*

On the basis of the results of the latter type of studies, a self-administered psychological survey was developed to enable the psychometric assessment of those prodromal feelings (Appels et al., 1979). This questionnaire was named - quite neutrally - the MAASTRICHT QUESTIONNAIRE (MQ). The MQ consists of 58 questions that were formulated on account of characteristic statements by MI-patients. They are asked either in retrospective or prospective form, and should be answered: Yes - ? - No. Some examples of questions are: Do you often feel tired lately ?; Do you sometimes feel like a battery going dead ?; Does the feeling that you might be ill contain your thoughts ?; Are you getting quickly irritated over minor hassles ?; Do you feel like you are not able to do something useful anymore ? (Appels, 1982). Recently, the MQ has been translated into English, German, Swedish, Italian, and Eastern-European languages, while other translations are in preparation.

1. The first version of the MQ was included in the Imminent Myocardial Infarction Rotterdam study, conducted some years ago. The main clinical purpose of this study was to develop some reliable means to identify - in general practice - middle-aged patients running a high risk of developing an acute coronary event in the near future. Patients who consulted their general practitioner about complaints of possibly cardiac origin were referred to this study for closer physical scrutiny, and were subsequently followe-up for a period of ten months. At the beginning of this interval they filled out the MQ. It was found that the 37 men (mean age 50 yrs.) who suffered a new coronary event (including MI and sudden cardiac death) within that period, had a significantly higher mean MQ score than the

remaining patient group of 405 men (mean age 49 yrs.) that did not develop new CHD during the follow-up interval (Appels, 1980). In addition, the mean MQ score from the 37 patients suffering MI or sudden death were also compared with the MQ scores from a healthy control group of 317 men (mean age 46 yrs.) participating in another screening study. In this instance, the difference in mean MQ scores was about two standard deviations. The 40 MQ items that discriminated between the first patient group and both comparison groups were subsequently analyzed twice by a hierarchical cluster analysis. The similar patterns that were identified by this procedure demonstrated that the MQ consists of item groupings denoting feelings of tiredness, helplessness and hopelessness, loss of vitality, depression and hypochondriasis, sleep disturbances, and exhaustion, or in short, the prodrome of vital exhaustion and depression.

2. A retrospective case-control study did confirm the validity of the MQ (Verhagen et al., 1980). In this investigation, the MQ scores of 58 middle-aged men with first MI were compared with those of 58 healthy controls, matched for age and socioeconomic status. Also, the SI assessment for identifying Type A behavior was included in this study, in addition to some other psychological questionnaires. The MQ followed by the SI proved to be the best discriminating instruments between cases and controls, both in univariate ANOVA, and in multivariate discriminant function analysis.

3. The definitive item selection for inclusion in the MQ will have to await the results from an ongoing prospective investigation. This study has begun recently among some 4,000 civil servants over 40 yrs. of age. This group is participating in a voluntary, periodical health screening with particular emphasis on the cardiovascular system. Data on the association between vital exhaustion and depression and MI occurring within three months after completion of the MQ are presented in *Table 1*.

	Exhausted/Depressed[1]	Vital	Total
MI	5	1	6
Well	557	3008	3565
Total	562	3009	3571

Fisher Exact Probability: .0005
Left Sided 95% Confidence Interval: 4.1

[1] MQ Score ≥ 85th Percentile Point.

These results hence demonstrate that those subjects who are identified as vitally exhausted and depressed, run a relative risk of at least 4.1 as compared with subjects scoring lower on the MQ to develop MI within three months (at a 95% confidence level). In an earlier analysis on some 2,000 subjects from this sample it was found that the association between the MQ and imminent MI is unconfounded by somatic risk factors, that is systolic/diastolic blood pressure, serum cholesterol, and smoking (Falger & Appels, 1982).

The findings from the studies in this section thus indicate that the MQ diferentiates psychometrically reliably and consistently between middle-aged men with various clinically assessed criteria for MI and healthy subjects from similar age brackets. Also, an enhanced MQ score appears to reflect an individual's feelings of vital exhaustion and depression as a psychologic precursor of MI or sudden cardiac death. However, these promising results as such do not explain yet sufficiently along what psychosocial and physiological pathways the prodrome of vital exhaustion and depression might be developed. Therefore, we have conducted some studies that may clarify the role of the Type A coronary-prone behavior pattern, and SLC, in the development of vital exhaustion and depression over the life course.

B. *Studies on the Development of Vital Exhaustion and Depression*

(This section is based on parts of an extensive article on factors that may influence the development of vital exhaustion and depression, to be published in the course of this year (Falger, 1983 b)).

1. First, a pilot study was performed as a preliminary step in clarifying interindividual differences in psychosocial factors that could bear upon the ontogenetic development of MI-patients and healthy subjects. In this investigation, some interactions were explored between reported SLC, the Type A behavior pattern, and manifestations of vital exhaustion and depression. To this purpose, we used a convenience sample of 136 healthy men, aged 39-41, that was randomly drawn from a larger sample of some 800 subjects. This particular age group was selected because the biographical investigations on personality development and mental health that were mentioned in section 4 strongly suggest that the transitional phase of mid-life (around age 40) may well be the most critical one with respect to the intensity of exposure to psychosocial stressors and with regard to subsequent psychobiological effects. All 136 participants completed three

questionnaires: The MQ, the JAS, and our self-constructed Middle Adulthood Life Changes Questionnaire (MALC).

With respect to the JAS, the authorized Dutch version was employed (Appels et al., 1979). This version has been cross-validated against the SI assessment (Appels et al., 1982). Seventy subjects scoring above the median JAS value for the population under study were classified as Type A, the remaining sixty-six as Type B. Those individuals who obtained a MQ score > one standard deviation above the mean for the population under study were characterized as vitally exhausted and depressed. Thirty-two subjects met this criterion; they were compared with an equal number of subjects scoring lowest on the MQ.

B.1. *The Middle Adulthood Life Changes Questionnaire* (MALC)

The MALC is a self-administered questionnaire that consists of twenty-two SLC items, divided into two domains. These are the Home & Social domain (12 items), and the Work & Career domain (10 items). With regard to each SLC item, the respondent should indicate: a) Whether or not this specific item has occurred (Yes - No), and if so: b) In what particular half year time interval over the last two years, and c) To what extent that item has been experienced as 'requiring adjustment', or as 'causing upsettingness', as indicated on a threepoints scale (Not at all - Average - Rather much). This latter procedure thus means that no attempt is made to assign differential weights to individual items, as in the Social Readjustment Rating Scale (Holmes & Rahe, 1967). Instead, the psychological impact of a SLC item that has been experienced over the last two years by a given subject is assessed through our procedure c).

An extensive description of the results of this pilot is reported elsewhere (Falger & Appels, 1982; Falger, 1983 b). In summarizing these results, it was concluded that in univariate ANOVA a positive association was found between Type A behavior and the amount of reported SLC in both domains. The prodrome of vital exhaustion and depression was also found to be positively associated in univariate ANOVA with the amount of reported SLC in both domains. These findings hence demonstrate that the Type A coronary-prone behavior pattern, as well as the prodrome of vital exhaustion and depression are associated with the occurrence of SLC in a particular phase in ontogenetic development, albeit with rather different emphases. Type A behavior is predominantly related to events in the Work & Career domain, as demonstrated by an analysis of single SLC that discriminate between Type A and Type B subjects (Falger, 1983 b). These items were: Problems with subordinates about their work; problems with supervisor about own work; working overtime; changes in social interactions at work; considerable increase in job responsibility; promotion to higher hierarchical level; and considered voluntary changing to other work. The onset of vital exhaustion and depression, on the other hand, seems to be much more related to SLC in the Home & Social domain (Falger, 1983 b). The items that discriminated between subjects that were characterized as vitally exhausted and depressed, and subjects who were not, are: Continuous educational problems with children; serious marital conflicts; effected considerable loan; financial troubles; lost close friends; and changes in personal life styles.

Our results further indicate a positive association between Type A behavior and the prodrome of vital exhaustion and depression. In addition, it should be noted that the thirty-two subjects characterized as vitally exhausted and depressed, consisted of twenty Type A subjects (63%), and twelve Type B subjects (37%).

This particular constellation of results strongly suggests that the Type A subjects not only demonstrate a larger degree of work involvement and conflict proneness. They also tend to exhaust themselves psychically to a considerable extent in the course of that (presumably chronic) struggle over their life course. Finally, it is worth noting that an analysis of the attributed psychological impact of the SLC reported by the subjects characterized as vitally exhausted and depressed, demonstrates that those SLC were rated as significantly 'more upsetting' or as 'requiring more adjustment' than in the comparison group (Falger et al., 1980).

2. After this pilot study on the interactions between Type A behavior, SLC and subsequent psychological symptoms of vital exhaustion and depression the MALC was used as an interview scheme with some 35 MI-patients with first MI. In these extensive interviews, also other stressful phases of the life course were explored, resulting in designing our Life History and Coping Strategies Interview.

B.2. *The Life History and Coping Strategies Interview*

This interview scheme currently contains 70 items: 15 items on childhood, adolescence and youth (that is, 3 items on place of birth, and familial composition, and 12 SLC), 8 items about formal education of subject and relatives, 20 items on work and career development (that is, 4 items on factual career development, present work status, and changes in personal life styles due to work and career, and 16 SLC), and 27 items about family and social history (that is, 3 items on single/marital status, number of children, and changes in personal life styles due to family and social life, and 24 SLC). With respect to each SLC reported by the interviewee, the following characteristics are being assessed, when appropriate: a) Respective frequency in different developmental phases, b) Conscious anticipation of, c) Psychological coping with, and d) Evaluation of each particular item (that is, whether in the opinion of the interviewee that item has contributed in any positive sense to his psychological and social development, or on the contrary, has proven detrimental), and e) Social support.

Although being more time-consuming than the MALC (that is, an average of $2\frac{1}{2}$ hours of interviewing), our interview scheme has some advantages over data collecting through postal questionnaires, as in the pilot study. The most obvious one is that each SLC can be discussed extensively, when necessary, with the interviewee in order to assess as precisely as possible the crucial features of anticipating, coping, and evaluating with respect to that item (Hultsch & Plemons, 1979; Brim & Ryff, 1980). This is the more important since the differences in perception of the latter features, as compared with controls, may constitute a distinguishing property in ontogenetic development in MI-patients, as we have discussed in section 4.
A second advantage is that in our interview scheme the SI assessment of the Type A behavior pattern is incorporated. This allows us to take into account the discriminating overt behavioral characteristics that the JAS does not (Jenkins, 1978; Rosenman, 1978). A last advantage constitutes the fact that the psychological contents of the prodrome of vital exhaustion and depression can be explored in further detail, since the MQ ought to be completed prior to interviewing.

This interview scheme, that next to the MQ also includes three concise questionnaires on cardiac symptoms, on previous illness episodes, and on predominant life styles (that is, eating, drinking coffee and alcohol, smoking, and sleeping habits and disturbances), is currently being employed in a retrospective study on 130 male MI-patients hospitalized with documented first MI, and two control groups. The first control group consists of 130 healthy males, without documented MI, living in the immediate vicinity of a MI-patient from our study. The second control group

consists of 210 men, without documented MI, currently hospitalized with a
non-stress related ailment (e.g., haemorrhoids, prostate hypertrophy,
inguinal hernia, cataract, etc.). MI-patients and the subjects from both
control groups are assigned equally to seven cohorts of five years each,
pertaining to the age range of 35 - 69 yrs. of age (that is, at least 15
MI-patients, at least 15 neighbourhood controls, and 30 hospital controls
of 35 - 39 yrs. of age, 40 - 44 yrs. etc.). In this manner, particular
cohort effects in the ontogenetic patterns of cases and controls might be
identified (Baltes et al., 1978). Also, in employing the above design,
covering a much larger segment of the adult life-span than our study in
a convenience sample, the inherent restrictions of a cross-sectional
approach can be alleviated to some extent (Schaie, 1973).

Since this study is still in progress, full-scale data pertaining to
the issues under discussion in the former sections cannot be presented yet.
However, preliminary analyses on 103 MI-patients and 101 neighbourhood
controls demonstrate that the patients score significantly higher on the
MQ than the controls. This result further corroborates the previous
findings, that manifestations of vital exhaustion and depression may fore-
shadow imminent MI. Other findings include the significant difference in
the prevalence of the Type A coronary-prone behavior pattern in cases and
controls. In the former group, 70 patients were assessed as Type A, and
33 as Type B, a ratio of 2:1. In the group of neighbourhood controls, 49
subjects were classified as Type A, and 52 as Type B, a ratio of 1:1. In
the younger age groups (35 - 49 yrs. of age), however, these ratios were
4:1 in MI-cases (26 vs. 6), and 1:1 in controls (27 vs. 26). Most important
with respect to our interest in ontogenetic developmental differences are
the obvious discrepancies between MI-patients and controls in exposure to
SLC in their respective life-span developmental patterns. In the combined
cohorts, aged 50 - 59, the 38 patients, as compared with 29 neighbourhood
controls, indicate significantly more often: Prolonged period(s) of
unemployment (odds ratio: 10.36); work place(s) closed down (odds ratio:
6.30); prolonged absenteeism (odds ratio: 2.81); and prolonged/serious
conflicts with supervisor (odds ratio: 2.54) on the interview section about
work and career development. Prolonged/serious conflicts with child(ren)
(in law) living on their own (odds ratio: 9.00); prolonged/serious marital
conflicts (odds ratio: 7.88); prolonged/serious educational problems with
child(ren) (odds ratio: 6.35); prolonged/serious financial problems (odds

ratio: 5.63); being necessitated to moving in with other people while married (odds ratio: 3.49), and prolongued/serious illness of subject (odds ratio: 2.81) are reported significantly more often in the section on family and social history. In the combined younger cohorts, aged 35 - 49, these differences are less outspoken, but are concerned with different single SLC, that might indicate a certain significance of SLC that were encountered in the youth of the 32 MI-patients involved, as compared with the 53 controls. These youth related events were: Prolongued living at other than elderly home (odds ratio: 2.57), and prolongued financial problems in elderly family (odds ratio: 2.29). Although these findings as such do not yet clarify the issues of peculiar timing and clustering of SLC in the life courses of MI-patients, that we have referred to in section 4, they do point to the importance of analyzing on cohort levels, necessary to capture the particular historical and environmental events that could have influenced the course of lives in a rather unpredictable manner.

B.3. Psychophysiological Profiles in Vital Exhaustion and Depression

3. A third area of investigation in our studies on the development of vital exhaustion and depression is concerned with psychophysiological characteristics of subjects who could be identified as such.

The basic question that we ask ourselves here is: Could individuals who display one or more of the psychosocial risk factors that have been discussed above, be characterized according to specific physiological adaptation reactions, as compared with subjects without those factors ? This question leads us to three additional problems: a) Which of those factors, the Type A coronary-prone behavior pattern or the prodrome of vital exhaustion and depression, do we investigate ?; b) What particular situations do we select in which it might be expected that clearcut adaptational reactions can be observed; and c) Which peculiar parameters of the adaptation profile do we measure ? Regarding the possible solutions to these problems we have to bear in mind that for instance the decision to select Type A behavior for further scrutiny will influence both the set-up of the experimental situations, preferably leading to a frustrating and moderately competitive setting, and the selection of the parameters of the adaptation response that are most sensitive to that particular situation. In this instance, these parameters should reflect β-sympathetic drive (Matthews, 1982).

Currently, we are engaged in some experiments in which we investigate the influence of both Type A behavior and manifestations of vital exhaustion and depression in adaptation to different tasks. These experimental tasks are: A binary choice reaction task under varying degrees of time pressure, in combination with either positive or negative feedback on task performance; a general knowledge test submitted under either easy or difficult instructions; an impossible comparison task, again under conditions of positive or negative performance feedback; and finally a vigilance task in which event-rate and stimulus/non-stimulus ratios are varied. As a general base-line condition we used a rest situation in which subjects may watch some videotaped documentaries on nature.

The subjects who participate in these experimental studies are selected on the basis of their scores on the JAS and the MQ from the healthy convenience sample, aged 39 - 41, that was described earlier in this section with respect to our pilot study on factors in ontogenetic development. In addition to the JAS, all subjects are interviewed and classified according to the SI assessment.

The parameters that are thought to reflect the adaptation response that we are interested in, belong to three different categories. These are: 1) Behavioral performance measures such as individual reaction time, percentage of correct reactions, and number of correctly answered questions; 2) Self-reported mood states; to this purpose, visual analogue scales are employed to measure degrees of involvement, concentration, effort, etc.; 3) Biological parameters. This last group could be divided into: a) Hormonal stress indicators (catecholamines, cortisol, etc.); b) Parameters reflecting thrombotic activity (platelet number, platelet activity, etc.); c) Neurophysiological indicators of the state of the cardiovascular system (heart rate, blood pressure, changes in blood vessel volume, systolic time intervals: PEP, LVET, TT; respiration rate, and under certain conditions: EEG-activity). Further detailed description of the experimental set-ups, including the mechanical and electronic devices that are employed, is beyond the scope of this paper.

Some preliminary results on the influence of Type A behavior pattern on the psychophysiological responses to a binary choice reaction task indicate that Type A subjects were less able to compensate for changes in blood pressure by varying their heart rate, probably because the sensitivity of the baroreflex was deminuished. Further, Type A subjects were found to be more sensible to variations in time pressure, as compared with Type B subjects.

With respect to the psychophysiological profile underlying manifestations of vital exhaustion and depression, it was found that in exhausted subjects platelet activity varied considerably during different experimental situations, while this variation was less in vital control subjects. This effect with respect to number of platelets is presented in *Table 2*.

	Exhausted/Depressed	Vital
Stress Condition	92,343	96,859
Rest Condition	101,257	96,211

Condition x State Interaction: $F_{(1,19)} = 3.05$; $p = .09$.

In this table, the total number of platelets was counted in a sample from which the irreversible aggregates were eliminated. Using the method proposed by Wu & Hoak (1974), we eliminated the amount of reversible aggregates. Then, the same tendency could be observed. Thus, we found a shift from reversible to irreversible aggregates during stress conditions in subjects who are identified as vitally exhausted and depressed.

In this manner, through epidemiological studies, biographical investigations, and laboratory experiments, we are engaged in investigating, from different vantage points, a wide array of psychosocial and psychophysiological factors that could contribute to 'breakdown in human adaptation' prior to myocardial infarction in different phases in individual ontogenetic development.

REFERENCES

Alonzo, A., Simon, A. & Feinleib, M.: Prodromata of myocardial infarction and sudden death. Circulation 52: 1056-1062 (1975).

Appels, A.: Psychological prodromata of myocardial infarction and sudden death. Psychotherapy & Psychosomatics 34: 187-195 (1980).

Appels, A.: Das Jahr vor dem Herzinfarkt. In: Koehle: Zur Psychosomatik von Herz-Kreislauf-Erkrankungen: 1-14 (Springer Verlag, Berlin 1982).

Appels, A., De Haes, W. & Schuurman, J.: Een test ter meting van het 'coronary-prone behaviour pattern' Type A. Nederlands Tijdschrift Psychologie 34: 181-188 (1979).

Appels, A., Jenkins, D. & Rosenman, R.: Coronary-prone behavior in the Netherlands: A cross-cultural validation study. Journal Behavioral Medicine 5: 83-90 (1982).

Appels, A., Pool, J., Lubsen, K. et al.: Psychological prodromata of myocardial infarction. Journal Psychosomatic Research 23: 405-421 (1979).

Baltes, P. & Goulet, L.: Status and issues of a life-span developmental psychology. In: Goulet & Baltes: Life-span developmental psychology: Research and theory: 3-21 (Academic Press, New York 1970).

Baltes, P., Cornelius, S. & Nesselroade, J.: Cohort effects in behavioral development: Theoretical and methodological perspectives. In: Collins: Minnesota symposia on child psychology 11: 1-63 (Erlbaum, Hillsdale 1978).

Baltes, P., Reese, H. & Nesselroade, J.: Life-span developmental psychology: Introduction to research methods: 1-27 (Brooks/Cole, Monterey, 1977).

Bateson, G.: Steps to an ecology of mind: 399-410 (Ballantine Books, New York, 1972).

Berkman, L.: Social network analysis and coronary heart disease. Advances in Cardiology 29: 37-49 (1982).

Bortner, R., Rosenman, R. & Friedman, M.: Familial similarity in Pattern A behavior: Fathers and sons. Journal Chronic Diseases 23: 39-43 (1970).

Brand, R.: Coronary-prone behavior as an independent risk factor for coronary heart disease. In: Dembroski, Weiss, Shields et al.: Coronary-prone behavior: 11-24 (Springer International, New York 1978).

Brand, R., Rosenman, R., Sholtz, R. et al.: Multivariate prediction of coronary heart disease in the Western Collaborative Group Study compared to the findings of the Framingham Study. Circulation 53: 348-355 (1976).

Brim, O. & Ryff, C.: On the properties of life events. In: Baltes & Brim: Life-span development and behavior 3: 367-388 (Academic Press, New York 1982).

Bühler, C.: The course of human life as a psychological problem. Human Development 11: 184-200 (1968).

Byrne, D.: Attributed responsibility for life events in survivors of myocardial infarction. Psychotherapy & Psychosomatics 33: 7-13 (1980).

Byrne, D. & Whyte, H.: Life events and myocardial infarction revisited: The role of measure of individual impact. Psychosomatic Medicine 42: 1-10 (1980).

Chesney, M. & Rosenman, R.: Type A behaviour in the work setting. In: Cooper & Paine: Current concerns in occupational stress: 187-212 (Wiley, New York, 1980).

Cobb, S.: Social support as a moderator of life stress. Psychosomatic Medicine 38: 300-314 (1976).

Cohler, B.: Personal narrative and life course. In: Baltes & Brim: Life-span development and behavior 4: 205-241 (1982).

Dean, A. & Lin, N.: The stress buffering role of social support. Journal Nervous and Mental Disease 165: 403-417 (1977).

Elder, G.: Age differentiation and the life course. Annual Review Sociology 1: 165-190 (1975).

Elder, G.: Historical change in life patterns and personality. In: Baltes & Brim: Life-span development and behavior 2: 117-159 (Academic Press, New York 1979).

Eyer, J.: Social causes of coronary heart disease. Psychotherapy & Psychosomatics 34: 75-87 (1980).

Falger, P.: Pathogenic life changes in middle adulthood and coronary heart disease: A life-span developmental perspective. International Journal Aging and Human Development 16: 7-27 (1983 a).

Falger, P.: Behavioral factors, life changes, and the development of vital exhaustion and depression in myocardial infarction patients. International Journal Behavioral Development 6: Accepted for publication (1983 b).

Falger, P. & Appels, A.: Psychological risk factors over the life course of myocardial infarction patients. Advances in Cardiology 29: 132-139 (1982).

Falger, P., Bressers, I. & Dijkstra, P.: Levensloopatronen van hartinfarctpatiënten en van controlegroepen: Enkele overeenkomsten en verschillen. Gerontologie 11: 240-257 (1980).

Friedman, M. & Rosenman, R.: Association of specific overt behavior pattern with blood and cardiovascular findings. Journal American Medical Association 169: 1286-1296 (1959).

Friedman, M. & Rosenman, R.: The key cause - Type A behavior pattern (1974). In: Monat & Lazarus: Stress and coping: 203-212 (Columbia University Press, New York 1977).

Glass, D.: Behavior patterns, stress, and coronary disease: 72-140 (Erlbaum, Hillsdale 1977).

Glass, D. & Carver, C.: Helplessness and the coronary-prone personality. In: Garber & Seligman: Human helplessness: Theory and applications: 223-243 (Academic Press, New York 1980).

Goldberg, E. & Comstock, G.: Epidemiology of life events: Frequency in general populations. American Journal Epidemiology 111: 736-752 (1980).

Gould, R.: Transitions: Growth and change in adult life. (Simon & Schuster, New York, 1978).

Grolnick, L.: A family perspective of psychosomatic factors in illness: A review of the literature. Family Process 17: 457-486 (1972).

Haynes, S., Feinleib, M., Levine, S. et al.: The relationship of psychosocial factors to coronary heart disease in the Framingham Study. II. Prevalence of coronary heart disease. American Journal Epidemiology 107: 384-402 (1978 a).

Haynes, S., Levine, S., Scotch, N. et al.: The relationship of psychosocial factors to coronary heart disease in the Framingham Study. I. Methods and risk factors. American Journal Epidemiology 107: 362-383 (1978 b).

Holmes, T. & Rahe, R.: The Social Readjustment Rating Scale. Journal Psychosomatic Research 11: 213-218 (1967).

Hultsch, D. & Plemons, J.: Life events and life-span development. In: Baltes & Brim: Life-span development and behavior 2: 1-36 (Academic Press, New York 1979).

Jenkins, D.: A comparative review of the interview and questionnaire methods in the assessment of the coronary-prone behavior pattern. In: Dembroski, Weiss, Shields et al.: Coronary-prone behavior: 71-88 (Springer International, New York 1978).

Jenkins, D., Hurst, M. & Rose, R.: Life changes: Do people really remember? Archives General Psychiatry 36: 379-384 (1979).

Kagan, J.: Perspectives on continuity. In: Brim & Kagan: Constancy and change in human development: 26-74 (Harvard University Press, Cambridge 1980).

Kuller, L., Cooper, M. & Perper, J.: Epidemiology of sudden death. Archives Internal Medicine 129: 714-719 (1972).

Levinson, D.: The mid-life transition: A period in adult psychosocial development. Psychiatry 40: 99-112 (1977).

Lulofs, R., Riedel, W. & Van der Molen, M.: Blood platelet activity in vitally exhausted subjects. Paper presented at the symposium: Cardiovascular psychophysiology: Theory and methods. Noordwijkerhout, The Netherlands, June 12-17, 1983.

Lundberg, U. & Theorell, T.: Scaling of life changes: Differences between three diagnostic groups and between recently experienced and non-experienced events. Journal Human Stress 2: 7-17 (1976).

Lundberg, U., Theorell, T. & Lind, E.: Life changes and myocardial infarction: Individual differences in life change scaling. Journal Psychosomatic Research 19: 27-32 (1975).

Machlowitz, M.: Workaholics. (Mentor Books, New York 1981).

Matthews, K.: Assessment and developmental antecendents of the coronary-prone behavior pattern in children. In: Dembroski, Weiss, Shields et al.: Coronary-prone behavior: 207-217 (Springer International, New York 1978).

Matthews, K.: Psychological perspectives on the Type A behavior pattern. Psychological Bulletin 91: 293-323 (1982).

Matthews, K. & Angulo, J.: Measurement of the Type A behavior pattern in children; Assessment of children's competitiveness, impatience-anger, and aggression. Child Development 51: 466-475 (1980).

Matthews, K. & Krantz, D.: Resemblance of twins and their parents in pattern A behavior. Psychosomatic Medicine 38: 140-144 (1976).

Mendelsohn, E.: The continuous and the discrete in the history of science. In: Brim & Kagan: Constancy and change in human development: 75-112 (Harvard University Press, Cambridge 1980).

Muhlenkamp, A., Gress, L. & Flood, M.: Perception of life change events by the elderly. Nursing Research 24: 109-113 (1975).

Neugarten, B. & Datan, N.: Sociological perspectives on the life cycle. In: Baltes & Schaie: Life-span developmental psychology: Personality and socialization: 53-69 (Academic Press, New York 1973).

Neugarten, B. & Hagestad, G.: Age and the life course. In: Binstock & Shanas: Handbook of aging and the social sciences: 35-55 (Van Nostrand Reinhold, New York 1976).

Orth-Gomer, C., Edwards, M., Erhardt, L. et al.: Relation between ventricular arrhythmias and psychological profile. Acta Medica Scandinavica 207: 31-36 (1980).

Rahe, R. & Romo, M.: Recent life changes and the onset of myocardial infarction and coronary death in Helsinki. In: Gunderson & Rahe: Life stress and illness: 105-120 (Thomas, Springfield 1974).

Rahe, R., Bennett, L., Romo, M. et al.: Subjects recent life changes and coronary heart disease in Finland. American Journal Psychiatry 130: 1222-1226 (1973).

Rahe, R., Romo, M., Bennett, L. et al.: Recent life changes, myocardial infarction, and abrupt coronary death. Archives Internal Medicine 133: 221-228 (1974).

Reese, H. & Overton, W.: Models of development and theories of development. In: Goulet & Baltes: Life-span developmental psychology: Research and theory: 115-145 (Academic Press, New York 1970).

Rissanen, V., Romo, M. & Siltanen, P.: Premonitory symptoms and stress factors preceding sudden death from ischaemic heart disease. Acta Medica Scandinavica 204: 389-396 (1978).

Rosenman, R.: The interview method of assessment of the coronary-prone behavior pattern. In: Dembroski, Weiss, Shields et al.: Coronary-prone behavior: 55-69 (Springer International, New York 1978).

Rosenman, R. & Chesney, M.: The relationship of Type A behavior pattern to coronary heart disease. Activitas Nervosa Superior (Praha) 22: 1-45 (1980).

Rosenman, R. & Friedman, M.: Association of specific behavior pattern in women with blood and cardiovascular findings. Circulation 24: 1173-1184 (1961).

Rosenman, R., Brand, R., Jenkins, D. et al.: Coronary heart disease in the Western Collaborative Group Study: Final follow-up experience of $8\frac{1}{2}$ years. Journal American Medical Association 233: 872-877 (1975).

Sands, J. & Parker, J.: A cross-sectional study of the perceived stressfullness of several life events. International Journal Aging and Human Development 10: 335-341 (1979).

Schaie, K.: Methodological problems in descriptive developmental research on adulthood and aging. In: Nesselroade & Reese: Life-span developmental psychology: Methodological issues: 253-280 (Academic Press, New York 1973).

Schroots, J.: Ontogenetische psychologie: Een eerste kennismaking. De Psycholoog 17: 68-81 (1982).

Silver, R. & Wortman, C.: Coping with undesirable life events. In: Garber & Seligman: Human helplessness: Theory and applications: 279-340 (Academic Press, New York 1980).

Theorell, T.: Life events and manifestations of ischaemic heart disease: Epidemiological and psychophysiological aspects. Psychotherapy & Psychosomatics 34: 135-148 (1980).

Theorell, T., Schalling, D. & Akerstedt, T.: Circulatory reactions in coronary patients during interview. A noninvasive study. Biological Psychology 5: 233-243 (1977).

Theorell, T., Lind, E., Fröberg, J. et al.: A longitudinal study of 21 subjects with coronary heart disease: Life changes, catecholamine excretion and related biochemical reactions. Psychosomatic Medicine 34: 505-516 (1972).

Thiel, H., Parker, D. & Bruce, T.: Stress factors and the risk of myocardial infarction. Journal Psychosomatic Research 17: 43-57 (1973).

Vaillant, G.: Adaptation to life. (Little, Brown and Company, Boston 1977).

Verhagen, F., Nass, C., Appels, A. et al.: Cross-validation of the A/B typology in the Netherlands. Psychotherapy & Psychosomatics 34: 178-186 (1980).

Williams, R., Haney, T., Lee, K. et al.: Type A behavior, hostility, and coronary atherosclerosis. Psychosomatic Medicine 42: 539-549 (1980).

Wu, K. & Hoak, J.: A new method for the quantitative detection of platelet activity in patients with arterial insufficiency. Lancet 2: 924 (1974).

Author's address:

Dr. Paul R.J. Falger,

State University of Limburg School of Medicine,
Department of Medical Psychology,

P.O.Box 616,

6200 MD Maastricht/The Netherlands.

PHYSIOLOGICAL ISSUES IN ESTABLISHING LINKS BETWEEN PSYCHOSOCIAL FACTORS AND CARDIOVASCULAR ILLNESS.

T. Theorell
National Institute for Psychosocial Factors and Health, Box 60210, S-104 01 Stockholm, Sweden.

In Sweden, as well as in many other European countries, there is growing awareness that psychosocial factors may be important not only to people´s general well-being but also to the maintenance of their bodily functions. A dramatic breakdown of human adaptation to psychosocial factors could be myocardial infarction and cardiovascular death. Whereas laymen have always accepted the possibility that psychosocial processes are linked with cardiovascular illness and death, the medical profession has been more hesitating in accepting the idea that there may be a causal link. Establishing causal links may be important both from a legal standpoint, for instance in work compensation cases, and from the standpoint of planning cardiovascular primary prevention. In the present article I do not intend to review all the efforts to establish causal links in this research field but rather ask a few questions of a general nature. Why does our cardiovascular system react as it does to certain challenges and what do some of the associations that have been seen mean in a physiological sense ?

The most important question may be: "What is a cause ?". Or in other terms: "Is it possible to prove a causal link between psychosocial factors and cardiovascular illness ?".

In the case of psychosocial factors we can make the assumption that our general life situation may influence our bodily functions either directly through the brain or indirectly through changes in our life style. This has great importance when we discuss causality. Most medical researchers in the field have focused their attention on direct links for instance between "stressors" in the work place and increased risk of coronary heart disease (CHD), that is neurogenic influence on relevant endocrinological and cardiovascular functioning. Indirect links have been disregarded since they have not been considered causal. By indirect links I mean the influence of "stressors" on accepted risk factors such as dietary habits, cigarette smoking and blood pressure. According to many researchers, indirect links should not be considered causal (1). But what happens if psychosocial factors cause a change for instance in smoking habits ? If that is the case no anti-smoking programme will be effective unless we take account of the psychosocial mechanisms involved in the smoking habit.

What is the overall physiological meaning of some of the pathophysiological processes relevant to CHD development ? How do these processes relate to psychosocial factors now and in previous generations ?

Every layman has heard of the fight and flight responses to threatening situations. The physiological meaning of these reactions is to provide the body with energy for muscular action and protection against loss of blood (platelet aggregation tendency increases) and salt and water due to sweating (mineralocorticosteroids) which takes place as a response to increased heat production. From a survival perspective the provision of energy is the primary goal. This is achieved by means of catabolism. If the state of alarm is longlasting the available stores of energy will be emptied and the catabolism will literally dissolve useful protein which is broken down into glucose and waste products. At the same time, anabolism is unfortunately inhibited. Anabolism restores tissues that are being worn out continuously and builds up useful protection against breakdown. Examples are replacement of worn-out epithelial cells in the skin and the gastrointestinal system and leukocytes that are important in the defence against infection. Thus, life could be seen as a continuous battle between catabolism and anabolism. If the catabolic reactions become too longlasting and intense, several of the body systems become "exhausted" (using Selye´s terminology (3)) and stress-sensitive. On the other hand intense catabolic periods may be harmless if periods of intense anabolism are interpersed between them.

Physicians frequently assume that cardiovascular disease processes are immune to "stress". In cases of work compensation the argument is often put forward that "stress" may precipitate episodes of myocardial infarction but is not causing atherosclerosis which should be regarded as the underlying cause. However, even if we make the cautious assumption that psychosocial factors are unrelated to atherosclerosis, the matter may not be quite as simple. Coronary atherosclerosis is neither a necessary nor a sufficient cause of myocardial infarction, although most persons with myocardial infarctions have significant coronary atherosclerosis. The state of the myocardium may modify the risk of myocardial necrosis dramatically. For instance, the catabolic effects of corticosteroids lead to depletion of myocardial magnesium and potassium, and this increases the risk of myocardial necrosis and disturbance of electrical conductance when the heart is exposed to severe strain (4). Of particular relevance is the observation that testosterone, growth hormone and insulin - hormones with predominantly anabolic functions in most conditions - favourably influence the protein synthesis of the myocardium itself (5, 6). This process may also disturb the coagulation mechanisms and increase the likelihood of clot formation in a coronary vessel. This further increases the likelihood of the development of a myocardial infarction (7, 8). Thus, when the organism is in a constant long-lasting state of catabolism and anabolism is not allowed to take place, a myocardial infarction or a life-threatening arrhythmia may occur even in the absence of severe coronary atherosclerosis. In such a case we may assume that the infarction may take place several years before it "should" do according to degree of coronary atherosclerosis. The theory is summarized in fig. 1 (9).

It has been assumed that in previous generations dietary and psychosocial habits have been more adapted to physiological and psychosocial demands than they are today. For instance, when a job is physically demanding excessive losses of energy, salt and water necessitate a diet rich in fat and salt. As has been pointed out by researchers dealing with dietary change, diets are very hard to change and may take a whole generation to effectively institute (19), and therefore those appropriate for a physically demanding life have been taken over by many people in our generation, whose life in general is _not_ physically demanding. A parallel phenomenon may take place with regard to psychosocial coping mechanisms. Harburg et al (11), on the basis of their studies of representative groups of people living in Detroit, have emphasized that coping with aggression-provoking situations exemplified by unfair aggression from boss, differs across social groups. In summary, the groups with little education eat more calories and salt (12), report that they react with more aggression in provoking situations and have higher blood pressure levels (13). One explanation of this may be that groups with a low level of education more frequently than other people have been prepared

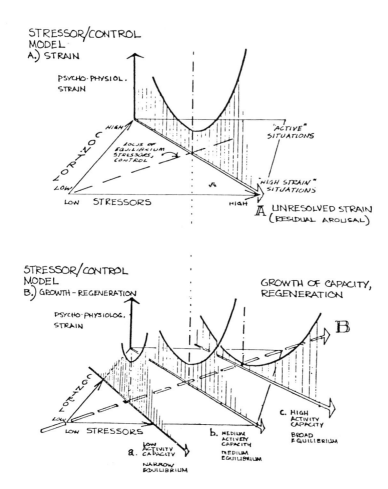

Fig. 1 Model describing associations between psycho-physiological strain and job structure.

(Karasek et.al., J. Human Stress, vol. 8, page 36)

to live a life that is physically strenuous and in which aggressive reactions may be functional. Unfortunately social segregation in modern societies makes improvement of this process difficult. In our own study of areas in the greater Stockholm region we found that in areas with low median income, high proportion of subjects on social welfare and in general pronounced social mobility, relative weights, muscular strengths and systolic blood pressures were on average significantly higher among 18-year old men than in other areas (14). In such socially "unresting" areas the school environment is known to deteriorate (13). There was also evidence of significantly lower average levels of verbal understanding in the socially deprived areas.

What are the difficulties in studying longterm effects of psychosocial factors on cardiovascular functions ?

I believe that the main difficulties in studies of possible longterm effects of psychosocial factors on cardiovascular illness are the following:

a) Subjects´ own reports of their psychosocial environment may be seriously distorted. Frese (15) has pointed out that part of this distortion may be caused by the psychosocial environment itself. Thus, a working situation that allows no development of skills may impoverish the worker´s ability to see problems in his work. An association of such conditions with development of cardiovascular pathology will be overlooked if the estimates of job conditions are based upon self-reports. Other strategies should be used, for instance objective descriptions of the conditions. A "middle way" may be the one used by Karasek´s and our groups:
"Standard weights" of psychosocial and physical job characteristics are obtained from national interview surveys for each one of all occupations on the labour market. These standard weights are subsequently applied on representative cases and non-cases distributed across different occupations (17, 18). In this way the importance of different job dimensions may be studied in a "macro-perspective".

b) Another difficulty arises because most of the pathological processes leading to chronic cardiovascular disease are asymptomatic. It has been pointed out that most symptoms of acute blood pressure elevation exist after persons become aware that their blood pressure is elevated (19). And even then, the symptoms are highly individualized. Some people report headache, others report sweating etc. (20) when the blood pressure is high.

Furthermore, in a prospective study, subjects who died a medically unattended sudden cardiovascular death have been described as overly optimistic and perhaps showing a denying life attitude more often than other subjects with or without manifestations of coronary heart disease (21). Perhaps such individuals tend to deny symptoms as well ? Due to this difficulty the researcher has to rely on physiological measures rather than self-reported symptoms in order to establish links with psychosocial factors.

c) Finally, <u>longterm</u> rather than acute physiological reactions (that are those most relevant to serious pathophysiological consequences) are not easy to record. The following possibly relevant cardiovascular and metabolic parameters may be seen as slow-reacting (within weeks).

1. The proportion of ractor A_1C in the blood hemoglobin (Hb) is very sensitive to fluctuations in carbohydrate metabolism. The HbA_1C/Hb ratio mirrors the average blood glucose level during the past 8-12 weeks. This means that longlasting psychosocial states resulting in catabolic conditions will be reflected in increased HbA_1C/Hb ratios. We have indeed been able to show in a controlled longitudinal study of elderly nondiabetic people that increased positive social activity is associated with significantly decreasing HbA_1C/Hb ratio (22). We therefore believe that this is a factor which   be used in studies of longterm consequences of psychosocial conditions.

2. The lipoprotein metabolism may also be used in the study of longterm effects of psychosocial conditions. Triglycerides and unesterified fatty acids react rapidly to challenges whereas serum cholesterol shows longterm variations that may be partly attributable to psychosocial factors. Rosenman et al showed that the cholesterol level rose during long periods of excessive overwork in bank accountants. These variations could not be explained by changes in dietary habits or physical activity (23). There is also indication that the <u>protective</u> high density lipoprotein part of cholesterol may be depressed during long periods of work strain (24) although this has not been sufficiently explored.

3. Various kinds of functional endocrinological and physiological tests could be used in order to follow longterm effects of psychosocial factors. The <u>response</u> to acute experimental mild challenges (psychological or physiological) in the laboratory varies in response to "chronic stress". One example is the standardized physical exercise test. Bank

accountants, going through periods of work overload, have been shown to respond with more heart rate acceleration when exposed to a standardized physical test than otherwise (Taggart, personal communication).

What do we know about interactions between physical and psychosocial factors ?

Research on the traditional risk factors for CHD has shown that the concurrent existence of two or more risk factors such as smoking and hypertension may increase the risk than the mere addition of these factors would do. The research tradition in the field of psychosocial factors and CHD so far has mainly tried to "isolate" the psychosocial factor from the physical one which may be unfortunate. In a prospective study of building construction workers five psychosocial factors, psychosocial work load, apartment small for family, unfavorable background and sick leave pay were used for prediction of myocardial infarction risk during a two year follow-up. At all levels of risk predicted on the basis of psychosocial factors, the category with the physically most demanding job, the concrete workers, had higher rates of CHD than other groups. An interaction between psychosocial and physical predictors was observed. In the 50-60 year old category, concrete workers with 4-5 psychosocial risk factors had four times higher risk than other construction workers with 0-1 psychosocial risk factor (25).

In a study of working men in Stockholm Alfredsson et al (18) divided subjects on the basis of 118 different job titles into jobs considered (by other representative workers in the same occupations) on average hectic vs on average non-hectic, on average noisy vs on average non-noisy etc. A quasi-prospective case-control study of all myocardial infarctions (dead and alive, cases n=334, controls n= 882) occurring during a three-year period within one area of Stockholm was performed. Table 1 shoes that sweating, physical monotony, heavy lifting, noise and vibrations are statistically unrelated to risk of myocardial infarction. However, when combined with hectic work practically all the combinations are associated with elevated risk of myocardial infarction. A multiplicative interaction seems to take place. This means that workers in occupations with for instance frequent heavy lifting and hectic pace run an elevated risk of myocardial infarction compared to workers in other occupation (26). Due to the design of the study (aggregate level analysis) the relative risks are probably underestimated (27). These results indicate that we must take a closer look at interactions between physical and psychosocial strains at the work places. The mechanisms underlying the associations are unknown.

Table 1. Calculated age-adjusted relative risks associated with occupation characteristics in Sweden. (Men, aged 40-64).

	Hectic work only 1.1	
	Only	And
Sweating	1.1	1.4^x
Physical monotony	1.2	1.3^x
Heavy lifting	1.2	1.4^x
Noise	1.0	1.2
Vibrations	1.1	1.4^x

What I have presented may illustrate some of the difficulties in establishing causal links between psychosocial factors and cardiovascular illness. Some of these difficulties are of a practical nature - how shall we measure the relevant factors ? - but some of them highlight the fact that we often do research without penetrating our own basic assumptions about the processes. For instance it is not possible to isolate a "psychosocial factor" from a number of physical factors in the study of these complex processes!

References

1. Prevention of coronary heart disease. Report of a WHO committee. Technical report series 678, World Health Organization, Geneva, 1982.
2. Rosenman, R.H., Brand, R., Jenkins, C.D., Friedman, M., Straus, R. and Wurm, M.: Coronary heart disease in the WCGS: Final follow-up experience of 8½ years JAMA 233:872-877, 1975.
3. Selye, H.: The physiology and pathology of exposure to stress. Acta Inc. Medical publishers, Montreal, 1950.
4. Selye, H. and Bajusz, E.: Conditioning by corticoids for the production of cardiac lesions with noradrenaline. Acta Endocrin. 30:183, 1959.
5. Hjalmarson, A., Isaksson, O. and Ahran, K.: Effects of growth hormone and insulin on amino transport in perfused rat heart. Am. J. Physiol. 217:1795, 1969.
6. Williams-Ashman, H.G.: Biochemical features of androgen physiology. Endocrinology 3:1527, 1979.
7. Cannon, W.B. and Mendenhall W.L.: Factors affecting the coagulation time of blood IV. The hastening of coagulation in pain and emotional environment. Am. J. Physiol. 34:251-261, 1914.
8. Haft, J.I. and Fani, K.: Stress and the induction of intravascular platelet aggregation in the heart. Circulation 48:164-169, 1973.

9.. Karasek, R.A., Russell, S. and Theorell, T.: Physiology of stress and regeneration in job related cardiovascular illness. J. Hum. Stress 8: 29, 1982.

10. Feinleib, M., Garrison, R.J., Stallones, L., Kannel, W.B., Castelli, W.P. and Mc Namara, P.M.: A comparison of blood pressure, total cholesterol and cigarette smoking in parents in 1950 and their children in 1970. Am. J. Epidemiol. 110:291, 1979.

11. Harburg, E,. Blakelock, E.H. and Roepen, P.J.: Resentful and reflective coping with arbitrary coping and blood pressure. Detroit. Psychosom.Med. 41:189, 1979.

12. Stunkard, A., D´Aquili, E., Fox, S. and Filion, R.D.L. Influence of social class on obesity and thinness in children. In Insel, P.M. and Moos, R.H. (eds): Hearth and the social environment. D.C. Heath, Toronto, 1973.

13. Langford, H.G. and Watson, R.L. Electrolytes and hypertension. In Paul, O.(ed.) Hypertension - Epidemiology and prevention, Stratton, New York, 1975.

14. Theorell, T., Svensson, J., Knox, S. and Ahlborg, B.: Blood pressure variations across areas in the greater Stockholm region: Analysis of 74 000 18-year-old men. Soc. Sci. Med. 16:469, 1982.

15. Löfgren, H.: Teachers´ working conditions (Swedish). Stenciled reports nr 23, Dep. of Education. University of Lund, Lund, 1980.

16. Frese, M., Greif, S. and Semmer, N.: Industrielle Psychopathologie. Hans Huben, Berlin, 1978.

17. Karasek, R.A.: Job characteristics, occupation and coronary heart disease. Final report - phase I, NIOSH, 1982.

18. Alfredsson, L., Karasek, R.A. and Theorell, T.: Myocardial infarction risk and psychosocial work environment: An analysis of the male Swedish working force. Soc. Sci. Med. 16:463, 1982.

19. Haynes, R.B., Sackett, D.L. and Taylor, D.W.: Increased absenteeism from work after detection and labelling of hypertensive patients. New Engl. J. Med. 14:741, 1978.

20. Pennebaker, J.W., Gonden-Frederick, L., Stewart, H., Elfman, L. and Skelton, J.A.: Physical symptoms associated with blood pressure. Psychophysiology 19:201, 1982.

21. Nirkko, O., Lauroma, M., Siltanen, P., Tuominen, H. and Vankala, K.: Psychological risk factors related to coronary heart disease. Prospective studies among policemen in Helsinki. Acta Med. Scand. Suppl. 660: 137, 1982.

22. Arnetz, B.B., Theorell, T., Levi, L., Kallner, A and Eneroth, P.: An experimental study of social isolation of elderly people. Psychosomatic Medicine, in press, 1982.

23. Friedman, M., Rosenman, R.H. and Carroll, V.: Changes in the serum cholesterol and blood blotting time in man subjected to cyclic variation of occupational stress. Circulation 17:852, 1958.

24. Chadwick, J., Chesney, M., Black, G.W., Rosenman, R.H. and Sevelius, G.G.: Psychological job stress and coronary heart disease. Stanford Research Institute, Menlo Park, 1979.

25. Theorell, T., Olsson, A. and Engholm, G.: Concrete work and myocardial infarction. Scand. J. Work, environment and health 3:144, 1977.
26. Alfredsson, L. och Theorell, T.: Psykosocial arbetsmilj- och hjärtinfarktrisk. Läkartidningen 79:4658, 1982.
27. Alfredsson, L.: On the effect of misclassification in case-control studies. Stenciled report, Dep. of Soc. Med., Huddinge Hospital, Huddinge, 1982.

WHITE-COLLAR WORK SETTING AND CORONARY-PRONE BEHAVIOUR PATTERN

J. Siegrist

Institute of Medical Sociology, Faculty of Medicine
University of Marburg, Marburg FRG

ABSTRACT

This contribution concentrates on the study of interactions between the coronary-prone behaviour pattern and specific characteristics of the work setting and of the occupational career. Findings from a large retrospective case-control-study on 380 male patients with first acute myocardial infarction and a healthy control group matched by age, sex and occupation suggest an analysis of coronary-prone behaviour in terms of a coping pattern in situations where efforts to personal control in a socially meaningful setting are threatened. In a more general framework results show the heuristic potential of a combined analysis of sociological, psychological and medical information in populations at cardiovascular risk.

INTRODUCTION

Every student of work-related cardiovascular epidemiology is faced with an obviously paradoxical finding: On the one hand, type A coronary-prone behaviour has been linked to the incidence of ischemic heart disease (IHD) as an independent risk factor (Dembroski et al., 1978). It has also been demonstrated that type A is prevalent among higher economic or educational groups (Rosenman et al., 1975). On the other hand, the incidence of IHD is especially high among lower socio-economic groups, even if one controls for somatic risk factors (Rose et al., 1981). How can we explain this discrepant finding? To some extent, it may be possible to solve the problem by detection of measurement bias or by re-analysis of sampling procedures. However, as differences are marked, we think a conceptual approach to the problem may be justified as well. Our main hypothesis states that different types of experiences of work stress prevail within these two socio-economic groups: In higher socio-economic groups, especially in white-collar occupations, several protective mechanisms such as social support or a healthier life style may lead to the fact that coronary-prone behaviour pattern becomes the main type of socio-emotional stress experience whereas among lower class working men, especially among blue-collar workers, heavy socio-environmental burden such as work load, including mental and physical stressors, increased social instability, and interpersonal conflicts are the types of stressors commonly experienced in this group. If this holds true, these socio-economically different areas of stress experience nevertheless should result in a common pathway of neuroendocrine changes which eventually affect the development of IHD.

We have suggested elsewhere that the concept of active distress might be a fruitful tool linking together these different types of stressful experience (Siegrist et al., 1982). In this context, we have also presented findings on blue-collar workers elsewhere (Siegrist 1983). This contribution will concentrate on the relation between white-collar work setting and coronary-prone behaviour. The coronary-prone behaviour can be described as an "attempt to assert and maintain control over stressful aspects of one's environment. Type A's engage in a continuous struggle for control and, in consequence, appear hard driving and aggressive, easily annoyed and competitive" (Glass 1981). These states of hyperresponsiveness can be followed by hyporesponsiveness in situations where individuals experience prolonged exposure to stressors which cannot be met by successful coping.

This approach has been elaborated and tested in a series of interesting experiments by D.C. Glass and coworkers (Krantz et al., 1981). Yet some intriguing conceptual problems persist, as one of the authors himself points out:"We must define more precisely those behaviours in pattern A that are risk inducing ... Efforts need to be directed toward a delineation of the classes of environmental stimuli that elicit the primary facets of the behaviour pattern" (Glass 1981).

Another problem is: How can we explain recurrence and relative stability of this response pattern, even in the presence of unpleasant and annoying experiences?

In order to answer this question we propose to concentrate on cognitive mechanisms. We hypothesize that type A's tend toward an unrealistic appraisal of demanding situations and their related internal coping resources. Such an unrealistic appraisal can function in multiple ways. It seems that two forms can be easily depicted:

1. Subjects overestimate given demands without being aware of their full coping resources. Inappropriately increased efforts are a probable behavioural consequence.

2. Subjects underestimate given demands and, by doing so, expose themselves to possible overload. Perceptions of exhaustion and tiredness are suppressed, and internal coping resources are overestimated again and again.

In the long run, underestimation as well as overestimation of demands lowers the threshold of critical experiences of active distress. Subjects who tend to overestimate challenges and obligations experience an imbalance or misfit between invested efforts and pay-off with concomitant feelings of irritation, frustration, dissatisfactions, and hostility. It would be interesting, by the way, to interpret hostility - one of the crucial features of CPB - in terms of reactive coping with disappointment and lack of reward.

Subjects who tend to underestimate demands and to overestimate their coping resources are likely to assume increased responsibility, to be work-addicted and to fulfill all kinds of obligations and expectations. For quite a while, they are successful and experience rewarding feed-back by significant others. But their burden increases, together with suppressed fatigue and exhaustion, and increased responsibility overwhelmes their adaptive efforts. In this situation, intense active distress is very likely, and it may even be followed by depressive states of exhaustion with the possible outcome of psychological and physiological breakdown (Appels 1980).

Our theoretical approach to CPB assumes that a specific link between cognition and performance may trigger psychoneuroendocrine reactions which, in the long run, are harmful to the cardiovascular system. This link is analyzed as an intraindividual coping technique in the presence of demanding situations - a technique which starts by unrealistic appraisal of challenges and elicits overactivity at the behavioural and at the physiological level. Long-term pay-off of this overactivity is assumed to be poor, and active distress is a probable emotional correlate.

Unrealistic appraisal of challenging situations is regarded as a relatively stable coping technique which is established over time. Three explanations which are not mutually exclusive are suggested:

a) unrealistic appraisal is the outcome of model learning during primary socialization. Children learn to appraise and behave in demanding situations the same way as their parents do (or one of them). Some experimental findings with coronary-prone children and their parents support this interpretation (Matthews 1977).

b) unrealistic appraisal can be understood as consequence of a specific motivation, i.e. a need for control and visible performance. This need may function as a compensation of experiences or fears of low self-esteem or marginal socio-emotional status. It is thought to evolve during primary socialization, as related to, but not identical with achievement motivation.

c) unrealistic appraisal of demands is learned during secondary or tertiary socialization as an adaptive technique in dealing with increased work load. Underestimation may be the more common reaction, given the routines of handling tasks and the perceptions of short-term reward associated with overcommitment and increased responsibility.

METHODS AND RESULTS

We present results on the hypothesis of coronary-prone behaviour as a coping technique in white-collar work settings from a large retrospective case-control study on 380 male patients with first acute myocardial infarction (AMI) (age 30-55) and a matched control group as well as from a follow-up study of the AMI-patients over 18 months. The healthy control group has been matched with the sample half by age, sex and occupational status. Clinical data included diagnosis of acute myocardial infarction on the basis of ECG and enzymes, information about somatic risk factors and family history of CHD. Sociological and psychological data included a structured interview with questions on sociodemographic characteristics, on occupational career, on experiences of workload, chronic difficulties and social support. Frequency and subjective impact of negative life events were assessed. Psychological measures concentrated on a special approach to the study of coronary-prone behaviour in terms "need for control" which was assessed by a self-administered questionnaire. Information about theoretical foundation and test-statistical analysis of this questionnaire can be found elsewhere (Dittmann et al., 1983). The follow-up questionnaire contained questions on the occupational and economic situations of cardiac patients after rehabilitation, on social activities and health behaviour including self-help groups, on information about the course of disease and subjective well-being. In addition, scales of coronary-prone behaviour were retested (r=.87) (Siegrist et al.,1982).

It must be emphasized that our sample includes a wide range of occupations. 20% of the MI subjects in the sample could be classified as unskilled or semiskilled workers, 23% as skilled blue-collar workers, 41% as white-collars, mainly employees, 10% as professionals and 6% as officials and higher ranks of civil service. 28% are in middle-echelon or in leading positions, and their work can be characterized by "coordination, organisation, disposition, control and leadership". An index of psychomental workload focussing on issues of work related control, coordination, decision and responsibility shows significantly higher mean values for white-collars than for blue-collars. The amount of CPB follows this pattern quite closely. In an analysis of very homogeneous small subgroups middle-echelon employees, supervisors and foremen in the AMI and control group were compared to blue-collar workers with simple repetitive tasks. Attitudes reflecting CPB are significantly stronger in the group of middle-echelon employees and foremen than in the group of workers with simple repetitive tasks ($p < .01$ in the disease group; $p = .06$ in the healthy control group). Of course it is possible to follow the argument of Mettlin (1976) and to interprete higher scores of CPB in middle-echelon employees as a result of professional competition, struggle and upward mobility. Thus, self-selection caused by personality traits such as job-involvement, competitiveness, hard driving and hostility might be the crucial variable. Yet longitudinal studies have not verified this interpretation to our knowledge. Another interpretation is consistent with an interactional approach to the study of CPB. The latter may be analyzed in terms of coping strategies which are elicited by challenges and demands of the work setting. Whereas much experimental research supports this perspective (Krantz et al., 1981, Chesney et al., 1980), only a few studies in real life settings are known to fit into this frame of reference (Howard et al., 1977, Burke et al., 1980). A closer view to dimensions of psychomental workload in our interview schedule revealed that threat to control and work autonomy, inconsistency of demands, time pressure and interruptions during responsible taskfulfillment were the most critical elements. These features in the work setting can provoke the well-known cognitive, emotional and overt behavioural reactions that constitute the core of CPB.

In our research group, the relation between forced occupational mobility and CPB in the sample described above was analyzed (Weber 1983). We not only found nearly twice as many AMI-subjects experiencing forced mobility - in most cases downward mobility - , as compared to matched controls, but also significantly higher degree of CPB among this group, in reference to a group with stable occupational positions. For example the percentage of AMI-subjects with high scores of CPB in the subgroup of forced mobility was 70% as compared to 51% in the group with stable positions. A similar, but weaker trend was present in the control group. A statistical test based on LOGIT-model showed significant main effects.

If it is true that coronary-prone persons react predominantly to those environmental stressors that threaten an individuals' sense of control, it can be concluded that experience of forced occupational mobility enhances cognitions, emotions and overt behaviours which try to seek control over this distressing situation.

As a final issue in the field of work setting and subjective coping we address ourselves to the question of intrapersonal change of CPB over time as a function of exposure to challenges and threats. A follow-up of our initial AMI-sample over 18 months gave us the opportunity to analyze some aspects of this issue. First, and very unexpectedly, we found a significant increase in mean scores of CPB in AMI-subjects after 18 months ($t=4.6$; $p<0.01$). The percentage of subjects with extremely high scores of CPB raised from 19 to 33%. A closer analysis of variables associated with an increase of CPB after rehabilitation showed that subjects who experienced more actual strain in the field of work and health also had higher mean scores of CPB ($C=.45$; $p<0.001$). A threefactorial analysis of variance with "age", "occupational status" and "actual strain in the field of work and health" as related to "amount of change in CPB" was carried out in 110 AMI-sub-subjects. Main effects were calculated on the basis of ANOVA procedure (Brockmeier 1980). Only 15,3% of total variance was explained by main effects (F=3.5; $p < 0.01$), but out of these, 13% could be attributed to the variable "actual strain" (F=7.4; $p < 0.001$). A multiple classification analysis shows this effect for adjusted and unadjusted CPB.

Figure 1: Multiple classification analysis: Change in coronary-prone behaviour over time (N=110 AMI-subjects)

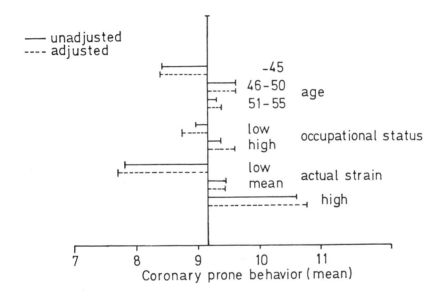

The most important items of the index "actual strain" were "being anxious about work and future" and "more vulnerable self-esteem at work as compared to time before disease onset". Thus, threat to one's occupational achievements may be answered by improving the risky attitudes and behaviours of the CPB pattern.

Taken together these results give some suggestive evidence for an interactional approach to the study of CPB at least in the white-collar work setting.

REFERENCES

Appels, A. 1980.
  Psychological prodromata of myocardial infarction and sudden death. Psychother. Psychosom. 34:187.
Brockmeier, R. 1980.
  Katamnesen psychosozialer Merkmale bei Herzinfarktpatienten. Diplomarbeit Soziologie. Unpublished.
Burke, R.J., Weir, T.1980.
  The type A experience: Occupational and life demands, satisfaction and well-being. J. Human Stress 4:28-38.
Chesney, M.A., Rosenman, R.H. 1980.
  Type A behaviour in the work setting. In: C.L. Cooper, R. Payne (eds.): Current in occupational stress. J. Wiley, New York: 187-212.
Dembroski, T.M., Weiss, S.M., Shields, J.L., Haynes, S.G., Feinleib, M. (eds.) 1978.
  Coronary-prone behaviour. Springer, Berlin, Heidelberg, New York.
Dittmann, Matschinger, H., Siegrist, J. 1983.
  Fragebogen zur Messung von Kontrollambitionen. In press.
Glass, D.C. 1981.
  Type A behaviour: mechanisms linking behavioural and pathophysiologic processes. In: J. Siegrist, M.J. Halhuber: myocardial infarction and psychosocial risks. Springer, Berlin, Heidelberg, New York: 77-88.
Howard, J.H., Cunnigham, D.A., Rechnitzer, P.A. 1977.
  Work patterns associated with type A behaviour: a managerial population. Hum. Relat. 30:825-836.
Krantz, D.S., Glass, D.C., Schaeffer, M.A., Davia, J.E. 1981.
  Behaviour patterns and coronary disease: a critical evaluation. In: J.T. Caccioppo, R.E. Etty (eds.):Focus on cardiovascular psychophysiology. Guilford Press, New York.
Matthews, K.A. 1977.
  Efforts to control by children and adults with the type A coronary-prone behaviour pattern. Child Developm. 50: 842-847.
Mettlin, C. 1976.
  Occupational careers and the prevention of coronary-prone behaviour. Social Science & Medicine C: 367-372.
Rose, G., Marmot, M. 1981.
  Social class and coronary heart disease. Brit. Heart J. 45: 13-19.
Rosenman, R.A., Brand, R.J., Jenkins, D., Friedman, M., Strauss, R. Wurm, M. 1975.
  Coronary heart disease in the Western Collaborative Group Study: final follow-up experience of 8 1/2 years. J. Amer. Med. Assoc. 233: 872-877.

Siegrist, J. 1983.
 Psychosocial coronary risk constellations in the work
 setting. In: H. Benson, W.D. Gentry, C. de Wolff (eds):
 Behavioral medicine: work, stress, and health. Elsevier
 Amsterdam (in press).
Siegrist, J.,Dittmann, K., Rittner, K., Weber, I. 1982:
 The social context of activ distress in patients with
 early myocardial infarction. Soc. Science & Med.
 16:443.
Weber, I. 1983.
 Berufstätigkeit, psychosoziale Belastung und koronares
 Risiko. Unpublished Ph.D. dissertation. University
 of Marburg.

PSYCHOLOGICAL METHODS: AN OVERVIEW OF CLINICAL APPLICATIONS

Andrew Steptoe
Department of Psychology
St George's Hospital Medical School
University of London
Cranmer Terrace
London SW17 ORE
U.K.

Psychological methods play a central role in the understanding of clinical disorders resulting from the breakdown of human adaptation. This paper provides a summary of investigative strategies, then focusses on methodological issues relating to the application of psychological instruments in the clinical field. These include the problems of subject selection and study design, and the questions of specificity and sensitivity of measures. Illustrations of these points are taken from research into essential hypertension, coronary-prone behaviour and adverse life events.

The part played by psychological methods in the investigation of clinical disorders must be seen in the more general context of evidence relevant to diseases of adaptation. A convincing argument in favour of psychological involvement in medical disorders requires the collation of data from diverse fields of research.

a. Evidence concerning biological mechanisms. These are the processes translating higher nervous stimulation into pathological dysfunction. The mechanism may be autonomic, neuroendocrine, neuroimmunological or behavioural. Unless these pathways are identified, the connection between social or psychological factors and a particular clinical disorder will remain incomplete. The fact that such mechanisms are present does not of course mean that psychological factors are necessarily important, as it is possible to delineate processes that are of no significance clinically.

b. Laboratory studies of responses to psychological stimuli. Such studies are valuable for two purposes. Firstly, experiments can be performed on animals that compress prolonged pathological sequences into a manageable time, showing processes that frequently take many years to develop in humans. This is made possible by virtue of the relatively short lifespan of laboratory species, and the degree of control which can be exerted over environmental conditions. Secondly, investigations of acute psychophysiological responses in humans permit a close analysis of those components of environmental stimulation and personal coping resource that are crucial in modulating physiological reactions.

c. Clinical/epidemiological data. Ultimately, the mechanisms and processes detailed in the laboratory have to be confirmed in the clinical perspective. Studies of patients, groups with genetic or constitutional vulnerabilities, and populations considered to be at risk psychosocially, are all relevant. Only when systematic evidence is available from all these levels of analysis can a coherent picture of disorders of adaptation begin to emerge.

The stage in the development of a disease at which psychological factors come to prominence may vary in different conditions. In the case of essential hypertension, the evidence suggests that psychosocial variables are important in the *initiation* of the disorder, promoting haemodynamic disturbance through the provocation of heightened sympathetic nervous system responses (Steptoe, 1981). In contrast, the role of psychological responses in cancer may be to *exacerbate* existing pathology, hastening tumor induction, growth and proliferation (Sklar and Anisman, 1981). On the other hand, research into bronchial asthma suggests that emotional factors are not significant in the aetiology of the condition, but influence or *modify* the severity of symptomatology and clinical status once the disease is established (Purcell and Weiss, 1970; Creer, 1979). Finally, it should be noted that different psychological factors may be relevant at different stages of a single disease process. Thus the conditions precipitating sudden cardiac death or acute episodes of coronary spasm and ischaemia may not be the same as those promoting atherosclerosis in the long term (Steptoe, 1981).

OVERVIEW OF PSYCHOLOGICAL METHODS IN CLINICAL SETTINGS

The literature concerning the measurement of psychological factors in the breakdown of human adaptation covers a range of procedures, techniques and disorders. Different methods are applied at various levels of social and psychological detail, each providing information appropriate to a particular type of analysis. These range from broad aspects of social activity and organisation down to precise studies of psychophysiological interactions at the individual level. Taking disease of

the cardiovascular system as an example, research has been carried out
into the influence of social class, the density of living conditions and
other major social variables (Marmot et al, 1978; Levy and Herzog, 1978).
Such investigations provide evidence that social and cultural forces are
relevant, but are non-specific; they do not attempt to account for wide
individual variations in response, or the vulnerability of different
people to breakdown. At the next level of detail are methods quantifying
the impact of chronic or acute lifestyles and experience. These fall
into two categories. Firstly there are investigations of life events,
which may involve studying events in general (Theorell and Rahe, 1975;
Connolly, 1976) or specific incidents such as redundancy or bereavement
(Parkes et al, 1969; Cobb and Kasl, 1977). Secondly, techniques have
been developed for recording chronic harassments, such as the difficulties
surrounding work conditions and job responsibilities (House et al, 1979;
Alfredsson et al, 1982). Analysis of this stratum enables variations in
life experience to be linked with clinical disturbances, but does not
take personal responses into consideration. The latter are treated more
fully in studies focussing on habitual behaviours and personality factors.
Research on Type A coronary-prone behaviour represents a major development in this field, but other elements have also been linked to heart
disease. Finally, psychophysiological experiments explore the links
between psychological stimuli and autonomic or neuroendocrine responses.

All these techniques are distinct but complementary, and it is
crucial in this area to resist the temptation of reductionism. Investigations directed at the psychophysiological interface do not supercede or
negate analyses of social influences on adaptive processes. Similarly,
not all social class and cultural variations in the prevalence of
ischaemic heart disease can be explained in terms of Type A behaviour
(Marmot, in press). Data from all levels are required in order to understand these phenomena fully.

STUDY DESIGN AND THE SELECTION OF SUBJECTS

Long-term prospective studies with multivariate pre-morbid evaluation of
psychological, biological and social factors represent an ideal for
researchers into the breakdown of human adaptation. They are however
logistically difficult to mount, and most investigators prefer to carry
out preliminary testing of their hypotheses using less elaborate designs.

The retrospective case control strategy is frequently utilised in the investigation of myocardial infarction and other severe disorders with a defined onset. Valuable information is provided by such studies through the use of sophisticated psychological and statistical methods (Siegrist et al, 1982). Nevertheless, the problems of retrospective designs must be borne in mind. They include retrospective reporting biases, and changes in reponse to psychological instruments with onset of the disorder. Thus the victim of a heart attack may re-appraise earlier experience, re-defining it to account for present ill-health. Likewise, newly diagnosed patients with malignancy may radically revise their ambitions, life goals and values, producing very different responses to psychological enquiry than would have emerged prior to diagnosis. A further problem is selective survival. Prospective studies of ischaemic heart disease indicate that the psychological profiles of survivors differ from those who die at early stages (Lebovits et al, 1967; Bruhn et al, 1969). Studies performed on survivors may therefore not be representative of the clinical group as a whole.

Chronic but sub-acute disorders such as essential hypertension and duodenal ulcer are generally examined in cross-sectional designs, comparing patient groups with age and sex-matched controls. In these cases, a major confounding factor is that diagnosed subjects may be atypical. Surveys of psychological factors in essential hypertension have produced very different results, depending on whether subjects are recruited through medical facilities or by screening of population samples (Steptoe, 1983). Subjects with diagnosed high blood pressure have higher neuroticism scores, may be more depressed, and complain of more symptoms than people with comparable blood pressure from the population (Robinson, 1964; Berglund et al, 1975; Goldberg et al, 1980). Although some of these effects are secondary to antihypertensive medication, reactions to the revelation of a long-term, potentially life-threatening disease may be significant. It was recently found in a community prevalence study that subjects mistakenly labelled as hypertensive (through errors in evaluation) reported more depression and poorer present health than the remainder of the normotensive sample (Sparacino, 1982). It is also possible that people with certain psychological characteristics are more likely to seek medical attention and have their high blood pressure detected.

A population screening strategy has been adopted by many

investigators in order to overcome referral bias. Psychometric data and blood pressure readings are collected from the study cohort prior to diagnosis. With this approach, the problem of blood pressure variability may arise, since levels at screening are frequently higher than those recorded subsequently (Armitage et al, 1966). Blood pressure measured on a single occasion cannot be considered a valid indicator of clinical status. Steptoe et al (1982) found that some 60% of male factory workers aged 18 to 65 with a blood pressure of 145/95mm or above on screening fell below this criterion on retest. A further contaminating factor is that people who refuse to participate in screening may differ in important characteristics from the remainder of the population (Theorell et al, 1982). Even the screening strategy may not therefore identify an entirely representative study population.

These factors have contributed to the mixed pattern of reports concerning personality characteristics and high blood pressure found in the literature (Harrell, 1980; Weiner, 1979). Monk (1980) has shown that none of the screening studies that demonstrate poor psychological status amongst hypertensives fulfil the methodological criteria of adequate repeated blood pressure determination, and administration of psychological tests before diagnosis. Investigations satisfying the criteria have tended not to show disturbances of hostility, anxiety or other psychological factors amongst hypertensives (Robinson, 1962; Davies, 1970; Steptoe et al, 1982). Unless these problems are considered in the design of clinical investigations, analyses of psychological aspects may be misleading.

Even when these criteria are satisfied, the causal significance of the resulting patterns cannot be assessed in a cross-sectional design. Modified responses to psychological or psychophysiological tests may arise after the emergence of essential hypertension or gastrointestinal disease. Recent years have witnessed a new approach to this problem, in the investigation of healthy subjects thought to be at high risk for a particular disorder. For example, since essential hypertension has a significant genetic component, children with hypertensive parents are at higher risk than their peers with normotensive parents (Julius and Schork, 1971). Thus psychophysiological experiments have been performed on the normotensive children of parents with high blood pressure (Falkner et al, 1979; Ohlsson and Henningsen, 1982). These offspring have normal blood pressure, so if they share patterns of psychological or physiological reactivity with hypertensives rather than normotensives, it can be inferred that the characteristics in question develop prior to the onset of the disorder.

The strategy is proving valuable in hypertension research, but its general applicability may be limited. High risk groups are difficult to identify for many clinical disorders. In the case of hypertension, it is possible that some psychological factors may be more important in people who are not at high genetic risk. Moreover, only a proportion of children of hypertensive parents will subsequently develop the disorder, so many false positives' are included in the risk group.

THE SENSITIVITY AND SPECIFICITY OF PSYCHOLOGICAL MEASURES

One of the most pressing needs in the assessment of psychological factors in diseases of adaptation is refinement of the predictive power of our techniques. At present, there are major limitations to the sensitivity and specificity of psychological instruments. It is possible that progress towards this goal will be assisted by greater emphasis on psychophysiological procedures.

When considering the clinical literature, it is striking that even amongst people considered to be at very high risk on measures of social or psychological dysfunction, the proportion who succumb to illness is not large. For example, although Type A coronary-prone behaviour is now established as an important risk for ischaemic heart disease, the majority of Type A individuals remain disease free (Brand, 1978; Zyzanski, 1978). Of course, this is attributable in part to the multi-factoral nature of disease; people exhibiting Type A behaviour may not be biologically vulnerable. But the incidence of disease is far from universal even in those at risk both physically and psychologically. It can be estimated from the Western Collaborative Group Study that of Type A men in the highest risk category on physical factors, only 26% would develop ischaemic heart disease over an 8.5 year period (Rosenman et al, 1976). The measurement of Type A behaviour seems to lack sensitivity, as far as prediction of disease is concerned.

Research on life events illustrates a related point. Much of the early work on life events and illness utilised 'dictionary' or respondent-based questionnaire approaches to measurement. The Holmes-Rahe Schedule of Recent Experiences was used extensively in this context, identifying associations between life changes and many different disorders (Holmes and Rahe, 1967). Such instruments have been criticised extensively on the basis of vagueness, low reliability and validity, and openness to bias (Brown, 1981; Paykel, in press). These shortcomings appear largely to be overcome by interview-based techniques which, despite some variations in methodology, may produce reliable estimates of life event incidence and impact.

Life events have now been implicated in the development of a whole host of disorders, from depression, anxiety and parasuicide to appendicitis, sudden cardiac death and non-fatal myocardial infarction (Brown and Harris, 1978; Finlay-Jones and Brown, 1981; Creed, 1981; Talbott et al, 1981; Siegrist et al, 1982). However, this wealth of evidence has its own drawbacks, since it indicates that life events have little specific predictive power. They cannot alert us to particular clinical consequences. The explanation of this effect commonly put forward is that life events lower the threshold for all disorders, so that people succumb according to their individual vulnerability. This is consistent with the diathesis-stress model of illness, in which psychosocial disturbances interact with biological predispositions (Kagan and Levi, 1974). A more intriguing possibility is that particular events and ways of coping with them affect the nature of the biological response. For example, recent research by Brown and his associates suggests that depressive illness may be linked with events involving loss, while anxiety is generated by events involving threat or danger (Brown and Harris, 1978; Finlay-Jones and Brown, 1981). Another case in which differences in reactivity may be due to variations in challenging conditions is in the development of essential hypertension.

There is considerable inconsistency in the literature concerning the cardiovascular reactions of hypertensive patients to psychological stressors. Some studies show that hypertensives produce greater heart rate and blood pressure reactions than normotensives (Nestel, 1969; Lorimer et al, 1971), but in many cases no differences have been observed (Hollenberg et al, 1981; Fredrikson et al, 1982). It cannot be assumed that different stimuli and psychological stressors will elicit the same reaction pattern. Obrist et al (1978) have shown that conditions involving active, effortful behavioural responses provoke a hyperkinetic cardiovascular adjustment, mediated by the sympathetic nervous system. This reaction pattern may be particularly significant in the early stages of essential hypertension (Steptoe, 1981; Obrist, 1981). Accordingly, it is possible that mild hypertensives will manifest pressor hyperreactivity only when exposed to conditions demanding active behavioural coping, rather than passively aversive stressors.

This prediction has been tested and confirmed by a comparison of mild hypertensives and age matched normotensive males in our laboratory (Steptoe et al, in press). The hypertensives showed exaggerated blood pressure reactions during tasks eliciting active behavioural coping, and not during exposure to an unpleasant cine film. In the latter condition, blood pressure reactions were unrelated to clinical status. It is possible that psychological involvement in the aetiology of essential hypertension is mediated in part by exposure to particular types of challenging environment. The individual's susceptibility will also depend on a propensity to engage in active behavioural coping. More generally, analyses of this kind may help to improve the precise predictive properties of psychological techniques.

Conclusions

This overview of clinical applications has not offered any detailed evaluations of particular psychological techniques. Rather the emphasis has been upon research strategy and methodology. I would like to argue that interview or questionnaire-based instruments that tap patients' experience at a psychological or emotional level must be supplemented by 'dynamic' procedures, in which physiological and biochemical responses are evaluated. Research in the rarified atmosphere of the psychophysiology laboratory can in turn help to refine hypotheses which are testable in clinical and epidemiological settings. The active interplay between these approaches will greatly enhance our understanding of clinical problems resulting from the breakdown of human adaptation.

References

Alfredsson, L, Karasek, R, and Theorell, T, 1982. Myocardial infarction risk and psychosocial work environment: an analysis of the male Swedish working force. Soc. Sci. Med., 16: 463-467.
Armitage, P, Fox, W, Rose, G A and Pinker, C M, 1966. The variability of casual blood pressure. 2. Survey experience. Clin. Sci. 30: 337-344.
Berglund, G, Ander, S, Lindstrom, B and Tibblin, G, 1975. Personality

and reporting of symptoms in normotensive and hypertensive 50-year old males. J. Psychosom. Res., 19: 139-145.

Brand, R J, 1978. Coronary-prone behavior as an independent risk factor for coronary heart disease. In: T M Dembrowski, S M Weiss, J L Shields, S G Haynes and M Feinleib (Editors), Coronary-Prone Behavior. Springer-Verlag, New York, pp 11-24.

Brown, G W, 1981. Life events, psychiatric disorder and physical illness. J. Psychosom. Res., 25: 461-474.

Brown, G W and Harris, T, 1978. Social Origins of Depression. Tavistock, London.

Bruhn, J C, Chandler, B and Wolf, S, 1969. A psychological study of survivors and nonsurvivors of myocardial infarction. Psychosom. Med., 31: 8-19.

Connolly, J, 1976. Life events before myocardial infarction. J. Human Stress, 2 (4): 3-17.

Creed, F, 1981. Life events and appendicectomy. Lancet, I: 1381-1385.

Creer, T L, 1979. Asthma Therapy: A Behavioral Health Care System for Respiratory Disorders. Springer Publishing Co., New York.

Davies, M, 1970. Blood pressure and personality. J. Psychosom. Res., 14: 89-104.

Falkner, B, Onesti, G, Angelakos, E T, Fernandes, M and Langman, C, 1979. Cardiovascular response to mental stress in normal adolescents with hypertensive parents. Hypertension, 1: 23-30.

Finlay-Jones, R A and Brown, G W, 1981. Types of stressful life event and the onset of anxiety and depressive disorders. Psychol. Med., 11: 803-815.

Fredrikson, M, Dimberg, U, Frisk-Holmberg, M and Ström, G, 1982. Haemodynamic and electrodermal correlates of psychogenic stimuli in hypertensive and normotensive subjects. Biol. Psychol., 15: 63-73.

Goldberg, E L, Comstock, G W and Graves, C G, 1980. Psychosocial factors and blood pressure. Psychol. Med., 10: 243-255.

Harrell, J P, 1980. Psychological factors and hypertension; a status report. Psychol. Bull., 87: 482-501.

Hollenberg, N K, Williams, G H and Adams, D F, 1981. Essential hypertension: abnormal renal vascular and endocrine responses to a mild psychological stimulus. Hypertension, 3: 11-17.

Holmes, T H, and Rahe, R H, 1967. The Social Readjustment Rating Scale. J. Psychosom. Res., 11: 213-218.

House, J S, McMichael, A J, Wells, J A, Kaplan, B H and Landerman, L R, 1979. Occupational stress and health among factory workers. J. Hlth. Soc. Behav., 20: 139-160.

Julius, S and Schork, M A, 1971. Borderline hypertension - a critical review. J. Chron. Dis., 23: 723-754.

Kagan, A R and Levi, L, 1974. Health and environment - psychosocial stimuli: a review. Soc. Sci. Med., 8: 225-241.

Lebovits, B Z, Shekelle, R B, Ostfeld, A and Paul, O, 1967. Prospective and retrospective studies of coronary heart disease. Psychosom. Med., 29: 265-272.

Levy, L and Herzog, A N, 1978. Effects of crowding on health and social adaptation in the city of Chicago. Urban Ecol., 3: 327-354.

Lorimer, A R, MacFarlane, P W, Provan, G, Duffy, T and Lawrie, T D V, 1971. Blood pressure and catecholamine response to 'stress' in normotensive and hypertensive subjects. Cardiovasc. Res., 5: 169-173.

Marmot, M G, Adelstein, A M, Robinson, N and Rose, G A, 1978. Changing social-class distribution of heart disease. Br. Med. J., II: 1109-1112.

Marmot, M G, in press. Stress, social and cultural variation in heart disease. J. Psychosom. Res.

Monk, M, 1980. Psychologic status and hypertension. Amer. J. Epidemiol., 112: 200-208.

Nestel, P J, 1969. Blood pressure and catecholamine excretion after mental stress in labile hypertension. Lancet, I: 692-694.

Obrist, P A, Gaebelein, C T, Teller, E S, Langer, A W, Grignolo, A, Light, K C and McCubbin, J A, 1978. The relationship among heart rate, carotid dp/dt and blood pressure in humans as a function of the type of stress. Psychophysiology, 15: 102-115.

Obrist, P A, 1981. Cardiovascular Psychophysiology. Plenum Press, New York.

Ohlsson, O and Henningsen, C, 1982. Blood pressure, cardiac output and systemic vascular resistance during rest, muscle work, cold pressure test and psychological stress. Acta. Med. Scand., 212: 329-336.

Parkes, C M, Benjamin, B and Fitzgerald, R G, 1969. Broken heart: a statistical study of increased mortality among widowers. Br. Med. J., I: 740-743.

Paykell, E S, in press. Methodological aspects of life event research. J. Psychosom. Res.

Purcell, K and Weiss, J H, 1970. Asthma. In: C G Costello (Editor), Symptoms of Psychopathology; a Handbook. John Wiley and Sons, Chichester, pp 597-623.

Robinson, J O, 1962. A study of neuroticism and casual arterial blood pressure. Brit. J. Soc. Psychol., 2: 56-64.

Robinson, J O, 1964. A possible effect of selection on the test scores of a group of hypertensives. J. Psychosom. Res., 8: 239-243.

Rosenman, R H, Brand, R J, Scholtz, R I and Friedman, M, 1976. Multivariate prediction of coronary heart disease during $8\frac{1}{2}$ year followup in the Western Collaborative Group Study. Am. J. Cardiol., 37: 903-910.

Siegrist, J, Dittman, K, Rittner, K and Weber, I, 1982. The social context of active distress in patients with early myocardial infarction. Soc. Sci. Med., 16: 443-453.

Sklar, L S, and Anisman, H, 1981. Stress and cancer. Psychol. Bull., 89: 369-408.

Sparacino, J, 1982. Blood pressure, stress and mental health. Nursing Res., 31: 89-94.

Steptoe, A, 1981. Psychological Factors in Cardiovascular Disorders. Academic Press, London.

Steptoe, A, Melville, D and Ross, A, 1982. Essential hypertension and psychological functioning: a study of factory workers. Brit. J. Clin. Psychol., 21: 303-311.

Steptoe, A, 1983. Communication to the editor. J. Psychosom. Res.

Steptoe, A, Melville, D and Ross, A, in press. Behavioural response demands, cardiovascular reactivity and essential hypertension. Psychosom. Med.

Talbott, E, Kuller, L H, Perper, J and Murphy, P A, 1981. Sudden unexpected death in women: biological and psychosocial origins. Amer. J. Epidemiol., 114: 671-682.

Theorell, T and Rahe, R H, 1975. Life change events, ballistocardiography and coronary death. J. Human Stress, 1 (3): 18-24.

Theorell, T, Svensson, J, Löw, H and Nerell, G, 1982. Clinical characteristics of 18-year-old men with elevated blood pressure. Acta. Med. Scand., 211: 87-93.

Weiner, H, 1979. The Psychobiology of Essential Hypertension. Elsevier,

New York.

Zyzanski, S J, 1978. Coronary-prone behavior and coronary heart disease: epidemiological observations. In T M Dembrowski, S M Weiss, J L Shields, S G Haynes and M Feinleib (Editors), Coronary-Prone Behavior. Springer-Verlag, New York, pp 25-40.

PSYCHOLOGICAL FACTORS IN THE BREAKDOWN OF HUMAN ADAPTATION
: SOME METHODOLOGICAL ISSUES

Colin Mackay[1]

Medical Division, Health and Safety Executive

London, United Kingdom.

ABSTRACT

This paper is in three parts. The first discusses some of the broader conceptual issues surrounding breakdown in human adaptation and the role of psychological factors. The nature of the scientific questions which are posed by this general area of study are briefly mentioned from the point of view of methodological problems concerned with causality. In the second part some specific issues are addressed including research settings, research designs and choice of dependent and independent variables including research tools used by the author and of relevance to studies of breakdown in human adaptation. In the third part some general conclusions are noted.

INTRODUCTION

The field of interest

These workshops are entitled 'Breakdown in Human Adaptation'. Whilst definitions and semantic considerations are generally uninteresting they may, on occasions, be illuminating. One of several definitions of 'adaptation' provided by the revised Shorter Oxford English Dictionary on Historical Principles (Friedrichsen, 1980), and dating from 1790, is 'the process of modifying so as to suit new conditions'. Clearly something more substantial than adjustment is implied and further properties are also indicated. First, it may be associated with a short-term (biological) cost. Second, it suggests a positive (evolutionary) gain in the long term. Third, it necessitates the achievement of 'fit' or 'fitness'. Although breakdown suggests a comprehensive failure of a mechanical system such an analogy is generally not appropriate. Breakdown in human adaptation suggests a gradual deterioration in functional effectiveness brought about by the chronic accumulation of costs. These in turn result from repeated unsuccessful or partially successful attempts at adapting to new conditions. Such a model suggests

[1] The views expressed are those of the author and not necessarily those of the Health and Safety Executive.

both positive and negative feedback loops. Self-correcting systems, the re-establishment of control over the environment, and the re-attainment of balance (homeostasis) are all cybernetic principles central to much of the current writing on stress and disease (Lumsden, 1975; Levi, 1974).

What are the scientific questions we wish to answer?

Very generally we are interested in delineating aetiological factors associated with the breakdown of adaptation and with the onset of pathophysiological disease processes, broadly conceived. In particular the role of psychosocial (Levi, 1974) and sociocultural factors (McQueen and Siegrist, 1982) will be of primary importance. Responses to breakdown processes which have a psychological component will also be of interest. In the past these factors have been subsumed under emotional or psychosomatic aspects of illness and illness behaviour. Broadly speaking if our aim is to establish reliable cause and effect relationships between the psychosocial factors of interest and relevant outcome measures (reversible and irreversible structural and functional changes, symptoms of disease and illness) we will wish to know the requirements for inferring cause. An eminent British epidemiologist Sir Austin Bradford-Hill has provided a list of criteria by which to judge the likelihood of a causal effect being present. Included in his list are strength of association, consistency, specificity, existence of relationships in time, presence of dose-response effects, coherence and biological plausability. In only a small number of the studies which ostensibly show cause and effect relationships between psychosocial factors and disease outcome have these criteria of validity been met.

How is failure to adapt manifested? The scope of such manifestations obviously defines the legitimate field of interest and, in particular, has implications for the choice of measurement and assessment techniques. A long term outcome may be disease. One epidemiologist has provided a working definition of disease as 'disability or failure in performance of a task' (Kagan, 1975). Structural impairments at the cellular level obviously have implications for organ dysfunction. Individual effectiveness or performance may also be degraded. Human reliability, human error and safety are therefore legitimate concerns. Indeed, there are empirical links between the experience of stressful life events and accidents (Connolly, 1980; Sheehan et al, 1981 Levenson et al, 1980). Certainly the quantification of human reliability presents measurement problems at least as difficult, if not more so, than those processes and outcomes concerned with 'disease'. (Swain and Guttman, 1980; Rassmussen, 1979; Moray, 1980). There exists a theoretical link between demanding life events, human reliability and psychological illness. Empirical links are, at the present, tenous but nevertheless promising (Broadbent et al 1982, Stuart and Brown, 1981). However these methodological issues are not specifically addressed in the present paper.

The wider context

The study of the extent to which psychosocial factors can elicit disease and illness cannot be seen in isolation. Their presumed causes and effects must be evaluated against other, and perhaps more reliably established, risk factors. In their recent review commissioned as a report to the Office of Technology, US Congress, concerned with determining cancer risks from the environment, Doll and Peto (1981) discussed 'unidentified' causes of the disease. They state:

> 'Two categories of environmental factors that we have ignored and may therefore be classed with "unknown factors" are that of psychological stress and that of some form of breakdown of immunological control, both of which have been suggested at intervals throughout this century to play some part in the production of cancer. It is possible, of course, that psychological factors could have some effect, eg by modulating hormonal secretions, but we know of no good evidence that they do nor that they affect the incidence of cancer in any other way, except insofar as they lead people to smoke, drink, overeat, or enjoy some other harmful habit'.

A similarly sceptical approach was taken recently by the Industrial Injuries Advisory Council (1981) in discussing occupational causation of illness and disease. The fact that Ischaemic Heart Disease (ICD Code 410-414) is excluded from the revised schedule of industrial diseases is not of particular relevance to our present concern except insofar as in reviewing the evidence they appear to be similarly unconvinced of the relevance of existing studies:

> 'Coronary disease is our leading cause of death (156,000 deaths in 1979), and was responsible for 185,000 spells of certified incapacity for work in the year 1978/79. It is quite commonly believed by the general public that stress at work can be responsible for the disease, but the epidemiological evidence in support of this belief is slender and still controversial. Far more important

components of its complex aetiology are
cigarette smoking, diet, lack of exercise,
raised blood pressure and obesity'.

The significance of these statements lies not in their unwillingness to consider purely psychological factors (relating to recognition perception and personality, for example) but in the erroneous conclusion that these (so-called) well-established risk factors are medical in nature, whereas clearly they are largely behavioural.

Although other policy making bodies have clearly recognised the need to consider psychosocial aspects of health and diseasé, for example the recent WHO Working Group on Health Aspects of Wellbeing in Working Places (WHO, 1980), the overall impact of research findings in this area has been generally of a low order. Successful intervention strategies (early identification of risk factors in the environment or, those at an elevated risk, job redesign, policies directed at social change) will depend upon the reliable interpretation of research findings. This in turn will depend upon sound methodology and it is in this area that problems have arisen in the past.

General methodological problems

Methodological development work is often unglamorous and unappealing. It is also time consuming. Also the fact that it often pays dividends in the long run is not for most sufficient motivation. Many of the methodological issues in stress research and in identifying psychosocial risk factors have been discussed elsewhere. Nevertheless they are crucial and therefore worthy of reemphasis. In the discussion that follows three general areas are covered: research settings, research designs, and measures. The third part of the paper offers some general recommendations.

Settings

The investigator wishing to study BHA (Breakdown in Human Adaptation) is faced with the choice between a number of contexts in which the study or investigation can be set. Four general types of setting generally appropriate for studying BHA can be distinguished; laboratory experimentation, simulations, field studies and clinical investigations. Each has both unique and shared advantages and disadvantages. Often the choice of setting is determined by availability of resources, expertise and personal preference. In many programmes of research only a single setting is used and typically in the area of BHA field studies have tended to dominate. The relative advantages and disadvantages of each approach has been extensively discussed with the respect to the literature on stress (Appley and Trumbull, 1967; McGrath, 1970). In general field studies tend to maximise realism but with a cost in precision and a threat to aspects of internal validity laboratory studies tend to maximise precision at a cost to realism and pose a threat to external validity.

Laboratory Studies

Two particular drawbacks are associated with laboratory studies of BHA. The first is the considerable limitation in both the severity and duration of imposed demands. Ethical considerations predominate although there are other difficulties. The second drawback concerns the problems of realism. Since current theories of BHA emphasise the perception and cognitive appraisal of demands, the ability of laboratory situations to successfully mimic perceived threat of any substantial degree must always be questioned. The necessarily short duration of experimental situations also severely limits the extent to which the temporal factors in the mechanisms which underlie BHA can be established. All of these drawbacks are well known. Nevertheless, clever experimental designs and manipulations can largely overcome some of these drawbacks. Moreover they have a number of advantages over other approaches. Random allocation of experimental units to conditions and systematic data collection means that all possible interaction between variables of interest can be established and reliable dose response relationships achieved. Within the present context, thoughtfully conducted and designed laboratory studies have an important role. Analogues of job characteristics such as task complexity or theoretical concepts such as 'underload' and 'overload' are particularly appropriate for laboratory investigation. The psychophysiological studies of Marrianne Frankenhaeuser and her colleagues is an example of a laboratory approach that can provide extremely valuable information (Frankenhaeuser, 1980). Studies involving the manipulation of physical environmental characteristics in exposure chamber studies (such as noise) are obviously relevant here. So are the laboratory analogues of concepts such as 'learned helplessness' in which experimental studies have played a major part (Seligman, 1975). Perceptions of imposed demands can also be manipulated. Laboratory studies of aversive stimulation (Averill, 1973;

Glass and Singer, 1972), where perception of control is manipulated, have proved particularly powerful techniques in examining theoretical constructs such as cognitive appraisal of threat.

Task demands have also been investigated with respect to subjective evaluations of workload. A laboratory study which differentiates between objective and perceived levels of demand has been described by Sales (1969). Using an anagram-solving task, Sales manipulated two conditions of objective workload: an underload condition in which subjects were kept waiting for anagrams for approximately 30% of the time, and an overload condition where 3.5% more anagrams were provided than could be decoded in the time allowed. Subjects were divided into those who reported high levels of subjective workload and those who reported low subjective workload. Data for two of the dependent variables are of relevance here: (a) reported interest in, and enjoyment of the task, and (b) changes in serum cholesterol. On both these variables an intereaction between subjective and objective workload occurred. Subjects who received the overload condition but reported low subjective workload and those in the underload condition who reported feeling overloaded both reported high levels of interest and enjoyment in the task. Those given the overload conditions and who felt overloaded and those given the underload conditions and who felt underloaded both had low levels of interest in the task. These latter groups also showed increases in serum cholesterol. The former groups, who showed greater interest and enjoyment in the task, exhibited decreases in serum cholesterol. Sales further analysed these data and found that levels of subjective overload were negatively correlated with scores on the verbal section of the Scholastic Apitude Test (SAT). Individuals with high levels of verbal ability reported low subjective workload, whilst high subjective workload was reported by those with relatively low verbal ability. A negative correlation between interest in the task and changes in serum cholesterol was also found. These data seem to be in general agreement with the contention that the effects of environmental demands (in this case from a task) are mediated foremost by perceptual factors. One of the primary determinants of the occurrence of psychological (interest in task) and physiological (serum cholesterol) correlates of stress is that of the individual's capability to deal with demand. These examples indicate that carefully conceived laboratory studies are of relevance in studies of adjustment. The investigation of task-inherent demands appears particularly appropriate for laboratory settings.

Experimental simulation

Experimental simulation appear to combine the precision demanded by laboratory studies with the realism provided by field studies but at the same time minimising some of the threats to the latter such as non-random allocation of experimental units and unsystematic data collection. However McGrath (1970) has highlighted three potential weaknesses of simulation strategies. First is the problem of fidelity or realism. Although simulators minimise the contrived nature of laboratory situations the extent to which they can mimic real-life situations, influence cognitive appraisals and hence appropriate levels of perceived threat is not clear. Similarly the extent to which motivation and observed behaviour are representative of actual conditions is difficult to assess. The second problem is one of resources, particularly cost. This in turn will depend very largely upon what is being simulated. Experience suggests that experimental simulation of a moderately complex system, yielding multiple outcome measures, with a number of variables and subject replications may be the most expensive strategy option. The third problem is the danger of over-concern with attempts to increase the realism of the simulation at the expense of data collection.

Nevertheless, laboratory simulations of stressful situations can be worthwhile. The present author has been involved in a series of studies at Nottingham concerned with the psychophysiological response to repetitive work (Mackay et al 1979; Cox and Mackay 1979; Cox et al 1982). These studies were based upon a simulation of a variety of repetitive tasks together with industrial field surveys of repetitive working practices and the responses to them. Two types of simulation were employed. The first was concerned with the simulation of paced repetitive work tasks. To create a realistic 'shopfloor' environment, a small workshop (approximately $90m^2$) was established, housing a conveyor belt. Using this belt, a machine-paced button sorting process was developed, which involved three separate tasks operating in sequence. The tasks, loading, sorting and minding, were chosen to reflect and combine various aspects of unskilled and semi-skilled industrial work. Loading and sorting were treated as 'repetitive' tasks and minding as a non-repetitive "at-work" control. Buttons were chosen as convenient material for this experiment: the project was not specifically concerned with work in the button manufacturing industry.

The first task in the sorting process was loading. This consisted of placing buttons, face upwards, onto the moving belt at predetermined rates. Black discs were marked at regular intervals (372 mm) towards the front of the belt as targets onto which the buttons had to be accurately placed. As the buttons passed down the belt, a second person was required to inspect them for faults: sorting. Faulty buttons were rejected by being pushed to the back of the belt. The operational definition of sorting was the detection and rejection of faulty buttons: 10% of all the buttons used had faults. The third task was machine-minding. The minder was required to sit at the end of the belt and watch over the collection of the buttons, tidying up any spillage, and reporting any breakdowns.

The repetitive tasks varied in two ways: in the level of activity of the motor response, and in the degree of attention or in the attentional strategy demanded. Loading required a more regular and active response than sorting where an active response was related to the detection of a faulty button. Although the minder's task appeared the least stimulating, unlike the other two it had no repetitive components, and it was less constrained by response requirements. All three tasks shared the same physical environment.

A period of training (15 min) was given on both repetitive tasks, to enable the loader to adapt to the speed of the belt and to the required pace of working, and to familiarize the sorter with the different faults present in the button used. Training appeared to overcome initial practice effects (Cox et al, 1981). Following the short training period, participants worked for two more 45 min periods during the afternoon. Participants were employed for 1 week, and worked for 1 day at each task (Tuesday to Thursday). Subjects were recruited through advertisements placed in the local press, which offered temporary unskilled work. Those selected were women, whom by self-report were in a good state of general health. Background information about each subject was obtained before work started.

This simulated working arrangement proved to be very useful for studying the effects of different types of repetitive task and for manipulating factors of interest notably degree of pacing and type of payment method, and for studying the psychophysiological responses of interest (urine catecholamines, continuously monitored cardiac activity, self-reported stress and arousal).

The second type of simulation carried out at Nottingham took a slightly different approach. This involved three women recruited from advertisements in the local newspaper working on a part-time basis for three months. An industrial task supplied by a local electronics Factory formed the basis for the work. It involved the assembly of simple components using a small hand operated press. The task was self-paced but each worker was required to assemble several hundred finished articles each day which would be checked for quality and dispatched to the factory. This type of approach whilst being based on only a few subjects allows moderately long-term adjustment to work to be investigated. Also by taking multiple measures from the same individual over many weeks, investigation of individual differences in pattern of responding can be assessed. Thus, both self-reported Stress and self-reported Arousal showed long term changes indicative of a gradual adjustment to the work routine (Mackay et al, 1978, Mackay, 1980).

Field Studies

Because BHA is essentially a complex multifactorial process concerned largely with chronic effects, field settings are often the only ones in which such phenomena can be studied. One is typically examining systems that are fundamentally dynamic. Individuals are growing and changing in interaction with one another and with the environment even in the absence of an imposed intervention. The complex interplay and links between environmental demands (life-events), individual perceptions of threat, conflict and harm, and their impact on pathophysiological processes cannot be satisfactorily handled by either experimental laboratory or simulation approaches. Nevertheless, good field studies are difficult to undertake.

Disadvantages with field studies are those concerned with procedural aspects and measurements problems. Often access to desired populations is difficult if not impossible, and the time and effort needed for this necessary part of field investigation should not be underestimated. Time to interview individuals or administer questionnaires is often linked to, for example, production constraints. In community or domestic settings this may be less of a problem. Because many of the available tests (see Section 3) have been designed for those with above average comprehension and reading ability they are often not appropriate for the lower socieconomic groups. Careful modification and piloting are made even more difficult if only limited access is possible. Moreover there is often a great desire to collect more data from the respondent than is possible within a given data collection session. Often the data are collected haphazardly and unsystematically. Analysis and interpretation are made even more difficult particularly when there are large numbers of missing data (physiological data are particularly at risk). The valid inferences which the investigator wished to make about cause and effect relationships cannot be made.

Study Designs

The discussion of appropriate study settings leads directly to issues concerned with study design. In many cases the choice of setting dictates the type of study design employed. Further, there are often other constraints on resources which hinder the investigator. When considering design decisions with respect to studies of BHA there are a number of key issues. The general principles of experimental design are well known to psychologists and epidemiologists and will not be discussed here except to emphasise that control is achieved by randomised assignment to treatments. Other generic design methods are available. Cook and Campbell (1979) have placed special emphasis on quasi-experimental approaches. These are experiments that have treatments, outcome measures and experimental units but do not use random assignment to create the comparisons from which treatment caused change is inferred. Instead, the comparisons depend on nonequivalent groups that differ from each other in many ways other than the presence of a treatment whose effects are being tested. Cook and Campbell state:

> The task confronting persons who try to interpret the results from quasi-experiments is basically one of separating the effects of a treatment from those due to the initial noncomparability between the average units in each treatment group: only the effects of the treatment are of research interest. To achieve this separation of effects, the researcher has to explicate the specific threats to valid causal inference that random assignment rules out and then in some way deal with these threats. In a sense, quasi-experiments require making explicit the irrelevant causal forces hidden within the ceteris paribus of random assignment.

There are a number of threats to the validity of such studies including differential learning, ageing and maturation effects. A further type of study which relies upon passive observation is also widely employed. Here, <u>passive</u> refers to the non-intrusive, non-interventionist nature of the study. Typically the design is a cross-sectional one with all the measures taken at the same time (although some may refer to past events in a retrospective sense). Cook and Campbell (1979) suggest:

> 'For all these techniques, it was assumed that statistical controls were adequate substitutes for experimental controls and that the functions served by random assignment, isolation, and the rest could be served just as effectively by passive measurement and statistical manipulation. The belief became widespread that random assignment was not necessary because one could validly conceptualize and measure all of the ways in which the people experiencing different treatments differed before the treatment was ever implemented. Also one could rule out any effects of such initial group differences by statistical adjustment alone. Similarly some researchers believed that all extraneous sources of variation in the dependent variable (that isolation and reliable measurement largely deal with) could be conceptualized, validly measured, and then partialled out of the dependent variable'.

Descriptive occupational epidemiology, which falls into this type of passive observational design can seldom offer definitive causal interpretations, and health differences in occupational groups cannot be simply interpreted by a posteriori intuitive judgments about the stressfulness of various occupations since later evidence may not quite support intuition (Kasl, 1978). Many occupational studies attempt to determine morbidity and mortality differences between groups of workers and then search for environmental factors in the workplace which may explain the differences. An alternative approach is to examine a population exposed to a hazardous agent (an organic solvent, a suspected carcinogen) and then to seek health differences between the exposed group and a non-exposed reference group. With some agents and observed diseases the causal link is comparitively easy to make (asbestos and mesothelioma, nickel carbonyl and nasal carcinoma). With others inferring a causal link with any degree of reliability proves difficult. The main reasons are well known, the most powerful include selection effects both into and out-of jobs particularly because of health effects; illness behaviour; difficulty of determining duration and degree of exposure; and ascribing observed occupational effects to occupational causes where in fact non-occupational factors are at play. Such confounding factors are legion. Even when occupational factors are known to play a causal role in differential occupational mortality and morbidity rates there may often be competing explanations.

An illustrative example of this type of study design is the study of occupational morbidity rates of mental health disorders carried out by Colligan, Smith and Hurrell (1977). It must be emphasised however that the investigators themselves were aware of the limitations of this approach. The aim of the study was to try to provide a basis for identifying and selecting specific occupations which may be at an elevated risk of mental health problems because of job stress. Using admission records of community health centres throughout the state of Tennessee (1972-1974), these authors were able to assess the incidence rate of diagnosed mental health disorders for 130 occupations. Occupations chosen were those employing 1000 or more individuals. Colligan and co-workers based their evidence upon the sample frequency of admission for each occupation relative to the working population frequency of that occupation. It was shown that health technicians exhibited the highest incidence of mental disorders, closely followed by waiters and waitresses, practical nurses and inspectors. A second analysis was carried out on the basis of a comparison of the observed frequency of admissions per occupation in the sample with the expected frequency based upon population (census) data. Those with scores significantly exceeding what would be expected by chance were waiters and waitresses, labourers and operatives, practical nurses, secretaries, nurses' aides and inspectors. The authors rightly point out that inferences regarding the causal direction of the relationship between occupation and incidence of mental health disorders are not justified by this type of study design. For example of the top 22 occupations six were related to hospital/health care operations. Whilst it is clear that there may be particularly severe stressful demands in the work of nurses (Parkes, 1982) other explanations of the observed differences in incidence rates are equally likely. Thus, sex sampling bias, willingness to report mental illness by health care professionals, earlier detection and treatment seeking, and predisposition to, or prexisting mental health problems are equally plausible explanations. Thus the obtained occupational incidence rates may be more reflective of occupational differences in reporting tendencies and accessibility to treatment than actual incidence of mental health disorders.

Examples of the pitfalls in retrospective, cross sectional designs are particularly common in the studies of psychosocial risk factors in the aetiology of cancer. These studies have recently been reviewed by Cox and Mackay, (1982). There appear to be several difficult methodological problems associated with studies on cancer; some unvoidable, others less so. In 1979, a Lancet editorial discussed some of these. All studies are difficult because of the likely multifactorial aetiology of the different types of cancer, and because of the 'relatively' low incidence of the disease. Hagnell's (1966) 'true' prospective study of 2500 Swedes provided only 42 cases of cancer after 10 years, while Thomas's study in America, begun in 1948, had yielded 55 cases by 1979 (Thomas & McCabe 1980). Furthermore, it is far from clear when the cancer process begins, and the disease may have been developing for many years before it becomes obvious and is diagnosed. Diagnosis itself is not a 'biologically' significant event, although it may be of critical importance psychologically. Some apparently cancer-free individuals, both in prospective and cross sectional studies, may be undiagnosed cancer cases. The process of diagnosis is very obvious to most people, and the perceived consequences of a positive diagnosis are such that the process itself must be anxiety producing. Laxenaire and his colleagues (1971) have described intense anxiety and suicidal preoccupations (without any reported suicide attempts) as the initial response to a positive cancer diagnosis. Further more, Craig and Abeloff (1974) report high levels of depression in one half of their cancer patients, and elevated anxiety levels in one third. Part of the cancer patients response must be related to general feelings about the diseases; cancers are an emotional subject within the population as a whole; more so than many other currently researched diseases. This does not make scientific study easy. Most of the problems discussed above apply to both cross sectional and prospective studies, but to varying degrees. Finally, there is the basic statistical caution that associations (correlations) do not imply causality. This point applies to data analysis and interpretation in all types of study, but also represents a design weakness in cross sectional studies.

Three types of 'clinical' study appear to exist. First, many early studies were cross sectional (retrospective), have typically concerned behavioural style or personality, and are greatly influenced in their interpretation by the fact of the disease being known to the patient group. They have either compared the past experiences and behaviours of different patient and control groups, or examined their present behaviours and attitudes. Some studies have recognised and discussed their limitations, but many imply some direction of effect (causality) where none is warranted. What these studies may well describe is not the role of behavioural style in the aetiology of cancer, but the impact of cancer diagnosis and the disease on behavioural style. Retrospective study designs, particularly in an area such as cancer where lifetime events need to be recalled accurately, are therefore subject to a number of biases and contaminations. Longitudinal prospective designs, whilst not free from all methodological faults are much more powerful in allowing causal relationships to be established. See Grossarth-Maticek et al (1982) for a recent example of a prospective study design in this area. Generally, Kasl (1978) has attacked 'the plethora of hopeless cross-sectional studies which attack extremely complex issues with the weakest of research designs; they are bound to yield almost uninterpretable findings (eg Ferguson, 1974; Shirom et al, 1973; Sinassi et al, 1974). in spite of the extensive effort they may go into collecting the data'.

Measures

The term measures refers to the class of psychological variables of interest to investigators of BHA. A number of attempts have been made to perfect a taxonomy to classify such variables (eg McGrath, 1970; Webb, 1966). None has been entirely successful. One basic approach, which emphasises the temporal nature of the breakdown process, has been to classify variables into three broad groupings: antecedent variables, concurrent variables and outcome measures. Antecedent variables for example include those such as personality traits which may enhance or buffer the susceptibility to stressful demands. Individual levels of skill and training may also be included here. Concurrent variables comprise those environmental events which may impinge on the person during a given time period. They may be subjectively or objectively defined. In many studies they are the main 'independent' variables of interest such as job demands, life events, social changes. Other concurrent variables include environmental resources such as social support mechanisms which may help to attenuate the effects of environmental demands. Outcome measures are those health related variables which are hypothesised to reflect the possible bodily responses to stressful demands (sometimes called stress indices or stress responses). Again a number of categorisations of such responses has been made. Thus Margolis and Kroes (1973) have suggested five dimensions of job-related stress reaction.

1. Transient short-term subjective reactions occurring in close temporal proximity to specific stresses. Examples of these may be feelings of anger, fear, tension and anxiety. Typically these are quantified by the use of checklists and rating scales (Mackay, 1980).
2. More chronic psychological responses that have become part of the individual's health status, rather than a reaction to specific work situations or events. These can be general malaise, constant fatigue, chronic depression or alienation. The General Health Questionnaire (GHQ, Goldberg, 1972) and the Middlesex Hospital Questionnaire (MHQ, Crown and Crisp, 1966) have been extensively used in the UK.
3. Transient clinical-physiological changes such as alterations in levels of catacholamines, blood lipids, blood pressure, gut motility etc, which may indicate in an 'objective' sense that the individual is under stress. These changes may also be the precursors of psychosomatic or physical illness.
4. Symptoms of physical and/or psychosomatic illness such as coronary heart disease, gastro-intestinal disorders, etc.
5. Decreasing work performance. Increased error rates and increased incidence of accidents.

In order to illustrate some of the issues concerned with the utility and validity of such response measures, examples of the first two response dimensions will be briefly described.

Transient short term subjective reactions : measures of mood.

One aspect of affective experience which has not been widely used in the stress literature is that of mood, although there is considerable anecdotal evidence that gross fluctuations in mood often follow life crises and stressful experiences in general. Much of the early work on the measurement of mood has been performed by Vincent Nowlis and colleagues (eg Nowlis and Green, 1957). An aspect of the mood-adjective checklist (MACL) which measured activation-deactivation was subsequently defined by Thayer (1963, 1967) into the Activation-Deactivation-Adjective Checklist (AD-ACL). This instrument consists of adjectives which are grouped into four separate, independent monopolar factors, each measuring an aspect of self-reported activation.

In a number of studies the present author (Mackay et al 1978; Mackay, 1980a) has examined the usefulness of this approach to self-reported arousal. These studies showed that whilst it was possible to approximate to Thayer's four-factor solution, the analyses described above clearly revealed the existence of two bipolar factors. The first was a combination of the two original monopolar factors High Activation and General Deactivation, and similarly, the second factor was a combination of General Activation and Deactivation Sleep. Based upon other re-analysis of Thayer's original checklist and the composition of the present factors these have been labelled STRESS and AROUSAL respectively.

A series of laboratory and field studies have been undertaken which have underlined the reliability and validity of self reported arousal measures. Laboratory studies have shown self-reported stress and arousal to be differentially sensitive to a range of noise and task effects and correlate substantially with biochemical changes (blood glucose) (Mackay, 1980b). Simulation and field studies have shown it to be a particularly useful measure of responses to repetititve work (Mackay et al 1979, Cox et al 1982). A clinical study by Ray and Fitzgibbon (1981) has shown that self-reported stress and arousal taken before surgery are correlated with subsequent measures of adjustment.

Symptoms of psychoneurotic illness

The need for a rapid and reliable method of quantifying levels of psychoneurotic illness arises both in clinical practice and in epidemiological research. In order to meet these requirements a number of self-administered symptom rating scales have been developed, including the Hopkins Symptom Checklist, (SCL) (Derogatis et al, 1974), General Health Questionnaire, (GHQ) (Goldberg, 1972), and the Middlesex Hospital Questionnaire (MHQ), (Crown and Crisp 1979). The last of these is an inventory composed of items designed to measure clinically recognised psychoneurotic illness and includes items referring both to illness and to personality, (Crisp, Gaynor-Jones and Slater, 1978). On the basis of clinical experience Crown and Crisp (1966) derived a series of sub-scales intended to measure a range of psychoneurotic and affective disorders. The six sub-scales, which each consist of eight questions, are as follows: Free Floating Anxiety (FFA), Phobic Anxiety (PHO) Obsessionality (traits and symptoms) (OBS), Somatic concommitants of anxiety (SOM), Depression (DEP) and Hysterical personality traits (HYS).

Several studies have investigated profiles of the six sub-scales in the general community (Crisp and Priest, 1971; Crisp, McGuinness and Harris, 1978) and in an industrial population (Crown, Duncan and Howell,

1970; Howell and Crown, 1971). Recent studies by the present author and his colleagues using a postal questionnaire approach have found the MHQ to be an appropriate instrument for assessing profiles of psychological illness in a general population (Alderman et al, 1983). Following a factor analysis study of responses from 8000 industrial workers the authors made the following conclusions. First, principal components factor analyses of the MHQ results in solutions which reflect the item composition of the original six sub-scales. Second, where differences do occur, they involve minor transpositions of items between sub-scales which, it is suggested, maintain reliability without adversely affecting validity. The third point is that the wording of some of the items may not be wholly appropriate for studies of working populations, for example those relating to a fear of going out or sleep patterns (particularly in shift workers). Some researchers have used a revised form of the MHQ for such studies (Broadbent and Gath, 1979), but rather than suggest the need for a revised version it would seem more appropriate for potential users to be aware of the small differences between the factors of the principal components solution and the sub-scales of the MHQ. Fifth, there appears to be two components of the obsessionality sub-scale one referring to symptoms and one to personality. These two scales show different patterns of correlations with other measures of response to the working environment.

GENERAL CONCLUSIONS

This general area of research and particularly the use of psychological methods presents a number of methodological problems. Some of these are shared with other branches of epidemiology. There are however some unique quantification difficulties. Many of these revolve around the use of subjective techniques aimed at tapping the perception and appraisal of stressful situations. The use of spouse or co-worker reports and the use of 'consensus' or group ratings or judgments may be useful in supplementing existing self-report techniques with investigations into the factors which determine the mapping of objective situations into subjective perceptions of them. A problem of the conceptual overlap between measures of demands and measures of responses exists. Item overlap between outcome measures may also occur (measures of job satisfaction and symptoms which refer to job characteristics; life events which refer to illness and subsequent illness reporting) General response bias and item overlap lead to 'inflated' correlations. Problems of poor recall and the phenomenon George Brown had called 'retrospective contamination' may reduce reliability further.

The plea for investigators to seek out opportunities for studying 'natural experiments' has been made before (Kasl, 1978) but is worth re-emphasising. Some occupational settings may also provide the opportunity for further control over non-random effects as in the study of Parkes (1982). Training situations where AB, BA allocation of trainees to jobs occurs are particularly appropriate. Job rotation is a further example. Where job-redesign is taking place, some experimental interventions may be possible if not too evasive? Such studies would also benefit from the use of multiple outcome measures (say self-report of symptoms and moods, psychophysiological measures, medical wastage).

REFERENCES

Alderman, K.J., Mackay, C.J., Lucas, E.G., Spry, W.B. and Bell, B. (1983). Factor analysis and reliability studies of the Crown-Crisp Experiential Index (CCEI). Brit. Journal. Medical Psychol. (in press).

Appley, M.H. and Trumbull, R. (1967). Psychological Stress. New York. Appleton.

Averill, J.R. (1973). Personal control over aversive stimuli and its relation to stress. Psychological Bulletin, 80, 286-303.

Broadbent, D.E., Cooper, P.F., Fitzgerald, P. and Parkes, K.R. (1982) The Cognitive Failures Questionnaire and its correlates. Brit. J. Clinical Psychol. 21, 1-16.

Broadbent, D.E. and Gath, D. (1979). Chronic effects of repetitive and non-repetitive work. In C.J. Mackay and T. Cox (eds). Response to stress: Occupational aspects. Guildford: IPC Science and Technology Press.

Colligan, M.J., Smith, M.J. and Hurrell, J.J. (1977). Occupational Incidence rates of mental health disorders. J. Human Stress. 3, 34-42.

Connlly, J. (1982). Life stress before accidents. J. Psychosomatic Medicine (in press).

Cook, T.D. and Campbell, D.T. (1979). Quasi-Experimentation: Design and Analysis Issues for Field Settings. Chicago. Rand-McNally.

Cox, S., Cox, T., Thirlaway, M. and Mackay, C.J. (1982). Effects of simulated repetitive work on urinary calecholamine excretion. Ergonomics, (In press).

Cox, T. and Mackay, C.J. (1982). Psychosocial factors in the aetiology and development of cancers, Soc. Sci. Med. 16, 318-396.

Cox, T., Mackay, C.J. and Page, H. (1982). Simulated repetitive work and self-reported mood. J. Occup. Behaviour, 3, 247-252.

Craig, T.J. and Abeloff, M.D. (1974). Psychiatric symptoms among hospitalized cancer patients. Am. J. Psychiat. 131, 1323-1327.

Crown, S. and Crisp, A.H. (1966). A short clinical diagnostic self-rating scale for psychoneurotic patients. The Middlesex Hospital Questionnaire. (MHQ). British Journal of Psychiatry. 112, 917-923.

Crown, S. and Crisp, A.H. (1979). Manual of the Crown-Crisp Experiental Index. London: Hodder and Stoughton.

Crown, S., Duncan, K.P. and Howell, R.W. (1970). Further evaluation of the MHQ. British Journal of Psychiatry. 116. 33-37.

Crisp, A.H., Gaynor Jones, M. and Slater, P. (1978). The Middlesex Hospital Questionnaire: A validity study. British Journal of Medical Psychology, 51, 269-280.

Crisp, A.H. and McGuinness, B. (1976). Jolly fat: relation between obesity and psychoneurosis in general population. British Medical Journal, 1, 7-9.

Derogatis, L.R., Lipman, R.S., Rickels, K., Uhlenhuth, E.H. and Convi, L. (1974). The Hopkins Symptom Checklist (HSCL). In P. Pichot (ed) Modern Problems in Pharmacopsychiatry. vol. 7, pp 79-110 Basel: Karger.

Doll, R. and Peto, R. (1981). The causes of cancer. Oxford. The University Press.

Ferguson, D. (1973). A study of neurosis and occupation. Brit. J. Indust. Med. 30, 187-198.

Frankenhaeuser, M. (1980). Psychobiological aspects of life stress. In: Levine, S. and Ursin, H. Coping and Health. New York. Plenum.

Friedrichsen, A. (1980). The Shorter Oxford English Dictionary. Oxford. Oxford University Press.

Glass, D.C. and Singer, J.E. (1972). Urban Stress. Experiments on noise and social stressors. New York. Academic Press.

Goldberg, D.P. (1972). The detection of psychiatric illness by questionnaire. Institute of Psychiatry. Maudsley Monographs. Oxford. The University Press.

Gossarth-Maticek, R., Siegrist, J. and Vecter, H. (1982). Interpersonal repression as a predictor of cancer. Soc. Sci. Med. 16, 493-498.

Hagnell, O. (1966). The premorbid personality of persons who develop cancer in a total population investigated in 1947 and 1957. Ann. NY Acad. Sci. 125, 846-55.

Howell, R.W. and Crown, S. (1971). Sickness absence levels and personality inventory scores. British Journal of Industrial Medicine, 28, 126-130.

Industrial Injuries Advisory Council (1981) Industrial Diseases: A review of the Schedule and the Question of Individual Proof. DHSS. London. HMSO.

Kagan, (1975). Epidemiology, disease and emotion. In Levi, L. (Ed) Emotions - their parameters and measurement. New York. Raven Press.

Kasl, S.V. (1978). Epidemiological Contributions to the study of work stress. In: Cooper C. and Payne, R. Occupational Stress. Wiley, Chichester.

Laxenaire, M., Chardot, C., Bentz, L. Quelques aspects of psychologiques du malade cancereux. La Presse Medicale 79, 2497-2500, 1971.

Levenson, H., Hirschfield, M.A. and Hirschfield, A.H. (1980). Industrial accidents and recent life events. J. Occup. Med. 22, 53-57.

Levi, L. (1974). Stress, distress and psychosocial stimuli. In McLean, A. (Ed.) Occupational Stress. Charles C. Thomas. Springfield. Illinois.

Lumsden, D.P. (1975). Towards a systems model of stress: feedback from an anthropological study of Ghanas Volta River project. In Sarason, I.G. and Spielberger, C.D. (eds). Stress and Anxiety Vol. 2. Hemisphere, New York.

McGrath, J.E. (1970). Major methodological issues. In: McGrath, J.E. (Ed) Social and Psychological Factors in Stress. New York. Holt, Rinehart and Winston.

McQueen, D.V. and Siegrist, J. (1982) Social factors in the etiology of chronic disease: an overview. Soc. Sci. Med. 16, 353-367.

Mackay, C.J. (1980a). The measurement of mood and psychophysiological activity using self-report techniques. In: Martin, I and Venables, P.H. (eds). Techniques in Psychophysiology, Wiley, Chichester.

Mackay, C.J. (1980b). Some experiments concerning the interaction between noise exposure and glucose loading on psychopysiological state. Unpublished Doctoral Thesis. University of Nottingham.

Mackay, C.J., Cox, T., Burrows, G.C. and Lazzerini, A.J. (1978). An inventory for the measurement of self-reported stress and arousal. Brit. J. Soc. Clin. Psychol. 17, 283-284.

Mackay, C.J., Cox, T., Watts, C., Thirlaway, M. and Lazzerini, A.J. (1979) Physiological correlates of repetitive work. In. Mackay, C.J. and Cox, T. (1979). Response to Stress. Occupational Aspects. Guildford. IPC.

Margolis, B.K., and Kroes, W.H. (1973). Occupational Stress and Strain. Occupational Mental Health, 2, (4).

Nowlis, V. and Green, R.F. (1957). The experimental analysis of mood. Technical Report No 3. Office of Naval Research. Contract No. NONR-668(12).

Parkes, K.R. (1982). Occupational stress in student nurses : A natural experiment. J. Applied Psychol. (in press).

Ray, C. and Fitzgibbon, G. (1981). Stress, arousal and coping with surgery. 11, 741-746.

Sales, S.M. (1969). Differences among individuals in attentive, behavioural biochemical and physiological responses to variations in workload. Unpublished PhD Thesis. University of Michigan.

Seligman, M.E.P. (1978). Helplessness: On depression, development and death. San Francisco. Freeman.

Stuart, J.C. and Brown, O.M. (1981). The relationship of stress and coping ability to incidence of disease and accidents. J. Psychosom. Research, 25, 255-260.

Sheehan, D.V., O'Donnell, J.O., Fitzgerald, A., Hervig, L. and Ward, H. (1981-82). Psychosocial predictors of accident/error rates in nursing students: a prospective study. Int'l. J. Psychiatry in medicine, 11, 1981-82.

Shirom, A., Eden, D. Siberwasser, S. and Kellerman, J.J. (1973). Job stress and risk factors in coronary heart disease among five occupational categories in a Kibbutzim, Soc. Sci. Med., 7, 875-892.

Sinassi, I., Crocetti, G., Spiro, H.R. (1974). Loneliness and dissatisfaction in a blue collar population. Arch. Gen. Psychiat., 30, 261-263.

Swain, A. and Guttman, R.S. (1980). Handbook of human reliability analysis with emphasis on nuclear power plant applications. Sandia Labs. Nuclear Regulatory Commission NUREG/CR 1278.

Thayer, R.E. (1963). Development and validation of a self-report adjective checklist to measure activation-deactivation. Unpublished Doctoral Dissertation. University of Rochester.

Thayer, R.E. (1970). Activation states as assessed by verbal report and four psychophysiological variables. Psychophysiology, 7, 86-94.

Thomas, C.B. and McCabe, O.L. (1980). Precursors of premature disease and death: Habits of nervous tension. The John Hopkins Medical Journal, 147. 137-145.

Webb, E.J., Campbell, O.T., Schwartz, R.D. and Sechrest, L. (1966). Unobstrusive measures, Skokie. Rand McNally.

World Health Organisation (1980). Health aspects of well-being in working places. Regional Office for Europe, EURO Reports and Studies 31.

MONITORING SIGNS OF DECREASE IN HUMAN ADAPTATION

- Use of quantitative measures available in official statistics

Töres Theorell, National Institute for Psychosocial Factors and Health, Box 60210, S-104 01 Stockholm, Sweden.

Summary:

The understanding of "breakdown of human adaptation" is greatly facilitated by access to good statistics. The Scandinavian experiences are given as examples and a critical evaluation is made.

Introduction:

An increased knowledge of means of monitoring signs of decreasing human adaptation is important in a period of economic crisis and threat of war. In several European countries vast sources of statistical information are available. They are not used to the extent that they deserve. In the following presentation I shall try to discuss limitations as well as utility and validity in some common sources of statistics available in many European countries. I shall also make some proposals regarding new uses of combinations of such sources.

In many instances, a social breakdown may be the first sign that something is going wrong, and in other instances distorted psychological defence mechanisms may cover a social breakdown which may become manifest much later as a biological breakdown (perhaps the extreme type A behaviour is such a mechanism ?). I think we should bear these thoughts in mind when we start discussing uses of official statistics available in most European countries. Figure 1 shows the theoretical interaction between biological, social and psychological breakdown.

Death Statistics:

Despite the fact that statistics of total deaths may provide important information (Table 1), the analysis of specific mortality in relation to possible psychosocial breakdown may provide even more information. Suicide for instance is an action that the person decides to take. The case of cancer is quite different. In that case we may be dealing with a chronic suppressed

Table 1. Number of living men/1000 live births (McKeown, 1976)

	Period				
Age	1693^x	$1838\text{-}54^+$	$1891\text{-}1900^+$	$1920\text{-}32^+$	1970^+
0	1000	1000	1000	1000	1000
5	582	724	750	870	976
20	481	652	712	837	968
50	275	456	531	699	909
80	-	80	82	150	236

(x) Germany

(+) England and Wales

---

cry for help as has been indicated in a recently published prospective study (Grossarth-Maticek, 1982). In the case of sudden coronary death we seem to be dealing with overly optimistic denying attitudes (Nirkko et al., 1981). Although there are similirities between these psychological syndromes the dissmililarities are bound to create interpretation difficulties if we group them together. The biological and physical environmental differences also contribute to the difficulties. Therefore, whenever possible, causes of death should be considered. On the other hand, it is obvious that the registries of specific causes of death must have pitfalls when we try to use them for comparisons between areas, periods or countries. Table 2 lists the sources of error I consider most important. Autopsy rates vary not only across countries but also across areas within countries. This is true also of most of the other factors. Perhaps the most difficult example is suicide. Historically the need for cover-up of this diagnosis has been more accentuated in catholic countries. Several of the pitfalls in suicide statistics could be discussed in Figure 2, which shows the age specific suicide rates in Stockholm during two periods, 1901-1910 and 1971-1975. The decline in suicide rates for men in their sixties and fifties during three quarters of this century may be due for instance to increasing rates of hidden suicides - fatal automobile accidents with a lonely driver were not possible in 1905 but may in many instances have been masked suicides in 1975. (Official Statistics, Stockholm 1978).

Despite all the difficulties everyone would agree that a skillful use of death statistics allows us to monitor gross changes in adaptation, also in a longer historical perspective. In Sweden the life expectancy of middle aged men is not increasing any more, and the female advantage in longevity is increasing. If we confine our analysis to Stockholm we observe that the age adjusted male/female mortality ratio for total illness has been slowly increasing during the century whereas the corresponding ratio for accidents has decreased dramatically. Obviously something is happening to our middle aged men in Stockholm. (Statistics Reports of the Stockholm City, 1979).

Table 2. Sources of error in registries of specific causes of death

Who writes cerfiticate ?

Routines for writing "primary" cause of death - and "other" causes

Autopsy rate

Skills in diagnostic routines

Variations in emphasis on and intensity of searching for various diagnoses

Cultural inhibitions in willingness to write certain diagnoses

Possibilities for cultural cover-ups - such as covert suicides which appear to be "single" motor vehicle accidents

---

Morbidity Statistics:

For morbidity statistics there are corresponding difficulties. The most serious ones arise due to variations in people's motivation to seek care for different illnesses. Some of this may be due to cultural or social attitudes to illness and some of it may be due to the quality and availability of care. Johnston and Ware (1976) have pointed out that there are pronounced differences between socioeconomically different groups in the relative importance persons attribute to somatic and psychiatric components of their health. Low income people put more emphasis on somatic components and vice versa.

A serious source of error in comparisons between European countries is the large variation in number of hospital beds per 1000 persons. For psychiatric care we have a variation between "high" countries such as Sweden (4.9/1000) and Norway (4.7) to "low" countries such as Bulgary (1.1), USSR (1.1), Poland (1.6) and Italy (2.2). (WHO official statistics, 1978). Differences in diagnostic procedure may also be important.

However, the sources of error that I have mentioned are less for life threatening somatic illnesses such as cancer and myocardial infarction. In these cases combined studies of death registries and hospitalization registries may be useful. It would certainly be of value for instance to use the infarct registries - which combine death and hospitalization registries - that are kept in several places throughout Europe for more systematic studies of "breakdown of adaptation". A special field of interest would be the study of young victims (below age 45) of myocardial infarction. Such a registry in Stockholm was used for the study of myocardial infarctions among middle aged building construction workers. By using a questionnaire to a large sample of them we were able to establish psychosocial predictors of increased infarction risk during a follow-up of two years. The table also shows a comparison with persons who suffered longlasting episodes of psychic morbidity or mortality during the first year of follow-up. In the latter case a third registry was utilized - the Swedish compulsory sickness registry of all episodes of sickness absence. All cases belonging to a geographically defined subgroup of the same population with more than 30 consecutive days of sickness, absence or deaths with a psychiatric diagnosis (but without such a diagnosis already during the preceding year) were identified. The study showed that quite different psychosocial predictors were found for myocardial infarction versus psychiatric diagnosis.

National Surveys:

Several European countries have national surveys of random samples of men and women. These surveys are based upon interviews with large samples of the population and describe various aspects of the level of living in various groups. There are several difficulties in interpreting results from such surveys as well. In Sweden we have observed increasing drop-out frequencies in these surveys. This may indicate increased feeling of threat among people randomly selected for interviews. Drop-out frequencies vary greatly between European countries. One other factor - changes in awareness - may be illustrated by Figures 3-6. Does the increased reporting of psychologically demanding work reflect a "true" difference or is it merely describing the fact that Swedes are becoming more aware of the importance of psychological factors at work ? The figures describe an interesting development in Swedish job descriptions - increasing psychological but decreasing physical demands as well as increasing influence and opportunities to learn something new at work. Results from comparisons between European countries have to be interpreted with due caution due to translation difficulties and differences in cultural attitudes and prevalence of denial etc. but may still be of value.

Social Statistics:

Another source of information is the official social statistics. It is probable that these vary greatly between countries and between regions in the countries. One of the difficulties is due to the fact that administrative regions may vary considerably in size. This means that very important variations may be drowned in large units and exaggerated in small ones. Table 3 gives an example from Stockholm. In this case the units are relatively comparable in size. Yet large differences are observed in proportions of people on social welfare. It is often said that that all social inequalities have been eliminated from our country. There is, however, a marked social segregation for instance in greater Stockholm. Cross-cultural comparisons of social statistics are very difficult to perform due to differences in administrative social routines in different countries. Still, cross-cultural comparisons of associations between data based on social statistics and illness may increase our understanding of breakdown processes.

Table 3. Proportion of subjects on social welfare in 63 different areas of the Stockholm community (parishes) % of all people living in the area. Only the five highest and five lowest are shown:

High			
Spånga	12.6	Västerled	1.7
Skärholmen	10.8	Essinge	1.7
Maria	7.4	Hedvid Eleonora	1.8
Skarpnäck	7.0	Kungsholm	2.4
Högalid, Hägersten	5.9	Oscar	2.5

Combinations of Sources:

One may sometimes have the opportunity to use socioeconomic or other kinds of data from one group of persons for the analysis of relationsships with somatic or psychological data in other persons belonging to the same group. This may sometimes be advantageous. It is important, however, to bear in mind what level of aggregation the two groups of data have. Table 4 shows results from the medical survey from the compulsory recruitment for military service in greater Stockholm. This was related to social statistics. The average blood pressure in 63 areas of residence was calculated. This was subsequently analyzed in relation to socio-economic characteristics of the same area. The table shows that there are high correlations between proportion of subjects on social welfare and median income of the area on one hand and average systolic blood pressure among 18-year-old men in the same area. It should be born in mind that the natural standard deviation has been almost eliminated in this study of 74000 men in 63 areas. This tends to increase the size of correlations. (Theorell et al., 1982).

Table 5a and 5b show the opposite phenomenon. In this case a national survey has been used in order to find which out of 118 occupations were above the median with regard to unemployment, overtime work etc. These standards have been used for calculation of relative risks of myocardial infarction associated with different types of jobs. These calculations were based upon other subjects, namely all male working subjects who had had myocardial infarctions within one area and one period (deaths and survivals) and controls. The occupational code (three digits international code) was translated into characteristics from the national survey. For instance, those working in occupations with more unemployment were compared with those with less unemployment. This crude method leads to misclassifications and underestimated relative risks (Alfredsson et al., 1982, Alfredsson, 1982). Despite this the method has advantages - the individual´s own denials or exaggerations are avoided. Occupations above the median for "hectic" and for "not learning anything new" have been compared with other occupations and age adjusted relative risks calculated. An analysis of the possible confounding effect of a number of factors was performed. Particularly in the age group below 55 the association between working in an occupation characterized by (other) workers in it as hectic and not giving any opportunities to lear anything new and risk of myocardial infarction held even when several confounding

Table 4. Area means from medical survey of 74000 18-year-old men (compulsory recruitment for military service) in 63 areas of greater Stockholm in relation to social statistics (product moment correlations).

Mean measures correlated with mean area syst. BP

Mean Height	$-.37^{xx}$
Mean Heart Rate	.11
Mean Muscle Strength	$.52^{xxx}$
Mean Verbal Understanding	$-.56^{xxx}$
Median Income	$-.43^{xxx}$
Mean Relative Weight	$.64^{xxx}$
Proportion on Social Welfare	$.27^{x}$

$^{x}p < 0.05$, $^{xx}p < 0.01$, $^{xxx}p < 0.001$

Table 5a. Single variables. Age-standardized relative MI risk of subjects belonging to 50% occupations with highest vs subjects belonging to the 50% occupations with lowest hypothesized risk.

	Relative risk	P Value (one-tailed)
Unemployment	1.14	0.15
Overtime Work	0.91	0.23
Shift Work	1.25	0.04
Piece Wage Salary	1.05	0.34
Hectic Work	1.06	0.33
High Accident Risk	0.96	0.38
Monotony	1.32	0.02
No Private Visits	1.20	0.08
Low Influence over Work Tempo	1.20	0.08
No Contacts	1.14	0.15
Not Learning New Things	1.19	0.08
Sweating	1.12	0.10
Physical Monotony	1.16	0.12
Heavy Lifting	1.19	0.09
Noise	1.03	0.40
Vibrations	1.11	0.21
Low Education	1.29	0.03
Immigration	1.07	0.28
Excessive Tobacco Smoking	1.28	0.03

Table 5b. Combined variables. Age-standardized relative MI risk of subjects beloning to the 50% occupations with highest percentages hectic work and concomitantly belonging to the 50% occupations with highest hypothesized MI risk with regard to other characteristics (vs all other subjects).

Rushed Tempo and	Relative Risk	P Value (one-tailed)
Monotony	1.26	0.08
No Private Visits	1.30	0.04
Low Influence over Work Tempo	1.35	0.02
No Contacts	1.24	0.07
Not Learning New Things	1.45	0.02
Sweating	1.43	0.02
Physical Monotony	1.31	0.04
Heavy Lifting	1.37	0.02
Noise	1.23	0.12
Vibrations	1.42	0.02

factors were considered. A multiple logistic regression analysis was also performed which verified Karasek´s hypothesis that a multiplicative interaction between demand (in this case tempo) and not learning new things ("strain" job, Karasek 1979) increases the risk of myocardial infarction even when other relevant factors such as average smoking habits in the occupation are being held constant. In this case, four sources were utilized: National Survey (standard weight)s, Population registry (occupation before myocardial infarction and in control cases), Death registry (case identification), Hospitalization registry (case identification).

Figure 7 shows an additional possible use of official statistics. In this case we have utilized the medical registries of 18-year-old men in order to find a risk group. We have selected the top 0.5% blood pressures. Intensive psychological (Schalling and Svensson, 1982) as well as psychophysiological (Svensson and Theorell, 1982, fig. 10) analyses were performed from three contrast groups in the survey. They seem to show that it is possible to find meaningful patterns by studying relatively small groups of subjects if they have been selected from the tails and the middle of a very large normal distribution. Anxiety proneness and aggression inhibition as well as slightly faster systolic blood pressure reaction to a provocative interview and a slower return to baseline as well as a much more accentuated systolic blood pressure reaction to the novel examination situation were observed in the high pressure group compared to the other groups. Finally, a follow-up study of the same population at the age of 30 has been started. Subjects have been characterized with regard to the kind of jobs they have had since the age of 18, using the classification system described by Alfredsson et al. (1982). Again we take advantage of the fact that "standard weights" for different occupations, although crude, allow us to avoid the individuals´ own denial and exaggeration.

Conclusion:

Despite many difficulties in the use of official statistics, the Scandinavian experience indicates that various combinations of statistical sources may provide new clues to our understanding of the breakdown of human adaptation. If these sources of information are used in a skillful way throughout the European community our societies may be able to discover early signs of breakdown and use this knowledge for improved planning.

REFERENCES

Alfredsson, L., Karasek, R.A. and Theorell, T. 1982.
  Myocardial infarction risk and psychosocial work environment. An analysis of the male Swedish working force. Soc. Sci. Med., 16: 463.

Alfredsson, L. 1982.
  On the effect of misclassification in case-control studies. Stenciled report, Dep. of Soc. Med., Huddinge Hospital, Huddinge.

Grossaeth-Maticek, R., Siegrist, J. and Vetter, H. 1982.
  Interpersonal Repression as a Predictor of Cancer. Soc. Sci. Med. Vol. 16: 493.

Johnston, S.A. and Ware, J.E. 1976.
  Income group differences in relationships among survey measures of physical and mental health. Int. J. Health. Serv. Res., 11: 416.

Kasarek, R.A. 1979.
  Job demands, job decision latitude and mental strain: Implications for job redesign. Admin. Sci. Quart., 24: 285.

Mc Keown, T. 1979.
  The Role of Medicine. Dream, mirage or nemesis? Basil Blackwell, Oxford.

Nirkko, O., Lauroma, M., Siltanen, P., Tuominen, H. and Vankala, K. 1982.
  Psychological risk factors related to coronary heart disease. Prospective studies among policemen in Helsinki. Acta Med. Scand. Suppl., 660: 137.

Scalling, D. and Svensson, J. 1983.
  Blood pressure and personality. Personality and Individual Differences, submitted.

Stockholm Community, 1974.
  Official Statistics.

Swedish Community, 1978.
  Statistical Reports.

Svensson, J. and Theorell, T. 1982.
  Life events and elevated blood pressure in young men. J. Psychosomatic Res., in press.

Theorell, T., Svensson, J., Knox, S. and Ahlborg, B. 1982.
  Blood pressure variations across areas in the greater Stockholm region: Analysis of 74000 18-year-old men. Soc. Sci. Med., 16: 469.

U.N. World Health Organization (WHO) Statistics, 1978.

INVENTORY OF STRESSFUL LIFE-EVENTS (ILE)

J. Siegrist, K.-H. Dittmann
Institute of Medical Sociology, Faculty of Medicine,
University of Marburg, Marburg FRG

ABSTRACT

An inventory for the assessment of number, time pattern and subjective impact of severe negative life-events is described. Conceptual approach, design and test-statistical properties are outlined. Selective results and overview over study populations are presented. Most of it's application until now has been in the field of cardiovascular epidemiology. Finally usefulness and limitations of the ILE are briefly discussed.

INTRODUCTION

Research on physiological and psychological consequences of critical life-events (LE) such as separation, loss of job, death of a close person has expanded in the field of psychosomatic medicine during the last 20 years (Dohrenwend et al., 1974, Gunderson et al.,1974, Katschnig 1980. Aagard 1983). A large amount of results was obtained, but much of this work has been shown to be too superficial from a methodological and conceptual point of view. Critics pointed out that careful assessment of quality, meaning and subjective impact of LE as well as of social contexts which increase vulnerability have been neglected (Brown 1974). Also, only a few studies have been carried out prospectively. As a consequence, a new generation of LE-inventories has been developed including subjective impact, coping processes, and social contexts in which LE occur.

METHODS AND RESULTS

The ILE is a combination of a structured interview schedule covering quality, quantity and time pattern of defined rather severe negative life-events (34 events) during the last two years, and a questionnaire with subjective ratings of impact administered after every event which has happened to the subject. Persons included are the subject him- or herself, husband/wife, closest friend, own children, parents, brothers/sisters. Items (5-point-scale) which measure subjective impact include the following dimensions derived from general stress theory: controlability of LE, predictability of LE, position of the LE in the subjective structure of relevance, interruption of everyday routines caused by the LE, situational vulnerability, active coping, experienced social support, psychological and social costs of adaptation after the LE had occured, and time period of impact. This approach enables us to calculate both the number of stressful events per selected subgroup or individual and the respective stress scores or their mean value. Assuming a cumulative nature of the scores, individuals can be assigned from 1 to 44 points per event, with 44 points indicating extreme stress. Accordingly, events can be classified following the degree of subjectively perceived stress (see Siegrist et al. 1981).

At present, more adequate multi-dimensional models of analysis of these subjective data are developed at our institute, in accordance with new approaches to the study of time sequence (e.g."discrete time models", "failure - time - data models") and of different a priori probabilities of events in different subjects or subgroups (Dittmann 1983).

Mean time of application of ILE is about 15 minutes. Of course, the ILE should be part of a broader inquiry including other areas of sociological, psychological and medical information. Initially, the ILE has been developed for the study of male working populations age 25-60, but an ILE-form for women has also been applied. The following epidemiological and clinical studies have been or are being performed with ILE (for details see Dittmann et al. 1983).

- several hundred male patients with first acute myocardial infarction and matched healthy controls
- male and female patients with functional heart disease
- male and female patients with recurrents of duodenal ulcer
- male and female patients with kidney stones

- female patients with parametropathic disorders and
- several hundred initially CHD-free industrial workers in a prospective study on cardiovascular risks.

A slightly modified english version is currently applied in a Johns-Hopkins study of women with alcohol problems.

The following short information on teststatistical properties and results of ILE concentrates on a retrospective case-control study of 190 male patients (sample half) with acute myocardial infarction (age 30-55) and a matched control group (N=190). Additional information from a small study including simultaneous application of ILE and London life-event schedule (LES, Brown 1974) is reported.

According to a general problem of life-event questionnaires (see Katschnig 1980), the test-retest reliability for sum of events was found to be better than reliability at the level of single event-items, indicating a somewhat insufficient interrater quality. On the other hand test-retest-reliability of subjective impact measures was quite high (f.e.: .81 for impact scores per single events, corresponding in test and retest).

Some aspects of validity are indicated in table 1:
- First, we found significantly more individuals with life events in the disease group (81%) than in the healthy control group (61%) and also significantly more events or more extreme event rates in the former group than in the letter.
- Second, sum scores as well as mean scores of subjective impact for every single stress dimension and for global indices were significantly higher in the disease group than in the healthy group. However it should be mentioned that the respective mean differences were diminished after controlling for rates of persons, who experienced life events, and the number of events per individual in each group.

Compared to other stress indicators, the global sum scores of number of events and subjective impact proved also to be quite good predictors or discriminators in differentiating between patients and healthy controls on the one hand and

also between patients with and without later myocardial reinfarct (stepwise multiple regression and discriminant analyses). In addition, analyses of variance between chronic social stressors and LE-measures were especially powerful with these parameters (see Siegrist et al. 1980).
- Third, a specific time pattern or clustering of events could be found marking a more significant increase in the disease group during the last three months before manifestation of myocardial infarction (u=2.6, p< .05).
- Forth, the mentioned mean differences in number and impact scores of LE remained significant in most cases after controlling for the following variables:
  1. age,
  2. occupational position,
  3. neuroticism (as measured by a widely used German personality inventory (FPI)),
  4. (non-systematic) interviewer effects.

Table 1: Comparison of mean values (t-test) for LE indices (number of LE, sum score and mean score of LE-impact) in subjects with myocardial infarction and healthy controls.

LE indices		all subjects				subjects with at least one experienced LE			
		IG 190	CG 190	t-value	p	IG 154	CG 121	t-value	p
number of LE	$\bar{x}$	2.08	1.09	7.20	.01	2.56	1.71	5.63	.01
	s	1.72	1.11			1.56	.93		
sum score of impact	$\bar{x}$	50.80	23.94	7.79	.01	62.67	37.60	6.03	.01
	s	47.19	25.66			44.75	22.79		
mean value of impact	$\bar{x}$	19.08	13.81	4.39	.01	23.54	21.68	2.75	.01
	s	10.64	11.50			5.84	6.08		

In a study on premorbid life-events in depressive women, conducted in Vienna, a simultaneous application of both instruments (the London life-event schedule (LES) and the ILE) could be obtained in a subgroup of 20 women with a total of 35 life-events rated identically by either istrument. The degree of correspondence between subjective impact ratings (four-point-scale on the basis of the total score from 1 to 44 points in the ILE) and objective threat ratings (short-term-threat and long-term-threat; both assessed on a four-point-scale in the LES) was calculated (Goodman-/Kruskal-coefficients). Table 2 demonstrates the rank correlations between both rating systems: (a) on the basis of experienced events (N=35) as observation units and (b) on the basis of subjects (N=20) as observation units. Coefficients are higher on the basis of subjects than of events, and they are higher for short-term ratings than long-term ratings. It can be concluded that there is substantial overlap between measures based on subjective impact rating (ILE) and measures based on objective threat rating (LES) - especially short-term-threat rating - of identical life-events.

Table 2: Coefficients of rank correlations between subjective and objective impact rating of life events in 20 depressive female patients.

Observation units	ILE rating with:			
	LES short-term threat	P	long-term threat	P
Events (N=35)	0.74	0.001	0.43	0.001
Subjects (N=20)	0.80	0.001	0.73	0.01

CONCLUDING REMARKS

Results so far show that a short-structured sociological schedule can be of use in epidemiologic and clinical studies of populations at risk. However, ongoing studies have to be analyzed in order to improve several test-statistical characteristics of the ILE and in order to elaborate its heuristic value under different conditions of disease. Also, prospective studies should be stimulated which relate ILE measures to other sociological and psychological characteristics as well as to neurohormonal and/or immunological parameters of adaptive versus non-adaptive coping.

REFERENCES

Aagard, J.
  Stressful life-events and illness. (In this volume of proceedings).
Brown, G.W. 1974.
  Meaning, measurement and stress of life-events. In: Dohrenwend, B.S., Dohrenwend, B.P. (eds.): Stressful life-events: their nature and effects. Wiley, New York.
Dittmann, K.H. 1983.
  Lebensverändernde Ereignisse: Lösungsansätze zu theoretischen und methodischen Problemstellungen. Ph.D. Dissertation, University of Marburg (in preparation).
Dittmann, K.H., Siegrist, J. 1983.
  Beschreibung des "Inventars zur Erfassung lebensverändernder Ereignisse" (ILE von Siegrist und Dittmann). (in press).
Dohrenwend, B.S., Dohrenwend, B.P. (eds.) 1974.
  Stressful life-events: their nature and effects. Wiley, New York.
Gunderson, G.K., Rahe, R.H. (eds.) 1974.
  Life stress and illness. Thomas, Springfield, Ill.
Katschnig, H. (ed.) 1980.
  Sozialer Streß und psychische Erkrankung. Urban & Schwarzenberg, München.
Siegrist, J., Dittmann, K.H. 1981.
  Lebensveränderungen und Krankheitsausbruch: Methodik und Ergebnisse einer medizinsoziologischen Studie. Kölner Zeitschrift für Soziologie und Sozialpsychologie 33, 115 .
Siegrist, J., Dittmann, K.H., Rittner, K., Weber, I. 1980.
  Soziale Belastungen und Herzinfarkt. Enke, Stuttgart.

THE NORWEGIAN FEMALE CLIMACTERIC PROJECT (NFCP)

Aslaug MIKKELSEN and A. HOLTE
Institute of Behavioural Sciences in Medecine
University of Oslo
P.O. Box 1111, Blindern
OSLO 3, NORWAY

THE NFCP
The problem studied by the NFCP is the female climacteric
and how social, psychological and biological factors contribute to the development of symptoms, lowered quality of
life and breakdown in human adaptation at the change of life

DESIGN
The design is a combination of a epidemiological cross-sectional study and a prospective longitudinal design. The
NFCP involved totally 3000 females in different studies:
The Drammen study (an epidemiological cross-sectional study)
Sample: N = 200 randomly selected on the following criteria:
- woman
- 45-55 years of age
- Norwegian citizen
- living in Drammen City
Method - postal questionnaire

Statistics: Factor analysis, analysis of variance, multiple regression analysis.

The Oslo study (an epidemiological cross-sectional study)
Sample: N = 2800 randomly selected on the following criteria:
- woman
- 45-55 years of age
- Norwegian citizen
- living in Oslo
Method - postal questionnaire
Statistics: Factor analysis, analysis of variance, multiple

regression analysis.

The prospective Oslo study (a longitudinal study)

Sample: N = 300 selected from the Oslo study on the following additional criteria:
- regular menstruation periods
- less than 2 months since last menstruation
- at least one intact overy
- no oestrogen substitution treatment last year

Methods
- Social background interview
- Symptom interview
- Psychological interview
- Gynecological examination
- Blood tests for endocrinological analyses

CONCEPTS AND INSTRUMENTS

The epidemiological studies

The mailed questionnaire in the epidemiological part contained questions about the traditional sociological background variables and social network, family, sex role identification, menstruation, menopause, expectations to the future and sexuality.

The longitudinal study

The social background interview

This part of the NFCP is aimed at adding information to the epidemiological study as to social background and registrate changes in the same variables. Below are some of the most important measures developed and used
- education - The Norwegian Standard Classification of education. It is based on length and type of education. The principles and definitions are translated to English
- work - The Nordic Standard Classification of Occupation. It is based on the International Standard Classification of Occupations
- socioeconomic status - It is a Norwegian measure developed by the Central Bureau of Statistics and based on the International Standard Classification of Occupations.

- sex role pattern and sex role identification
- life events and life problems
- quality of life. The measure contains questions about happiness, life satisfaction and a semantic differential about the respondents feelings about the life in general. The measure is developed by Moum (1982) and translated to English. The measure is based on the international traditions in this field.
- social network. The measure contains both quantitative and qualitative aspects of social network, social support and confidance.

The psychological interview

The interview is on this part self-administered and includes:
- the "Body Catexis"-method (Jourard et al 1955) for the registration of experience of the women's own appearance and body (Lazare 1970)
- Torgersen's personality scale based on Lazare (V, Torgersen, 1980)
- Bem Sex Role Inventory translated and adjusted to Norwegian circumstances (Bem, 1974)
- Spanier's Marital Adjustment Scale (Spanier, 1976)

The symptom interview

The interview includes:
- The General Health Questionnaire (Goldberg 1979) modified for interview by Svenn Torgersen
- A symptom check list from the epidemiological part based on the 24 most mentioned climacteric symptoms in the literature
- Other actual symptoms and diseases
- The symptom measurement is further developed to registrate:
- first occurence
- frequency
- duration
- subjective discomfort
- changes since last interview of each symptom

Gynecological examination

In addition to an ordinary gynecological check the following are systematically registrated: menstruation, menopausal status, births, abortions, preventives, sterilization, gynecological diseases and problems, sexual problems, climacteric complaints, blood pressure, heart examination, vaginal mucous membrane, descens, exploaration.

Blood tests

A double set blood tests are collected and deep-frozen (-80° C). The following analysis shall be conducted:
- total oestrogens
- gonadotrophines
- progesteron

References

Bem, S., 1974. The measurement of psychological androgyny. J. of Consulting and Clinical Psychology 42: 155-162.

Goldberg, D.P., Cooper, B., Eastwood, M.R., Kedward, H.B., Shephard, M., 1970. A standardized psychiatric interview for use in community surveys. Brit.J.Prev.Soc.Med. 24:18-23.

Jourard, S.M., Secord, P.F., 1955. Body Catexis and the ideal female figure. J.Abnorm. Soc. Psychol. 50: 243-246.

Lazare, A., Klerman, G.L., Armor, D.J. 1970. Oral, obsessive and hysterical personality patterns: An investigation of psychoanalytic concepts by means of factor analysis. Arch. Gen. Psychiat. 14: 624-630.

Moum, T., 1982. The role of values and life goals in quality of life. Unesco, Paris/Institute of Social Research, Oslo.

Spanier, G.B., 1976. Measuring dyadic adjustment: New scales for assessing the quality of marriage and similar dyads. J. of Marriage and the Family 38: 15-38.

Torgersen, S., 1980. Hereditary-environmental differentiation of general neurotic, obsessive and impulsive hysterical personality traits Acta Genet.Med.Gemellol. 29: 193-207.

QUESTIONNAIRE FOR ORGANISATIONAL STRESS (VOS)

Nico van Dijkhuizen
SWO/DPKM, Ministry of Defence
P.O.Box 20702, 2500 ES Den Haag, Netherlands

ABSTRACT.
The following regards information on a Dutch questionnaire for measuring organisational stress, based on a stress model originally developed at the University of Michigan (French and Caplan, 1972). It measures a number of work-related stressors, personal characteristics and strains. The questionnaire, thoroughly tested before being used in research, possesses satisfactory psychometric qualities. It has, by now, been used on a variety of populations in a number of studies, both cross-sectional and longitudinal. The results may be used both in programmes intended to improve employees' health (by using the questionnaire as screening device) and in changing and developing organisational structures appearing to have detrimental effects.

INTRODUCTION.
The VOS is based on a variables-model dealing with the effects of psychosocial stress on health and illness; this model is an elaboration of the one originally developed by French and Caplan (1972, see Van Dijkhuizen and Reiche, 1976; Van Vucht Tijssen et al., 1978; Van Dijkhuizen, 1980). The questionnaire itself is an advanced version of the one used by Caplan et al. (1975).
In the stress-model stress refers to those characteristics in the job-environment which are perceived by the individual as threatening. Two kinds of stress may threaten the person: 1. he may not be able to live up to environmental demands; 2. the environment does not provide sufficient opportunities to fulfill the person's needs.
Strains refer to deviations from the person's normal response patterns. Three kinds of strains are discerned: psychological strains (related to the job as well as in general), behavioural strains (p.e. smoking and absenteeism), and physiological strains (as high blood pressure).
In the model stress is studied within a sequence of events. The sequence consists of the objective environment, the subjective environment, the individual's reactions to these and, finally, the consequences for both the individual and his environment. These four elements are supposed to be causally related.
Personal characteristics are thought to have a conditioning influence on the sequence. In some studies social support variables are treated as conditioning variables as well; others treat lack of social support as a stressor.
Stress in work situations is best studied using the role set technique (Kahn et al., 1964) in which both focal persons and the most important persons in their environments are included in the research. Moreover, it is advised to use depth-interviews with at least a subsample of respondents in order to elaborate on specific problem areas. An

interview guide is available for this purpose (Van Vucht Tijssen et al., 1978).

THE QUESTIONNAIRE.

The questionnaire measures 14 stressors, 11 strains and 2 personality characteristics, apart from biographical data such as age, sex, marital status, time being employed, and education and training.

Modules are optional, depending on the research-questions, regarding physiological strains (blood pressures, cholesterol-level, catecholamines, pulse rate. length, weight, 17-ketosteroïds, etc.), caffeïne-intake, use of drugs (sedatingdrugs as librium, valium, nobrium, seresta, temesta, but also anti-depressiva, beta-blockers, aspirines, etc.), regarding the interaction between the work- and the home-situation, and regarding physical stressors (temperature, noise, humidity, air pollution, lighting, vibration, etc.).

In the following the concepts in the standard questionnaire are listed, together with the number of items used to measure the concept and its reliability (Cronbach's alpha) as found in the various studies (range).

Concept.	No.items	Reliabil.
Stressors:		
- role ambiguity	6	.72 - .85
- responsibility for persons	4	.66 - .76
- work load	12	.75 - .91
- underutilisation of skills and abilities	3	.55 - .76
- tensions in relations with superiors/subordinates	2	-
- tensions in relations with other departments	1	-
- lack of participation	3	.70 - .84
- role conflict	3	.66 - .84
- lack of support from superior	4	.77 - .88
- lack of support from colleagues	4	.73 - .82
- lack of support from others at work	4	.71 - .83
- lack of support from partner	4	.71 - .83
- lack of support from relatives	4	.76 - .83
- job future ambiguity	4	.70 - .76
Strains:		
- job dissatisfaction	3	.52 - .71
- psychosomatic complaints general health	8	.66 - .83
- psychosomatic complaints regarding heart	3	.65 - .85
- anxiety	4	.80 - .87
- depression	6	.70 - .86
- irritation	3	.77 - .84
- loss of self-esteem	3	.64 - .73
- job-related threat	4	.61 - .79
- smoking	2	-
- absenteeism	1	-
- alcohol consumption	2	-
Personal characteristics:		
- rigidity dogmatism	4	.68 - .79
ordening	6	.76 - .82

- job involvement (A/B typology)   9   .73 - .79

Figures for content and construct validity are satisfactory. The average time to fillout the paper and pencil questionnaire ranges from 20 to 90 minutes, depending on the respondent's degree of education. No specific problems were found, however, in lower levels of hierarchy.

STUDY-POPULATIONS.

The VOS has been used in the following populations:

- 254 professional soldiers (officers and non-commissioned officers Royal Netherlands Army): Warlicht (1977);
- 578 employees in industry (middle managers, their immediate subordinates and superiors, shopfloor-workers and staff-specialists): Van Vucht Tijssen et al. (1978);
- 260 production-workers in chemical industry: Penders (1979);
- 68 student-nurses in middle management training: Driessen (1979), Van Oss (1980);
- 177 ex-mining supervisors: Denteneer (1980);
- 100 nurses and physicians: Prins (1981);
- 600 hospital employees (head-nurses, nurses, student-nurses, medical specialists and superiors): Van den Bergh-Braam (1981);
- 46 building construction workers: Borrie et al. (1982);
- 1246 industrial employees: Winnubst et al. (1982);
- 304 personnel-officers in industry: Van Bastelaer et al. (1982);
- 120 personnel-officers in the Ministry of Defence: to be published.

A study with about 1000 all ranks personnel in the Royal Netherlands Navy is currently being set up (Van Dijkhuizen, in preparation).

REFERENCES.

Bastelaer, A. van and Van Beers, W., 1979. Vragenlijst Organisatiestress (VOS): psychometrische kanttekeningen bij vergelijking van data over vijf onderzoeksgroepen. Nijmegen, Kath.Univ. (internal report 79A001/CG01).

Bastelaer, A. van and Van Beers, W., 1980. Vragenlijst Organisatiestress: testhandleiding deel 2, konstruktie en normering. Nijmegen, Kath.Univ. (internal report 80A008).

Bastelaer, A. van and Van Beers, W., 1982. Organisatiestress en de personeelsfunktionaris. Lisse, Swets and Zeitlinger.

Beers, W. van and Van Bastelaer, A., 1979. Interimrapport onderzoeksproject organisatiestress bij personeelsfunktionarissen. Nijmegen, Kath.Univ. (internal report 79A002/CG02).

Bergh-Braam, A.H.M. van den, 1981. De hoofdverpleegkundigen: een onderzoek naar hun positie en problemen in de Nederlandse ziekenhuizen. Nijmegen, Kath.Univ. (internal report 81A003).

Borrie, H., Kamp, T., Moonen, S. and Versteeg, F., 1982. Vooronderzoek langdurig ziekteverzuim in de bouw. Nijmegen, Kath.Univ. (internal report 82A001).

Caplan, R.D., Cobb, S., French, J.R.P., Van Harrison, R., and Pinneau, S.R., 1975. Job demands and worker health. Washington, HEW

Publication (NIOSH) 75-160.
Denteneer, H.M.C., 1980. Gedwongen beroepswisseling: de lotgevallen van de mijnopzichters van Staatsmijnen die wegens bedrijfssluiting van beroep moesten veranderen. Nijmegen, Kath.Univ.
Driessen, J.G.M., 1979. Stressoren, sociale steun en strains. Nijmegen, Kath.Univ. (internal report 79A005).
Dijkhuizen, N. van, 1980. From stressors to strains: research into their interrelationships. Lisse, Swets and Zeitlinger.
Dijkhuizen, N. van, 1981a. Measurement and impact of organisational stress. In: J. Siegrist and M.J. Halhuber (Editors), Myocardial infarction and psychosocial stress. Berlin, Springer.
Dijkhuizen, N. van, 1981b. Towards organisational coping with stress. In: J. Marshall and C.L. Cooper (Editors), Coping with stress at work: case studies from industry. Aldershot, Gower.
Dijkhuizen, N. van, 1983. Towards a sequential model of stress. In: I.G. Sarason, C.D. Spielberger and P. Defares (Editors), Stress and anxiety Vol. 11. New York, Wiley.
Dijkhuizen, N. van and Reiche, H.M.J.K.I., 1976. Het meten van organisatiestress: over de bewerking van een vragenlijst. Leiden, Univ. of Leyden (internal reports 001-76, 002-76 and 003-76).
Dijkhuizen, N. van and Reiche, H.M.J.K.I., 1978. Stress en strain bij middenkader. Nijmegen, Kath.Univ. (internal report 78A006).
Dijkhuizen, N. van and Reiche, H.M.J.K.I., 1980. Psychosocial stress in industry: a heartache for middle management? Psychotherapy and Psychosomatics, 34: 124-134.
French, J.R.P. and Caplan, R.D., 1972. Organisational stress and individual strain. In: A.J. Marrow (Editor), The failure of success. New York, AMACOM.
Kahn, R.L., Wolfe, D.M., Quinn, R.P., Snoek, J.D., and Rosenthal, R.A. 1964. Organizational stress: studies in role conflict and ambiguity. New York, Wiley.
Kleber, R.J., 1982. Stressbenaderingen in de psychologie. Deventer, Van Loghum Slaterus.
Oss, P.J.A.S. van, 1980. Spanningen bij middenkader in de verpleging: een onderzoek bij een groep cursisten van de Kath. Hogere School voor verpleegkundigen te Nijmegen. Nijmegen, Kath.Univ. (internal report 80A001).
Penders, H.J.P., 1979. De bruikbaarheid van de vragenlijst survey of organizations voor onderzoek naar stress in organisaties. Nijmegen, Kath.Univ. (internal report 79A003).
Prins, J., 1981. Stress in het ziekenhuis: een vergelijkend vooronderzoek bij verpleegkundigen en artsen. Nijmegen, Kath.Univ. (internal report 81A002).
Reiche, H.M.J.K.I., 1982. Stress aan het werk: over de effecten van de persoonlijkheid en sociale ondersteuning op strains. Lisse, Swets and Zeitlinger.
Reiche, H.M.J.K.I. and Van Dijkhuizen, N., 1979. Bedrijfsgrootte, hiërarchie en persoonlijkheid: beïnvloeden zij het ervaren van stressoren en strains? Gedrag, 7: 58-75.
Reiche, H.M.J.K.I. and Van Dijkhuizen, N., 1980. Vragenlijst organisatiestress: testhandleiding deel 1 testafname. Nijmegen, Kath.Univ. (internal report 80A008).
Ros, J.G., 1982. Ongelijkheid in werk en gezondheid. Nijmegen, Kath.Univ. (internal report 82A003).
Vucht Tijssen, J. van, Van den Broecke, A.A.J., Van Dijkhuizen, N., Reiche, H.M.J.K.I. and De Wolff, Ch.J., 1978. Middenkader en stress.

Den Haag, COP/SER.
Warlicht, E.E., 1977. Stressoren, strains en het hart. Leiden, Univ. of Leyden.
Winnubst, J.A.M., Marcelissen, F.H.G. and Kleber, R.J., 1982. Effects of social support in the stressor-strain relationship: a Dutch sample. Soc.Sci.Med., 16: 475-482.

Internal reports from Nijmegen University may be obtained from:

Stressgroup Nijmegen Verkoopcentrale
Psychologisch Laboratorium
P.O.Box 9104
6500 HE Nijmegen, Netherlands

Information and documentation: tel. 080-512660

A SCALE FOR MEASURING THE MARITAL RELATIONSHIP AMONG MALES

M. Waltz
Universitaet Oldenburg
Westerstr. 2
Oldenburg - FRG

Abstract

The marital relationship can be a source of both social support and chronic role stress. Since these factors are not independent of one another, they should be conceptualized in relation to the latent variable marital relationship. A factor analysis of 32 items measuring various dimensions of positive and negative marital interaction produced one latent factor. Items of emotional closeness and concern had the highest factor loadings. The whole scale may be used as well as single-item, crude indicators of the general appraisal of the marital relationship. These variables were shown to be correlates of important dimensions of outcome of the stress process.

Introduction

The conjugal bond is a major interpersonal relationship in the social network of adult persons. The spouse is frequently the most important attachment figure and the closest emotional relationship that individuals possess. Core human requirements for affection, intimacy, social approval and belongingness are met by every-day positive social interaction between the marital partners. In crisis situations, such as serious illness, the spouse is frequently the chief source of socio-emotional and tangible support. Turner et al. (1982) have defined experienced social support as follows: "Social support thus refers to the clarity or certainty with which the individual experiences being loved, valued, and able to count on others should the need arise" (15). Much social support is the product of every-day family life of which the receiver is not aware until some exigency, such as widowhood or divorce, separates him or her from the other marital partner. The measurement of this unexperienced social support poses a major problem for the adequate operationalization of social support as the "provisions of social relationships" (Weiss, 1974).

The marital relationship can also be a source of chronic

life stress. Marital conflict, non-reciprocity in give and take and emotional isolation are important dimensions of chronic role strain in the marriage (Pearlin et al., 1978). A lack of common interests and emotional closeness, as well as the absence of genuine involvement and concern between the marital partners, lead to what Weiss (1974) has termed "the loneliness of emotional isolation". Socio-emotional support and emotional isolation are thus two sides of the same coin. They are the gratification or non-gratification of basic needs for affection, security, trust, intimacy, belongingness, affiliation and approval (Kaplan et al. 1977). The marital relationship in it's twofold function as a source of stress and of social support is thus a major aspect of the social context in which adaptation to a life event occurs. Stress and support variables are not independent of one another.

Individuals can be categorized in three groups: 1. those possessing a close marital relationship, 2. those married but not possessing such a relationship and 3. the unmarried. Whether the social institution marriage or the possession of an emotionally close heterosexual relationship are decisive here is an empirical question for further investigation. The marital relationship can be conceptualized relative to key dimensions of social interaction and social support. These include the following:
- emotional closeness and affection
- involvement and intimacy
- adequacy of the sexual relationship
- common interests and gratifying leisure activities
- degree of conflict and the non-reciprocity of give and take

MATERIALS AND METHODS

These theoretical dimensions of conjugal interaction were operationalized with such items as:
- my spouse showed me in the hospital that I meant very much to her
- she is always there for me and my problems
- I could talk over my inmost feelings and fears with her in

the hospital
- she understands my problems and helps me to master them
- my spouse is affectionate toward me
- my spouse is someone who is a good sexual partner
- my spouse appreciates me just as I am
- my spouse tries to force her will on others

Three other items were developed to measure the general quality of the marital relationship. The <u>general appraisal of the marriage</u> as good was a distinct correlate and thus validation of these separate items. Sexual and emotional closeness, experienced social support following the illness and chronic marital strain were all significant correlates of <u>appraisal of the marriage</u>. (The measure of association was GAMMA.)
- emotional closeness (.692)
- adequacy of social support (.675)
- domineering, Xantippean (-.674)
- good mate (.643)
- esteem support (.660)
- marital conflict (-.541)
- lack of common interests (-.520)
- ability to confide (.381)
- good housekeeper and mother (.459)

A general appraisal of the marriage would thus seem to be an adequate surrogate for and crude indicator of both measures of social support and chronic role strain. Three such measures were adapted from Pearlin et al. (1978):
- How often do you feel that your marriage is a strain on you?
- How often do you feel that you married the wrong person?
- How often do you feel that you're not the husband you would like to be?

RESULTS

As part of a longitudinal study on adaptation to life with a myocardial infarction, a large representable sample was interviewed on their marital life by means of a written questionnaire. A factor analysis was performed on 32 items measured at T2 in the second wave of the study (N=682). One latent factor emerged, with the highest loadings being on measures of emotional closeness, experienced socio-emotional support and general appraisal of the marriage. The eigen-value was 16.1, the factor explaining 50.4 % of the variance in the item set. Items with high factor loadings included the following:

	Factor loading
- warm-hearted, affectionate	.847
- social approval from spouse	.843
- involvement and concern in adjustment to illness	.835
- reliable alliance	.817
- experienced love and affection	.808
- moral support following illness	.797
- appraisal of marriage	.767
- support in changing life-style	.754
- adequate sexual relation	.683
- Xantippean	-.665
- lack of common interests	-.652
- lack of involvement and concern	-.635
- lack of common friends	-.581

CONCLUSIONS

A good marital relationship would seem to be characterized by high benefits in the form of experienced and non experienced social support and low emotional costs. A poor marriage, on the other hand, would be characterized by high costs in chronic role strain and low social support. On the operational level it is difficult to distinguish between the supportive and stress-inducing aspects of an interpersonal relationship. For this reason the quality of the relationship can be used to measure both the social support and the chronic role strain originating in the relationships. Crude indicators of the appraisal of the relationship as well as a large scale, encompassing central dimensions of interaction, can be used to operationalize the latent construct. The marital relationship was seen to be a determinant of major outcomes of the stress process, such as negative affect (-.437), self-derogation (-.574), sense of control/mastery (.516) and depression (-.323), (Vid. Badura/Waltz, 1983; Badura et al., 1983).

REFERENCES

Badura, B., Bauer, J., Kaufhold, G., Lehmann, H. and Waltz, M.,in press. Zur Bedeutung von medizinischer Versorgung und von Selbsthilfe im Laiensystem für die Herzinfarktrehabilitation. In: C. von Ferber and B. Badura (Editor), Selbsthilfe und Selbstorganisation im Gesundheitswesen, Oldenbourg-Verlag, München.
Badura, B. and Waltz, M., in press. Social support and the quality of life following myocardial infarction. Soc.Ind. Research.
Pearlin, L. and Schooler, C., 1978, The structure of coping, J. Hlth. Soc. Behav., 19:2-21.
Turner, R.J., Frankel, B.G. and Levin, D., in press. Social support: conceptualization, measurement and implications for mental health. Res in Comm. Ment. Hlth., 3.
Waltz, M., 1982. Neue Trends in der methodologischen und theoretischen Erfassung sozialer Unterstützung und Netzwerke. Mimeographed.
Weiss, R., 1974. The provisions of social relationships. In: Z. Rubin (Editor), Doing unto others, Prentice Hall, Englewood Cliffs, N.J., pp. 17-26.

This scale was developed as part of the Oldenburg Cardiac Rehabilitation Study, funded by the West German Federal Ministry of Research and Technology.

PART 2

Human performance and breakdown in adaptation

edited by

H.M. Wegmann

HUMAN PERFORMANCE IN TRANSPORT OPERATIONS
INTRODUCTORY REMARKS

H.M. Wegmann

DFVLR-Institute for Aerospace Medicine
D-5000 Köln 90, Federal Republic of Germany

Abstract

The author reviews the general problem of research of human performance decrease in stressful conditions and indicates some issues common to transport operators of air, road, sea and rail systems.

In principle, investigations on human performance in transport operations are dealing with man-machine systems that are designed for the special purpose of movement and of carrying people or goods from one place to another. Fig. 1 presents the basic elements of such a system. By certain rules and through various interfaces the main components (man and machine) correspond and interact with each other. The "behaviour" of both is influenced by a variety of factors from the environment. In contrast to most other man-machine systems, the operator is moved when the system is set in motion, and in addition he is able to move in the inside of the vehicle. The specific task of the operator in this system is to control and guide the vehicle in accordance with mission demands. Particular skills in perceptual, cognitive, and motor functions are necessary to cope with this task. Optimal performance of

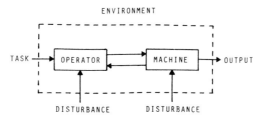

Fig. 1 Basic model of transport system.

the system requires high-level performance of the operator. Human errors can induce malfunctioning of the system with serious risks not only for the system itself but also for the environment where it moves. Characteristic for transport systems is the considerable potential of damage, a compulsary consequence of the high energy level that is needed for movement and transportation. Any degradation of operator performance from whatever cause can lead to an increase in the probability of accidents. From to-day's statistics it must be concluded that human factors are far more causes of accidents than technical failures. This is true for all transport systems, and it follows that operator performance is of central significance for traffic safety and reliability.

The integration of operators into transport systems takes advantage of the human as a major source of flexibility and adaptability. Many "internal" and "external" factors can affect operator performance, as outlined in Fig. 2. The complex interaction of man- and machine-dependent conditions determines coping or non-coping adaptation. Special attention must be paid to the natural capacities and limitations of the human organism. Physiological, psychological as well as organisational and environmental aspects have to be considered to prevent break-down in performance. Some key properties of modern transport give rise to frequent conflicts with those limits. There are indications that these conflicts and a permanently

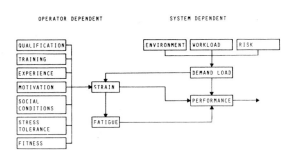

Fig. 2   Operator- and systems dependent factors influencing performance

high pressure on operators may lead in some areas to a higher rate of absenteeism or disease.

From the many issues pertinent to the various transport systems, some appear to be of particular relevance and of common interest for all parties involved. In the following we will briefly mention a few and it is hoped that the reader will understand that to present a complete list of all problems arising from operating transport systems of today would be beyond the purpose of introductory remarks.

As mentioned above inadequate applications of human factors are a major source of accident causes. It is, therefore, all the more surprising that so little is done with respect to training and education in this area as F.H. Hawkins pointed out in his excellent chapter on this topic for the field of air transport. There is conformity among experts that conditions are quite similar in other traffic systems and much needs to be done to improve the situation.

The problem of irregular work-rest schedules is a matter of unique concern for all traffic systems. The acute effects upon operator performance and sleep as well as their impact for his social and domestic life are widely recognized, but not thoroughly investigated. Serious long-term consequences for health and well-being are yet to be studied. Items such as extended work periods and their maximally tolerable duration are also part of this problem.

Medical and psychological care with respect to selection and surveillance has been observed with different emphasis in the past. In particular in the area of aviation this aspect has been taken seriously and sufficiently considered. In other traffic systems the situation is unsatisfactory. Long-term epidemiological studies are needed. There are indications that the higher loads on operators and specific problems of nutrition can be the cause for higher rates of disease and absenteeism.

Future trends in vehicle design indicate a considerable increase in automation. Consequences for operator performance are poorly understood. "Encephalization" of the task, monotony and underload, boredom, arousal and vigilance, reliable functioning in emergencies are some of the key aspects of the problem. Parameters and methods need clarification. Relevant simulation studies could be also part of this topic. Since simulators are extremely expensive and difficult to operate, cooperative efforts could be of particular advantage and the use of unique facilities could be maximized, if their utilization would be shared by several research groups.

Alcohol and drugs have been realized in the past to be a major problem for transport operations. Much attention has been paid to the exploration of the negative effects upon performance. In future research, emphasis will shift more to investigations on measures to prevent abuse. An important objective in this category will also be the safe applicability of certain medicaments. Short-acting hypnotics for sleep disturbances and antihistamines for allergic states are examples of major implications. But also analgetics, anti-hypertensives, antidepressants and stimulants are drugs that may be of some relevance in this context.

In the past human factors researchers have generally ignored essential segments of the population, as for instance the female the very young and the older operator. Much basic research is still needed with respect to differences in stress tolerance, adaptibility and behaviour. The age limitation for operators is presently a matter of intense discussion and needs scientific clarification.

Finally, environmental factors such as vibration, heat and noise, are also problems that have general relevance for performance and health of the operator in the different traffic systems. Although an extensive body of data already exists, little has been done with respect to norms and standards, at least not on the European level.

AIRCREW WORKLOAD

S.R. Mohler and H.D. Nichamin
Wright State University School of Medicine
Dayton, Ohio 45401 - U.S.A.
(with R. Sulzer, Aviation Psychologist, Linwood, N.J. and
W.J. Cox, Aviation Consultant, Arlington, Virginia)

ABSTRACT

Advances in biotechnology, psychology, physiology, aerospace medicine, bioengineering and computer sciences (especially the development of microprocessors) have made possible increasingly refined studies of aircrew workload requirements. The workload required by a given aircraft flying a specific mission can be quantified in regard to each crewmember. The categories for quantifying workload assessments are: mental functions ("cognitive workload"), perception and motor. This paper addresses the workload elements within each category. In addition, the overall summation of workload characterizing the total mission workload is discussed.

DEFINITION OF WORKLOAD

The workload imposed by a particular aircraft and flight mission is comprised of perceptual (visual, aural, and tactile senses), mental (cognition) and motor (musculoskeletal activity) requirements. These are covered in a previous report (Mohler, 1981). All aircrew workload aspects can be assigned to one or another of the above categories depending upon the primary characteristics of the specific workload aspect under consideration. Within the categories, demands on the operator at any given time can be quantified. Communications demands between crew members or between the crew and external stations can be listed under mental workload.

THE AIR CREW MEMBER

Self-perceived and actual workload are directly related to the characteristics of the individual crew members. These include achieved skill level, recent proficiency, general physical fitness, and the state in the rest/fatigue cycle, including degree of anxiety (if any) and presence of illnesses or chemical dependencies (alcohol, nicotine, drugs - these constituting additional self-imposed stresses due to cyclic withdrawal phenomena).

MODERN DEVELOPMENTS

Cathode Ray Tube (CRT) attitude and station displays, with information isolation and display magnification capabilities, accurate time to station readouts with True Air Speed (TAS), Ground Speed (GS), winds, and other pertinent data, including fuel remaining information, flight duration capability, and continuous weight and balance details, have decreased cockpit cognitive and sensory workload tasks. Magnifiable terminal and enroute CRT displays, plus cockpit displays of key vicinity traffic, will further aid the crew in mission accomplishment.

In effect, the modern trend in aircraft development is toward "encephalization" of flight management, that is, increasingly delegating monitoring, "motor" and "housekeeping" tasks to automated systems. The crew member roles are evolving as computer and systems managers, retaining, of course, basic fall-back skills in the event of various sytesm failures.

MEASURING WORKLOAD

Figure 1 illustrates example classifications of studies that assess workload. Some detailed breakdowns of the subcategories are available (Wierwille and Williges, 1978). It is necessary that workload measures undertaken in the laboratory, in simulations or during flight, be repeatable, non-interfering, readily scored and characterized by cause and effect relationships. The mental factors relating to perception, decision making, recent and past memory, stage of alertness, command capability, emotional state, and motor capability, are amenable to measurement (Mohler, 1979).

Individual aircrew factors that determine workload level for a specific flight are skill level, practice (proficiency) level, motivation, fatigue status, emotional state (including anxiety level) and general health status.

```
FIGURE 1 - Workload Study Classifications

    I.   System Factors
         A. Task Demands
         B. Aircrew Performance
            1. Aircraft
            2. Simulator

    II.  Task Characteristics
         A. Perceptual Category
         B. Cognitive Category
         C. Motor Category

    III. Measurement Areas
         A. Subjective Grading
         B. Objective Measures
         C. Psychophysiologic Data
         D. Available Spare Capacity

    IV.  Accident Analyses

    V.   Mathematical Models
```

Examples of workload methodologies for measuring various key aspects are given in Figure 2 as adapted from Wierwille and Williges (who cite Berliner, Angell and Shearer, 1964, as a prior source). Some of these have been applied in assessing the crew station requirements of present aircraft

```
FIGURE 2  Dimensions of Operations

    I.   Perceptual Processes
         A. Searching for, and receiving,
            information
         B. Identifying objects, actions,
            events

    II.  Mediational Processes (cognition)
         A. Information Processing
         B. Problem Solving and
            Decision Making

    III. Motor Processes
         A. Simple and Discrete
         B. Complex and Continuous

    IV.  Communication Requirements and
            Processes
               (cuts across all categories
                above)
```

(Boeing 747, McDonnell Douglas DC-9, DC-10 and Lockheed 1011). Documentation of the applied workload studies in civil air transport aircraft to date are reported (Sulzer et al, 1981).

ALERTING SYSTEMS

Figure 3 describes the increases in alerts (cautions, advisories and warnings - visual and auditory), that have occurred in jet transport aircraft since turbojet introduction (Veitengruber, 1977). These increases cannot continue indefinitely as they can potentially result in information overload to the crew. A new approach toward consolidation and integration is in process.

New flight displays (some pictorial), whether "head-up" or panel, of CRT, plasma or other non-mechanical constructions, will in the future off-load present perceptual demands. In addition, second-order cognitive extrapolations as now required by conventional mechanical displays will be lessened if not essentially eliminated.

FIGURE 3 Jet Transport Alerting
Systems Since Introduction

Increase

200%	in Cautions and Advisories
100%	in Wide-body compared to Narrow-body
100%	in Visual Alerts
60%	in Warnings
50%	in Auditory Alerts

SPECIFIC WORKLOAD MEASURES

Aircrew workload measures fall into the categories of Perception, Cognition, and Motor. Perceptual processes are those of information gathering, the primary in-flight receptions occurring through visual, auditory and tactile receivers. Cognitive functions are those that require central cerebral processing, understanding, verbalizing, remembering, emoting, monitoring, commanding, cooperating, alerting, and related brain activities. Motor functions include the conscious or unconscious musculo-skelatal and organ movements that carry out the nervous system dictates, constituting the human/machine actuator interface.

Workload methodologies are primarily classified as either subjective or objective. Examples of subjective methods include conjoint measures of workload, concurrent discrete or continuous reports, polar adjective scales, semantic differential scale ratings as to preference, three-attribute rating scales of mental workload and physiologic index rating scales. Objective measures include embedded secondary task measures (communications, time estimation), activity intensity measures, steady state evoked potential responses, transient evoked potential responses, times required for conscious mental processes, EEG correlates of attentional load (event related potentials), self-adaptive and cross-adaptive secondary tasks, speech pattern analyses, electromyographic measures, cross-coupling appendage measures, communications load measures, measures of aural and visual warning discretes, oculomotor eye movement measures, and cockpit display analyses in terms of bits of information presented (information theory).

The above methods permit each aircraft to be characterized by workload demand in regard to any point in a flight mission profile. An important aspect of these measures is the opportunity they present to assess the "spare capacity pools" in each crew member that exist in the perceptual, cognitive and motor categories, respectively, at any given point in the flight profile. The implications of certain operational regimens for diminishing these capacities can be identified and quantified, with remedial steps prescribed.

In addition to the above, a given crew schedule can be assessed quantitatively for its "physiologic demand potential" by a physiologic index (Mohler, 1976). This index scores the physiologically demanding aspects of multiple night flights, 24 hour layovers after an evening arrival, multiple time zones, and, or multiple transits, contained in a given schedule pattern. It identifies the "hot spots" in a schedule that are apt to be the most fatiguing.

REFERENCES

Mohler, S.R. 1976. Physiological Index as an Aid in Developing Airline Pilot Scheduling Patterns. Aviation, Space and Environmental Medicine. 47:3. March.

Mohler, S.R. 1979. Mental Function in Safe Pilot Performance. Human Factors Bulletin. Flight Safety Foundation. January/April.

Mohler, S.R., Sulzer, R.L. and Cox, W.J. 1981. Elements of Aircrew Workload. Human Factors Bulletin. Flight Safety Foundation. Arlington, Virginia. January/April.

Sulzer, R.L., Cox, W.J. and Mohler, S.R. 1981. Flight Crewmember Workload Evaluation. DOT/FAA RD/82/81, ASF-82-1. Federal Aviation Administration. Washington, D.C. April.

Veitengruber, J.E., Boucek, G.P., and Smith W.D. 1977. Aircraft Alerting Systems Criteria Study. Report FAA-RD-76-222. I. Federal Aviation Administration. Washington, D.C. May.

Wierwille, W.W. and Williges, R.C. 1981. Survey and Analysis of Operators Workload Assessment Techniques. NATC. SED. U.S. Navy. Patuxent River, Maryland. September.

SAFETY, INDIVIDUAL PERFORMANCE & MENTAL WORKLOAD IN AIR TRANSPORT:
OEDIPUS AS ICARUS

P. Shipley

Department of Occupational Psychology
Birkbeck College, University of London
Malet Street, London WC1 E7HX
U.K.

ABSTRACT

The increasing practical need to investigate the mental content of jobs to maintain performance quality is illustrated with examples from air transport. The state of knowledge in workload is reviewed in a re-statement of the debate about psychology as a natural or social sciences discipline. A synthesis of the two approaches is recommended for mental workload. Studies of the natural sciences kind have produced some interesting findings and should continue. The interdisciplinary attack on this complex subject needs to be broadened, however, to gauge the influence of social factors important for stress and performance at the level of the individual operator. The paper then returns to the issue of generalisation, and asserts the need for more naturalistic studies conducted either in field settings or good simulations of them. Problems of generalisation to and from specially selected and trained groups in transport are also raised.

INTRODUCTION

Much has already been written about the subject of mental load. An exhaustive treatment of such complex literature is outside the scope of this paper and a personal view is given instead. The physical stresses of flying have been well documented elsewhere, (see Ernsting, 1978, for example). There are also useful texts available for lay readers on general 'human factors' aspects of aircraft pilotage. In particular, see Reason (1974). The bias of this paper is more towards the social contributors to performance and safety, which have so far received little attention in the field of air operations.

WORKLOAD : MEANING AND DEMAND

Like 'stress' workload is a fuzzy, ambiguous concept. Attempts to define it exactly have been consistently defeated. Reasons for this include two which fit the argument of this paper. One is that such concepts are borrowed from mechanics, physics and physiology. The other is that human beings are exceedingly complex and we are a long way from understanding ourselves. Yet we continue to think 'there must be something in it' and if the scholar in us protests at this lack of precision, our human nature finds the idea intuitively appealing. We all know what it is like as private persons, to feel stressed or strained, and we often feel our workload has something to do with this.

Despite recognition of this definitional problem among scholars for years, the concepts are now part of our everyday language and requests continue to be made, for help with practical problems cast in such terms. Take for example the controversy in the airline industry over the issue of reducing the flightdeck crew complement from three to two members as a general norm (see Europilote and US ALPA, 1981). Manufacturers and employers claim that automation of the modern flightdeck will make a third crew member redundant. Pilots on the other hand, refute this claim on grounds of safety. They call for 'workload studies' (op cit, page 2), and state that: "There is going to be more stress in the cockpits of all future airliners." (ibid, page 15). Reference is made to heart disease as a stress disease and "....a real threat of shortened careers due to the excess stress in a two-pilot cockpit." (ibid, page 16). ...."mental workload is the real crux of the issue." (ibid, page 36).

The author was a member some years ago of a 'human factors' team project based at the University of Loughborough set up to study the contribution of navigation and communication to flightdeck workload in airliners. Radio telephony was recorded into a standard domestic tape recorder directly from the flightdeck's electrical system, and activities of the crew recorded using a code agreed with operators and a manual (32 switch) event recorder, carried on the observer's lap. Both streams of data, radio-telephony and coded events, were analysed and integrated on to a 2 or 3-pilot multiple activity chart.

We found considerable variability in workload type and intensity on the same routes on different occasions even in relatively similar daytime conditions. The interrelation between ATC and pilots in their membership of a larger system meant that flightdeck workload could not be viewed as if in a vacuum. ATC load led to changes in strategy in that environment which would affect flightdeck load as a result. The average length of communications between flightdeck and ATC was about 4½ seconds in a sample of flights on one home sector, and congestion at one terminal area in that sector on one particular flight led to halving of that average message length. Instead, content analysis of messages for the flight showed that flight level clearances given to the captain were double the average found in this sector, and frequent level checks were requested with fewer acknowledgements given. Other aircraft in the vicinity queried several times the communication frequencies they had been given, since many messages had been drowned by simultaneous transmission. Traffic congestion invariably led to increased time spent 'looking out' by all pilots, and the sharing of the load among the crew was illustrated in the multiple activity charts.

Norms, and rules of thumb exist for containing physical workload and physiological effort but it has not so far been possible to do this with mental workload despite the need for it and the increasing mental content of jobs in modern industry. There are no such codes of practice for mental stress and mental health, at the workplace. Because linear relationships between mental load and performance impairment do not hold up even in tightly controlled laboratory studies, the notion of dose-response relationships appears misplaced in the field of mental workload.

In their account of the mid-air collision in 1976 between a British Airways Trident and an Inex-Adria DC9 over Zagreb, Yugoslavia, when all 176 people on board these aircraft were killed, Weston and Hurst (1982, page 171) state there were no ATC regulations in Yugoslavia stipulating maximum workload, and that the shift supervisor thought they were undermanned at Zagreb with 30 men doing the work of 40. When the Darmstadt team directed by Rohmert studied the Frankfurt controllers a few years ago, the simultaneous processing of four aircraft was thought to be about the maximum for a trained controller, (Laurig et al, 1971), whereas the safe limit is now considered to be about three times more, thanks to technological advances. One controller (Tasic) at Zagreb took sole blame for the accident in 1976 despite (we are told) the existence of Swedish radar at the Centre over the previous two years, but which had not been put into service because of missing ancillary equipment. A witness in the court case stated that the Centre had to rely heavily on procedural control. At the time, Tasic, who was sentenced to seven years in prison, was said to be handling alone eleven aircraft simultaneously. Regulations for hours of work did exist in Zagreb. We are also told that Tasic had been scheduled to work 50 hours in the week of the crash, 8 hours more than should have been prescribed.

AN OLD DEBATE : PSYCHOLOGY AS NATURAL OR SOCIAL SCIENCE

Two general approaches inform psychology. The 'nomothetic' or natural sciences approach is grounded in the scientific method and is exemplified by many activities in the field of 'human factors'. This approach seeks to establish general laws of human behaviour based on aggregated findings taken from homogenous groups of subjects, and its attractiveness lies prime facie in the possibilities it allows for providing engineers and other designers with solutions to practical problems. An antithetical view is taken by others who fill a variety of professional roles including the counselling and clinical. Their tradition is 'ipsative', their interest is in individual differences in behaviour and their embedment in particular cultures and societies at certain times. The former view dwells mainly upon people as biological organisms, whereas an interest in the social influences on behaviour prevails in the latter which adopts a more historical perspective.

The view taken here is that these contrasting approaches need not be conflicting, but complementary, and this synthesis could lead to a better understanding of performance and safety problems. The natural sciences approach looks for causes, whereas the social approach is interested in reasons. The 'eclectic' middle-ground position takes from both, proposing that it is helpful to assume that people can be both self-determined and other-determined. Social determinants of behaviour have tended to be neglected in the 'human factors' literature because they are less visible, less easily controlled for or measured in experimental laboratories. A broadening of the frame of reference to encompass these social influences will require a widening of skills and knowledge in the inter-disciplinary 'human factors' teams of the future.

A current humanistic view is that of Taylor (1981) in his article applying hermeneutics to the field of road safety. While it is agreed with Taylor that there are limits to mechanistic explanations of human error, and that over-subscription to models and analogies borrowed from the natural and engineering sciences can blind us to other explanations, it is not proposed here to replace the mechanistic with the humanistic, nor the nomothetic with the ipsative. We have so far managed to establish some useful facts and generalisations about behaviour. No two people are exactly alike but people also have things in common. Rather it is proposed that the knowledge we have accumulated so far of human behaviour and cog-

nition in mechanistic, nomothetic studies should be enriched with more
systematic investigation of the processes of motivation, emotion and
attitudes, and related ideas of subjective meaning, personal plans and
intentions. Individual differences are not necessarily random differences
lying beyond the process of abstraction.

SOCIAL SOURCES OF REGULARITY IN BEHAVIOUR

Putting people together in homogeneous groups of air traffic controllers, professional drivers and divers, aircraft and ships pilots leads to
behavioural regularities. It is at the level of a particular individual
that irregularities and uniqueness assert themselves. A person's mind
not only has biological propensities and limitations, it is also a mirror
of the society of which it is part. It is also capable of influencing
that society through our main medium of communication, language, and
through behavioural example. Being human is to have the capacity for
thought, for language, for self-consciousness manifest in comparisons of
personal standards and performance, with those exhibited by others in
reference groups important to us.

Several 'human factors' specialists have referred to the need to
study motivation and personality, but hardly any have done so in depth.
The competences of 'human factors' specialists typically lie elsewhere,
dealing with biological properties in nomothetic studies. They are more
inclined to see pilots as components of man-machine systems rather than
as persons and social beings. The application of ergonomics to the field
of aviation should not be curtailed now however, since technology continues to make demands on our biological adaptive capacity. But rarely are
behavioural and attitudiual factors studied together. An exception is
Bainbridge's work on mental load and mental modelling in plant process
control, (Bainbridge, 1974). She drew on her own field studies of
process controllers, also on work done in Europe at the time of ATC. She
noticed how performance differed with external task demand or load, with
individual capacity, with practice and experience, with fatigue and with
motivation. She described motivation, as 'the operator's willingness to
use his potential capacity optimally', (op cit, page 289) and emphasises
the non-linear and discontinuous relationship between task demands and
performance. Operators will change strategy as a function of task and
environmental constraint. The skilled and experienced can be relied upon

to use efficient strategies when demands increase. Flexibility and strategic range or repertoire is acquired with skill, and capacity appears limitless.

Studies of mental workload rooted in the nomothetic tradition emphasise what operators have in common, using non-human, especially engineering analogues, for describing such properties. A popular example in lay and ergonomic circles alike is the single-channel model of information-processing; and the information 'bit' is the metric for quantifying mental load. More recent models of higher cognitive functioning, such as serial and parallel-processing, are variations on the same theme. It would be interesting however to know what social factors influence changes in strategies. Behavioural, motivational and task factors interact to confound variables rendering the general prediction of performance inexact. Biological and social constraints together hold standards up. Unfortunately, factors like fatigue, circadian and ultradian rhythms, practice and ageing, limit performance from one direction; while social factors influence it from another.

Social factors, on the other hand, must not be emphasised to the exclusion of other important contributions to performance breakdown. In a later paper Bainbridge (1978) points out that experienced controllers have lower workloads because of enlarged individual capacity for the task, but unexpected demand may break down performance because unfamiliar decisions have to be made. Such demand can be expressed in qualitative information-theory terms of uncertainty rather than numbers of variables to be controlled per unit time. Habitual responses inappropriate to the situation, rather than over-loading, could be the reason for error.

We know that decisions and perceptions are not made in an all-or-none manner, and that previous experiences and other expectations result in the operator putting a probabilistic value on events. It is not unusual for performance breakdown to stem from an incorrect hypothesis. Perhaps the best operators always anticipate the unexpected, but even many of them may not do so under intense emotional stress, or when fatigued or drugged. Flexibility means the ability to switch from open-loop to closed loop modes when either internal or external factors require it. Simulators can be useful tools for allowing practice of alternative strategies with impunity.

But the required flexibility may be of a different order. We know the upper limits to human performance in tracking studies for example; typically a second-order level of control and a 5-seconds lead time constant, (Moray, 1982). No doubt learning to control such multi-variable processes and plants, especially under unstable conditions, can be stressful to the trainee, especially if that trainee cares about actual performance being below what could be achieved. We need to know more, however, about what makes people care or not, especially when skill and knowledge are at about the same level within a population.

Motivation has both a directional and an intensity component, factors which become salient when few obvious constraints apply. We need to know on what basis choices are made: to vary speed or accuracy; to adopt this or that strategy to attain a given goal; to change one goal for another. Perceived failure or anticipated failure to meet the goal set may lead to emotional stress (anxiety, fear, worry) in the susceptible. Long-term criteria may be sacrificed, temporarily or permanently, for shorter-term expedients. Some individuals will be less bothered by this than others. Social group membership also affects how work is done, and the experience of stress in itself, or its anticipation, may impair performance even further once performance degradation has set in.

SOCIAL FACTORS AND SAFETY STANDARDS

Experts have concluded recently (see for example, Singleton 1982), that accidents occuring in developed countries, including accidents to airliners, have reached an asymptote somewhere above zero. Safety rates may appear good given the complexity of the systems and technology involved, primarily because of the extent of performance control achieved in such societies, including the quality of the performance expected at the level of the individual. The latter, it is proposed in this paper, is broadly a function of two factors intrinsic to the individual, i.e. factors of skill and attitude, as well as many external factors impinging on human performance. Frequency rates and objective probability risks for such events to complex systems are exceedingly small, but when they do occur they are dramatic and costly. They are costly in both material and non-material terms. The extent of the subjective costs of these events is not known. A consideration of hitherto somewhat neglected social aspects may help to bring down the accident rates of the developed countries nearer to the null or zero point.

Singleton (ibid) speculates that some kind of social norm, a level of acceptable risk, is responsible for this status quo, and that the attempt to decrease levels in one area such as the occupational, is compensated by people satisfying their need for risk in another area, in leisure pursuits say. The problem with this view is that the explanation given for a social phenomenon is a mechanistic one, and it has something in common with the homeostatic notion of 'danger compensation' in road driving risks discussed by Brown (1980). The implication is that there are fixed needs in drivers like the need for essential nutrients, incapable of modification. Why should risk-taking in the air or on the road not be seen instead as a value? Rules governing driving are interesting. They can be externally and internally induced; overt, e.g. speed control,and covert, such as courtesy for other road users and a respect for their lives. It would appear that in order to reduce accident costs still further we need to understand why individuals vary in their awareness and compliance of such rules. Unsafe, young male drivers could be understood partly in terms of limited skill and awareness, and partly in terms of other personal attributes. 'Macho' values of a parent culture may be reflected in exhibitionist behaviour at the wheel of the car and the identity of the male ego with aggressive control of a potentially lethal weapon. Allnut (1982, page 15) takes a similar line on risk taking and flying. "Our culture and training teach us that men are brave; that it is natural, desirable and indeed heroic to struggle through adversities such as pain, bullies, strong teams of opposing sportsmen, and enemy troops. Man's lowly instincts therefore often generate a contempt for bad weather, or a risky flight, as just another hazard which will yield to bravery. This distortion of logic is compounded by social attitudes offering a combination of censure and sneaking admiration for the man who breaks the rules and wins; while of course, the wrath of righteous indignation is called down upon the man who breaks the rules and loses." ...."Pilots are trained to control the urge to express 'manhood' by flying dangerously, and the urge *is* controlled, but not entirely destroyed." Similarly, Roscoe (1982, page 153) remarks: "There is a giant pride associated with breaking a wild horse or the masterful manoeuvring of an unruly aeroplane at the edges of its performance envelope. The 'right stuff' is required and, like manhood, must be defended."

Meaning can be unique or shared. Fears can be idiosyncratic, as are many phobias and obsessions (such as the phobia a ship's pilot may develop about going onto the bridge of a ship). Threats are also shared and common-place such as the threat to life and limb and the more subtle social fear of public-speaking which entails no obvious objective danger to life. Values permeating a whole society, or one of its sub-cultures or organisations and institutions, percolate down to the level of the group or the individual. Sometimes these matters are relevant for safety, stress and health. Edwards (1981) has taken a speculative interest in the topic of professional discipline for aircraft pilotage. Systematic research is also needed.

Professional groups have their own codes of practice, written and unwritten. In a study of ships pilots Shipley & Cook (1980) noticed that the norms, habits and lifestyles of seafarers inculcated whilst at sea as bridge officers were carried ashore by the pilots and may have had some part to play in their health status. Flying instructors are important role models for young aircraft pilots, as embodiments of attitudes towards safety and professional discipline. They also have authority and constitute a source of praise or criticism. Social pressure may have operated in the case of the fatal crash on the runway of London airport of a Vanguard airliner 'Double-Echo' in 1965, on overshoot. This was a domestic night flight when a marginally incorrect runway visual range value was communicated to the crew as was a landing made successfully just previously by another airliner. Couching such judgements in pseudo-rational terms about 'pay-off', borrowed from econometrics, or in engineering terms like 'signal-detection', may take our attention off important social considerations. Such terms, like 'workload', are merely convenient metaphors.

In 1972 a British Airways Trident airliner crashed on climb outside London airport killing all on board. Some of the recommendations made in the report of the 'Papa India' enquiry were of the conventional 'human factors' kind, such as the introduction of a speed-operated baulk to prevent premature retraction of leading-edge slats or droops. Others required thought to be given to a consideration of broader socio-organisational factors to improve communications. It is not known, for example, how far the junior pilot's status on the flightdeck inhibited his monitoring and checking duties. He may have felt intimidated in the presence of an angry senior captain who had just left an argument in the crew room over industrial action and whose reputation included identification with management.

To assert that social rules will influence behaviour to produce assessable regularities is not to imply that people do not take responsibility in principle for their behaviour. Indeed, our legal system is predicated on this and the rightness of it is a matter for debate in seminars on moral philosophy. At Zagreb, Tasic took over the controller's responsibility for the upper sector without assistance, was reputed to prefer to work alone, and was described by a Supervisor as one of their best controllers. It is clear that Weston and Hurst believed that a conscientious operator had been scapegoated to cover up a 'ramshackle structure' (op cit, page 166). However we describe Tasic's personal style he paid dearly for it. Shutting down a chemical or nuclear process plant is expensive and rule-bending may be encouraged and safety margins adjusted to save costs. Such rule-bending may actually be 'rewarded' by less safety-conscious bosses.

PACING AND WORKLOAD VARIATION

Researchers of mental workload in the field of air transportation have shown a predilection for seeing all load problems as overburdening; see for example the report of the evaluation of the New Jersey electronic journal using mental workload as its theme (Sheridan et al, 1981). A subjective rating scale for pilot mental workload is given (op cit, page 108) with three categories of 'satisfactory-acceptable-unacceptable'. The description against the 'satisfactory' rating is interesting for uncovering its designer's assumptions: "There is plenty of time to perform all tasks with generous idle time between them. There is little or no mental effort required, and no emotional stress."

Coping with workload variation rather than load as such may be a major stress. Human beings make notoriously poor watchkeepers and are easily distracted from the task by irrelevant external stimuli or inner thoughts, reveries and ruminations. Perhaps the glamorous image of flying blinds us to its less heroic side, and we overlook the long hours of sustained vigilance on a routine flight. ATC has its own stress jargon among which is 'bounce back'. the ability to recover alertness in conditions of sudden demand after a lengthy spell of low workload; i.e. the flexibility to adapt quickly and easily when workload varies between high and low levels. We need a methodology for studying workload variation.

The three year study by Rose and his associates (1978), of some 400 controllers from a range of centres in the North Eastern States of America has provided strong evidence of an association in the susceptible hypertensive controller between blood pressure variation and workload (variation in complexity of pattern in the active separation of aircraft). A screening examination at the start of the programme revealed a high hypertension prevalence of 32.5%, using a criterion of 140/90 and above, from readings taken on two successive occasions. This rate had risen to 55% by the end of the three-year period of monitoring. An earlier study, (Cobb and Rose, 1973) compared over 4000 US controllers with 8000 second-class airmen. Licensing bias was allowed for, and the authors found a four-fold higher prevalence of hypertension in the controllers, a statistically significant finding. The disease was found generally to have begun a few years earlier in the controllers, who also had a greater number of ECG abnormalities and higher plasma serum lipid levels than the airmen. Controllers from the higher density centres had significantly higher prevalence rates. We need more research on the effects of workload variation and the external and internal pacing dimensions of pilotage and ATC control, on performance safety and health.

In quiet periods, controllers will go in search of more work, sometimes combining sectors. Another piece of ATC jargon is 'losing the picture'. Weston (1982) describes this as a controller's trauma, and uses the analogy of a pyramid of cards built up in the controller's mind, with the edifice collapsing because of one card too many. The extra card may be an additional aircraft exceeding cognitive processing limits, or it may be one aircraft already in the pack doing something unexpected. The event can wipe out everything in the controller's short-term/working memory store. We need to know how far situations of underload also could bring about a loss of picture. Caplan et al (1975) studied job stress in some 2000 men from a range of American blue collar and white collar occupations including ATC. Controllers did not report having a different level of workload (either higher or lower) compared to other groups, but within the whole sample they had the highest workload variation and highest demand for concentration on job tasks. Controllers at small sites were amongst the most satisfied groups in the sample.

The men from the higher density centres (larger centres) reported levels of role conflict and ambiguity which were significantly higher than those reported for smaller sites. The former also reported significantly lower levels of social support from others at work.

Using Karasek's model (Karasek, 1981) for predicting coronary heart disease from the interacting variables of job demands and decision latitude, we would expect ATCs to be at risk at larger centres, given the reported role conflict and ambiguity, the workload variation and external pacing factors under high load, and low social support. Amongst the occupations sampled by Caplan et al (op cit) controllers were low down the league table for reported cardiovascular disease, but these authors point out that low rates for disease are not surprising for men who are relatively young and who have to pass stringent screening tests to remain in the job. Physiological measures were taken in seven of the 23 occupations looked at in this study and the controllers were highest for serum uric acid (independent of site). FAA scientists have also conducted several studies of US controllers developing a biochemical (stress hormone) index which distinguishes the more demanding (numbers of aircraft) from the lower workload sites. They do not regard the former as typical of the job, however, (Melton et al, 1977); particular controllers may be at risk on particular sites only. O'Hanlon (1981) compared catecholamine excretion rates for various jobs arguing that machine-paced assembly line workers suffer the greatest stress given this criterion, a working group predicted to be at risk for CHD using Karasek's model.

The susceptibility of particular people emerges again, in a British study which distinguished paced assembly line workers from matched controls (repetitive work) in British car plants, for mental ill-health symptoms. Broadbent and Gath (1981) went a step further than usual in seeking salient individual differences within groups. Short-cycle time was less relevant for strain than pacing, and among paced workers those at risk appeared to be more anxious. Of the more anxious it appears that obsessional personality and job involvement are important. The more obsessional people get more anxious it is suggested in pacing conditions which prevent them from frequently checking their own performance.

Explanations of excessive or insufficient load can be psycho-physiological in kind in terms of over-excitation or depression of the cortex, perhaps disadvantaging anxious or depressed people respectively. Another view is more psychosocial, in that some individuals may be more worried about compromising their own high standards through lack of time to check results constantly, as in overload, or through lapses of attention and mind-wandering, when underloaded. When high costs and penalties are attached to human error (as in air transportation) these load problems outside the operators' control may potentially be more distressing to them.

PERSONALITY AND WORKLOAD

Considerable interest has developed since the formulation of the concept in the latter part of the fifties of the notion of a personality at risk for heart disease labelled 'Type A', described as competitive, ambitious and hard driving. Useful reviews of research on Type A are Glass, (1977) and Glass and Contrada, (1982). Price (1982) interprets Type A in social learning theory terms and suggests that it may be a product of a culture (Anglo-Saxon Protestant, say) which emphasises hard work, ambition, personal responsibility and achievement. The disposition may be dormant until drawn out in Type A conditions of work, i.e. challenging and competitive. Several studies have confirmed a link with heart disease. There is also a 'denial' psychic defence component to the personality. For example, in self-paced tasks comparing As and Bs, the former repressed feelings (and possibly symptoms) of fatigue until the task goal was accomplished. The 'achievement' explanation suggests that the ambitious motive of such people to win at all costs is responsible for such denial, while the 'psychoanalytic' view is that As have a subconscious neurotic need for control and any kind of failure challenges a deep-seated sense of insecurity.

Steptoe (1981) gives a review of the relationship between cardiovascular pathology and psychological variables, including Type A. Since the seminal work by Cannon (1915) on the body's flight-fight defensive response to threat and the stress of fear and anxiety, considerable attention has been paid to the assessment of stress, especially using physiological and endocrinological measures. Although the establishment of a clearcut link between organic disease and psychological variables continues to elude us, there have been quite consistent findings of Type A responses to challenging conditions. Typically the reaction is a marked elevation of the sympathetic nervous system compared to Bs, in particular blood pressure (especially SBP), heart rate and catecholamine (stress hormone) excretion rates. We do not know the performance and health implications of responses which are short term in duration. Interesting hypotheses are that CHD links with Type A may reside in the marked fluctuation in levels of circulating hormones, and that physiological response specificity may be a reflection of _psychological_ response style; i.e. the subjective meaning and interpretation of events.

Few studies have been done on Type A personality style in the field of air transportation, and these are confined to ATC. We did not know if the captain of Papa-India was Type A, neither do we know if the crew room argument that immediately preceded the flight had anything to do with his possible heart attack. Caplan et al (1975) assessed the ATC members of their sample of 2000 workers for Type A, finding their scores to be close to the mean for the sample, although controllers from larger sites had higher scores than those from small sites. Interestingly, the controllers disposed toward hypertension in the Rose et al (op cit) study were more Type B than the normotensives. A team at Stockholm University directed by Frankenhaeuser, has done much to clarify the psycho-endocrinological stress response. This team has distinguished a 'cortisol' from a 'catecholamine' factor. Type As demonstrate the elevated catecholamine response to the stress of challenge, but they also do so when required to be inactive. Type As may feel distressed under paced, loaded conditions where a trade-off between speed and accuracy is forced on them. We would expect As to prefer to go at their own pace and to sacrifice personal comfort if necessary to maintain performance levels. Johannson (1981) summarised the Stockholm team's research on paced and unpaced work and neuroendocrine responses. She concluded that compared with self-paced work machine-paced jobs may have better performance but a greater subjective cost, including biological cost.

Studies in the fields of aircraft pilot selection and counselling have been done using a psychoanalytic model. The psychoanalytic approach looks for weaknesses of ego formation as the source of personal problems. Although this author prefers a level of explanation which reaches beyond the Freudian categories implicit in the psychoanalytic approach, the reference to pre-conscious and sub-conscious motivations and emotional instability having implications for safety is a useful one. An early laboratory study, seminal in the field of perceptual defence, was that reported by Bruner (1957) and this was followed by a later confirmatory study by Broadbent (1973). These researchers were able to 'beat' the defences of their subjects by the subliminal presentation of emotional stimuli, and impair the efficiency of cognitive processing.

One of the first applications of this approach to flying was that of Davis when he was a member of the Cambridge Cockpit Studies team headed by Bartlett (Davis, 1948). Davis differentiated two kinds of extreme control movement behaviour in wartime pilots who took part in the studies: the overactives and the inerts. The former were anxiety-prone and it was thought that anticipatory tension contributed to the 'fatigue', a source of skill breakdown and disorganisation, reported by Bartlett (1943). Tension was thought to be related to fear of social disapproval if good performance was not maintained. Since this early work various sub-systems of arousal have been postulated; high emotional arousal, for example, of anxious people, as well as the low activation of fatigued individuals, has been implicated in the 'tunnel vision' thought to be responsible for much pilot error in stressful conditions, when peripheral signals in the perceptual field get overlooked.

Davis' pilots are reminiscent of Hayward's 'anancastic', obsessional pilots whose attention on flying he theorises, can be distracted by inner ruminations, leading to dangerous omissions in behaviour. For Hayward mental workload has to be broken down into an external demand (cognitive stress) and inner subjective component (emotional stress). As a clinician he has counselled many aviators and helped to rehabilitate them professionally. He has also conducted stress research on aviators, parachutists and air controllers. He argues that emotional stress can seriously impair flying performance (Hayward 1976) and that the source of this stress often resides outside the job in a pilot's private, domestic life. People have a stress threshold and the relationship between stress intensity and reaction is not linear. To know this threshold for any one individual requires intimate psychological knowledge of that person. 'Anancastic' pilots are driven by habits and easily distracted by stress. These pilots are to be distinguished from dysthymic pilots who are more like unstable introverts. Excessive rumination, or excessive feelings (rage, anxiety, fear, frustration), are the two broad categories of unsafe behaviour. Overcontrolling and errors of commission are typical of the more tense pilot and errors of omission are associated with ruminators. Disorganisation of skill occurs rapidly when emotional intensity is high. The obsessional, however, has negligible arousal problems in that sense, but may become close-minded under stress and be worse on vigilance tasks; more like unstable extraverts. A recent study has shown a statistical relationship between stable extraversion and success on the RAF flying course, (Bartram and Dale, 1982). Perhaps such personalities get on better with their instructors.

Stoker (1982) failed to find predictive ability in the Kragh defence mechanism test for RAF pilots, and this may be because it was administered as a group test. Perhaps personality can only be assessed realistically by a clinician. This projective test presents stimuli in a subliminal way. There is a visible central 'hero' figure and a less visible peripheral 'parental' figure. Responses are interpreted clinically using Freudian categories. It is used for pilot selection by the Swedish and Danish Air Forces, and it is claimed that it can predict accident behaviour. The less suitable applicant is said to have a fragile ego, personal identity problems, and strong psychic defences. Dixon (1982) believes that ego defences and personality disposition may become intensified, and behaviour irrational under stress. The stress may arise from social conflict not just objective danger. Adaptation in flying reaches to the deepest layers of personality, therefore. In the professional instructor too, institutional norms overlie particular ego structures.

THE ART OF GENERALISATION IN APPLIED PSYCHOLOGY

It is clear from the foregoing that individual performance is not in itself a sensitive and sufficient criterion, not least because of differences in individual strategies under conditions of varying load and constraint, for comprehensive studies of stress and workload. The new synthesis will require of its adherents the investigation of relevant phenomena in more naturalistic settings truly empirical for the study of motives and emotions, in conjunction with a variety of unobtrusive and non-invasive observational techniques and procedures, covering longer time-scales than heretofore. Contrived situations involving artificial manipulations of motivational states which are unrealistic to the people who are subject to such manipulations merely add to our stock of artefactual findings. Exceptions exist in the case of simulations designed to be sufficiently real in psychological terms as to induce a sense of illusion. We are more interested in the natural reactions of stressed workers than the preconceptions of experimenters in laboratories.

The 'new look perception' wave in psychology in the sixties accompanied the growing awareness of the unintended 'experimenter effect'. People are not only passive at times. They actively search for stimulation and meaning in their environment. Coping, however, is so often construed in reactive terms, as if individuals spend most of their time scanning their worlds for potential threats to their existence. To be vigilant against threat is certainly important, but so is the setting of goals and plans and the monitoring of one's position in relation to such goals. Similarly, so few studies look at the whole picture, to include what is brought into the job from peoples' private worlds outside, and their history before taking up the job. Naturalistic studies which take the whole person into account are scarce in studies of stress and safety in the world of aviation.

There are not only dangers in generalising from the laboratory to real life. Airline pilots are subject to stringent medical examinations. Epidemiologically-speaking, therefore, they are an atypically fit occupational group. Fitness is at least as great in military aviators. A biochemical field study of French military pilots (Escousse, 1969) found greater stress hormone responses among student pilots compared with trained pilots, and greater responses during demanding flights (attack and night flights) compared with more routine trips, especially for the leader who carried more responsibility even though his experience was greater than the rest of the crew. Overall the author concluded that even in the more stressful flights the stress hormone secretions were generally low which testify to the level of fitness. The levels were lower than those found in car drivers and passengers. With these highly selected and trained groups there are clear dangers in over-generalisations based on the nomothetic approach. Military pilots may be reliably different from airline pilots and general aviation pilots are likely to have over all the groups of pilots the lowest standards of fitness.

British controllers will also be different from American controllers. Even if selection and training standards are comparable working conditions will differ between the two countries. Within one country important differences were found between controllers at large and small sites. We do not know how far differences of a psycho-social kind, specifically different attitudes, values, standards and norms and the prevailing climate of control (socio-organisational patterns of sanction and reward) have a bearing for safety and health. The art appears to be that of knowing when to generalise and when not. In the meantime it is wiser to define workload in broad terms to include both internal and external factors relating to people's capacities and limitations of both a biological and social kind. Ultimately, the workload of an identical task will mean different things to different people even in the same control room.

REFERENCES

Allnut, M. 1982. Human Factors : Basic Principles. In R. Hurst & L.R. Hurst (Eds). Pilot Error : The Human Factors. Second Edition. London:Granada.

Bainbridge, L. 1974. Problems in the assessment of mental load. Le Travail Humain, 37 (2) p.279-302.

Bainbridge, L. 1978. Forgotten alternatives in skill and work-load. Ergonomics, 21 (3) p.169-185.

Bartlett, F.C. 1943. Fatigue following highly-skilled work. Proceedings Royal Society, B, 131 p.247-257.

Bartram, D. & Dale, H.C.A. 1982. The Eysenck Personality Inventory as a selection test for Military Pilots. Journal of Occupational Psychology. 55 (4) p.287-296.

Broadbent, D.E. 1973. In Defence of Empirical Psychology. London: Methuen.

Broadbent, D.E. & Gath, D. 1981. Symptom levels in assembly-line workers. In: Salvendy, G. & Smith, M.J. (eds) Machine Pacing and Occupational Stress. London: Taylor & Francis Ltd.

Brown, I.D. 1980. Error-correction probability as a determinant of drivers' subjective risk. Swansea, International Conference on Ergonomics and Transport.

Bruner, J.S. 1957. On perceptual readiness. Psychological Review 64 p.123-52.

Cannon, W.B. 1915. Bodily Changes in Pain, Hunger, Fear and Rage. London: Routledge and Kegan Paul

Caplan, R.D. Cobb, S. French Jr. J.R.P. Van Harrison, R. and Pinneau Jr. S.R. 1975. Job Demands & Worker Health. Washington: National Institute for Occupational Safety & Health. Publication No. 75-160 (NIOSH). U.S. Department of Health, Education & Welfare.

Cobb, S. and Rose, R.M. 1973. Hypertension, peptic ulcer and diabetes in air traffic controllers. Journal of American Medical Association 224 p.489-492.

Davis, D.R. 1948. Pilot Error. London: H.M.S.O. Air Ministry Publication A.P. 3139 A.

Dixon, N. 1982. Some Thoughts on the nature and causes of industrial incompetence. Personnel Management, December

Edwards, E. 1981. Safety via Discipline. Birmingham: University of Aston, Applied Psychology Department, publication no. 258.

Ernsting, J. 1978. Aviation Medicine : Physiology & Human Factors. London: Tri-Med Books Ltd.
Escousse, A. 1969. Flight and adrenosympathetic reaction. Flight Safety 3 (2) p.3-5.
Europilote & U.S. ALPA . 1981. No Compromise with Safety : The Crew Complement Question.
Glass, D.C. 1977. Behaviour Patterns, Stress and Coronary Disease. New Jersey : Lawrence Erlbaum Associates.
Glass, D.C. & Contrada, R.J. 1982. Type A Behaviour and Catecholamines: A Critical Review. In: Lake, C.R. & Ziegler, M. (eds) Norepinephrine: Clinical Aspects. Baltimore : Williams & Wilkins.
Hayward, L.R.C. 1976. Impairment of flying efficiency in anancastic pilots Aviation, Space and Environmental Medicine, February
Johansson, G. 1981. Psychoneuroendocrine correlates of unpaced and paced performance; In: Salvendy, G. & Smith, M.J. (eds) Machine Pacing and Occupational Stress. London: Taylor & Francis Ltd.
Karasek, R.A. 1981. Job decision latitude, job design, and coronary heart disease. In: Salvendy, G. & Smith, M.J. (eds) Machine Pacing and Occupational Stress. London: Taylor & Francis Ltd.
Laurig, W. Becker-Biskaborn, G.U. and Reiche, D. 1971. Software problems in analysing physiological and work study data. Ergonomics 14 (5) p.625-632.
Melton, C.E. Smith, R.C. McKenzie, J.M. Wicks, S.M. & Saldivar, J.T. 1977. Stress in Air Traffic Personnel : Low-Density Towers and flight Service Stations. Springfield, Virginia : National Technical Information Service.
Moray, N. 1982. Subjective mental load. Human Factors 24 (1) p.25-40.
O'Hanlon, J.F. 1981. Stress in short-cycle repetitive work : general theory and empirical test. In: Salvendy, G. & Smith, M.J. (eds) Machine Pacing and Occupational Stress. London: Taylor & Francis Ltd.
Price, V. 1982. What is Type A? A cognitive social learning model. Journal of Occupational Behaviour 3 p.109-129.
Reason, J. 1974. Man in Motion : the Psychology of Travel. London : Weidenfeld & Nicolson.
Roscoe, S.N. 1982. Neglected Human Factors. In: R. Hurst & L.R. Hurst (eds) Pilot Error : The Human Factors. Second Edition. London : Granada.
Rose, R.M. Jenkins C.D. and Hurst, M.W. 1978. Air Traffic Controller Health Change Study: A Prospective Investigation of Physical, Psychological and Work-Related Changes. Springfield, Virginia : National Technical Information Service. (FAA-AM-78-39).
Sheridan, T. Senders, J. Moray, N. Stoklosa, J. Guillaume, J. and Makepeace, D. 1981. Experimentation with a Multi-Disciplinary Teleconference and Electronic Journal on Mental Workload. Massachusetts : Institute of Technology, Report to the U.S. National Science Foundation.
Shipley, P. and Cook, T.C. 1980. Human factors studies of the working hours of UK ships' pilots. Applied Ergonomics 11 (3) p.151-159.
Singleton, W.T. 1982. Accidents and the progress of technology. Journal of Occupational Accidents, 4 p.91-102.
Steptoe, A. 1981. Psychological Factors in Cardiovascular Disorders. London : Academic Press.
Stoker, P.J. 1982. The Validity of the Swedish Defence Mechanism Test as a Measure for the Selection of Royal Air Force Pilots. London : Ministry of Defence.

Taylor, D.H. 1981. The hermeneutics of accidents and safety. Ergonomics 24 (6) p.487-495.

Weston, R.C.W. 1982. Human Factors in Air Traffic Control. In: R. Hurst and L.R. Hurst (eds) Pilot Error : The Human Factors. Second Edition. London: Granada.

Weston, R. and Hurst, R. 1982. Zagreb One Four : Cleared to Collide? London: Granada.

STRESS MANAGEMENT IN AIR TRANSPORT OPERATIONS:
BEYOND ALCOHOL AND DRUGS

F.H. Hawkins
Schiphol Airport
P.O. Box 75577
1118 ZP. AMSTERDAM
The Netherlands

ABSTRACT

In occupations where lives depend on the maintenance of a consistently high level of performance on the part of the operator, the use of alcohol and drugs as tools for stress management must give rise to concern. There is, however, evidence to suggest that non-pharmacological approaches can be effective in providing the operator with a safe means of controlling stress. One such method, Autogenic Training, is noted, together with a discussion of some of the practical aspects involved with setting up an Autogenic Training course in the air transport industry and the results which may be expected.

STRESS

The semantic jungle

Any extensive work on "stress" must inevitably involve setting forth into the semantic jungle surrounding the word itself; this in turn leading to similar hazardous expeditions with motivation and fatigue. Such an adventure, however, is beyond the scope of this short paper, and in any case has been bravely undertaken elsewhere (Murrell 1978). For the purpose of this paper a stressor may simply be defined as an event or situation which induces stress. It may be seen as a pressure being applied to an individual. The extent to which this pressure and the resulting stress has adverse results, such as dissatisfaction, reduced work effectiveness, behavioural changes or health damage -- one might say the extent to which the human system becomes unable to cope and begins to break down -- depends on the individual's response; on his adaptive capability. Neither will this paper attempt to discuss the chemical anatomy of stress which has also been covered elsewhere (Carruthers 1977, etc.).

Stress in industry

More than a century ago people were writing of "the stresses of modern living". At that time aviation was confined to balloons, so we must be cautious in identifying stress with aviation, or any other particular industry or occupation for that matter. Or even with the nature of modern society.

Nevertheless, the aviation industry does involve a cocktail of stressors which is unique when combined with a critical need for a high level of human performance. Those in the industry responsible for safety and efficiency are often reminded of the problem when trying to explain dramatic accidents resulting from less than optimum human performance. But profound discussion of human performance in accident investigation reports is regrettably rare, as illustrated in the use of the term "pilot error" as a common manner in which to close the investigation file.

Several areas in the aviation industry are unremarkable in terms of occupational stress. Aircraft manufacturing is really little different in this respect from other manufacturing industries, except, perhaps, that work levels peak with new projects, then fall with staff redundancy following. Management stress in an airline is probably no more severe than in, say, an insurance company (though management stress itself may be something of a mythical factor). But other branches of the industry can be seen as rather more individual.

Air Traffic Control has been the source of numerous industrial stress studies (Hopkin 1982) and while the problem appears from recent research to be somewhat less severe than was thought earlier, it nevertheless merits attention. ATC officers are engaged in a very critical activity in which, in spite of a generally high degree of automation, small human errors can have catastrophic consequences. A high level of vigilance is required **continuously and a momentary** weakening of this vigilance can bring disaster. Furthermore, air traffic control is normally a round-the-clock activity involving shift work. This disruption of the natural circadian rhythms of the body introduces unfavourable influences on human performance with loss of motivation and sleep disturbance as well as the inevitability of sometimes having to work during the low phase of the circadian performance curve (Klein & Wegmann 1980).

Flight crews -- in particular, long-range flight crews -- also face a unique combination of stressors. Perhaps the earliest stressors recognised were those created by the immediate environment -- noise, vibration, temperature and humidity extremes and acceleration forces. Many of these now have less significance than formerly, and they have been replaced by more complex factors, the implications of which are not always yet fully understood; not least, their impact on motivation and other behavioural characteristics. Working/resting patterns while away from base are now totally irregular; not even the degree of regularity involved in shift work is maintained. Upon this irregular working/resting pattern is superimposed the disturbing influence of transmeridian flying on the circadian rhythms. To these physiological circadian rhythm disturbances is also added the emotional disturbance of family separation for approaching 2/3 of the working life. This can be particularly unpalatable during the years when the children are growing up and a wife needs support, and may appear at times as putting career before family. Other psychological stresses are also present, such as the feelings of insecurity induced by 6-monthly medical and proficiency checks; four times each year the pilot's licence and thus his career are put under challenge. Unlike most other professional people, he is routinely required to perform, in perhaps hostile conditions, with a company check pilot or state aviation inspector looking over his shoulder, constantly posing a psychological threat to his security. Role overload and other role pressures are certainly not confined to flight crews but are additional elements with some crew members, and occasionally, a fear of flying may also be present.

The pilot lives a life of deadlines; one might say "passions that poison and deadlines that kill" (Carruthers 1976). He is under constant pressure to maintain a public relations image. He is exhorted endlessly to be professional, responsible, vigilant, dedicated, avoid complacency, be economically conscious and so on. He works under the threat of immediate media spotlight in the event of a deviation from routine operational circumstances which could influence safety -- or at least which could make a good story. Yet to admit to suffering from the pressures upon him may be seen, in a society which extols achievement and competitiveness, as an admission of failure. And so, too often, the existence of the related symptoms are denied by the individual and ignored by the company. Until, that is, they become apparent through behavioural changes, sickness or reduced performance. This might be too late to avoid an accident and maintain operational and commercial efficiency. It may also be too late to save the individual's career.

An analysis of safety reports (Lyman and Orlady 1981) has shown significant human performance decrements to exist associated with what was described as fatigue. The majority of the incidents concerned involved such unsafe events as altitude deviations, take-offs and landings without clearance and the like. Monitoring and vigilance tasks suffered severely. Fatigue has been the subject of much study, but its origins have not been clearly established. The relationship between stress and fatigue, while warranting serious discussion, is beyond the scope of this short paper.

In organised industry, the limiting of fatigue has generally been accomplished by means of work rules -- usually restrictions applied to the number of hours worked -- though attention to chronobiological factors has usually been inadequate. Even government controlling agencies are sometimes reluctant to become involved due to the controversial nature of the problem and the possible implication in terms of commercial cost (FAA 1981). It is unrealistic to imagine that in the aviation environment the stressors can be totally removed, though with proper chronohygiene their severity can be reduced. Neither is it realistic to suggest that "those who cannot stand the heat should get out of the kitchen". Even though we may have techniques which would enable us to determine those who are better able to adapt to circadian rhythm disturbances, there may be sociological and economic problems in being able to select or schedule staff taking such criteria into account. Adaptation to occupational stressors is thus likely to remain a problem for a significant part of the relevant population.

In discussing occupational stress it is not possible to ignore the interaction of work and non-work factors. It is delusory to instruct staff to leave their domestic problems at home; the brain and body suffering from the effects of stress at home cannot simply become healed on arriving at work. Family separation due to work can be expected to induce stress at home. Frustrations at home can be expected to be reflected in attitudes at work. Insomnia at home, often a symptom of stress, may be reflected in reduced performance at work. Studies of occupational stress often fail to take this interaction into account.

CURRENT STRESS MANAGEMENT TECHNIQUES
Alcohol

As stress may be seen as a source of adaptive energy, which becomes a problem only if it becomes excessive, its management, rather than its total avoidance, may be the appropriate objective. Perhaps the most widely used technique for managing stress today -- and for centuries past -- is alcohol. It is now believed that physiological damage from alcohol occurs at daily consumption levels considered as moderate and far lower than those leading to neuropsychiatric symptoms (Pequignot 1979). Recent scientific research shows that in a given country alcohol pathology is linked to the average per capita alcohol consumption and that the proportion of excessive drinkers is growing faster than the per capita consumption. For the 20 years up to 1972 this consumption rose by 276% in the Netherlands, by 182% in Germany and by 133% in Denmark (Godard 1979).

But in a skilled occupation where optimum performance is required for the protection of life, the use of alcohol has dimensions beyond purely medical ones. Alcoholism can be the basis for permanent loss of a pilots's licence. More significantly in the context of this paper, alcohol impairs performance -- reaction times increase, tracking performance is lowered, distance judgement is degraded, visual acuity is decreased and so on. A second effect, highly dangerous in the flying environment, is that it creates an illusion of improved performance; "you think you are doing fine".

Dedicated research has been carried out to demonstrate the adverse effect of alcohol on the performance of specific flying tasks, such as instrument approaches (Billings et al 1973).

In spite of all that is known in research circles of the impairment of performance due to alcohol, some 50% of general aviation pilots in the USA responding to a survey stated it would be safe to fly within 4 hours after drinking alcohol. Analysing the answers further it was estimated that 27-32% of the respondents considered flying after drinking, within a time period that would result in a 15 mg % blood alcohol concentration (BAC) or higher, to be a safe behaviour. The relationship between these attitudes and actual behaviour can only be hypothesised. However, it is noted that about 20% of general aviation pilots killed in accidents are found to have a BAC of 15 mg% or higher (Damkot et al 1978).

In commercial aviation, evidence of working under the influence of alcohol is usually difficult to obtain and in any case it is no doubt less than in general aviation. Nevertheless, dramatic reminders occur from time to time to suggest that the problem cannot be ignored (e.g. Anchorage, DC8, Jan. 1977, all aboard killed, pilot under the influence of alcohol).

Drugs

The 20th century's tool for the management of stress is drugs -- tranquilisers, sedatives, hypnotics. And then, sometimes, amphetamines to counteract the resulting drowsiness. In road transport a highly significant statistical correlation has already been shown to exist between the use of tranquilisers and accidents (Skegg et al 1977). More recently it was demonstrated that car-driving behaviour was changed in such a way as to increase the risk of an accident on the road 12 hours after (the morning after) taking a single dose of temazepam (Normison, Euhypnos) or flurazepam (Dalmane, Dalmadorm) (Betts et al 1982). This is particularly interesting in the case of temazepam as it has a relatively short half-life (less than 6 hours) and has previously been shown to have little effect in psychomotor tests the following morning (Bond et al 1981). Flurazepam has a half-life of 47-100 hours and its influence on a tracking task was included in the study noted below (Nicholson et al 1976).

Commercial airlines, however, are perhaps understandably reluctant for various reasons to speak out on either the existence of a stress problem or the techniques currently used to manage it. Even after research showing the effect of certain benzodiazepines and barbiturates on an adaptive tracking task (Nicholson et al 1976) one of the world's major airlines announced to its crews, in response to concern expressed by some of its captains, that the use of one nitrazepam (Mogadon) a day was not harmful (the half-life of nitrazepam is 25-30 hours). A survey carried out amongst flight crew members of a European airline indicated that almost half the crew members interviewed occasionally used pharmacological aids to sleep when away from home on long-range flying duties. At home the figure dropped to 13% (see table 1). The drug most commonly used at that time was nitrazepam. This survey was carried out independently of company and staff union (Hawkins 1980). In a few cases crew members admitted to taking such drugs as little as 6 hours before departure. One must be cautious, however, in too hastily criticising crews for their actions in this respect. Each individual crew member is faced with the difficult task of making a differential assessment of the loss of performance -- and motivation -- resulting from sleep deprivation compared with that resulting from the use of drugs. They generally have nothing upon which to base this assessment apart from subjective feeling, which can, of course, be very misleading.

TABLE 1 Use of pharmacological aids to sleep by airline flight crew members while away from home and at home.

a) Within the working environment (i.e. on the line, away from home)

Crew category	Ss	Av. age	Never	Rarely	Sometimes	Frequently
Cockpit crew	81	48	54%	16%	27%	3%
Cabin crew(M)	41	45	56%	12%	25%	7%
Cabin crew(F)	236	26	54%	24%	18%	4%
Total crew	358	33	54%	21%	21%	4%

b) Outside the working environment (i.e. at home)

Crew category	Ss	Av. age	Never	Rarely	Sometimes	Frequently
Cockpit crew	83	48	90%	6%	3%	1%
Cabin crew(M)	40	45	75%	15%	3%	7%
Cabin crew(F)	225	26	89%	7%	3%	1%
Total crew	348	33	87%	8%	3%	2%

The incidence of drug taking may be expected to vary widely between different ethnic groups, and staff from different countries are likely to favour different drugs. In the USA, for example, where nitrazepam is not marketed, flurazepam is likely to be favoured.

Even if it were possible to produce a drug with a hypnotic or chronobiological effect which had no side effects to influence performance for more than a short period, the problem remains of how to police any safeguards placed on its use. And the wisdom of relying on any drugs for a lifetime management of occupational stress is debatable.

Non-pharmacological approaches

Proper management of stress involves intervention at the level of the individual and the organisation.

As far as the individual is concerned it is likely that both genetic and cultural backgrounds will influence the strategy which is found most suitable. Individual strategies may include physical activities such as exercise, particularly sport, and diet. They may involve psychophysiological methods such as meditation, Yoga, biofeedback, hypnosis as well as other techniques for "relaxation" such as the neurophysiological method of Progressive Relaxation. They may implicate outside intervention through counselling, guidance or psychotherapy.

Little systematic work appears to have been done by organisations in the air transport environment to examine the different individual strategies available, and to make recommendations and offer facilities to staff, though some private thought has been given to the problem (Hawkins 1980). One such programme in Holland in 1982, privately initiated and financed, utilised Autogenic Training and involved nearly 50 air transport staff, some 75% of whom were flying personnel (Hawkins 1982). This initial programme has since been followed by another, but with a somewhat lower percentage of active flying staff, and demand is accumulating for a third programme. The rest of this paper outlines some aspects of such a programme.

AUTOGENIC TRAINING PROGRAMMES FOR AVIATION STAFF

The technique

Autogenic Training (AT) is a psychophysiological technique prescribed by a trained therapist and carried out by the subject himself. It is designed to promote the subject's own homeostatic mechanisms. In an industrial application it has the objectives of improving the quality of working and domestic life and of providing a personal tool for the long-term management of stress. It can be expected to have a favourable influence on working efficiency as well as personal well-being. In air transport a relationship with safety can be clearly inferred.

AT appears suitable for this particular industrial application for several reasons. Once the technique is learned it is in the subject's own hands; he needs no further booster or assistance from the therapist for lifetime use of the tool which he has acquired. He becomes self-reliant, participates in the maintenance of his own well-being and does not develop dependency upon a therapist. No equipment or drugs are needed in either the acquisition of the technique or its application. It enhances the performance of the body's own homeostatic, self-regulating mechanisms rather than interfering with them. It is easy to learn and has no ritualistic, religious, mystical or folkloric connotations. It can be utilised effectively by the majority of people. Finally, its application, particularly in the clinical field, has been very well documented over more than half a century and these reports also include special applications in highly skilled occupations such as that of surgeons (Labhardt 1980). It must, however, be taught by a skilled practitioner and for optimum results must be industry-adapted and case-adapted. Furthermore it has physiological and psychological implications which must be well understood by the therapist.

The vast majority of literature available on AT is concerned with its clinical application, with a relatively small amount devo-

ted to application in sports and education and very little to industrial use. Yet increasingly, large organisations are introducing AT as a facility for their staff. It has been used, amongst others, by Renault and SNECMA. By I.G. Farben. By IBM, Marks & Spencer and the British National Coal Board. By railway workers and telephone operators. The observations reported from several industrial programmes may be summarised as-:
- improvement of health and efficiency
- long-term protection against industrial/professional stressors
- decrease of absenteeism, errors and accidents
- improvement in interpersonal relations

In air transport, one European airline offered AT to its staff in 1973 and in 1981. Another now offers AT in its evening school programme. A number of staff members of a third European airline have individually supported 2 programmes without them being initiated or sponsored by the company, during 1982.

Establishing a programme

For an effective AT programme to be established in an airline or other industrial environment, proper planning and organisation are important-:
- Staff requirements

    A fully trained therapist or instructor is required and he may be expected to hold degree level or equivalent qualifications in psychology or medicine. He should have undertaken an AT therapist's course in a centre approved by the International Committee of Autogenic Therapy (ICAT). He must be fully aware of the physiological and psychological implications (contra-indications, etc.) of AT. There must be a routine consulting line of communication with a physician very familiar with AT, in connection with the screening of trainees and the differential diagnoses which may be required during the course. For various reasons it has been found preferable that the therapist not be an employee of the organisation.

- Training location

A quiet, and if possible centrally located, comfortable, correctly furnished room must be available. This could be in the medical department of the company, but a case could be made for choosing a restful location remote from the working environment.

- Consultation

Full cooperation can only be assured by adequate consultation with management, the medical department and relevant staff associations. It may be necessary to arrange informative talks or lectures to groups of staff to dispel prejudices and misunderstandings. Some scheduling cooperation may be required from the company.

- Training logistics

AT lends itself well to group instruction. However, groups should be small, ideally about 5, though some flexibility is possible with careful screening. The group should be homogeneously constituted taking into account such factors as status, and psychological and medical condition. The course consists of 8-10 weekly sessions of $1-1\frac{1}{2}$ hours each.

- Motivation

Benefits of AT often do not appear to the trainee until after some 3 or 4 weeks of training although the internal therapeutic process begins much sooner. It is **therefore necessary to utilise** various techniques to enhance motivation so as to avoid drop-outs during these early weeks.

Effectiveness

The programmes which have been used in airlines were practical, industrial applications of AT and not controlled experimental research studies. It would be extremely difficult in the commercial aviation environment to carry out such controlled studies in the field. The variables are so extensive -- flight schedules, crew functions, domestic situations and so on -- that establishing matched control groups would seem an unrealistic objective. Not least, because commercial airlines do not readily accept the role of research establishments. However, controlled studies of many therapeutic applications of AT have already been carried out, and so, after more than 60 years of clinical use, AT is not without justification (Luthe 1970). There are, nevertheless, certain specific laboratory research studies which could usefully be done on aspects of AT directly relevant to long-range air transport operations.

Analysis of results of 2 of the airline programmes have been made. The Swedish (1973) analysis was particularly related to the effects of AT on sleep in the disturbed circadian rhythms of long-range flying. Improvement in sleep was reported for all the different categories of flying staff. The first programme in Holland analysed several behavioural and performance aspects as reported by the trainee and, independently, by the peer (wife, husband, etc.) as well is recording blood pressure changes (Hawkins 1982).

In this Dutch 8-week programme, too, special reference was made to the subjective reporting of sleep both in the working environment and in the home environment. The most common cause of complaint of long-range flight crews relates to the disturbance of sleep and body rhythms (Hawkins 1980) and this problem can be directly related to human performance (Klein & Wegmann 1980). In this programme, 43% of the trainees had indicated sleep problems as one reason for wanting to participate in the programme. After the course 70% of trainees reported an improvement in sleep in the working environment. Interestingly, 77% of peers reported that the trainee slept better in the normal, home environment after the course. The mean of the sleep latency times reported by all participants was reduced from 26 minutes before the course to 14 minutes after the course. Before the course,

13 participants reported a sleep latency time of more than 30 minutes in the normal regular waking/sleeping environment. After the course, only 2 reported more than 30 minutes.

While all of these sources reflect only subjective reporting of sleep improvement, there is a consistency in the information which suggests confirmation of previous conclusions that sleep tends to improve following the use of AT.

AT has normally been shown in clinical application to have favourable influence on hypertension. Analysis of the first course in Holland revealed that the blood pressures of the 10 participants (all flying staff) having the highest blood pressure of the group at the beginning of the course, had all fallen by the end of the 8 weeks. All were originally mild hypertensives.

These programmes did not, of course, establish a causal relationship between AT and the therapeutic benefits achieved, as they were not designed as controlled experimental studies. The programmes in Holland, involving an industry- and case-adapted application of AT did, however, suggest that-:

a) a technique such as AT can be made acceptable to a sophisticated, technically oriented, population of this kind.

b) motivation in such a population can be maintained at a level which results in minimal drop-out.

c) the majority of participants can be expected to report benefits in several behavioural and performance aspects, reflecting an improvement in personal stress management.

d) highly individual personalities often represented in flight crews, can be integrated homogeneously enough for group training.

e) an effective programme can be established in a transportation organisation, involving scheduled travel away from base, without disruption of those schedules, and with only a small degree of logistical cooperation.

f) a "fixed package" course with a cut-off date and without an open-ended clinical therapeutic facility, appears acceptable, in spite of the rather less than clear distinction between industrial and clinical application. Nevertheless, a referral facility should be available for use in isolated cases.

g) the most common complaint reported as being amenable to such a programme appears to be sleep disturbance, but many other aspects, such as concentration, memory, anxiety and personal relationships, often variable with stress, were also reported as changing favourably.

h) various disorders of a psychosomatic nature, not specially related to the occupation, may be expected to show improvement during the programme.

i) most participants believe that their company or organisation should support such stress management schemes for staff, although confidentiality is considered important.

While these programmes were not considered as clinical applications of AT, clinical symptoms, in addition to insomnia, may be expected to appear occasionally. In the Dutch programmes, for example, such cases included tinnitus, bulimia nervosa, bruxism, amnesia, and several cases of hypertension; all responded favourably to AT. In addition, case-adaptation may be required, even in an industrial programme, for conditions such as stomach disorders, migraine, pregnancy, and individual disturbing responses.

CONCLUSION

In air transport operations, as in other activities where safety depends upon a high level of human performance, the use of alcohol and drugs as tools of stress management entails risks which make their use questionable. Necessary safeguards are difficult to apply effectively.

There are, however, non-pharmacological methods available which appear to be effective in many cases when carefully planned, industry- and case-adapted and taught with skill.

As the benefits accrueing have been demonstrated to involve improved efficiency in addition to improved well-being, it would seem to be in the interest of organisations to facilitate the acquisition of such techniques by interested staff members. Such facilitation involves disseminating information, and preferably some logistical and financial support.

REFERENCES

Betts, T.A. and Birtle, J. 1982. Effect of two hypnotic drugs on actual driving performance next day. Br. Med. J., 285, 852.

Billings, C.E., Wick, R.L., Gerke, R.J., Chase, R.C. 1973. Effects of ethyl alcohol on pilot performance. Aerospace Med. 44 (4), 379-382. April 1973.

Bond, A. and Lader, M. 1981. After-effects of sleeping drugs. In "Psychopharmacology of Sleep" (Ed. D. Wheatly) (Raven Press, New York). pp. 177-197.

Carruthers, M.E. 1976. Passions that poison and deadlines that kill. New Society, 303-304. 11 Nov 1976.

Carruthers, M.E. 1977. The chemical anatomy of stress. In "Beta blockers and the central nervous system" (Ed. P. Kielholz). (Huber, Berne/Stuttgart/Vienna). 1977.

Damkot, D.K. and Osga, G.A. 1978. Survey of pilots' attitudes and opinions about drinking and flying. Aviat. Space and Environm. Med., 49 (2), 390-394. 1978.

FAA 1981. Operation review program notice no. 7. Flight crewmember flight and duty time limitations and rest requirements. Notice of withdrawal of supplemental notice 78-3B. Fed. Reg. 46:119. 22 Jun 1981.

Godard, J. 1979. Alcohol consumption in Europe and its consequences. In "The medico-social risks of alcohol consumption". Working party report", Commission of European Communities 1979. ISBN 92-825-1007-7.

Hawkins, F.H. 1980. Sleep and body rhythm disturbance amongst flight crew in long-range aviation. MPhil thesis, University of Aston, Birmingham, England.

Hawkins, F.H. 1982. Autogenic Training programme for aviation staff: an analysis of results. Unpublished report. F.H. Hawkins, P.O. Box 75577, Schiphol Airport(C), 1118 ZP Amsterdam, Holland.

Hopkin, D. 1982. Human factors in air traffic control. NATO-AGARD-AG-247.

Klein, K.E. and Wegmann, H.M. 1980. Significance of circadian rhythms in aerospace operations. NATO-AGARD-AG-247.

Labhart, F. 1980. The influence of Autogenic Training or beta-blockade on stress in surgeons -- a preliminary report. In "Psychosomatic cardiovascular disorders -- when and how to treat" (Eds. Kielholz, Siegenthaler, Taggart, Zanchetti). (Hans Huber, Berne/Stuttgart/Vienna).

Luthe, W. 1970. "Autogenic Therapy: Vol. IV; Research and Theory" (Grune and Stratton Inc., New York).

Lyman, E.G. and Orlady, H. 1980. Fatigue and associated performance decrements in air transport operations. NASA report CR 1666167.

Murrell, H. 1978. Work stress and mental strain: a review of some of the literature. UK Dept. of Employment. Work Research Unit, Occasional paper No. 6. Jan 1978.

Nicholson, A.N., Borland, R.G., Clarke, C.H. and Stone, B.M. 1976. Experimental basis for the use of hypnotics by aerospace crews. NATO-AGARD-CP-203.

Pequignot, G. 1979. General assessments of the risks. In "The medico-social risks of alcohol consumption". Working party report. Commission of European Communities. 1979. ISBN 92-825-1007-7.

Skegg, D.C.G., Richards, S.M. and Doll, R. 1977. Minor tranquilisers and road accidents. Br. Med. J., $\underline{1}$, 917-918, 7 Apr 1979.

REASONS FOR ELIMINATING THE "AGE 60" REGULATION FOR AIRLINE PILOTS

Stanley R. Mohler
Wright State University
School of Medecine
Dayton, Ohio 45401
U.S.A.

ABSTRACT

The calendar age of 60 is no longer medically justifiable as an upper age limit for airline pilots. Advances in gerontologic studies, clinical medicine, and operational flight proficiency evaluations, now allow individual pilot assessments for health status and performance capability. Individualizing the career duration of pilots by eliminating the present age 60 upper limitation will enhance flight safety and efficiency as the highly qualified, experienced, and proficient older healthy pilots continue their productive careers.

There is today no medical, physiological, psychological or operational justification for retaining the calendar age of 60 as a mandatory career cut-off for an airline pilot. Age alone, as is the case with race or sex, gives no information about an individual's competency or health.

The three critical determinants of pilot fitness are: freedom from impairing disease, ability to perform and desire to continue flying.

It is now clear that the evolution of understanding and knowledge about the normal developmental aging process and the pathophysiology of specific acquired diseases, has given us new insight concerning individual health. This and the new diagnostic and disease prevention techniques have revolutionized the thinking about who is, or isn't, healthy and, consequently, fit to fly. Parallel advances have occurred in flight technology, including flight simulator fidelity. This latter field has progressed to the stage (especially in the aspect of high quality optical displays) that it is now feasible to give an airline pilot the necessary training for a new type of aircraft in a simulator (Mohler, 1981). The first time a captain actively flies the aircraft can be in revenue, passenger-carrying flights, if the new procedures are followed.

THE AGING PROCESS

The normal, genetically programmed development of an individual is a life-long continuum that results ultimately in the terminal involuntional subtraction of the aged individual from the population. The entire life-span is a developmental process - a functioning, normal process totally distinct from acquired diseases.

Human lifespan potential in the 20th century is reaching 100-120 years, depending upon lifestyle, environment, and genetic strain and diseases. The changes with age involve modifications of functions and structures, but to describe, for example, graying hair in pejorative terms is to fall victim to arbitrarily adopted sociocultural concepts. Graying of the hair, as with all other normal aging changes, is a developmental change, no more, no less. Until the ultimate involutional changes of advanced age occur, these aging changes have no bearing upon an individual's ability to perform.

Many things improve with age, including judgment and intelligence (Comfort, 1977). Experience enhances judgment and older healthy persons tend to be less impulsive, and consequently, have better safety records. Sonnenfeld points out that factual evidence of older workers' performance rejects the "stereotyping" and prejudices that link age with senility, incompetence, and lack of worth in the labor market (Sonnenfeld, 1978).

Improvements from 1901-75 in survival into old age as seen in U.S. males shows a marked continuation of greater numbers of individuals extending into the older age area. For individuals whose lifestyle and environment promote the full potential of their genetic endowment, healthy survival into the 90s and even into the early 100s is feasible.

Fries has recently shown that the average length of life in the U.S. has risen from 47 to 73 years in this century, and that this length is moving toward an ideal average lifespan of 85 years for the present cohort of young adults (Fries, 1980). Postponement of the disease phenomena previously identified with "aging" by present social interactions, health promotion, and personal autonomy, is highlighted by Fries. Life in the older age brackets has continued, for those who realize their potential, to be physically, intellectually, and emotionally vigorous and productive. This achievement, of course, is inhibited by externally imposed constraints, as typified by arbitrarily imposed upper age limits for employment or other activities. Many of the conditions previously construed as "aging" - for example, atherosclerosis, emphysema, and other acquired conditions - simply take a number of years to develop in susceptible persons; hence the confusion by some with the normal aging process.

There is wide individual variation in changes with age at all age groups. This is one of the primary findings in the longitudinal age study conducted by Shock and associates in Baltimore (Shock, 1980). Although certain physiologic functions based on population averages show a change toward less capacity in older age groups, the important factor is that there is a tremendous variation among individual capabilities in a given age group. For example, the cardiac stroke index group mean declines somewhat with age but, by data obtained by Landowne, show there are healthy 70-year-olds (and older) who outperform others in this respect in their 30s (Landowne, 1957). Present capacity to perform by an individual is the signigicant point, not the chronological age.

MENTAL FUNCTIONS AND AGING

It is now understood that senility (senile dementia) is not a part of the aging process (Butler, 1978). Only about one in five persons of advanced age will become "senile" and there are tests available that determine this syndrome. A proportion of these victims suffer from Alzheimer's disease, an entity receiving considerable attention. Some have hypothesized a possible viral etiology.

Spirduso and Clifford have shown that 70-year-old physically active persons (racquet ball sports or jogging, for example) can have simple, complex, and choice reaction times equivalent to those of sedentary 30-year-olds (1978).

The Framingham study has shown the greater likelihood of coronary artery disease in persons with untreated hypertension, glucose intolerance, or who smoke and are overweight (Kannel, 1977). A risk profile for a primary candidate for coronary artery disease includes a steady rich diet, smoking, little exercise, obesity, high blood pressure, high blood lipids, a marked sense of "time urgency" and a tendency toward diabetes. One in three with this profile will have a cardiovascular event by the age of 60.

Kannel points out that multiple marginal abnormalities, if not tended to, can, by multiplexing, become the equivalent of a mjor risk factor. These risk profiles developed by the Framingham study powerfully enhance the ability to predict an individual's risk of developing cardiovascular disease. High-risk individuals can be readily identified today.

Captain ages in scheduled airline accidents for 1970-77 as contained in National Transportation Safety Board Reports show the peak ages for these accidents is in the 40s, with a rapid fall-off to age 60. The older captains have the greater experience, ability, and judgment, and have become older captains by not having fatal accidents while younger. These are the safer pilots, and this is why command of the larger, more sophisticated, wide-body, high-density passenger aircraft is univerally entrusted to them when they bid on them, as they almost invariably do.

Table I provides an analysis of pilot-in-command accidents by age during 1978-1979 involving those pilots holding Air Transport and Commercial Certificates operating in the general aviation area. Note that the progressive decrease in observed accidents per 1,000 pilots with age further substantiates the accident experience on scheduled airlines. Increasing pilot age and experience is correlated over and over with a decrease in accidents.

Older healthy captains have the experience, judgment, problem-solving ability, and rapid response capability to avert emergencies of all types. Two examples of this are given in Table II. In both cases, the captains were commended by high authority for averting catastrophe. Unite Airlines gave a major cash award to the B-747 captain who, a few weeks later, was forced out prematurely by the age 60 rule. The Department of Transportation and the Federal Aviation Administration gave a major award to the DC-10 captain, who was making his last flight before being forced out by the age 60 rule when he encountered the potentially catastrophic take-off event, reacting within 1.2 seconds and averting disaster. Many other examples on all airlines could be cited. There is no evidence that healthy competent older pilots are susceptibel to degradation in performance capability compared with younger pilots. There is no evidence that older pilots have any greater difficulties transitioning to new aircraft than younger pilots.

The continuing premature loss today of these and hundres of other experienced airline captains is no longer medically or operationally justified.

CONCLUSION

As illustrated in this paper, there is today no physiological, psychophysiological, or medical justification for the "age 60" airline pilot rule. Some of the many reasons why this is so follow:
1. U.S. morbidity and mortality data of 20 years ago, when the regulation was established, have markedly changed for the better;
2. Longitudinal age studies, including those on pilots, have exploded outmoded concepts of inevitable declines of capability prior to very advanced ages;
3. Dramatic advances during the past 10 years in disease de-

tection, understanding, and treatment have achieved practical application throughout the U.S. and many other countries;

4. Improvements in predicting the development of disease and the availability of preventive measures provide a powerful tool for health maintenance;

5. Aircraft simulators and flight performance provide detailed information on a specific individual's cognitive and perceptual flight skills and overall performance capabilities, including handling emergencies. Together with a longitudinal record of performance, a total history and record of the individual's capabilities are clearly defined and enable a reasoned decision concerning future performance.

By 1980, the point was well passed for biomedical justification of an age 60 rule. Individual assessment of airline pilots, irrespective of age, is within practical reach now, and actually is being practiced by various airlines today that are not subject to the rule. Indeed, elimination of the age 60 rule can only enhance air safety, as companies will be able to continue utilizing the advanced skills and experience of their older healthy pilots.

REFERENCES

Butler, R.N. 1978. Senile Dementia. Goldfarb Memorial Lecture. National Institute of Aging. Bethesda, Maryland.
Comfort, A. 1977. A Good Age. Mitchell Beazley. London.
Fries. J.F. 1980. Aging, Natural Death and the Compression of Morbidity. NEJM. 303, 130-135.
Kannel, W.B. 1977. Preventive Cardiology. Postgrad. Med. 61:1-85.
Landowne, M. 1957. Methods and Limitations in Studies of Human Organ Functions. Ciba Colloquium on Aging. 3:73-91.
Mohler, S.R. 1981. Reasons for Eliminating the "Age 60" Regulation for Airline Pilots. Aviation, Space, and Environmental Medicine. August.
Shock. N.W. 1980. Presentation to Longitudinal Studies Workshop, Johnson Space Center, Houston, Texas. April 10.
Sonnenfeld, J. 1978. Dealing with an Aging Work Force. Harvard Business Review. Nov/Dec.
Spirduso, W.W. and Clifford, P. 1978. Replication of Age and Physical Activity Effects on Reaction and Movement Time. J. Geront. 33. 26-30.

TABLE I

Pilot Age and Accidents

Age	Active Pilots 1978	No. Accidents Expected 1978	No. Accidents Observed 1978	Accidents Per 1000 1978	Active Pilots 1979	No. Accidents Expected 1979	No. Accidents Observed 1979	Accidents Per 1000 1979
16 - 19	374	3	8	21.4	468	4	7	15.0
20 - 24	10,839	92	167	15.4	11,839	90	160	13.5
25 - 29	26,102	220	312	12.0	25,755	196	294	11.4
30 - 34	45,011	379	414	9.2	44,606	341	359	8.0
35 - 39	41,742	352	321	7.7	42,520	324	309	7.3
40 - 44	35,270	297	236	6.7	35,031	267	209	6.0
45 - 49	28,012	236	214	7.6	29,585	225	191	6.5
50 - 54	19,660	166	164	8.3	18,803	143	149	8.0
55 - 59	22,499	190	131	5.8	23,073	176	123	5.3
60 +	12,205	103	71	5.8	14,069	107	72	5.1
	241,714	2,038	2,038		245,749	1,873	1,873	

Source: NTSB and FAA Statistical Handbook, Calendar Years 1978-79. Pilots-in-Command having Commercial and Air Transport Certificates, General Aviation Accidents

TABLE II

OLDER CAPTAINS

	Good Judgment Fast Response Time Best Safety Record Captain Age	Date	Location
United 747	59	May 21, '78	Pacific (Honolulu)

(Engines 1, 2, and 4 failed at 22,000 feet due to ice. Captain got No. 2 going at 300 feet above surface and made safe landing.)

Continental DC-10	59	March 1, '78	Los Angeles

(Reacted in 1.2 seconds when tires blew on take-off roll.)

HUMAN FACTORS EDUCATION IN EUROPEAN AIR TRANSPORT OPERATIONS

F.H. Hawkins

Schiphol Airport, P.O. Box 75577, 1118 ZP. AMSTERDAM, The Netherlands.

ABSTRACT

Human Factors or ergonomics is concerned with the effectiveness and well-being of man in his working environment. It has for long been an applied technology in its own right, with academic qualifications available up to doctorate level.

For nearly half a century it has been recognised that most aircraft accidents are associated with the role of the human in the system. Yet in spite of this recognition, Human Factors as a technology is frequently not understood and the use in air transport operations of staff properly qualified in Human Factors is the exception rather than the rule.

It seems unlikely that adequate progress in the better application of Human Factors can be expected until these educational gaps have been filled. This must be accomplished at all levels in the industry and appropriate courses are now available for each level.

HUMAN FACTORS AND SAFETY

Statistics and good intentions

Nearly half a century ago it was calculated (Meier Muller 1940) that some 70%-80% of aircraft accidents involved inadequate application of Human Factors (HF). More recently, the commercial air transport industry has confirmed this calculation (Fig. 1) and has called for action to meet the challenge (see Appendix I). A conclusion of that conference (IATA 1975) was that "the wider nature of Human Factors and its application to aviation seem still to be relatively little appreciated. This neglect may cause inefficiency in operation or discomfort to the persons concerned; at worst it may bring about a major disaster". This ominous statement was followed 17 months later by the double Boeing 747 disaster at Tenerife in which 583 people died and which cost about $150 million. It is significant that this accident -- the greatest in aviation history -- resulted entirely

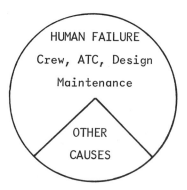

Fig. 1 A diagram illustrating the important role of Human Factors in aircraft accidents (IATA 1975).

from a series of HF deficiencies. It is also highly significant and typical of most other accidents, that the crew members concerned, if tested, would have been found to have possessed at the time the capacity and the skill to have performed the task safely. Little more than a year after that IATA conference the other side of industry, the pilots, also called for action to improve the application of HF in their working environment (see Appendix II). They concluded that "there is widespread concern at the frequent failure to apply HF knowledge .... and a current lack of expertise in HF in many areas of the commercial aviation industry" (ALPA/IFALPA 1977).

It may not now be unreasonable to assess, after several years have passed since these conferences, where the European air transport industry stands with respect to its knowledge and application of HF, which for our purpose is synonymous with ergonomics (Murrell 1958). Professor Elwyn Edwards of the University of Aston in England has defined HF as the technology concerned to optimise the relationship between man and his work by the systematic application of the human sciences. From proceedings of the conferences of both the International Air Transport Association (IATA) and International Federation of Air Line Pilots Associations (IFALPA) referred to above it was clear that the term Human Factors was understood by very few of the delegates present. Some interpreted the expression to mean the pathology applied to accident victims. Others considered that it was mainly related to flight technical procedures and operational policies applied on the flight deck. Yet others saw it as almost synonymous with aviation medicine, which could be left to the physicians. Collected together at these two major conferences were most of those in the world's air transport operations primarily responsible for operational safety and efficiency and for the application of HF in their organisations. The inevitable conclusion was that a fundamental educational gap existed -- at all levels -- and action to rectify this situation was apparently a primary need.

Do-it-yourself approach

The consequences of the apparent lack of adequate HF education in the air transport industry should not, perhaps, create surprise. We would not expect too much in the way of medical advice from practitioners without formal education in medicine. Neither would we consider being represented in court by an official without formal legal training. In air transport itself, we would not be content to fly behind a pilot who had not been formally trained and properly qualified.

Yet it seems that in the specialised field of HF, or ergonomics, we have in the past generally been content to accept placing the responsibility for its application on staff with no formal education whatsoever in the applied technology which forms the subject of their responsibility. Much practical operational experience, perhaps, but no basic education. This applied to a considerable extent to the manufacturing industry as well as the air operators.

Examples of HF deficiencies in air transport cover a wide range of applications. The realisation that wartime equipment design was advancing faster than man's ability to utilise it stimulated much early work, particularly in the UK and USA. But the scope now extends far beyond these hardware developments based on military effectiveness. The reassurance given by an airline to its flight crews that the daily use of nitrazepam by them was not harmful is an example of one kind of deficiency related to human performance. Another is reflected in the limited attention which appears to have been devoted to the effects of stress and fatigue on human performance in aviation and to techniques for managing these (Hawkins 1983). The case of the damaged doppler panels in one major airline is another simple illustration, in this case of the economic cost of ignorance of HF. These panels, mounted on the centre control pedestal in the aircraft, were being routinely damaged by being used as a footrest by pilots on long-range flights. It took more than a year of fruitless exhortation and threats of disciplinary action (and thousands of $ of damage) before the airline was prepared to follow Human Factors guidance and instal a simple, inexpensive foot-rest at the base of the instrument panel. This provided the pilot with a means of relieving the pressure under his thighs and the blood pooling which are the inevitable result of long, continuous hours immobilised in a pilot's seat. Examples of such deficiencies are manifest, and they normally reflect ignorance at the point of application rather than a lack of research or unavailability of research data.

Why has this "do-it-yourself" attitude to HF developed? And why has it been allowed to persist? Firstly, HF or ergonomics was only fully recognised as an applied technology in its own right in 1949, although serious aviation HF research was already being carried out earlier, for example, in Dayton (USA) and Cambridge (England). Few of the current senior operational staff will have acquired any such knowledge as a part of their own basic education and most will have been since then too busy, or not motivated, to fill the gap at some later stage. Yet specialised education has been available from a short course level up to a doctorate qualification for many years.

Other reasons can also be suggested. The aircraft and equipment manufacturer may have felt reluctant to invest much in Human Factors education of its staff on the grounds that the customer -- the airline -- will in any case specify what it wants, and the manufacturer will simply have to produce it. Even where certain major aircraft manufacturers do have HF departments, they tend to have an advisory function only and it is not unusual for them to be overridden on grounds of cost or production deadlines. The aircraft operator, on the other hand, may feel that as far as hardware is concerned the manufacturer should have provided HF optimisation before supply. Responsibility has thus often been unclear, resulting in neglect. Paradoxically, the growth of the litigation industry, particularly in the USA, has sometimes served to inhibit progress in some areas of HF application. Neither has air transport received the enlightened support from professional and other influential bodies which it might have expected. In 1971 one of the world's most distinguished aeronautical societies declared that the science of ergonomics "was not yet sufficiently advanced for the society yet to take it seriously". This quite remarkable declaration was made about a quarter of a century after the classical research studies on human performance in connection with reading errors on 3-pointer altimeters (by Fitts, Jones and Grether) which played such an important role in focussing attention on cockpit system-induced errors in the early postwar years. Incidentally, as recently as December 1980, it was authoritatively declared that "despite findings from this and other studies, it is unlikely that 3-pointer-type altimeters will be replaced .... in older operational aircraft" (NASA 1980). This again points to a lack of application rather than any lack of research, and may not be totally unrelated to widely held attitudes as reflected by the declaration of the aeronautical society.

For whatever reason, the air transport industry has for long felt it could survive well enough without the benefit of much formal HF investment or expertise. There is a growing feeling, however, that with the cost of a breakdown in the man-machine system reaching very substantial proportions -- in both human and financial terms -- a more professional and educated approach to HF is now required.

THE MEANING OF HUMAN FACTORS

The SHEL concept

Before discussing the educational requirements to ensure an effective application of HF in air transport it would perhaps be wise first to define the scope of this applied technology so that all are tuned to the same frequency. It is concerned with the study of man in his working environment and to understand this better it might be helpful to utilise a model which has been developed by the author from a basic concept originally proposed by Professor Elwyn Edwards. This has been called the SHEL concept, an acronym derived from the initial letters of the components of the man-machine system with which we are concerned, Software, Hardware, Environment and Liveware (Edwards 1972).

Fig. 2   The L component

In the centre of the model is man, or the Liveware in the system (Fig. 2). It is necessary to have a sound understanding of the characteristics of the liveware in order that a proper match can be made with the other components. To use terms familiar to engineers, it could be said that it is necessary to understand the-:

. input characteristics, in other words, the human sensory systems.
. output characteristics, for example, limb movements and speech.
. power potential, the forces which can be applied, for example.
. fuel requirements, such as food, water and oxygen needs -- including the effects of contamination.

. information processing, and factors which are known to influence its accuracy and speed, perception and so on.
. environmental factors, such as the effect of heat, noise, vibration, ozone, time-zone changes, pollution, etc.
. physical size and shape, varying as they do between individuals, men and women and between ethnic groups.

These headings are, of course, rather general and do not totally or specifically cover the full range of knowledge of man's behaviour and performance required. Attention must, of course, be applied to an understanding of motivation -- the difference between what a man CAN and WILL do -- and to the adverse effects of stress on different aspects of performance and the various ways of minimising these effects.

Man is not a simple, standardised component; people differ in many respects from their size and shape to their response to stressors. In this model, then, the block representing man is not symmetrical, thus emphasising the need for matching of components in order to achieve an effective system.

Fig. 3   The L-H interface

Perhaps the most commonly discussed interface in the system is between man (the Liveware) and the Hardware (Fig. 3). It is necessary to design seats, for example, with a proper understanding of man's size and shape. Displays need to be designed taking into account the characteristics of man's sensory and information processing systems. A warning system which failed to discriminate adequately between, say, an engine fire and an APU fire, would reflect a deficient L-H interface. As would a cabin emergency door which was too heavy for a stewardess to open in an emergency, or an ill-fitting oxygen mask.

Fig. 4   The L-H-S interfaces

Another component which must be recognised in the system is the Software (Fig. 4). This includes, for example, the procedures which are used in flight and on the ground, the concepts utilised in operating manual design, checklists and similar areas. Some aspects of the recent 2- or 3-man crew argument on short-medium range commercial aircraft might be seen as being related to this interface.

Fig. 5   The L-H-S-E interfaces

Somewhat easier to recognise and perhaps more extensively researched, is the interface between man and the Environment (Fig. 5). The effects of heat, cold, noise, vibration and pollution on man's performance and physical well-being are now widely recognised. Perhaps less so are other environmental factors such as time-zone changes and irregular work patterns, with their resultant disturbance of sleep and body circadian rhythms and the consequent adverse influence on human performance.

Fig. 6   The L-H-S-E-L interfaces

The final component to complete the model is, once again, man or Liveware (Fig. 6). Air transport operation involves team work -- cooperation between flight crew members, for example. Personality differences, aspects of authority, questions of leadership, discipline and loyalty may all be seen to be related to the L-L interface. Relations between flying staff and specialised groups such as the medical department, involving confidence and trust, also reflect effectivity of this interface.

Ergonomics and medicine

It will be apparent from this discussion that HF calls upon several disciplines as sources of information. Perhaps the two most significant are psychology and physiology, but others are also involved such as anthropometry, biomechanics and biology (and its sub-discipline, chronobiology). Neither should we forget statistics and experimental methodology, without which HF studies would frequently prove impossible to complete effectively.

It is perhaps the involvement of physiology which has led to the common misconception that full responsibility for HF within an organisation lies with the medical department. It is clear that there are certain common areas of interest between the HF specialist and the physician in an air transport unit (Fig. 7).

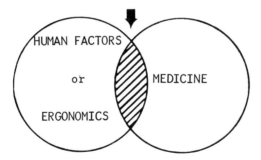

Fig. 7 Area of common interest between ergonomics and medicine

Collaboration between the the HF specialist, leaning towards applied psychology and the physician leaning towards physiology may be a useful element in an organisation. An example may be seen in the use of medication and stress management by crews and their effect on performance and behaviour in the operational environment. It would be as inappropriate, however, to expect a physician to make an authoritative contribution on error classification, motivation at work or computer keyboard design as it would be to expect a HF consultant to select the appropriate drug for the treatment of, for example, myasthenia gravis.

EDUCATIONAL REQUIREMENTS

Required expertise within an organisation

All staff in positions of even low responsibility in air transportation will have had their basic mathematics, languages, and science at school. Very few indeed are likely to have had any formal background of HF education. It is apparent, then, that the deficiency must first be handled within the industry, and at all levels. In order to facilitate an assessment of what is required in any organisation, another model, the HF "educational wigwam", may be found useful (Fig. 8).

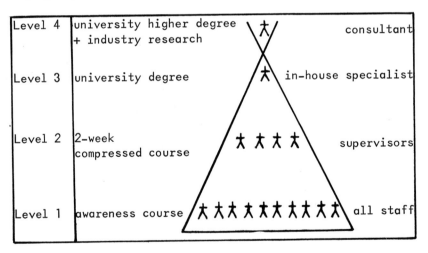

Fig. 8 The HF "educational wigwam" illustrates the different levels of HF education and expertise required within an organisation.

Level 1: all staff

It is reasonable to assume that without a basic awareness of HF by all staff in skilled or semi-skilled tasks little progress is likely. Operations management are able to publish a notice to pilots on idiosynchrasies in the performance of a particular engine and this is likely to be an effective communication because all have had the original education in aero engines upon which to base their comprehension and response. A similar crew notice on HF is likely to have less comprehension and response if there is no original education upon which to draw.

Many deficiencies in the operational environment are corrected as a result of feedback from the operators. The quality, and thus the effectiveness of that feedback will be greatly enhanced if it is made with some basic knowledge of the technology involved. For example, a pilot may be able to make a meaningful and productive report on the performance degradation of his aircraft in icing conditions because he has had an educational background in both aerodynamics and meteorology. He is far less able to make a constructive report on the human performance degradation of his crew resulting from circadian rhythm disturbance, because of a lack of relevant education.

The contribution of usable and educated feedback to safety and efficiency is not confined, of course, to pilots. It is applicable, to any operator in the transport system in the aviation industry from the airline employee at the ticket desk to the instrument repairer at his workbench; from the cargo handler to the stewardess in the aircraft galley.

It would be comforting to believe that in the very near future such a level of instruction will be incorporated into other basic educational programmes. This, regrettably, would be an unrealistic assumption, taking into account the slow progress noted earlier. One state flying school (Netherlands) has indeed now moved in this direction, but it will be many decades before we can expect to see the air transport industry fully populated with operational staff so qualified.

What type of education, then, is likely to be adequate to meet this "awareness" level of knowledge and in what form should it be presented? There are a number of criteria which should be taken into account in designing such an educational programme.

- Exhortation by itself is not enough. Neither is simply the issue of more paper or manuals in an industry already loaded -- some would say overloaded -- with documentation, unless they are directly designed to support other forms of education.
- The programme must be flexible and broad enough to cater for staff of widely different specialised interests within air transport.
- With recurring periods of economic strains in the air transport industry, the educational programme should be cost-effective; that is, a large number of staff should be educated at a low cost per head. This necessarily involves in-house education, or at least a programme not attracting hotel and other such overhead costs. It also favours a method which will not require the routine presence of a qualified instructor. From a logistical point of view it must also be possible to provide education to large numbers of staff without costly loss of production.
- In an industry employing large numbers of highly skilled, technical staff, the programme must carry high credibility, displaying expertise and trustworthiness and be of a high technical quality.

These criteria were taken into account in the design of a Human Factors Awareness Course (KHUFAC), now acquired by airlines such as KLM, Aer Lingus, Lufthansa and Indian Airlines, as well as being in use in the Netherlands State Flying School. The course in its basic format uses a twin-screen, tape-slide technique and is presented in

15 separate, self-contained units, each on a different aspect of HF and each lasting 30-40 minutes (see Appendix III). Optional single screen and video-tape versions are available. In KLM, which sponsored the course design and started this educational programme in 1980, more than 7000 man/units had been presented by the end of 1982 within the Flight Operations Division alone. Plans are now being introduced to expose staff from other company divisions to relevant units and individual units are being integrated into other company courses.

A feature of the course is that having stimulated interest of staff in what to many is a new applied technology, easily readable and, as far as possible, inexpensive books are made available for purchase by course participants. To date, some 700 such books have been purchased by KLM staff. It is an encouraging reflection of the interest of staff in the subject, and perhaps unique in a company course, that so many have chosen to carry out further study of the subject in their own home and at their own expense.

The amount of information contained in each unit is such that not more than 2 units are presented in any one day. As the units are self-contained, require no instructor and only last 30-40 mins. each, they lend themselves to adding on to other scheduled assignments or as lunch-time presentations, thus minimising logistical costs. They are easily transportable from one location in a company's system to another.

This audio-visual course has been designed to meet the educational needs of Level 1 in Fig. 8. It is the only course of its kind available in the air transport industry. An HF authority in the USA, currently particularly concerned with the study of flight crew incident reports (Air Safety Reporting System), has commented that more airlines must be exposed to this kind of education in order to make any significant progress in reducing human error induced accidents (Lauber 1982). It has also been welcomed by representatives of airlines, the manufacturing industry and IFALPA. Enquiries have been made from outside air transport (e.g. shipping and the oil industry) to assess whether the course can be adapted for these other industries. The course is currently available in the English language only.

Level 2: supervisory staff

Such an awareness course (KHUFAC), consisting of a total of only some $8\frac{1}{2}$ hours of instruction, is not adequate for supervisory staff, although it is, of course, far more than most can currently claim. For this second level in the "wigwam" a longer course is required and this will probably entail something like a 2-week participation in a programme made available by a university or technical institute. General, non-aviation, HF (ergonomics) courses of this kind have long been available (e.g. that offered by the West of England Management Group and conducted by the University of Wales, Institute of Science and Technology. However, such courses, covering a wide range of working environments, including mass production manufacturing industries, are not really cost-effective for air transport staff. For this purpose an industry-dedicated course is needed.

The only such 2-week course totally dedicated to air transport operations is that conducted at the University of Aston in Birmingham in the UK. The course is the responsibility of the Department of Applied Psychology of the university and the course consultants represent the airlines and the operational engineering side of the aircraft manufacturing industry. About 15 lecturers are used and these are drawn from airlines, the manufacturing industry, government agencies and acadaemia. While this is certainly too short to produce a "specialist", it is about the maximum time that most companies are prepared to release supervisory staff for educational purposes. In smaller countries, with no other similar expertise available, the graduate from this course may, relatively speaking, be seen locally as a specialist.

The course title "Human Factors in Transport Aircraft Operations" (HFTAO) reflects its scope. Subjects covered represent a practical approach to HF in air transportation (see Appendix IV). In addition to conventional lectures, workshops are held on various topics of operational interest.

The course was founded in 1972 and in the first 10 years 178 participants from 28 different countries attended. The distribution of participants between countries is not, however, even, with the Netherlands, for example, sending to the course more than double the number of participants sent by any other country outside the UK (see Appendix IV). As a further illustration of policy differences, it is interesting that while many of the major international scheduled airlines provide little or no specialised HF training for any of their staff, Britannia Airways, an independent, non-scheduled carrier (based at Luton, England), has for long maintained a policy of having all its senior supervisory operational staff educated in HF at this HFTAO course (see Appendix IV). This particular company is commercially directed and successful, currently operating about 30 aircraft. The total of 178 HFTAO participants, of course, represents only a very small proportion of the supervisory staff of the world's air transport industry, but very slowly a nucleus of aviation staff is growing with some knowledge of what is meant by HF. However, without the existence of a general level of education as described for Level 1, their task in applying HF after they return home from the course may be a daunting and frustrating one.

A $4\frac{1}{2}$-day HF course ("Schnittstelle Pilot/Flugzeug") was founded in 1979 and is now provided annually at the Carl-Crans-Gesellschaft, Heidelberg, Germany. Major areas covered in this course, which is given in the German language, are pilot-aircraft interface, display and control design, the role of the pilot in semi-automated systems and methodology of system evaluation with respect to system performance and pilot workload. The course is designed specifically for those working in the design, development and evaluation of aircraft systems rather than those involved in operating aircraft and so it covers only a part of those envisaged in Level 2 of the "wigwam".

As there are so few short aviation-dedicated HF courses, reference should also be made to such a course outside Europe. This is available at the University of Southern California as a part of the Aviation Safety programme offered by the Institute of Safety and Systems Management of that establishment. It lasts 5 days and is less comprehensive than the HFTAO course at Aston. Its objective is to teach selected current principles of behaviour science particularly appropriate to the advancement of safety. It allocates proportionally more time to physiology and biomedicine than the Aston course, a difference which reflects Aston's background of degree level ergonomics educational programmes. It may be questioned whether the scope of this 5-day course is adequate for supervisory staff as intended in Level 2 of the wigwam model (see Appendix V).

Level 3: the in-house HF specialist

A participant in the 2-week compressed, air transport operations (HFTAO) dedicated course described above will be able, back in his own industrial environment, to recognise most HF problems as such, be able to solve the more straightforward ones and know when and from where to seek further assistance. And, indeed, with the knowledge acquired after only such a short exposure, he will certainly find on many occasions that he does need more information and assistance. He would not feel qualified to handle extensive experimental studies on, for example, human performance, without such help, and would also wish to obtain information on studies conducted elsewhere.

The next level in the wigwam, therefore, is that of the in-house source of such assistance -- the degree-qualified HF specialist or ergonomist. His degree will be in applied psychology or industrial/occupational psychology or ergonomics in Europe; in the USA, engineering psychology, Human Factors or Human Factors engineering is likely to form the title of his studies.

A number of universities in Europe provide such degree courses (see Appendix VI). However, very few provide facilities for specialisation in aviation HF. In the USA certain universities, such as Ohio State and Embry-Riddle, do provide aviation orientation and the US Air Force Academy, for example, is graduating approximately 30 students a year with a degree in psychology with a specialisation in Human Factors.

Within the organisation, the in-house specialist will be available on demand to solve HF problems wherever they occur. He should also be able to carry out research studies to find solutions to more complex problems where answers are not readily available from elsewhere. Experience suggests, however, that it may take some years of practical exposure to the specialised problems of the flight deck, before the operations department will be able to make adequate use of his skills. Those who have been applying HF on a "do-it-yourself" basis for decades may possibly see the HF specialist as a threat to their security. Even when the HF man also has operational experience there still appears to be an entrenched attitude to avoid asking his advice or assistance. This situation emphasises the need for adequate education at the lowest level in the "wigwam", as discussed earlier.

Level 4: the HF consultant

The highest level in the model is the consultant. He is nominally outside the company or organisation structure, though permanent lines of communication to him should be maintained.

Some companies may not feel large enough to employ an in-house ergonomist on a full time basis. However, this should hardly apply to the major national airlines, most of which have a workforce of over 10000 personnel. In the case where the in-house specialist is omitted, then the consultant must play a far more active role in monitoring company HF affairs. In order for the company to know when to call for his assistance, it is obvious that Level 2, the supervisors, must be properly educated to make such a call, if no in-house specialist is available.

Even when there is an in-house specialist at Level 3, he will occasionally find it necessary to seek guidance on research work which has been done elsewhere (to avoid duplication), experimental methodology, and so on.

There are various advantages for a company in using a consultant. He is not on the normal company payroll, and is called upon only when needed. He is able to make comments and give advice free from internal company pressures or biases. He will also have routine contact with industry and research developments which may be of benefit to the company he is advising.

He will hold a higher degree (Masters or Doctors) in ergonomics, applied psychology, engineering psychology or related discipline and will have a good knowledge of the HF problems of the air transport industry, including flight operations.

This, then, is the kind of structure which should form the basis for progress in providing a more effective application of the HF knowledge which is now available. It is unreasonable to expect adequate progress without some basic education in what is to many a new applied technology. There may come a time when such education does form a part of original schooling and industry can recruit new staff secure in the knowledge that this educational gap has been filled. However, this day is far away and in the meantime responsibility for ensuring an adequate level of HF education will remain within the air transport industry itself.

RELATED EDUCATIONAL ACTIVITIES

This paper has referred to courses covering the full scope of HF which provide a sound foundation for HF application.

Nevertheless, without this foundation some airlines and organisations have attempted to tackle individual HF aspects of which they have become aware without perhaps always fully comprehending the manner in which different aspects interact with each other. Initiative is often due to a local incident or accident having drawn attention to one particular facet of HF. Often the programme is related to the airline's own operational procedural background or particular operational environment and so may not have universal applicability. In this connection it may be appropriate to recall that accidents occur from a combination of factors which are unlikely ever to combine again in exactly the same way. Therefore, for accident prevention it may not be fruitful to tackle the specific combination of circumstances that obtained in a particular case. Rather the basic weaknesses and deficiencies which would still cause an accident when combined differently should be addressed. This objective may involve a more general educational process.

One US airline has successfully introduced an extensive programme to improve, in particular, the working relationship between the captain and first officer in its own company by using video-taped simulator sessions. This can be seen as just one example of the L-L interface discussed earlier. "Captain's" or "command" training in a number of airlines aims at developing the desired behavioural pattern in its aircraft captains -- good leadership, commercial consciousness and so on. In Europe, one such course is offered, in France, by Air France. A 5-day module of the captain's development training applies particular attention to communication between people and spends much time on human behaviour and interfaces within a team or group. Another module includes aspects of group leadership. In Lufthansa (DLH), supervisory staff are given training in motivational theories and strategies, leadership theory and behaviour, communication theory and communication on the flight deck. Psychologists are used in the training of instructors. DLH, which has acquired the KHUFAC programme (see Appendix III), is planning to introduce a one-week HF course for all captains. Other in-house courses attempt to influence favourably cabin staff attitudes to passengers, yet another L-L interface question. Sometimes outside consultants are involved in establishing and conducting such courses. A number of airlines are moving towards integrated crew training and qualification rather than seeing each crew member as a separate entity. Again, an aspect of the L-L interface.

Such contributions are admirable but it seems very probable that they would be more effective if constructed upon a sound overall foundation of HF expertise in the company, as discussed earlier. And, of course, when considering overall safety, they tackle only a part of the Human Factors areas which are relevant to air transport operations.

It is noted that other aspects of HF education are often neglected. The L-S interface, involving, for example, manuals, forms, questionnaires and other publications is one. There is widespread HF criticism of the maps and charts most commonly used on board commercial aircraft. An airline may design and use some 3000-5000 forms -- mostly questionnaires of one kind or another. It is certainly unusual to find company staff involved with the design of such documentation having any formal education in the HF of documentation. The effectiveness of documentation, which includes safety bulletins and other publications directly influencing operational efficiency, can be profoundly influenced by its HF quality. There is now much known about layout, printing and language which should be applied.

As noted earlier, and largely as an L-E or an L-L interface question, educated advice on stress management and serious consideration of the effects of stress on performance are uncommon in the air transport industry. The use of fatigue and stress as bargaining elements in management/staff negotiations does not encourage objective study of the subject, though it does not reduce the need for it.

HF is slowly being introduced into aeronautical courses in the more progressive academic institutions. Only a few years ago it would have been surprising to find as much as 10% of accident investigation and prevention courses devoted to HF. But at the 8-week Aircraft Accident Investigation course at the College of Aeronautics, Cranfield, England, for example, the HF instruction time was approaching 15% by 1982. The Institute of Aviation Safety in Stockholm allocated 26% of the 77 hours of instruction to HF in the 1982 2-week Aviation Accident Prevention Investigative Techniques course. The KHUFAC programme (see Appendix III) is being incorporated into the Rijksluchtvaartschool pilot training curriculum in the Netherlands. At the level of the International Civil Aviation Organisation (ICAO) the situation is less encouraging. No significant HF role appears to be recognised in accident investigation courses being offered at regional ICAO training centres. This situation is also reflected in the omission of HF expertise from ICAO technical assistance programmes for developing countries. While this is not of direct concern to Europe, aviation is international in nature and the effects of such safety measures are not confined within national boundaries.

Some 6 years after their Istanbul Conference which called for urgent action to promote HF (see Appendix I), IATA published a small "Airline Guide to Human Factors". This booklet included chapters on a few selected aspects of HF and it was anticipated that in due course the scope would be extended. However, it has been questioned whether simply adding one additional manual to the heavy load of documentation already carried by air transport operators will make a significant impact on achieving the educational objectives proclaimed by IATA members in Istanbul in 1975. Nevertheless, a well-written and comprehensive aviation-dedicated HF manual which provided a back-up for and which was coordinated with properly constructed courses would, no doubt, make a useful contribution to the required educational process. Such a project has yet to be tackled.

CONCLUSION

In view of the public and industry attention which has been focussed on aviation HF, in particular in connection with human performance related accidents, it appears paradoxical that so little on a world-wide scale has been done to fill the basic educational gap which has been revealed. It is also surprising that the two major international aviation bodies which spoke out so strongly several years ago for action do not yet appear to have followed up their call with appropriate and effective measures.

In isolated cases, substantial progress is now being made in establishing a sound structure of HF expertise in air transport operations. In most cases, however, progress has been disappointing and activity unsystematic.

Sources of HF education dedicated to air transport operations are now available. However, in view of the slow progress generally being achieved nationally and internationally, a strong case could be made for an initiative at a CEC level. This would encourage activity and disseminate information in the field of HF education, with the aim of producing long-term benefits for the safety and economic health of European aviation.

-oOo-

"Not only will men of science have to grapple with the sciences that deal with man, but -- and this is a far more difficult matter -- they will have to persuade the world to listen to what they have discovered."
Bertrand Russell (1872 - 1970)

REFERENCES

ALPA/IFALPA 1977. Symposium on Human Factors, emphasising human performance, workload and communications. Washington, D.C. Feb 1977.

Edwards, E. 1972. In proceedings of BALPA symposium: Outlook for safety. Man and machine: systems for safety.

Hawkins, F.H. 1983. Stress management in air transport operations: beyond alcohol and drugs. In: Human performance in transport operations . Commission of European Communities. Köln, Germany 28-30 Jan 1983.

IATA 1975. Safety in flight operations. 20th Technical Conference. Istanbul, Nov 1975.

Lauber, J. 1982. Private correspondence between NASA and KLM. May 1982.

Meier-Muller, H. 1940. In: Flugwehr und technik . No. 1, 412-4, No. 2, 40-42.

Murrell, H. 1958. The term "ergonomics". The American Psychologist. $\underline{13}$, 10, 602.

NASA 1980. NASA Aviation Safety Reporting System: quarterly report. Prepared by NASA, Ames Research Center, Mountain View, Calif. in cooperation with Battelle Columbus Labs., Dec 1980. NASA-TM-81252.

APPENDIX I

IATA 20th Technical Conference

Extracts from the summing up of the International Air Transport Association (IATA) 20th Technical Conference in Istanbul, November 1975.

-oOo-

- ".... the statistical data presented to the Conference clearly indicated that the Human Factor problem not only comprehends the vast majority of accidents now occurring but also constitutes the area where the industry's knowledge is least perfect."

- "Vigorous and intensive programs are necessary to keep flight crews continuously conscious of the safety problem, ..."

- "There is an urgent need for the inclusion of Human Factors experts (ergonomists) in all accident investigation teams. Airlines should consider the use of the same type of expertise in their "in-house" safety programs."

- "The wider nature of Human Factors and its applications to aviation seem still to be relatively little appreciated. This neglect may cause inefficiency in operations, or discomfort to the persons concerned; at worst it may bring about a major disaster."

- "In order to facilitate the applications of Human Factors principles, much more requires to be done .... to make available more short courses to provide an appreciation of Human Factors for relevant personnel."

- ".... advantage could be taken of the availability of short-term courses for airline personnel. This would at least broaden awareness of the Human Factor problems faced by the industry ...."

- "Specific training for flight crews in their behaviour and limitations is not an easy task. .... but an effort must be made to educate."

- "To assist pilots in fighting complacency, it has been suggested that some education in their own human limitations would be useful."

- ".... emphasized the need for pilot understanding and awareness of the limitations and defects of the human sensory perception mechanisms."

- "Steps are needed to promote an increased awareness on the part of accident investigating authorities of the importance of accounting for Human Factors aspects."

- "... attention was drawn to the need for airlines to recognize the importance of education and training in Human Factors and the availability of literature and professional expertise in this specialised field."

- "Increased attention by airlines to HF .... implies the availability of personnel with knowledge and training in this specialised field."

APPENDIX II

ALPA/IFALPA Symposium on Human Factors
Extracts from the final position paper of the Airline Pilots Association (ALPA/IFALPA) symposium on Human Factors, Washington, Feb. 1977.

-oOo-

."The symposium .... remains concerned at the lack of expertise in the applied technology of Human Factors (ergonomics), worldwide, in many key centres of the air transport industry and associated bodies which control the quality of flight production, and thus the safety and welfare of the travelling public and flight crews ...."

.".... IFALPA be requested to study availability of training facilities in aviation Human Factors (ergonomics) and advise all member associations."

.".... IFALPA advises all member associations .... to encourage the local airlines and aviation agencies to ensure adequate levels of Human Factors expertise within their own organisations."

.".... IFALPA recommends that each member association ensures that at least one of its officers receives some formal training in Human Factors."

.".... in all pilot/manufacturing industry contact, information shall be requested on the Human Factors expertise which has been involved in the product design and development."

.".... there is widespread concern at the frequent failure to apply Human Factors knowledge already available and at the current lack of expertise in Human Factors in many areas of the commercial aviation industry."

.".... that Human Factors aspects of accident/incident investigation should be more deeply pursued."

.".... notes the frequent failure to apply knowledge which has long been made available from past Human Factors research, often a reluctance to take specialised Human Factors input into consideration of operational problems and a widespread lack of understanding of even the nature of this applied technology."

APPENDIX III

HUMAN FACTORS AWARENESS COURSE (KHUFAC)

Objectives
The course is designed to provide a broad awareness of aspects of Human Factors important for the safe and efficient conduct of air transport operations. The audio-visual format provides a very cost-effective means of presenting such information in a compact and concise manner to large numbers of personnel. The style is entertaining as well as educational and is intended to stimulate individual interest and initiative.

Course content

Unit 1 The Meaning of Human Factors
This ensures that all are talking the same language, clarifies some misconceptions and provides practical illustrations of current deficiencies.

Unit 2 The Nature of Human Error
In order to cope with errors it is necessary to understand their nature and be able to classify them. This unit is conceptually associated with Unit 3.

Unit 3 Meeting the Challenge of Human Error
A two-pronged attack is explained to minimise the occurrence of errors and to reduce the consequences of those that remain.

Unit 4 Fatigue, Body Rhythms and Sleep
The nature of sleep and body rhythms is described together with current and potential means of relieving associated problems in aviation.

Unit 5 Vision and Visual Illusions
Geometric illusions and illusions from flying situations in this unit make clear that all can be fooled sometimes, with potentially fatal results.

Unit 6 Fitness and Performance
The latest information of the effects of exercise, drugs, alcohol, smoking and diet is here related to performance and to health.

Unit 7 Motivation and Leadership
Motivation is concerned with the way one behaves and performs when not under close supervision. This is directly relevant to safety and efficiency.

Unit 8 Communication: Language and Speech
Some of the most dramatic accidents in aviation history have been caused by inadequate application of Human Factors to communications.

Unit 9   Attitudes and Persuasion
Many company publications are forms of persuasive communication, often aimed at modifying attitudes so as to increase safety or raise efficiency.

Unit 10   Training and Training Devices
The effectiveness of training in producing skills and reducing errors depends largely on an understanding of human learning and teaching processes.

Unit 11   Displays and Controls
Information on board must be presented in the optimum way for the human brain to utilise it. This is right in the mainstream of Human Factors.

Unit 12   Space and Layout
Efficiency and safety in any workplace -- and the flight deck is a workplace -- are fundamentally influenced by aspects of space and layout.

Unit 13   Documentation
No airline can function, it seems, without a mass of documentation. Research has shown that Human Factors plays a vital role in its effectiveness.

Unit 14   Passengers: the Human Payload
From seat and galley design to cabin crew leadership in emergencies, Human Factors has a major impact on safety and efficiency in the passenger cabin.

Unit 15   Awareness and Application
A look back over the course and "where we go from here".

Course duration
15 x 30-40 minutes (audio-visual format, tape-slide and video-tape options).

Information
KLM, Royal Dutch Airlines, Crew Training Manager (SPL/NT), Schiphol Airport, Amsterdam, Holland

APPENDIX IV

HUMAN FACTORS IN TRANSPORT AIRCRAFT OPERATION (HFTAO)

Objectives
　　The discipline of Human Factors is concerned to contribute to the Safety and Effectiveness of systems by the integration of scientific knowledge about human beings into total system design. The aim of this specialized course is to provide, for senior personnel in aviation, a review of human factors in relation to the design and operation of transport aircraft.

The course will be suitable for:
　　Accident Investigators
　　Airline, Project and Test Pilots
　　Design and Development Engineers
　　Flight Safety Officers
　　Medical Officers
　　Operations Officers
　　Technical Managers
　　Training Captains
and other personnel closely concerned with the design or operation of aircraft, or with the performance and welfare of flight crews.

Course content
　　The first course in this series was held at Loughborough University of Technology in 1972. Further courses have since been held at annual intervals and the venue has been changed to the University of Aston. The course is essentially concerned with assisting people to recognise the nature of human problems in civil aviation, and indicating the ways in which these problems may be handled. Several well-known experts contribute to the course as visiting speakers.

The following topics are included in the presentations:
　　Automation
　　Display and control systems
　　Documents: their design and use
　　Fatigue and sleep
　　Flight safety
　　Human error
　　Stress and workload
　　Training
　　Warning systems
Experience has shown that the international and multi-disciplinary background of the course students is a valuable feature of the course, facilitating the interchange of views and information.

Course duration
2 weeks

APPENDIX IV (continued)

Information
Professor Elwyn Edwards, Department of Applied Psychology, University of Aston in Birmingham, College House, Birmingham B4 7ET, Great Britain.

Country of origin of participants

Country	'72	'73	'74	'75	'77	'78	'79	'80	'81	Total
Australia	-	2	-	-	2	1	2	1	1	9
Brazil	-	-	-	-	-	1	-	-	-	1
Canada	1	-	-	1	3	4	1	4	2	16
Columbia	-	1	-	-	-	-	-	-	-	1
Denmark	-	-	-	-	-	1	-	-	1	2
Eire	-	-	-	-	-	-	2	-	-	2
Finland	-	-	-	1	-	-	-	-	-	1
France	-	-	1	-	-	-	-	2	-	3
Ghana	-	1	-	-	-	-	-	-	-	1
India	-	1	-	-	-	-	-	-	-	1
Indonesia	1	-	-	-	2	-	-	-	-	3
Israel	-	1	-	-	-	1	-	-	1	3
Italy	-	-	-	-	-	-	-	-	1	1
Japan	-	-	-	-	-	1	-	-	-	1
Jugoslavia	-	-	-	-	-	-	1	-	-	1
Lebanon	-	-	1	-	-	3	-	-	-	4
Netherlands	3	1	1	-	3	7	5	9	6	35
New Zealand	-	-	-	-	1	1	-	-	1	3
Norway	-	-	-	-	-	-	-	-	2	2
Portugal	-	-	1	-	-	1	1	1	-	4
Spain	-	-	-	-	-	-	-	1	-	1
Sweden	2	-	-	-	-	1	-	2	-	5
Switzerland	-	-	-	-	1	-	-	1	-	2
Syria	-	1	-	-	-	-	-	-	-	1
West Germany	-	2	-	2	2	5	-	2	-	13
United Kingdom	9	11	8	8	4	5	6	1	3	55
United States	1	1	1	-	1	1	1	-	-	6
Zambia	-	-	-	1	-	-	-	-	-	1
Totals	17	22	13	13	19	33	19	24	18	178

APPENDIX V

HUMAN FACTORS (AVIATION SAFETY PROGRAMMES)

Objectives
To teach selected current principles of behaviour science particularly appropriate to the advancement of safety. The course is designed around the MAN-PRODUCT-ENVIRONMENT system and will provide the students with a working knowledge of hazard control techniques within this area.

Course content
1. Physiology/Biomedicine
   a. Maximum human capabilities and limitations
   b. Medical selection standards and procedures
   c. Health maintenance and training
   d. Effects of habit, stress and disease
   e. Aging

2. Psychology
   a. Social effects on human factors: historical and contemporary
   b. Sensory capabilities
   c. Auditory capabilities
   d. Involuntary/voluntary attention
   e. Perceptual and problem-solving capabilities

3. Human engineering
   a. System design concept formulation
   b. Preliminary design
   c. Detail hardware design
   d. Personnel system development

4. Communication
   a. The nature of language -- yours and others
   b. Interpersonal communication

Course duration
Five days

Information
Institute of Safety and Systems Management, University of Southern California, Los Angeles, California 90089-0021, USA.

APPENDIX VI

EUROPEAN UNIVERSITIES OFFERING DEGREEE OR DIPLOMA COURSES IN ERGONO-
MICS, APPLIED PSYCHOLOGY, INDUSTRIAL/OCCUPATIONAL PSYCHOLOGY OR RE-
LATED DISCIPLINES

NOTE

This list should be considered as a general guide only, as programmes are subject to change and degrees/diplomas are not standardised internationally. Further details should be obtained from the
appropriate university. At additional establishments, some ergonomics
or applied psychology education may be incorporated into other programmes. The International Ergonomics Association is currently pre-
paring a new and comprehensive international directory of educational
programmes in ergonomics/Human Factors.

* = full degree course in ergonomics or related disciplines
** = air transportation Human Factors/ergonomics short course

Country	University	* / **	Programme
Austria	Technische Universiteit Wien		industrial engineering
Belgium	Université Libre de Bruxelles	*	ergonomics, work psychology
	Katholieke Universiteit te Leuven	*	occupational psychology
Denmark	University of Copenhagen	*	ergonomics, work physiology
Eire	University College, Dublin	*	applied psychology
	University College, Cork	*	applied psychology
Finland	Helsinki University of Technology	*	industrial psychology
France	Centre Interuniversitaire d'education permanente	*	study of work conditions
	Institut National de Recherche et de Sécurité		ergonomics and safety
	Laboratoire de Psychologie du Travail	*	work psychology, ergonomics
	Université Louis Pasteur		ergonomics of the physical environment

APPENDIX VI (continued)

Country	University	* **	Programme
	Université Jean Moulin (Lyon III)	*	applied ergonomics
	Université de Paris 1 (Panthéon-Sorbonne)		ergonomics and human ecology
	Rene Descartes (Paris V)	*	ergonomics
	Université Pierre et Marie Curie (Paris VI)		ergonomics, work psychology
	Université Paris Nord (Paris XIII)	*	ergonomics
	Université des Sciences et Technique de Lille	*	ergonomics, work physiology
Germany (FRG)	Technische Hochschule Aachen	*	industrial engineering
	Institut fur Arbeitsphysiologie		ergonomics, occupational physiology
	Universität Bremen	*	industrial engineering
	Carl-Crans-Gesellschaft Heidelberg	**	aviation Human Factors
	Technische Hochschule Darmstadt	*	industrial engineering
	Universität Hamburg	*	ergonomics, work/industrial/environmental psychology
	Technische Universität München	*	work sciences
	Technische Universität Stuttgart	*	ergonomics, work psychology
Greece	National Technical University of Athens		industrial engineering
Hungary	Technical University of Budapest		ergonomics
	TUC Institution for Labour Safety Education of Engineers		labour safety engineering
Italy	Instituto Superiore per le Industrie Artistiche	*	ergonomics
	Societa Italiana di Ergonomia (sie) e Regione Lombardia		ergonomics

APPENDIX VI (continued)

Country	University	* / **	Programme
Netherlands	Agricultural University Wageningen		ergonomics
	Catholic University Tilburg	*	techno-psychology
	Delft University of Technology	*	man-machine systems
	Twente University of Technology		ergonomics
	Utrecht University	*	ergonomic psychology
Norway	Agricultural University of Norway		ergonomics
	University of Trondheim -- Norwegian Institute of Technology	*	ergonomics
Poland	Politechnika Wroclawska Instytut Organizacji i Zavriadrania	*	ergonomics, work physiology
	Warsaw Technical University		ergonomics
Roumania	University "Babes-Bolyai"	*	industrial psychology and ergonomics
Spain	Universidad Complutense		industrial psychology
Sweden	University of Lulea	*	human work sciences
Switzerland	Swiss Federal Institute of Technology		work physiology, work and psychology
USSR	Moscow Institute of Transport Engineers	*	engineering psychology
	Yaroslavl State University		work psychology and engineering psychology
United Kingdom	University of Aston	* / **	ergonomics / air transport HF
	University of Birmingham	*	work design and ergonomics
	Cranfield Institute of Technology		work organisation and industrial ergonomics
	University of Hull	*	industrial psychology
	Birkbeck College, University of London	*	occupational psychology

APPENDIX VI (continued)

Country	University	* **	Programme
	University College London	*	ergonomics
	Loughborough University of Technology	*	ergonomics
	Nottingham University	*	occupational psychology
	Napier College of Commerce and Technology	*	ergonomics
	University of Wales Institute of Science and Technology	*	applied psychology
Yugoslavia	University of Zagreb		work physiology, ergonomics, work ecology

BEHAVIOUR RESEARCH IN ROAD TRAFFIC

J. Moraal*, J.B.J. Riemersma*

*Institute for Perception TNO
3769 ZG Soesterberg, The Netherlands

ABSTRACT

A review is given of some of the main research areas on human behaviour in road traffic, the purpose of which is the application of research findings in order to contribute to the solution of problems in road safety and efficiency. The subjects are discussed under three main headings; methods and instrumentation, basic research on human behaviour and applied research areas. The first two of these are seen as being in support to the third one. For a large part the review reflects the work done in the Institute for Perception TNO. Some future trends are marked and research priorities proposed.

INTRODUCTION

In daily road traffic the traffic participant is an element of the man-vehicle-roadway system. Safety and efficiency in road traffic, like in other transport systems, can only be obtained when the three system elements are mutually adapted. This means that, first, vehicle and road environment should be designed according to man's capabilities and limitations and, second, the traffic participant should be well trained to perform his tasks and obey the traffic rules. Research into human behaviour in road traffic aims to contribute to the solution of problems of safety and efficiency, i.e. is applied in nature. Its results should lead to conclusions how to design safe roads, legible and comprehensible road signs, visible road markings, education and training programmes, knowledge and skill testing procedures, rules which can be learned and applied easily, etc. However, questions with regard to safety, legibility, comprehensibility, visibility, training, testing and rule learning, all are related to human functions. For that reason, the problem-oriented or applied research necessarily should be supported by basic research on human functioning, i.e. investigating the ways in which people generally adapt to and cope with their environments in which they have to perform tasks. Basic research seeks to understand the ways in which people learn to look and search for information and how they process this information in order to decide to safe and meaningful responses and actions, whether it be under normal or adversive conditions. Besides support from basic research, the solution of road traffic problems asks for specific research methods and instrumentation. For instance, results from laboratory studies will have to be validated under real-world circumstances to enable the generalization of research findings. Otherwise, the evaluation of specific design features asks for well adapted observation techniques in order to gather relevant traffic data.

In the following sections some of the most relevant problems of human behaviour in road traffic will be reviewed, respectively with regard to methods and instrumentation, basic and applied research areas. In the final section some future trends will be marked and research priorities be proposed.

METHODS AND INSTRUMENTATION

The wide variety of problems in transport operations requires also a variety of research methods. The methods in use differ according to a number of dimensions which will be referred to as the (1) amount of experimental control of the circumstances, (2) the unit of measurement: from process outcome to the underlying behavioural processes themselves, (3) whole- versus part-task studies, and (4) the amount of preselection of the data (Riemersma, 1982).

Accident data

The first step in quantifying behaviour in traffic is by *registration of accident data*. However, relevant data under controlled conditions, for instance taking exposure into account, are seldom available. Besides, because of reasons of preselection these data usually are incomplete, not always easily accessible, and mostly lacking in statistically representative numbers. When complementary data are available, e.g. of exposure, environment or weather conditions, consecutive *analysis of accident data* is possible but still is concerned with the *outcomes* of the transport system as the unit of measurement.

Observation and registration methods

The second step, *observation and registration of traffic events* by human observers, video or film, provides data on actual traffic behaviour and, therefore, is more oriented towards the underlying transportation *processes*. This kind of data enables the analysis of interactions and conflicts between traffic participants and of behaviour of single traffic participants, in as far as this can be observed from outside (Van der Horst, 1982).

In-car observation enables the acquisition of data which are even more related to behavioural processes, like visual scanning behaviour of drivers, but the observer's presence may affect the behaviour under observation. Before this method can be used successfully, the question of observer reliability has to be answered.

Instrumented vehicles exclude some of the disadvantages of human observers, by using instrumentation for the registration of vehicle- and driver-related variables (Blaauw and Burry, 1980). When instrumented vehicles are used, more rigorous experimental designs can be followed like, for example, when the influence of drugs on driver behaviour is investigated. However, one of the disadvantages of the use of real-world experiments with instrumented vehicles is that strict replication of conditions is not guaranteed, whereas "testing-to-the-limits" is considered as being too dangerous.

Simulation

The next alternative, therefore, is *whole-task simulation*. Nowadays, many driving simulators are in use for research as well as for training purposes. Simulators offer the possibility for rigorous experimental control, replication of conditions is guaranteed, and they can be regarded as safe research tools. Questions remain, however, with regard to the validity (or fidelity) of the simulator, the role of the level of motivation of the subjects, and the choice of relevant task conditions (Blaauw et al., 1978; Blaauw, 1982; Moraal and Poll, 1979).

Instead of using whole-task simulators, however, most laboratory studies only use *part-task simulations*. For example, in measuring threshold values of lateral speed perception, subjects do not have to respond by actually operating a vehicle, but only to decide on the presence of a nonzero value of the independent variable. The part-task simulation consists of a driving scene with the corresponding information processing cycle (Riemersma, 1979). Part-task simulations are useful for the detailed analysis of task processes under experimental control. Again, however, the question of the generalizability of the research findings remains.

The optimal choice of - combinations of - methods is a research problem in itself. Sometimes theoretical considerations dictate what method should be chosen, whereas otherwise these are mere practical problems, e.g. availability of time and space. Still, not all problems with regard to the applicability of the methods have been solved. Whether simulators can be used in training people in dangerous situations, in which they have to use the full capabilities of the car, still has to be investigated. Video techniques for registration of traffic events seems to be very promising, e.g. for the quick evaluation of design features. However, the analysis of the pictures by human observers is, despite advanced methods, still very time-consuming and, therefore, waiting to become fully automated.

BASIC RESEARCH

Basic research in road traffic focusses on human perceptual and cognitive capabilities and limitations. The results are basic in the sense that performance limitations have to be overcome by the design of the traffic environment and the management of the traffic system.

Basic research has its starting point in an analysis of the tasks of the traffic participant. Three levels of task performance can be distinguished. The first, *strategic* level, is related to trip planning, route choice and guidance processes. The second, the *tactical* level, is related to the perception of other traffic participants and the consecutive manoeuvres. The third, *operational* level, is related to basic skills in the control of one's own vehicle by perception and maintenance of course and speed. The task levels are ordered hierarchically. Their main difference lies in the time delay by which information is gathered, processed and translated into actions.

Route choice and guidance

Trip planning normally takes place on the basis of partial knowledge of the road network, supplemented by information on road maps. Detailed knowledge of these processes of the use of mental maps of the road network could help in formulating the relevant information requirements to be met by route guidance systems (Janssen, 1979). Current research focusses on the weight drivers may give to the expected stress on particular routes.

Perception of motion

Overtaking and controlling the heading distance to a leading vehicle depend on the perception of relative speeds of cars on the same road. The change in optimal angle between taillights or headlights was found to be the optimal cue for this information (Janssen et al., 1976). Also, thresholds for relative speed differences for this situation could be established. Current research is concerned with the basic cues for the perception of impending collision at nighttime on intersections, under various circumstances of relative position, speeds and geometric situation (Janssen and Van der Horst, 1980).

Perception of egomotion

The optical cues used in course holding and control of one's own speed are still not exactly known. Perceptual thresholds for several candidate cues have been determined in laboratory simulations of driving scenes, and their effectiveness in real driving is demonstrated. It is investigated whether a further distinction can be made between position- and rate aspects of the optical cues for lateral position and speed. Results could be applied in the design of guidance systems, such as used in road-work zones (Riemersma, 1981; 1982).

Non-visual information

The driver's knowledge about the characteristics of his vehicle and the course to be followed may lead to control strategies in which non-visual information plays a role. The effectiveness of these strategies could be improved by appropriate design of steering systems and is investigated in curve driving and overtaking. Registration of steering behaviour during periods in which visual information is occluded has proven to be a valuable technique (Milgram et al., 1982).

Driver modelling

The aim of mathematically modelling driver behaviour is to replace drivers as subjects, in order to enable consistent search for optimal interfacing of drivers, vehicles and roadway environments. Starting from system theory and optimal control models, the usually less consistent knowledge on human perception, information processing and control behaviour is incorporated. In a combination of methods, computer modelling as well as experiments with instrumented vehicles, the human operator is described as a supervisor of system behaviour (Blaauw, 1981).

APPLIED RESEARCH

Traffic environment

Most of the relevant information for the traffic participant is provided by the environment, mostly by special purpose information systems such as route information, traffic lights, signs and guidance systems. However, the design of traffic facilities, the perceptibility of characteristics of other traffic participants and the impression of the total environment provide information as well. Physical characteristics determine visibility as such, but with competing information also cognitive factors become involved. Much is still unknown about the comprehensibility and the consistency of the total information stream and its impact on behaviour. Even less so about the more uncoded types of information from design elements and the wider structure of the environment.

Traffic signs and lights

Apart from visibility aspects research in this area now also includes cognitive factors as well and their effect on behaviour. Even very conspicuous signs are not always able to evoke the desired traffic behaviour. Questions have to be asked, then, with regard to comprehensibility, motivation and decision making (Riemersma et al., 1982). Knowing the meaning of traffic signs is no guarantee that drivers will also know what they have to do. In this respect manipulation of the duration of the yellow phase of traffic lights seems to be a means to improve the "red light discipline" (Van der Horst and Godthelp, 1982).

Guidance systems

The visibility of road markings, especially during nighttime, leaves much to improve. Road markings are elements of guidance information. Temporary changes in guidance, like in road-work zones, sometimes impose conflicting information. Path following through road-work zones often is hampered because the height of beacons and lights prevent curve anticipation by the driver (Godthelp and Riemersma, 1982). Furthermore, the design of tunnel walls and entrances have a clear effect on the lane position of cars which easily can lead to dangerous situations (Blaauw and Van der Horst, 1982).

Intersections

A considerable number of accidents occur at intersections. Drivers need various types of information when approaching and negotiating intersections: presence of an intersection, route guidance information, the geometry of the intersection, right-of-way rules and motion information of other road users. The quality of each of these should be related to accident and exposure data. Research findings show that certain types of accidents are more related to information quality than to traffic density. Nonlinear multivariate data analysis techniques have proven to be valuable research methods in this case (Janssen and Van der Horst, 1982).

Bicycle facilities and residential areas

With bicycle paths and low-speed residential areas designers of traffic facilities try to influence traffic behaviour according to certain standards. Evaluation, by means of before/after studies, can be based on the registration of actual traffic behaviour such as choice of path, interactions between traffic participants, speed, and the like, in order to enable comparison with intended behaviour. Analysis of these data may lead to suggestions for a more optimal design (Van der Horst, 1982).

Analysis of accidents and conflicts

As a first screening of safety aspects of the environment accident statistics may be helpful. However, exposure data are very often lacking and usually the numbers are statistically not representative. As an alternative, the registration and analysis of traffic conflicts seems more promising, since exposure data can be gathered simultaneously. Methodologically, however, there is no general agreement yet on the definition of what a conflict exactly is or should be regarded as such. Therefore, calibration studies should be undertaken and various methods of conflict registration should be compared (Kraay, 1982).

Vehicle ergonomics

Optimal task performance also depends on the adaptation of the vehicle characteristics to man's capabilities in respect of the task demands. Seating posture, location of displays and controls, outside view and mirrors, steering characteristics, aspects of comfort such as noise and vibration, interior lighting, etc., all may contribute to safe task performance of the traffic participant. Nowadays, many manufacturers seem to incorporate ergonomic standards in car design. However, several ergonomic principles still lack meaningful theoretical considerations and there is no agreement yet with regard to the relative importance of design features. In this respect special attention should be paid to design for the handicapped and elderly (De Waal, 1983).

Traffic participants

To improve the efficiency of training and testing procedures one needs to understand those aspects of tasks and skills which determine the smoothness and safety of performance: Basic understanding of principles of skill learning is still incomplete. Research in this area should concentrate on laboratory as well as on real tasks and manipulate in both the same task variables in order to enable generalization of research findings. Driver knowledge tests can be regarded as part-task simulations and, therefore, should have a high validity and not measure, for instance, verbal abilities (Veling, 1981; 1982).

Most of the research discussed sofar concentrates on the "normal" alerted traffic participants. However, there are considerable differences between people. Several categories of traffic participants are being recognized as problematic, although not necessarily less safe. Decreased safety also may result from transient states of the normal traffic participant when his capabilities and skills are impaired by fatigue (Riemersma et al., 1978), sleep deprivation, drugs, and the like. Research in this area requires a somewhat different methodology. Questions are in particular concerned with the capabilities of older drivers and the handicapped. In no way their right to participate in traffic can be denied. However, each driver will face the moment when he or she is no longer capable to drive, at least under certain adverse conditions. Basic research on information processing, functional visual field and attention allocation will have to support research with regard to measures to be taken to facilitate traffic participation for these people as long and safe as possible.

Handicapped drivers ask for special facilities. New technologies, like voice input control, seem to be promising but still have to be further developed and tested. With rehabilitation patients, like people with permanent diffuse cerebral dysfunctions, the question inevitably will arise whether they are able again to participate in traffic. It seems that in this case the best methodology lies in a combination of laboratory tasks, driving skill tests and cognitive testing of the subjects (Ravestein et al., 1982).

Subnormal functioning of traffic participants as affected by drugs (diazepam) or fatigue has been investigated (O'Hanlon et al., 1982). Although substantial deterioration of performance has been demonstrated, the results are not yet sufficient to allow generalizations, particularly because most studies did not include habitual drug users. The research effort is directed towards the development of suitable measuring techniques for task performance and physiological indicators of stress.

FUTURE TRENDS

Until now, the main research effort has been on the investigation of information gathering or input processes. It is realized, however, that a shift is necessary towards more cognitive processes, like interpretation of information and decision making. Studies on the perception, acceptation and taking of risk are coming more into focus, since they seem to cover the gap between, on the one hand, what people are able to do according to their basic capabilities and, on the other hand, how they actually behave in traffic (Benjamin and Somers, 1982). Once traffic participants have become involved in risky operations, recovery from threat is necessary.

Sofar, however, most of the emphasis was laid upon avoidance of risky situations. It should be investigated whether it is feasible to use simulators for training people in recovery behaviour. Much priority should be given to studies to improve training efficiency in general. A second shift to be recognized is from the relatively simple skill-based operation of vehicles towards the more complex task levels of manoeuvring and trip planning. This undoubtedly will raise questions about the interactions between task levels and different stages in the information processing chain. Third, there is a growing interest in the effects of degraded conditions under which traffic participants have to operate. Although one is beginning to understand the effects of stressors, such as fatigue, drugs and alcohol, yet no substantial research effort has been invested in their combined effects on behaviour in road traffic.

The growing numbers of older people in our societies ask for a concerted action to investigate their declining capabilities and to look for ways in which the traffic environment can be adapted, in order to enable them to participate in traffic as long and safe as possible (Planek, 1981; Sivak et al., 1981). For the same reason attention should be paid to people who have become handicapped in one way or another.

A final remark should be made concerning instrumentation. The developments of automated unobtrusive observation techniques should be highly prioritized. Video techniques, in use for the registration and analysis of traffic events such as conflicts, seem very promising. However, the analysis of the recordings by hand is very time-consuming, and therefore should be automated.

REFERENCES

Benjamin, T.E.H. and Somers, R.L. 1982. (Eds.) Road user exposure and risk. Accident Anal. and Prev., 14, 335-424.

Blaauw, G.J., Van der Horst, A.R.A. and Godthelp, J. 1978. The driving simulator of the Institute for Perception TNO - A validation study in straight road driving. Report no. 1978-16, Institute for Perception TNO, Soesterberg, The Netherlands.

Blaauw, G.J. and Burry, S. 1980. ICARUS, an instrumented car for road user studies. Journal A, 21, 134-138.

Blaauw, G.J. 1981. Drivers' internal representation and supervisory control; a first model verification in relation to driving experience, task demands and deteriorated vision. Proceedings 1st European annual conference on human decision making and manual control. Delft, The Netherlands, 315-327.

Blaauw, G.J. 1982. Driver experience and task demands in simulator and instrumented car; a validation study. Human Factors, 24, 473-486.

Blaauw, G.J. and Van der Horst, A.R.A. 1982. Lateral positioning behaviour of car drivers near tunnel walls; final report. Report no. 1982 C-30, Institute for Perception TNO, Soesterberg, The Netherlands.

De Waal, B. 1983. Road vehicle ergonomics; a review of the literature. Report no. 1983 C-., Institute for Perception TNO, Soesterberg, The Netherlands.

Godthelp, J. and Riemersma, J.B.J. 1982. Vehicle guidance in road-work zones. Ergonomics, 25, 909-916.

Janssen, W.H., Michon, J.A. and Harvey, L.O. 1976. The perception of lead vehicle movement in darkness. Accident Anal. and Prev., 8, 151-166.

Janssen, W.H. 1979. Route planning and guidance; a review of the literature. Report no. 1979-C13, Institute for Perception TNO, Soesterberg, The Netherlands.

Janssen, W.H. and Van der Horst, A.R.A. 1980. The perception of impending collision in night-time driving. Report no. 1980 C-17, Institute for Perception TNO, Soesterberg, The Netherlands.

Janssen, W.H. and Van der Horst, A.R.A. 1982. Predicting the accident potential of intersections from a consideration of information processing aspects. Proceedings 1st Nordic congress on traffic medicine. Linköping, Sweden, 7-9.

Kraay, J.H. 1982. Proceedings 3rd int. workshop on traffic conflicts techniques. Institute for road safety research SWOV, Leidschendam, The Netherlands.

Milgram, P., Godthelp, J. and Blaauw, G.J. 1982. An investigation of decision-making criteria adopted by drivers while monitoring vehicle state in the temporary absence of visual input. Proceedings 2nd European annual conference on human decision making and manual control. Bonn, FRG, 104-116.

Moraal, J. and Poll, K.J. 1979. The Link-Miles driver training simulator for tracked vehicles; a validation study. Report no. 1979-23, Institute for Perception TNO, Soesterberg, The Netherlands.

O'Hanlon, J.F., Haak, T.W., Blaauw, G.J. and Riemersma, J.B.J. 1982. Diazepam impairs lateral position control in highway driving. Science, 217, 79-81.

Planek, T.W. 1981. The effect of ageing on driver abilities, accident experience, and licensing. In "Road safety, research and practice" (Eds. H.C. Foot et al.). (Praeger, New York).

Ravestein, R., Veling, I.H. and Gaillard, A.W.K. 1982. The driving skill of rehabilitants with a diffuse traumatic injury of the brain; a preliminary approach. Report 1982-16, Institute for Perception TNO, Soesterberg, The Netherlands.

Riemersma, J.B.J., Biesta, P.W. and Wildervanck, C. 1978. On the effects of prolonged night driving. In "Driver fatigue in road traffic accidents". (C.E.C., Brussels). pp. 72-81.

Riemersma, J.B.J. 1979. The perception of deviations from a straight course. Report no. 1979-C6, Institute for Perception TNO, Soesterberg, The Netherlands.

Riemersma, J.B.J. 1981. Vehicle control during straight road driving. Acta Psychol., 48, 215-225.

Riemersma, J.B.J. 1982a. Applied Psychology and Traffic Safety: Research at the Institute for Perception TNO. Report no. 1982 I-2, Institute for Perception TNO, Soesterberg, The Netherlands.

Riemersma, J.B.J. 1982b. Perceptual cues in vehicle guidance on a straight road. Proceedings 2nd European annual conference on human decision making and manual control. Bonn, FRG, 127-136.

Riemersma, J.B.J., Van de Hazel, A.T., Buist, M. and Moraal, J. 1982. Design and evaluation of pictograms adapted for a matrix panel. Report no. 1982 C-25, Institute for Perception TNO, Soesterberg, The Netherlands.

Sivak, M., Olson, P.L. and Pastalan, L.A. 1981. Effect of driver's age on nighttime legibility of highway signs. Human Factors, 23, 59-64.

Van der Horst, A.R.A. and Godthelp, J. 1982. Drivers' obedience to the red traffic signal in relation to the termination of the green phase; a review of the literature. Report no. 1982 C-11, Institute for Perception TNO, Soesterberg, The Netherlands.

Van der Horst, A.R.A. 1982. The analysis of traffic behaviour by video. Proceedings of OECD seminar on short-term and area-wide evaluation of safety measures. Institute for Road Safety Research SWOV, Leidschendam, The Netherlands, 198-205.

Veling, I.H. 1981. Driver performance evaluation. In "Guidelines for driver instruction". (OECD, Paris). pp. 31-38.

Veling, I.H. 1982. Measuring driving knowledge. Accident Anal. and Prev., 14, 81-85.

SOME THEORETICAL CONSIDERATIONS ON ACCIDENT RESEARCH

M.L.I. Pokorny and D.H.J. Blom

Netherlands Institute for Preventive Health Care/TNO
P.O. Box 124, 2300 AC LEIDEN, The Netherlands

INTRODUCTION

Since many years the problems of load and load capacity of task performers have been raised with respect to scheduling, duration and nature of the work and the possible effect thereof on the behaviour and well-being of the task performer.
In 1977 a start was made at the Netherlands Institute for Preventive Health Care/TNO with a project*, of which the purpose could roughly be defined as follows: To develop a useful measuring instrument capable of demonstrating the (assumed) effect of the performance of a task on the task performer and, assuming that this would prove possible, to ascertain to what extent the task effect changes under different task conditions.

The majority of research has been concentrated on describing short-duration effects of physical components of a task on the individual performing that task. The investigations aiming to demonstrate more lasting effects of the task performance on the one hand and the effects of an invariably present mental component of a task on the other, are faced with the absence of a good measuring method enabling interpretation of the to be determined effect in a quantitative way (Broadbent, 1979).
Some known indicators of changes in the condition of the organism can be used under properly controlled laboratory conditions, but their indicator value greatly diminishes when they are used in a real field situation (Johnson, 1970; Lacey, 1967).
The spectrum of influences under which such indicators can change is actually so wide that it is difficult to represent them quantitatively and to interpret the effect of the actual task examined.
The conclusion that must be drawn from the foregoing is obvious: measuring results in respect of the effect of the task on the task performer must actually represent this effect and not all kinds of other, possible, underlying factors.
It was decided to set up a project for examining the effect of an occupational task on the task performer. For practical reasons the task of a

---

* Members of the project-team are Drs C.H.J.M. Opmeer and both authors.

busdriver was chosen.

Measurements performed in this investigation can be roughly divided into two categories, viz., measurements with respect to the individual (physiological and psychometric measurements) and with respect to the bus (involving two components: speed of the bus and steering-wheel movements). These measurements are performed under different task conditions (routes, shifts) with respect to task performers in two age groups. In order to improve interpretation of the data, events in and around the bus were observed during the work. In determining the research strategy it was assumed that the effect of the task performance is a cumulative effect. Thus, we assume that the condition of the organism changes under the influence of the task to be performed by this organism. This consideration led to the decision to measure the assumed effect on physiological and psychometric variables not during performance of the task but before, during some rest intervals, and after the performance of the task. These measurements are carried out in a specially equipped mobile laboratory, designed for the purpose. The task-specific measurements are, of course, carried out during task-operation.

Apart from the field study just described, the project contained a methodologically different but in view of the purpose integrated part: the accident analysis.

THE ACCIDENT STUDY

At the beginning of the above-mentioned project during a general orientation on the busdriver's profession an investigation was started of the archives of accident reports of busdrivers in the company where the project was to be carried out. It appeared that the contents of the archives together with other available information about the accidents, the busdrivers and their work-organization, might be interesting for a more detailed analysis relevant for the purpose of the project. It was assumed that the occurrence of a traffic-accident could contain information about possible effects of the task performance of a busdriver, and of the conditions of this task, on the man. However, also in the field of accident research one is faced with the absence of a generally applicable methodology (Hale and Hale, 1972).

The accident study contains two parts because the initial explorative investigation of the above-mentioned accident archives could be followed by a second study of the accidents of a different establishment of the bus company. In this second study the hypotheses derived from the first part were tested.

In this paper a brief outline will be given of a conceptual model on which accident research could be based. Some practical and methodological problems encountered during the study will be considered in another paper (Blom and Pokorny, 1983).

The results of the first part of the study are published (Pokorny et al., 1983) and the results of the second part will be published in the near future.

THEORETICAL CONCEPTS

Introduction

The following hypothesis was the basis for the project: The performance of an occupational task has a (cumulative) effect on the individual performing this task; the nature of this effect is determined by characteristics of the task itself, of the environment (of which the task forms an integral part) and by individual qualities of the task performer.

With specific reference to the accident study this means that the occurrence of accidents in task situations might give indications concerning existence and nature of the effect of the performance of the task on man.

The basic principle applied was the Stimulus-Organism-Response (S-O-R) concept, which has become a classical concept in psychology. This choice is associated with the assumption that changes in the response to various stimuli can lead to statements on changes within the intermediating variable (the organism); the changes can be interpreted as effects of, i.a., the task performance.

This concept should not be taken as a chain of separate reactions on solitary stimuli, but only serves as a schematic representation of a very complicated interactive proces. With regard to the accident study this proces can also be approached in terms of a man-environment interaction, well-known in epidemiological research (figure 1).

In general it can be assumed that the nature of the response on external stimulation is associated on the one hand with characteristics of the environment, interpreted as sources of stimulation; and on the other with qualities of the individual, the organism.

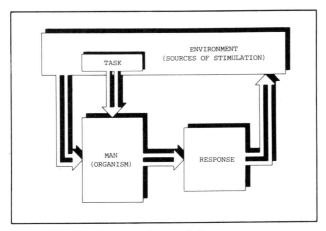

Fig. 1   Scheme of the interaction model.

It is important for the interpretation of the response to have available as much information as possible concerning the stimuli and their sources and concerning the individual reacting to these stimuli.

The use of this concept leads to our conceptual definition:

"An accident is a possible result of a hazardous situation with or without damage and/or injury and/or death".

This is illustrated in figure 2.

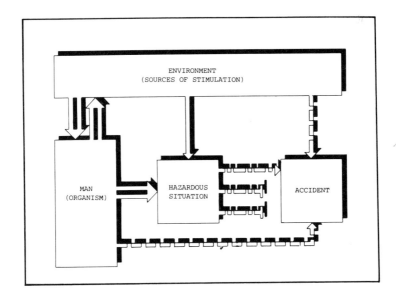

Fig. 2   Scheme of Accident Aetiology.

Two main elements can be discerned in this diagram (figure 2).
1. A central position is occupied by the "hazardous situation". This is defined as follows:

"A hazardous situation is a possible result of certain interactions between man and environment, which may or may not result in an accident".
2. A differentiation is made between the hazardous situation on the one hand and the possible result thereof, the accident, on the other.

It should be emphasized that the consequences of the accident in the form of the extent of the damage or the occurrence of injury or death are in this approach more or less independent of the process leading to the hazardous situation, since: "..... the same error can have many different consequences, ranging from injury, through damage to equipment, to momentary dislocation of production. The factors which cause errors are not necessarily the same as the factors which cause errors to produce injuries" (Hale and Hale, 1972).

This approach, partly inspired by the critical reports on accident research by, i.a., Kerr (1959), Haddon et al (1964) and Hale & Hale (1972), the epidemiological approach of Mc Farland et al (1955, 1958), the emphasis placed on the process character of accidents by Arbous and Kerrich (1953), can in our view be applied in many task situations in two ways:
1. For analytical-epidemiological investigations of accidents in certain risk groups (as is the case in this study).
2. In any systematic recording of hazardous situations, as done in the work of Hildebrandt et al (1973), and Prokop and Prokop (1955).

In the two components of the man-environment interaction various factors can be distinguished, which to a different extent and in different ways can be important in the process leading to an accident. This will be discussed in more detail further on.

The man-environment interaction

The factors discussed in this section have been subdivided into three categories. This classification (with regard to the individual factors after Häkkinen, 1958) is of an arbitrary nature. In effect the categories gradually merge. In addition, it should be pointed out that only some factors will be mentioned for purpose of illustration.

The following three categories are distinguished with respect to the human component of this interaction:
1. Constant factors or factors showing relatively little change during the

survey period. Factors that would come to mind in this respect are the sex of the individual or his (her) personality structure.
2. Factors that change in the course of time, but which are, in principle, controllable, such as age and experience.
3. Rapidly and (or irregularly changing factors, such as an intercurrent disease, fatigue, etc.

This division into 3 categories has of course consequences for the analysis. It can be said that these categories show a diminishing degree of constancy over time, and conversely an increasing difficulty for controlling these factors in an analytical design, from the first to the third category. Another consequence is that the relevant factors, depending on, i.a., the length of the period surveyed, the composition of the population examined and the questions to be answered by the analysis, will form an uncontrollable source of variability at the level of analysis. The factors of the first and partly the second category can be conceived as personal "qualities", those of the third category as temporal "attributions".

With respect to the environmental component of the interactive process three categories of factors can be distinguished in the same way, both as regards the physical and the social dimensions of the environment:
1. Constant factors or factors showing relatively little change. Factors that come to mind here are static environmental characteristics, ergonomic quality of task instruments or the socio-cultural context of the individual.
2. Factors that change in the course of time, but which, at least in principle, are controllable e.g. secular factors (years, months, etc.) or the alternation of light and darkness, changes in the technical properties of task instruments or organization of the work.
3. Rapidly and/or irregularly changing factors, such as weather conditions, behaviour of other individuals, type of shift, duration of service, etc.

When analysing the accidents allowance must be made for diminishing constancy and/or controllability of the relevant factors from category 1 down to category 3.

Only some of the numerous factors that can come to mind, both in the individual and in the environmental component, are of varying importance in the interactive process that can lead to an accident.
With some of the factors from the first and second categories, which are relatively constant, readily definable and detectable, it may be possible to determine the course of the interaction and the extent to which these

factors contribute to the process causing an accident, while other factors cannot or hardly be implied in the research design. In the case of factors from the third category (both with the human and with the environmental component) at least three possible forms of influence on the interactive process can be distinguished:

a. The influence on the interaction is only of marginal importance (in the accident literature sometimes referred to as 'Act of God'). On the individual side this could be for instance a sudden heart attack or stroke, suicide attempts and the like; with respect to the environmental component an example would be a suddenly occurring breakdown of equipment or a natural disaster.
b. The influence on the interaction is completely unsystematic or cannot be determined. Examples: incidental excitement or a light form of hypoglycemia.
c. The influence on the interaction can be established in principle, but is difficult to quantify, at least in an epidemiological study such as the present. Examples: Fatigue and the like (individual); behaviour of other individuals and the like (environment).

Whereas in the case of eventualities mentioned under a. there is actually no question of interaction and a definite cause of the accident can often be demonstrated, the possible causal contribution of factors mentioned under b. will probably remain undetected. The factors mentioned under c. will be discussed further.

The foregoing discussion concerned some categories of individual- and environment-related factors and their influence on the interactive process, which may lead to an accident, arranged according to their degree of constancy. It proved possible to draw up an analogous order for both the individual and the environmental components. Another approach to these factors, which in our view is also of great importance, is to draw up an order according to the degree of (mutual) influence. With this approach, however, a difference can clearly be observed between the two components. Whereas with a number of factors of the individual component the influence by situational aspects may be assumed to increase as the constancy of the relevant factors is diminishing, a comparable reasoning cannot be followed in the case of the environmental component. It would after all be true to say that the momentary situation in the time preceding the accident cannot be influenced by the individual concerned.

As mentioned before, the influence of individual factors is reversely associated with the constancy of these factors. It need hardly be pointed out that age or sex for instance are not at all dependent on the interaction, while hazardous behaviour for example and to an even greater extent such factors as "fatigue" and speed of information processing can be influenced by the interaction with the environment.

Selection problems

In the approach outlined above one should, ideally, wish to have information concerning all events which could be defined as "hazardous situations", i.e. information concerning the distribution of the probability of such an event taking place. In the analysis of (road) accidents one is, however, confronted with the fact that the information usually available only relates to registered accidents. Quite apart from representativity problems in connection with the completeness of the various registration systems, the implication is that the analysis of factors that may have contributed to the occurrence of the situation from which an accident resulted is influenced by selection bias right from the start.

Hence, only those data are available for which the results of these events have acquired a given form, viz. damage and/or injury and/or death, i.e. information concerning the distribution of the probability of an accident. The two probability distributions are theoretically not identical. One could say that the probability of an accident is determined by the probability of a given event and some other factors. This means that the probability of an accident is a selection from the distribution of the probability of a given event taking place.

Nevertheless, we take the view that this selection need not necessarily distort the analysis results to any serious extent, if the hypothesis is correct that the above-mentioned other factors determining the result of a hazardous situation (hence, whether or not an accident, with appertaining damage and/or injury, takes place) follow mainly a random distribution. This hypothesis can be illustrated with the following example: A driver of a car applies the brakes suddenly on a wet road - the hazardous situation. The result of this situation can differ widely, ranging from simply driving on after braking; a fright for the driver if the skids, but manages to regain control over the car; a dent in the wing of the car, if a small post happens to be in the way; serious material damage with possible injury to the driver, if there happens to be a tree in the way during skidding; to a complete catastrophy, if there is no post or tree in the way during skidding, but a class of infants happens to be walking there.

Only in some cases the results might still be influenced systematically. Such influence could take the form for instance of an association between the extent of the damage and the size of the vehicle (heavy trucks; Bygren, 1974), or the suspicion expressed in the literature that during the night shift fewer, but more serious accidents take place (Andlauer and Metz, 1967; Fröberg, 1974). Hence, it is important when analysing possibly causal influences on the occurrence of accidents to make allowance for this selection and, if possible, to imply these factors in the analysis by checking the results of categories of accidents with different extent of damage.

The interaction

In the analysis it will generally be easier to include data from categories 1 and 2 of both the individual and the environmental components of the interaction model described in the preceding section than from the categories 3. In order to improve the possibility of explaining the variability in the occurrence of accidents, it would seem necessary not just to involve the more or less constant factors from these categories in the analysis, but also as much as possible factors from the two categories 3. It must be endeavoured to establish a possible systematic influence on individual factors in this category by situational aspects. Since these rapidly and/or irregularly changing individual factors are hard to define and to operationalize (Muscio, 1921; Broadbent, 1979), a description of the conditions under which the individual must function might be helpful. The results of these analyses could then be interpreted under specified conditions as effects of momentary attributions of the individual. A certain amount of variability will for that matter probably remain unexplained. For the sources of such variability reference may be made to the previously mentioned non-systematic influences from the two components of the man-environment interaction, as well as to possible systematic influences from the environment on the individual, of which the source of stimulation cannot be detected or be implied in the analysis. Examples of this would be changes emanating from the environment to which individuals can, of course, react quite differently, but which may nevertheless have a systematic effect on the occurrence of accidents.

SUMMARY

- In accordance with the purpose of this study it is assumed that the oc-

currence of an accident might present indications of the existence and nature of effects of the task performance on man.
- An accident is defined as a possible result of a hazardous situation with or without damage and/or injury and/or death.
- A hazardous situation is defined as a possible result of an interaction between man (the organism) and his environment (sources of stimulation).
- The consequences of the accident (extent of the damage, etc.) are more or less independent of the process that led to a hazardous situation.
- In both components of the man-environment interaction three analogous categories of factors can be distinguished within the analytical-epidemiological design used:
  1. Constant factors or factors showing relatively little change during the survey period.
  2. Factors which change in the course of time, but which are, at least in principle, controllable.
  3. Rapidly and/or irregularly changing categories.

From category 1 to category 3 one can speak of a diminishing constancy, an associated diminishing controllability, and in the case of the individual component of an increasing possibility to be influenced in the interaction.

REFERENCES

Andlauer, P. and Metz, B. 1967. Travail en équipes alternantes. In "Physiologie du travail; Ergonomie vol. 2" (Ed. J. Scherrer). (Masson, Paris).
Arbous, A.G. and Kerrich, J.E. 1953. The phenomenon of accident proneness. J. abn. soc. Psychol., $\underline{48}$, 99-107.
Blom, D.H.J. and Pokorny, M.L.I. 1983. Accidents by busdrivers - practical and methodological problems. In "Human Performance in Transport Operations; Proceedings of a CEC-Workshop" (Ed. H.M. Wegmann). (Commission of the European Communities, Luxembourg).
Broadbent, D.E. 1979. Is a fatigue test now possible. Ergonomics, $\underline{22}$, 1277-1290.
Bygren, L.O. 1974. The driver's exposure to risk of accident. Scand. J. soc. Med., $\underline{2}$, 49-65.
Fröberg, J.E. 1974. Scheduling hours of work. (Rep. Working Party to Swedish Work Environment Fund, Stockholm).
Haddon, W., Suchman, E.A. and Klein, D. 1964. Accident research; methods and approaches. (Harper, New York).
Häkkinen, S. 1958. Traffic accidents and driver characteristics; a statistical and psychological study. Sci. Res. no. 13 (Finland Inst. Technol. Helsinki).
Hale, A.R. and Hale, M. 1972. A review of the industrial accident research literature. (HMSO, London).
Hildebrandt, G., Rohmert, W. und Rutenfranz, J. 1973. Über Jahresrhythmischen Häufigkeitsschwankungen der Inauspruchnahme von Sicherheitseinrichtungen durch die Triebfahrzeugführer der Deutschen Bundesbahn.

Int. Arch. Arbeitsmed., 31, 73-80.
Johnson, L.C. 1970. A psychophysiology for all states. Psychophysiol., 6, 501-516.
Kerr, W. 1950. Accident proneness of factory departments. J. appl. Psychol., 34, 167-170.
Lacey, J.I. 1967. Somatic response patterning and stress; some revisions of activation theory. In "Psychological stress" (Eds. M.H. Appley and R. Trumbull). (Appleton-Century-Crofts, New York). pp. 14-42.
Mc Farland, R.A., Moore, R.C. and Warren, A.B. 1955. Human variables in motor vehicle accidents; a review of the literature. (Harvard School Publ. Hlth., Boston-Mass.).
Mc Farland, R.A. 1958. Health and safety in transportation. Publ. Hlth. Reports, 73, 663-680.
Muscio, B. 1921. Is a fatigue test possible? Brit. J. Psychol., 12, 31-46.
Pokorny, M.L.I., Blom, D.H.J., van Leeuwen, P. and Opmeer, C.H.J.M. 1983. Ongevallen bij buschauffeurs - Een epidemiologische studie. Deel 1. (Accidents of Busdrivers - An epidemiological study. Volume 1. With Summary in English). (NIPG/TNO, Leiden).
Prokop, O. und Prokop, L. 1955. Ermüdung und Einschlafen am Steuer, Dtsch. Z. gerichtl. Med., 44, 343-355.

ACCIDENT OF BUSDRIVERS - PRACTICAL
AND METHODOLOGICAL PROBLEMS

D.H.J. Blom and M.L.I. Pokorny
Netherlands Institute for Preventive Health Care/TNO
P.O. Box 124, 2300 AC LEIDEN, The Netherlands

INTRODUCTION

An important discovery at the onset of the project (as mentioned in the previous paper, Pokorny and Blom, 1983) was that because of insurance regulations all bus accidents, whatever the damage may be, with or without injury, had to be reported to the company. This fact, together with the existance of an archive containing information about accidents which had happened in previous years, formed the beginning of this part of the study. These findings can be considered as some of the basic conditions for an accident analysis to yield satisfactory results. The operational definition of an accident was therefore:

"An accident is an event of damage to a bus, or by a bus, that needed to be reported to the insurance company".

In practice this means all accidents of busses of that company, including the very minor ones. Because of the liability of the busdrivers in case of not reporting one can reasonably assume completeness of the data. The accident data collected represent that part of the hazardous situations as meant in the previous paper, in which the result has taken a given form.

Part of the information about the accidents and the circumstances in which they occurred was available on the accident forms. The greater part of the context information, however, had to be sought in other sources, like time-tables etc. (figure 1).

In this paper the various sources used and the problems encountered therewith will be discussed. A brief outline is given of the method to assess accident risk under different conditions of work, and of the consequencies for the analysis.

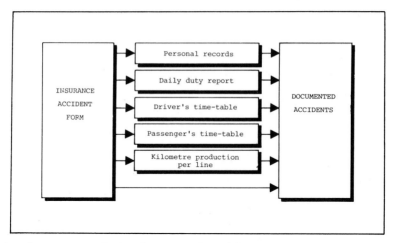

Fig. 1 Sources of the documented accidents.

SOURCES OF INFORMATION

The accident form

This form, the basis of the data used in this study contained the following information:
- Name of the busdriver;
- Date, hour and place of the accident;
- A number of other relevant data as type of bus, weather, road condition, collision-object, extent of damage or injury, etc.

The name of the driver, date, time and place of the accident gave the necessary connection to context information from other sources.

Not every form was completed correctly, resulting in missing data, but also in the impossibility to identify the exact shift the driver was working on the day of the accident.

Personel records

Information about the bus drivers was available from the personel department of the bus company. This included date of birth, date of entering into employment, if applicable termination of employment not only of bus drivers involved in one or more accidents, but of all drivers of the company. These data are necessary to describe the population under study with regard to the possibility to incur an accident. In other words: the population at risk. A problem encountered in this respect is, that sometimes it is difficult to establish the date of a person entering into employment as a bus driver - not as an employee in other functions.

Another source of data for possibly important individual factors influencing accident risk could be the results of entry-examination, periodic health examinations etc. These data are however collected in a nonstandardized way and were therefore left out of consideration.

Daily duty reports

Daily duty reports are forms containing day to day information about the bus drivers on duty, their names, the number of the shift they were working, the number of the bus they were taking from the garage etc. These reports which are kept in the archives of this particular bus company for at least five years proved to be almost essential to connect a certain accident to the necessary context information. Using the name of the driver, date, time and place of the accident from the accident form it was possible to reconstruct exact information about the actual shift the driver was working on the day of the accident.

Problems: Besides missing reports, sometimes the information from the accident form differed from the data on the daily duty report, due to irregularities in the execution of the services (illness of driver on duty, etc.).

Driver's Time-tables

Every time a new passengers' time-table of bus-line operations is issued and sometimes in between, due to changes in the drivers' schedule-organization, new drivers' time-tables are made with more or less changes, or completely renewed. Such a time-table is valid for some months up to more than one year. They contain information about all line-runs to be made by all drivers on the different buslines on every day of the week during the period of validity. This includes time of departure and arrival of every run, and, because a certain number of runs is combined into a shift, the time of commencement and ending of each (numbered) shift. Multiplied by the number of weeks of validity, these tables contain the basic information about the *exposure* of all drivers of that company to their task, viz. when and how long each of them was on duty, and therefore also their exposure to the risk of incurring a bus accident.

Complementary to this information the time-tables also contain the shift schedule system for the current period of validity, i.e. the order in which (groups of) drivers were supposed to be on duty in each of the numbered shifts described, during the successive days, weeks, months etc. of that period of validity. This shift schedule forms the basis of the daily duty report already mentioned.

Some of the problems in this part of the data collection can be described as follows:

1. All information in the time-tables is related to the *scheduled* line-runs and shifts, not the *actual* time or distance covered by each driver;
2. One has to decide which unit of exposure should be used (see below);
3. As routes, frequencies of runs and therefore schedules change from time to time it is necessary to assess this exposure apart for each period of validity of a schedule.

In figure 2 some of the changes from period to period are shown.

	PERIODS OF VALIDITY								
	1	2	3	4	5	6	7	8	9
number of shifts per week	415	486	528	578	578	549	554	537	525
number of weeks per period	22	31	20	49	28	28	11	40	32
year of registration	1973			1974		1975		1976	1977

Fig. 2 The various periods a time-table was valid, length of the period of validity and number of shifts per week in each period (first study).

Not only the length of the period and the number of shifts per week as shown in figure 2 are varying but also the relative frequency of the different types of shift (figure 3). In the company at issue three types of shift can be discerned:

- Early shift: shifts with a mean duration of 8 hours, starting between about 5.30 hours and 10.00 hours (before noon);
- Late shift: shifts with a mean duration of 8 hours, starting between about 13.00 hours and 17.00 hours (after noon);
- Split shift: compound shifts, consisting of two parts. The first part starting between 6.00 to 8.00 hours and lasting for about 3 hours; the second part starting between 13.00 to 15.00 hours and lasting for about 5 hours. This means that a person on a split shift starts working early in the morning for a shorter part of his shift, is off duty for 3 to 4

hours and then starts working again in the afternoon for the second (longer) part of his shift.

The relative frequency of these shifts, in the successive periods of validity of a time-table are shown in figure 3.

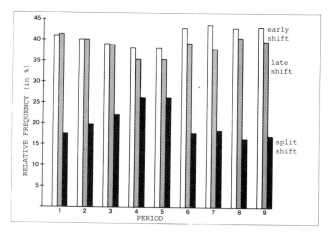

Fig. 3 Relative frequency of early, late and split shifts in the successive periods of validity (first study, 1973 - 1977).

Passenger's time-tables

These time-tables contain basically the same information on all line-runs as the driver's time-tables, only arranged in a different way, viz. per line (route) number and not per shift. They are used for reference and clarification only.

Kilometre production per line

From time to time the accounting office of the bus company makes an assessment of the number of passengers-kilometres (as unit of production), based on the known length and frequency of each line-run, during a certain period of validity of a schedule.
A problem is that this assessment is not made for every period.
For the missing periods one has to assess the kilometre-production by extrapolation from the known periods using the information on changes of line-runs (frequency, length) from the time-tables. Alternatively this assessment can be made by computing from the time-tables after addition of the number of kilometres to each line-run.

EXPOSURE

An important feature of the shift-organization of this particular bus company is that all drivers perform all shifts on rotating schedules, therefore each driver has a comparable exposure regarding shifts, bus line, kilometres driven, hours of service etc. This can be described as a 'found experiment' (Boyle, 1980) because in this population at risk a kind of randomization has taken place with regard to the mentioned variables of interest. The comparability of the exposure of all members of the population is a 'conditio sine qua non' for a valid analysis.

As to the *unit of exposure*, the use of the number of days (or weeks etc.) worked in various shifts seems not to be justified due to large differences between the contents and length of each shift and therefore its exposure to the risk of the driver to be involved in an accident. The same applies for line-runs.

This is illustrated in figure 4 in which a scheme is drawn of a part of the shift-schedule with three shifts (No. S_1, S_2 and S_3) and in each shift a varying number of line-runs of various length (above the line). The first run covers a distance of x_1 km and has a duration of y_1 hours etc. Below the line are shown the rest periods between the line-runs, also of variable duration (sometimes non-existent) the first r_1 min. etc.

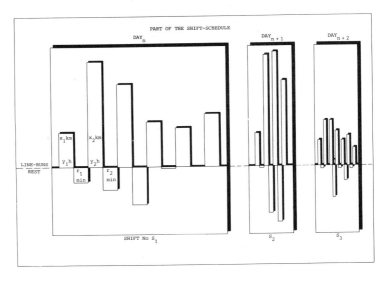

Fig. 4   Schematic representation of a part of the shift-schedules*

---

*   The diagram of shift No. S_1 has been magnified to show the contents more clearly.

The best, well-defined, estimator of the actual exposure to the task is, in our opinion, the driving-distance and/or driving time scheduled in the various shifts on each bus line. The use of this unit of exposure gives the possibility to assess risk in terms of the number of accidents per 100.000 km or per 1000 hours driven, the most commonly used accident rate in this study. Where applicable also the number of accidents per man-year or comparable rates are used. The choice of the accident rate is an important one, a different rate can give a different picture because of a different assessment of the exposure to risk (Bygren, 1974; Borkenstein, 1977). Thus accident rates can be computed per bus line, type of shift, year, month, day and hour etc., and the statistical testing of differences between various rates can be based on expected numbers of accidents, derived from the number of kilometres driven under each condition (see Appendix).

THE TWO PARTS OF THE STUDY

Because of the nature of the available data, knowledge about the population at risk and an estimation of their exposure to the risk of incurring an accident, the possible influence of various factors of the man-environment interaction as mentioned in the previous paper could be assessed and in the explorative part of the study hypotheses could be derived about the strength of their association with the occurrence of accidents and with hazardous situations. In the first part only accident rates based on the number of kilometres driven were used. These numbers were estimated from the kilometre-production data of the company for some periods and our extrapolation for the other periods.
Data were available of 944 accidents in 5 years (1973 - 1977) of 197 bus drivers from one establishment of the bus company, which covered a total of more than 27 million kilometres. A complete report on the results is available (Pokorny et al., 1983) and a series of papers is in preparation. In the second part of the study on accidents of busdrivers of another establishment of the same company, all data, not only from the accident forms and personel records, but also from all time-tables with addition of the number of kilometres per line-run, were put into computer-files. Accident rates will be computed based on both the number of kilometres and the number of hours driven. The procedure is illustrated in figure 5.

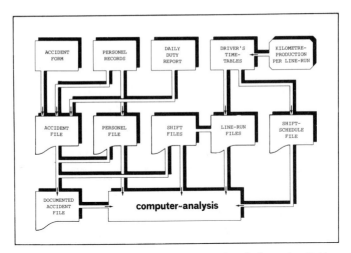

Fig. 5 Scheme of the data-handling in the 2nd part of the study.

To test the hypotheses derived from the first part of the study, data are available of 2130 accidents of 427 drivers in a period of 8 years (1973 - 1980). During this period they covered more than 42 million kilometres in 1,762,000 hours.

APPENDIX

Format of statistical testing

Suppose for a certain variable A the material is distributed in i categories $A_1, A_2, \ldots A_i$, and let $K_1, K_2, \ldots K_i$ be the corresponding numbers of kilometres driven, $O_1, O_2, \ldots O_i$ the observed numbers of accidents.

Given the null hypothesis that the probability of an accident is only depending on the number of kilometres driven, the expected number of accidents can be calculated using the following expression:

$$E(O_{A_i}) = \frac{K_i}{\sum_{i=1}^{i} K_i} \sum_{i=1}^{i} O_i \qquad (1)$$

The value of

$$X^2 = \sum_{i=1}^{i} \frac{\left[O_{A_i} - E(O_{A_i})\right]^2}{E(O_{A_i})} \qquad (2)$$

approximates a χ^2-distribution with i-1 degrees of freedom.

REFERENCES

Borkenstein, R.F. 1977. An overview of the problem of alcohol, drugs and traffic safety. In "Proceedings of the Seventh International Conference on Alcohol, Drugs and Traffic Safety. Melbourne, Victoria, 23-28 January 1977". (Austr. Govt. Publ. Service, Canberra).

Boyle, A.J. 1980. 'Found experiments' in accident research; report of a study of accident rates and implications for future research. J. Occup. Psychol., 53, 53-64.

Bygren, L.O. 1974. The driver's exposure to risk of accident. Scand. J. soc. Med., 2, 49-65.

Pokorny, M.L.I., Blom, D.H.J., van Leeuwen, P. and Opmeer, C.H.J.M. 1983. Ongevallen bij buschauffeurs - Een epidemiologische studie. Deel 1. (Accidents of Busdrivers - An epidemiological Study. Volume 1. With Summary in English). (NIPG/TNO, Leiden).

Pokorny, M.L.I. and Blom, D.H.J. 1983. Some theoretical considerations on accident research. In "Human Performance in Transport Operations; Proceedings of a CEC-Workshop" (Ed. H.M. Wegmann). (Commission of the European Communities, Luxembourg).

EFFECTS OF ALCOHOL ON DRIVING PERFORMANCE
A CRITICAL LOOK ON THE EPIDEMIOLOGICAL, EXPERIMENTAL AND PSYCHOSOCIAL APPROACHES

E.A. Sand

Free University Brussels
Université Libre de Bruxelles
Laboratoire d'Epidémiologie et de Medecine Sociale
Campus Erasme CP 590
808, route de Lennik
1070 - Bruxelles, Belgium

ABSTRACT

While some effects of alcohol and drugs on driving performances have been widely investigated, a more comprehensive understanding of these effects has not yet been achieved. Little is known mainly in two important fields of research :

- Due to methodological constraints, studies on <u>real-life situations</u> including intoxicated drivers and control groups <u>are yet insufficiently</u> conclusive.
- The increased risks of being involved in an accident does not depend only on the actual level of alcohol intake but as well on various <u>personality factors</u> which may deeply modify the behaviour of the driver, <u>be it in a negative</u> way (attenuated risk-perceptions etc), be it in a positive way (e.g. the intoxicated driver may refrain from driving).

More should be known about these psychosocial factors.
In this paper some recent experimental studies will be examined as well as the endeavours to study real-life situations and the influences of some personality-components.

INTRODUCTION.

The objective of this paper is to place recent informations on the effects of alcohol on driving performance in a broader context, a "system", including :

- the vehicle in the real-life situation on the road.
- the driver and some important components of his, her, personality.

Indeed it is not possible to reach a comprehensive perception of the effects of alcohol on reaction-time etc.

1. SOME PRELIMINARY REMARKS ABOUT THE SCOPE AND ORIENTATION OF THE PAPER

1.1. Only topics relevant to epidemiological and experimental research carried out about alcohol and car-driving performance shall be included. The scientific literature concerning the epidemiology of alcohol use by pilots and crews of commercial aircraft appears as rather limited.

We could not spot reliable data on the prevalence of alcohol use (e.g. proportions of members of the personnel, levels of qualification, quantities of alcohol used etc.)

The same is true, but to a lesser extent, in the field of commercial navigation. Recently in a Scientific meeting this subject was considered (see : Verkehrs-medizinische Probleme der See-und-Binnen-Schiffahrt- 1975).
At this meeting, E. Zorn, presented an interesting view of clinical observations ("aus der Praxis") about the use of alcohol on board of ships.
He mentioned 27 references (1961-1974) and, while presenting some statistics, finally, stated that it is difficult to situate the use of alcohol by seamen in comparison with the use by a comparable group of population, and that "it is impossible, to answer this question in a general manner for the seamen".
A. Backhaus completed E. Zorn's contributions by a view on psychological and social aspects of alcohol use on board (no references), whilst W. Naeve analyzed the possible influence of alcohol on the occurence of accidents on board and in the harbour (cfr. also A. Low).
Clinical and anecdotical informations are apparently available in all these fields, but they contribute little to a broader understanding of the problem.

    1.2. In the field of road traffic and of the effects of alcohol on driving performance, research has been active and a great variety of dates are available.
However, even in this field, little is known about the prevalence of alcohol use by bus or coach-drivers etc.
This paper concerns thus informations stemming from research about alcohol and driving-performance.

    1.3. As a third preliminary statement one must stress the fact that there are apparently no acceptable justifications for the use of alcohol by drivers, whilst one should consider that psychotropic drugs may be used for medical or psychiatric reasons.

    1.4. Drug use will not be considered in this paper, since this problem is being reviewed by A.N. Nicholson.
Moreover, most of the research concerning the effects of drug use on driving performance show that these effects vary considerably from one drug to another. Even for some specific drugs such as marijuana, the effects are not clearly determined yet (Moskowitz, 1976, 1 and 2).

1.5. Four aspects of the problem will be looked at :

- Epidemiological research
- Experimental research
- Comprehensive approaches, combining these two orientations : "bridging the gap"
- Psychosocial factors, the personality of the driver.

2. EPIDEMIOLOGICAL EVIDENCES

Numerous projects have been carried out in this field.
As soon as 1962, Norman L.G. collects data indicating, for example, that above a certain level of blood alcohol concentration ("B.A.C.") the risk of being involved in a serious or fatal accident increases considerably. Norman mentions a study carried out in Bratislava (C.S.S.R.) in 1969 by Vamosi.

TABLE 1 Accident risk by various BAC's (Vamosi, cit Elbel H., 1960) (Bratislava, C.S.S.R.)

(a) BAC (mgr/100 ml)	(b) Number of persons with accidents	(c) Number of persons examined not implicated in accidents	(d) $\frac{b}{c}$	(e) Increase of risk
≤ 30	123	370	0.33	1
100	89	37	2.4	7
150	82	8	10.0	30
>150	124	3	41.3	124
Total	418	418		

Most of the earlier studies were based on the retrospective analyses of drivers involved in accidents. Hence a systematic comparison with control-groups was rarely, if ever, at that period, achieved.

More recently research yielded more complete informations, namely through studies comparing intoxicated drivers (DWI = driving while intoxicated) with others who were not intoxicated (DWNI).

In this respect, various results could be presented a few years ago in a comprehensive document of U.S. Department for Health, Education and

Welfare (1974).

Chapter IV presents data on alcohol and highway safety. As a synthesis of the research mentioned in that chapter, the authors give the following statement :

> "The risk of having a traffic accidents rises steeply as a person's blood alcohol concentration rises. Arrests for driving under the influence of alcohol, as well as alcohol-related deaths on the highway, are connected with very heavy drinking, far above the limits of legal impairment. The indications are strong that these alcohol-traffic-trouble people are problem drinkers."

Many epidemiologic analyses confirm the relation between B.A.C. and the probability of occurrence of various kinds of accidents.

P.M. Hurst (cit : Perrine M.W. 1974) pooled sets of data stemming from five U.S. studies. He calculated (Fig 1) the relative probability of involvement in crashing. In this study the level "1" is applicated to drivers who are not intoxicated by alcohol. Probability of involvement increases slowly from the level of .05 to 06 mgr %; there is a sharp increase from .08 to 09. mgr % and more.

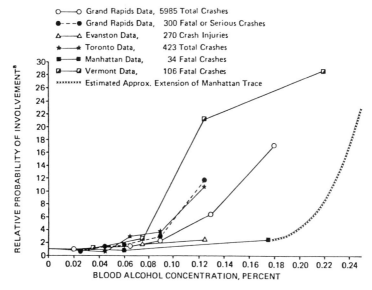

Fig. 1

Perrine M.W. has shown that the relative risk of being responsible for a fatal crash is increasing exponentially. The curve presented by this author indicates that from 0.08-0.09 mgr % the risk increases rapidly.

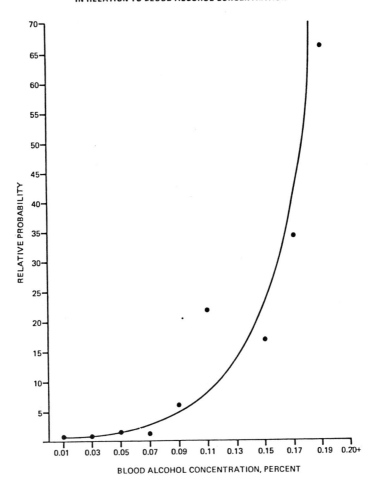

Fig. 2

A comprehensive study was published in 1970, by Laessig R.H. on the involvement of alcohol in fatalities of Wisconsin drivers.
The author mentioned four levels of BAC's :

TABLE 2   Blood alcohol concentrations (after Laessig L.H. and K.J. Waterworth)

A. ———	000	010 mgr %
B. "drinks"	010	079 mgr %
C. "under influence"	080	149 mgr %
D. "intoxicated"	150 +	

The study concerned 507 drivers who were killed in road accidents between the 1 February 1968 and the 30 May 1969, mainly in non-urban areas. Informations have been collected on 17 variables, (Ex : Age, Sex, BAC, hours of the day, type of vehicle, meteorological conditions etc.).

In his conclusions Laessig R.H. pointed out that approximately 50 % of the fatalities were DWI at level C or D, two thirds of the drivers were at level B, C or D.

Moreover young drivers were particularly prone to be involved in fatal accidents : "Young people, 16 to 20 years old, had a fatality rate of 0.50 per 1000 population compared with 0.22 for the entire United States". Laessig R.H. indicated as well that "the drinking driver usually was killed in a single-car accident".

The originality of Laessig's work was that he completed epidemiological data with biological (age) and sociological characteristics, which, taken jointly, offer better ways of understanding the "drinking driver's" phenomenon.

However the majority of actual research projects remain confined to some limited fields of observation. This applies to most of the above mentioned studies :

" However no controlled study has yet been conducted to obtain systematic data on the actual influence of alcohol upon real-world driving behaviour in its natural environment " (U.S. Dept. H.E.W. 1974 p.97).

3. EXPERIMENTAL RESEARCH

In this field numerous studies have been carried out, e.g., on the influence of alcohol (and drugs) on perceptions (visual etc.), on reaction-time etc. (e.g. Moskowitz, 1976,1, 1976,3).

Most of these have been achieved in laboratory-settings. Whilst they add undoubtedly interesting elements to the understanding of the DWI problems, their relevance to the "real-world driving behaviour" is not clear.

Recently, Perrine M.W. (1976) repeated studies of that type, as the first step in an attempt to go beyond the experimental approach, to "bridge the gap".

Instead of measuring only fractions or elements of performance, Perrine selected variables which have direct relevance to the real-life driving situation, i.e. braking, stopping the car "in time".
This performance was first submitted to evaluation in an experimental environment.

The task assigned to two male volonteers in an experiment built on a 2X2X2 factorial design was to :
- keep car at a designated speed
- respond to (dim) light signals (foveal and extra-foveal) as fast as possible by braking.

The observations were achieved in driving and parked conditions. The results are presented in table 3.

TABLE 3   Reaction-time (braking) in m.sec under influence of alcohol and placebo. (Perrine M.W., 1976)

Condition		Light : Foveal	Extrafoveal
Driving	Alcohol	909	1152
	Placebo	744	868
Parked	Alcohol	678	791
	Placebo	572	660

(In m.sec = means of simple reaction-time)

As Perrine M.W. stated, "in response to an extrafoveal signal (e.g. a child or a car coming out from the side of the road), it would take (the driver) at least 2 sec. longer to stop than if he were sober. During this time, his car would have travelled (...) 28 m (...) further than if he had been sober."

Perrine M.W. (1976) made observations on brake use carried out on instrumented cars. These also are close to the real-life situation. However some important features do limit the possibility of applying the conclusions to real-life. Indeed the experimental drivers (volunteers under observation) cannot be considered as similar neither in their driving situation, nor in their personalities to a group or a sample of real-life drivers.

In quite a different orientation of research one shall mention studies on some important physiological effects of the use of alcohol (or use of drugs, or tobacco) on various hormonal systems. In this field, curiously, a great number of data are indeed available. However relatively little is known about the changes induced by alcohol use on such fundamental phenomena as levels of catecholamines, corticoids etc, as well as on the duration of these modification and their influence on, say, vigilance, blood pressure, psychological reactions (anxiety, fatigue) etc (Levi, 1972, p.148). The same applies to the interaction of these effects with psycho-social stimuli (ibid).

4. A WAY TO COMPREHENSIVE APPROACHES : "BRIDGING THE GAP"

Perrine M.W. reports the results yielded by a previous research carried out on the road. The design of this research allowed the observation of drivers (control-group). The objective of the study, was to compare the driving behaviour of motorists with high BAC's and those of low or zero BAC's. (1976)

The author presents the "smoothed plots of speed" observed on drivers with various BAC's. The braking patterns of drivers with the highest BAC's are represented in fig. 3C and 3D. The risks of losing the control of the car during the deceleration-phase are clearly increased in these cases (risk of skidding etc.) : "Actual braking responses made under high BAC's are conducive to losing control of the car".
Perrine M.W. points out that alcohol impairs real-life driving abilities more seriously than in experimental situations.

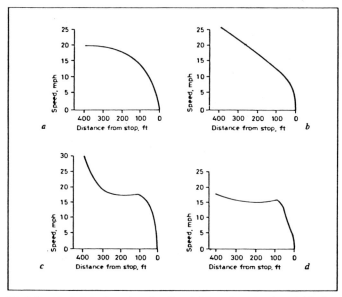

Fig. 3. Smoothed plots of speed as a function of distance from stopping point for four motorists at two roadside surveys. *a* BAC, 11 mg%; roadside survey number (RB), 32; car number (C), 93. *b* BAC, 30 mg%; RB, 16; C, 10. *c* BAC, 130 mg%; RB, 32; C, 89. *d* BAC, 120 mg%; RB, 16; C, 11.

Another and quite different approach to the on-the-road situation has been worked out, already several decades ago, by Platt F.N. (in Norman L.E. 1962).

This author calculated the probability of taking wrong decisions in driving situations on the road. Such errors may lead, by a stochastic phenomenon, to accidents. Platt F.N. showed that, on average, when driving at a speed of, say 60 km p.h., the driver has to realize 25 observations per minute, and has to take one decision every 8 seconds. It can be estimated that, on average, drivers with a zero BAC level will make one error approximately per forty decisions.

One should consider the significance of these calculations in the light of the earlier mentioned research on drivers with high BAC's. The frequency of observations and the probability of the occurrence of decision-to-be-taken will be similar for these drivers. But at each stage of the observation-integration-decision-making-reaction process the intoxicated driver will be probably slower than the sober driver.

5. PSYCHOSOCIAL FACTORS - THE PERSONALITY OF THE DRIVER

A set of psychological and psychosocial factors or variables have to be considered jointly with the data on age, sex, BAC's etc, taken into account in the above mentioned studies (WHO 1981). This is true clearly, for any driver (or pilot etc.) whether he is using his car for private or professional reasons and probably even more for adolescents and young adult drivers. (Beck K.H., 1981, Rutherford D., 1977).

The following figure summarizes the cycle leading from an actual driving situation (result of a preceding reaction of the driver), through the perception of the current situation of the car and its environment (road, meteorological conditions, traffic, hour of the day), through the integration of these data (i.e. a weighted comprehensive evaluation) to finally, an appropriate decision and the consecutive reaction (braking, accelerating, steering). This action puts the vehicle in a new situation etc.

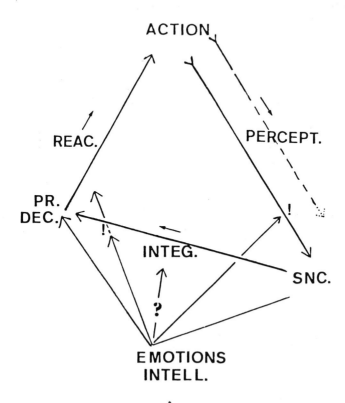

Fig. 4

In most circumstances the cycle includes the interventions of a) sensorial organs and proprioceptive mechanisms (perception), b) of cognitive mechanisms (cerebral cortex) and c) neuro-muscular performances at the level of action.

However, a whole <u>ensemble</u> of other psychological factors may, and often will add their weight to these mechanisms, accelerating them and eventually increasing their efficiency or, on the contrary, acting in a negative way.

Whilst research on chronic alcoholics has shown that there are no specific "alcoholic personalities", some symptoms or tendencies are frequently observed, e.g., high level of anxiety, and at some stages of the disease, troubles in the cognitive field etc. A reduced ability to cope with stress situations may be the main factor leading to alcohol-intake (Jellinek E.M., 1960). On the other hand, in alcoholics as well as in non-alcoholics, personality-characteristics, such as aggressive tendencies, psychopathic traits, "accident proneness", etc, may play an important role in the decision-making process on the road.

These more or less stable psychological characteristics are under the influence of momentary situations which, in turn, may modify the reaction of a given individual driver at a given moment : worries, stress, conflicts, depression....

In this respect Schuman S.H. and Pelz D.C. (1967,1970) have shown the possible effects of conflict situations leading to increased and not expressed aggression, the "blowing-off steam" phenomenon where car-driving may be used to lessen psychological tensions.

These mechanisms should be investigated in a more systematic way, because of the high prevalence of neurotic symptoms (and personalities) in average populations (Srole L., 1962, Tsung-Yi Lin, 1963) and because of the high frequency of occurrence of stressful life situations at least in some social groups (Levi L., 1971).

Adolescents and young adults deserve particular attention in this respect due to their normal tendencies towards risk-taking behaviour (Jeanneret O., Sand E.A., 1983).

Whilst Browning J.J. and Wilde G. could not demonstrate in adults a clear relationship between alcohol-use and risk-perception in simulated and real car-driving situations (1977), the higher tendency towards risk-taking behaviour is apparently a specific characteristic of adolescents and young adults.

The high proportional death rates at age levels between 15 and 25 must be underlined (Levy C., 1980). Suicidal tendencies are not unfrequent and in most European countries the trends are increasing.

Adolescents have to cope not only, if they use alcohol, with its effects on their performances, but with several learning-programs.

This concerns, of course, perception and understanding of the traffic-situation. But simultaneously, adolescents and young adults may still be in a learning situation relative to their own body and their neuro-muscular system. The physical growth may go on for some time during late adolescence and e.g. the mere dimensions of their limbs have to be fully integrated in all the manoeuvres to be undertaken when driving a car (Jeanneret O., Sand E.A., 1983).

One should recall at this point the Manhattan Study mentioned above where the authors indicated, at least in Manhattan, N.Y., the high prevalence of neuroses and neurotic symptomatology. Various other studies confirm the observations of Srole L. and his co-workers, even if the observed prevalence-rate may be somewhat lower (Tsung-Yi Lin, 1963).

The influence of psycho-emotional variables cannot be omitted in any approach of the epidemiology of accidents, neither or rather - a fortiori - in the analysis of the influence of the alcohol on driving.

An extensive analysis of these influences should include distinctions between alcohol-intake by chronic alcoholics and occasional alcohol-use by non-alcoholic drivers. But these distinctions, however important, would lead to longer developments, which are not really relevant to our point, and it has been mentioned that the serious "alcohol-traffic-trouble people are problem drinkers" (U.S. Dept. H.E.W. 1974).

Table 4 indicates the main characteristics of the personality which :
a) may be associated with a tendency to alcohol use or abuse, whether occasional or regular.
b) may be associated with the level of driving performances.

TABLE 4 Personality characteristics and driving performance.

⎧ INTELLIGENCE
⎨   ↕ ↕
⎩ EMOTIONS

A. STABLE CONDITIONS

    LEVEL OF MATURATION ⎧ RISK-TAKING BEHAVIOUR (ADOLESCENTS)
                                   ⎨   ⎧ DRIVING
                                   ⎩ LEARNING SITUATIONS ⎨ ALCOHOL
                                                               ⎩ SOMATIC DEVELOPMENT

    "STRENGTH" OF PERSONALITY (EGO)

    NEUROTIC SYMPTOMS ⟶ NEUROSIS, ETC.

    ⎡ DEPRESSION
    ⎢ ANXIETY
    ⎣ ACCIDENT PRONENESS
    SUICIDAL TENDENCIES
    MASOCHISM ⟵
                    VERSUS
    ⟶ AGGRESSION ⟵
    ⟶ SOCIABILITY   VERSUS   EGOCENTRISM ("PSYCHOPATHY")

B. EPISODIC CONDITIONS

    ILLNESS
    STRESS
    FATIGUE
    "WORRIES"
    "BLOWING-OFF STEAM"
    USE OF PSYCHOTROPIC DRUGS

6. CONCLUSION

Hence, alcohol use, being a multivariate phenomenon, associated to a variety of personality factors, may not be considered as a one-dimensional variable (e.g. B.A.C.) in a study of its effects on driving performance.

Studies like those achieved by Schuman S.H. and Pelz D.C. (1967), by Perrine M.W. and others are oriented in this broader sense.
Scales indicating the relative impact of life-events, may provide another tool to implement such research (Rahe R.H., 1971).
Studies of alcohol-use by professional drivers, pilots etc, should be based on a comprehensive approach, stating for ex., not only "how much" alcohol could have been used, but in what cricumstances, by whom (personality) etc.

REFERENCES

Beck, K.H. 1981. Driving while under the influence of alcohol : relationship to attitudes and beliefs in a college population. Am. J. Drug Abuse 8,3, 377-388

Browning, J.J. and Wilde, G.J.S. déc. 1977. L'effet de l'alcool sur la perception du risque dans des situations simulées et réalités de conduite en automobile. Toxicomanie 10, 235-267

Jeanneret,O. and Sand, E.A. , Manciaux M., Deschamps J.P. 1983. Les adolescents et leur santé.(Flammarion, Paris).

Jellinek, E.M. 1960. The disease concept of alcoholism. College and University Press, Hillhouse Press.

Laessig, R.H. and Waterworth, K.J. June 1970. Involvement of alcohol in fatalities of Wisconsin drivers. Public Health Rep. 85,6,535-549.

Levi, L.(Ed) 1971. Society, Stress and disease. Vol 1. Oxford Medical Publications, The Oxford Univ. Press.

Levi, L. (Ed) 1972. Stress and distress in response to psychosocial stimuli. Acta Medica Scandinavia. Suppl. 528-166.

Levy, C. 1980. La mortalité des enfants et des adolescents dans huit pays développés. Population, 35, 291-319.

Low, A. 1981. Arbeitsstress - Alkohol. Verkehrsmedizinische Tagung, Hamburg 7. - 9.Mai 1981. Hansa Schiffahrt - Schiffbau - Hafen - 118 Jahrgang, nr 17, 1255.

Moskowitz, H. (1) 1976. Drugs and driving : introduction. Accid. Analysis Prevent. 8,1,1-2.

Moskowitz, H. (2) 1976. Marihuana and driving. Accid. Prevent. 8,1, 21-26.

Moskowitz, H. and Ziedman, K. and Sharma, S. (3) 1976. Visual search behaviour while viewing driving scenes under the influence of alcohol and marihuana. Hum. Fact. (Balt) 18,5 417-432.

Norman, L.G. 1962. Les accidents de la route. Cahiers de Santé Publique 12 OMS Genève.

Perrine, M.W. 1974. Alcohol, drugs and driving. Nat. Highway Traffic Safety Admin. Washington D.C. Technic. Dep. DOT HS-801-096, (in U.S. Dept. HEW 1974).

Perrine, M.W. and Waller, J.A. and Harris, L.S. 1971. Alcohol and highway safety : behavioural and medical aspects. Washington DC : Nat. Highway Traffic Safety Admin. Techn. Dep., DOT HS-800-599 (in U.S. Dept HEW 1974).

Perrine, M.W. 1976. Alcohol and highway crashes. Closing the gap between epidemiology and experimentation. Med. Probl. Pharmacopsycho. 11,22-41 Karger, Basel.

Platt, F.N. 1958. Rev. Internat. circ. Séc. rout. 6,2, p8, cit : Norman, LG. 1962.

Rahe, R.H. 1971. Psychotropic drug response. Advances in prediction : in

Levi, L; (Ed), p.39
Schuman, S.H. and Pelz, D.C. and Erlich, N.J. et al. 1967. Young male drivers : Impulse expression, accidents and violations J.A.M.A. 200: 1026 - 1030.
Schuman, S.H. and Pelz, D.C. 1970. A field trial of young drivers. Arch. Environm. Health 21, 462-467. Sept 1970
Srole, L, and Langner, T.S. and Michael, S.T. et al. 1962. Mental health in metropolis : The Midtown Manhattan study.(New York, Mc-Graw-Hill).
Tsung-Yi Lin and Standley, C.C. 1963. La place de l'épidémiologie en psychiatrie. Cahier de Santé Publique 16. O.M.S., Genève.
U.S. Dept. of Health Educ. and Welfare, Second special report to the U.S. Congress on Alcohol and Health June 1974. Chap. VI : Alcohol and Highway Safety-97-110.
Vamosi, (cit Elbel, H.) 1969. Ciba Symposium 7, 242.
Verkehrsmedizinische Probleme der See-und Binnenschiffahrt. Jahrestagung der Deutschen Gesellschaft für Verkehrsmedizin. Hamburg 11.-13. April 1975.
Backhaus, A. Alkoholkonsum an Bord. Psychologische und soziale Aspekte p. 17-21.
Naeve, W. Bedeutung des Alkohols bei Unfällen an Bord und im Hafen p.22-33.
Zorn, E. Alkoholkonsum an Bord. Beobachtungen aus der Praxis p.7-16.
W.H.O Euro. The influence of alcohol and drugs on driving. Rep. on a WHO ad hoc Technical Group. Monaco 30 Oct. - 2 Nov. 1978. Euro Rep. and Studies 38. WHO Euro Copenhagen 1981.

INVESTIGATIONS ON THE INFLUENCE OF CONTINUOUS DRIVING
ON THE MOTION ACTIVITY OF VEHICLE DRIVERS [+]

M. Lemke

Volkswagenwerk AG, Nutzfahrzeug-Entwicklung
D3180 Wolfsburg 1, FRG

ABSTRACT

Changes of the motion activity of car drivers as a result of monotonous driving tasks should be investigated under the aspect of decreasing vigilance. The electro-myogram was supposed to be the indicator for decreasing vigilance of the driver. This could not be confirmed under the actual test situation with motorically active subjects. The myo-integral was rather found to be an indicator of instantaneous changes in the driver-vehicle-environment control-circuit caused by changes of the traffic situation or the road quality.

The motion activity was measured at the back of the driver seat. It was not only transformed into time segmented RMS-values but also counted as frequencies of pre-adjusted level crossings. A summary comparison of both 3 hour test sections of continuous highway driving showed a transfer of the total motion activity and the motion frequency from the thorax to the pelvis area. In the second test section an increased motion frequency was noticed and motions with high amplitude were transferred from the thorax to the pelvis area.

INTRODUCTION

The area of "Active Safety" in road traffic has gained in importance through the last years. While before this time efforts mainly had been made towards a reduction of accident consequences ("Passive Safety"), now intensified precautions were taken to avoid accident or accident relevant situations from the beginning. To achieve this in the driver-vehicle-environment control-circuit, special care had to be taken of the human element, which, of course cannot be optimized (Helander 1974 and Küting 1976).

On the one side the human controller is remarkably adaptable, on the other side his control quality can be changed by numerous extraneous disturbances (like emotions, fitness, fatigue) in a hardly predictable way. Especially his fatigue or vigilance level is changed by the mere performing of the control task - if performed over sufficient time. At the same time an alteration of his control activity can be observed in the sense

---

[+] Investigations were carried out at the Technical University of Berlin (Institut für Fahrzeugtechnik, Prof. H.-P. Willumeit) and were sponsored by Deutsche Forschungsgemeinschaft (Le 418/1).

of a reduction of control quality and therefore an increase of accident
hazard (Lemke 1980 and 1982, Walz 1976). This has a special impact for
commercial vehicles in the long distance road haulage and for passenger
cars in the holiday traffic: In both cases drivers often perform long distance travels with high monotonous stress. A decreasing vigilance level can
generally be determined by means of physiological variables. For the practical application of a warning device, however, indicating variables have
to be restricted to those which are gained directly from the vehicle. This
means that the driver must not be irritated by electrodes. In former experiments in real vehicles and in a driving simulator at the University of
Berlin correlations between physiological signals of subjects and their
actions in the driver-vehicle control-circuit had been investigated (Lemke
1980 and 1982). Besides the intentional control actions of the driver in
the control-circuit his motion activity in the seat could give information
about his actual control capacity (or state of "fitness" or "fatigue").
Therefore vehicle experiments were carried out to investigate alterations
of the motion activity of drivers in their seats as a result of continuous
driving.

PRELIMINARY EXPERIMENTS

In two preliminary experiments the applicability of the electro-myogram
(EMG) of the right trapezius had been investigated to indicate muscular
tensions as a result of monotonous actions: The EMG-signals of the right
trapezius of two subjects were recorded as a myo-integral (EMI). The first
subject, seated in a chair, performed technical drawings over a time of
100 minutes (fig. 1).

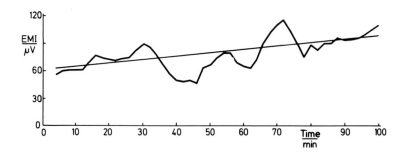

Fig. 1   Myo-integral of a sitting subject; activity: drawing.
EMG of the right trapezius, integration time 2 minutes, moving
average of n=5 values.

The second subject drove a passenger car on the highway Berlin-Nürnberg, a four track highway with stringent speed limit of 100 km/h and high monotonous strain on the driver, followed by two sections of a country road (total duration 360 minutes, fig. 2).

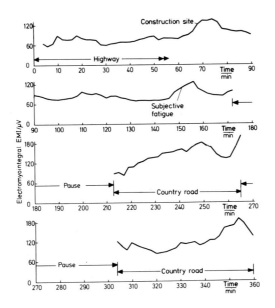

Fig. 2 Myo-integral of a sitting subject; activity: driving. EMG of the right trapezius, integration time 2 minutes, moving average of n=5 values.

The myo-integral of the drawing subject had a monotonous increasing tendency superimposed with periodical variations which had a period of approx. 20 to 30 minutes. Compared with the driving task, the drawing task had only little external stimuli. The myo-integral of the driving subject had a generally increasing tendency over test duration but had additional situation-dependent overshoots caused by construction sites or convoy traffic. At a certain moment (after 145 minutes of driving time) the subject noticed a feeling of high drowsiness: At that time the myo-integral showed a high increase. In this test no superimposed periodical variations like in the drawing experiment were recorded. After the transition from the highway section to the country road the myo-integral increased with a larger slope and after a decrease in a pause again increased with a large slope over driving time. It was concluded, that for the driving task the absolute value of the myo-integral and its gradient over time depended on the difficulty of the actual control task. The myo-integral obviously did not allow to differentiate clearly between situation dependent variations of the control task (e.g. by changing traffic situations) and spontaneous vigilance decrease (e.g. sudden feelings of drowsiness). The preliminary experiments showed that the EMG-signal had limited applicability as a physiological control variable and that care had to be taken that the test conditions were largely constant. Therefore the test track was limited to the monotonous highway section and the tests were performed at nearly the same daytime.

EXPERIMENTAL SET-UP

The test vehicle was a Volkswagen Passat (Dasher). The driver seat had an additional seat-back with inserted spring steel sheets equipped with strain gauges. Two signals analogous to the driver's motions against the seat-back could be recorded and evaluated, one from the upper and one from the lower section of the seat-back (fig. 3). There were two additional

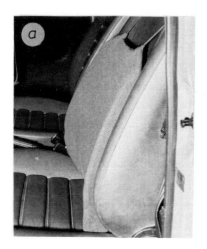
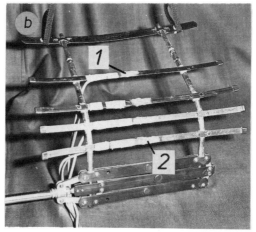

Fig. 3 a, b   Additional seat-back for the registration of the motion activity of the driver (a). Active strain gauges on steel springs (b). Top strain gauge: 1, bottom strain gauge: 2).

threshold indicators and counters in the signal paths which recorded the frequency of pre-adjusted level crossings of the motion activity.

The electro-myogram of the right trapezius was taken as a physiological control variable. The signal was reduced in the vehicle to a myo-integral with 5 minutes integration time and recorded for the whole test duration as numerical values. Additionally the original EMG-signal and the motion activity signal from the strain gauges were recorded on magnetic tape every 30 minutes for 5 minutes.

The tests were carried out with 11 subjects (aged 22 to 32); the test track was the Autobahn Berlin-Hof and back, which is a four track highway with a speed-limit of 100 km/h. The subjects had to drive the whole distance of 600 km without any interruption. An overview of the whole test program and a recording schedule is shown in fig. 4.

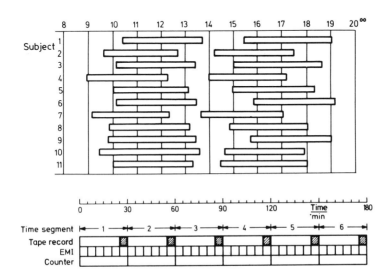

Fig. 4   Test program (top) and recording schedule (bottom) for one section of each test.

RESULTS

Electro-myogram

During the test every 30 minutes the electro-myogram was recorded on magnetic tape for 5 minutes. These records were reduced after the test to their respective RMS-values. A representative diagram of one subject's RMS-values is shown in fig. 5: In spite of the 5-minute integration time there were large variations from record to record which means a high portion of intraindividual dispersion. As additional interindividual dispersion was found, the RMS-values were normalized with each subject's average, thus achieving a certain level adjustment and the possibility of proper averaging between subjects. A time history of the average RMS-values of all subjects is shown in fig. 6; the EMG is strongly influenced by arte-

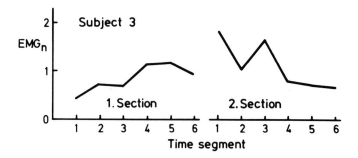

Fig. 5   EMG-RMS-values of subject 3 during the test, normalized with his average EMG-value.

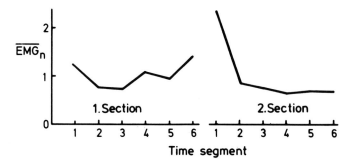

Fig. 6   Average of the normalized EMG-RMS-values of all subjects.

facts: The first time segment of the return trip has an exceedingly high RMS-value which coincides with a bad road condition in that segment.

Besides the interindividual differences in the EMG-amplitude there were also found interindividual differences in the trend. Therefore the concordance of all values was checked: The result was non-significance for the first test section but high concordance for the return trip ($W = 0,403$, $\alpha = 0,01$).

Myo-integral

The electro-myo-integral was continuously calculated and registered during the tests:

$$EMI = k \int_{t}^{t+\Delta t} abs(U) \, dt$$

The integration time was $t = 5$ minutes. The representative time histories of the myo-integrals of 3 subjects are shown in fig. 7. They clearly show

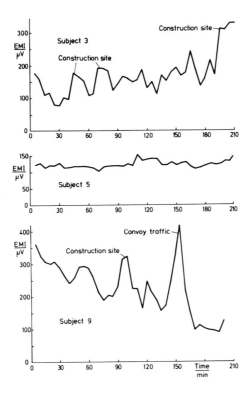

Fig. 7   Time histories of the myo-integral EMI of 3 subjects, showing three totally different tendencies. Integration time 5 minutes.

opposite trends over time and they additionally show the more situation-dependent behaviour of the EMI. For level adjustment the numerical values were normalized with the individual averages and afterwards the total averages of all subjects for each time segment were calculated (fig. 8). Their time histories show slightly decreasing tendency for both track sections. The tendency is not very pronounced, which is a result of the above mentioned interindividual trend dispersion and they have also an overlay of variations caused by situation-dependent influences.

Motion activity (RMS)

The strain gauges of the spring steel insertion of the seat-back gave, after amplification, a signal analogous to their deflection and therefore analogous to the motion of the driver against the seat-back. The recorded signals were reduced to their RMS-values, normalized with the individual averages (as was done above) and then averaged for all subjects. Fig. 9

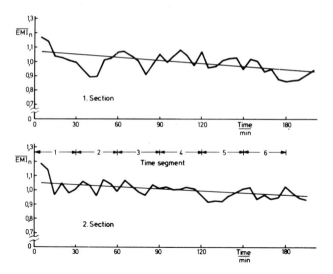

Fig. 8 Total average of the normalized myo-integral EMI of all subjects; normalization with the individual average of each subject. Integration time 5 minutes.

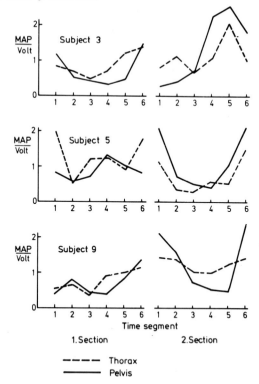

Fig. 9 RMS-values of the motion activity of 3 representative subjects.

shows the representative time histories of 3 subjects, fig. 10 shows the total average of all subjects separately for the upper and the lower portion of the seat-back (corresponding with the thorax and the pelvis). The individual time histories of the motion activity have interindividual differences in amplitude and trend. The total average values show a higher motion activity for the return trip (second section) than for the first section. The separate graphs of the signals from the upper and the lower strain gauges (fig. 10) show no significant variation of the thoracical motion activity in the first section; in the second section it has a high value at the beginning, then decreases, and again increases towards the end of the test. The initial high value of the second section could have been influenced by a bad road surface (as was presumed with the EMG-signals).

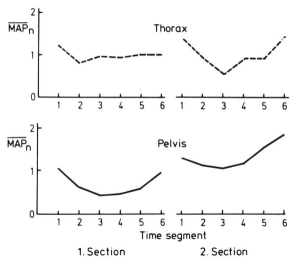

Fig. 10   Total average of the normalized RMS-values of the motion activity MAP of all subjects; normalization with the individual average of each subject.

The signals from the lower portion of the seat-back (pelvis area) show a 'U'-shape for the motion activity for both test sections but an increase in the average amplitude from the first to the second section. These 'U'-shaped time histories were also often found in the individual records of the subjects. This could be explained by an initializing phase in each test section in which the subjects tried to find their comforftable seating position therefore having a high motion activity. The motion activity then decreases in the first 3 or 4 time segments of the respective test section and turns into a positive slope with increasing test duration which could be caused by fatigue effects. Towards the end of the return trip a so-called 'Coming Home'-effect was noticed which has been found in other experiments: The knowledge about the approaching end of the test increases the activation level and the motion activity of the subjects.

The more pronounced tendency of the motion activity of the pelvis against the thorax could be explained by less influence of artefacts from vertical vehicle vibrations and from the operation of the vehicle on the pelvis area; signals from the lower strain gauges rather had to be the result of intentional variations of the driver's seating position. The summary comparison of both test sections shows a decrease for the motion activity of the thorax and an increase for the pelvis area (fig. 12, next paragraph).

Motion activity (frequency)

In the signal paths of the seat-back sensors additional threshold indicators and counters were provided to achieve a continuous registration of the motion activity over the whole test duration. They recorded the accumulative frequency of the crossing of pre-adjusted levels every 30 minutes. Although the threshold levels were adjusted to each subject's weight and pressure on the seat-back, high interindividual dispersion had to be noticed. Therefore the values were normalized as above. Nevertheless a clear tendency was not found: The total average of all subjects varied only little about the value of 1.0 (fig. 11). Therefore the time histories were not analysed further. A summary comparison of the motion frequency, however, yielded the following results: The motion frequency increased in the pelvis area and pronouncedly decreased in the thoracic area from the first to the second test section (fig. 13). These results corresponded with the RMS-values of the periodical records of the motion activity shown in fig. 12.

DISCUSSION

The results of the EMG-variables were in contradiction to the hypothesis stated in the preliminary experiments; the electro-myogram was not applicable as a physiological control variable for 'fatigue' or a decreasing vigilance level (confer Radl, 1969): Between the subjects large differences in trend were noticed; the EMG-signals increased, decreased, or

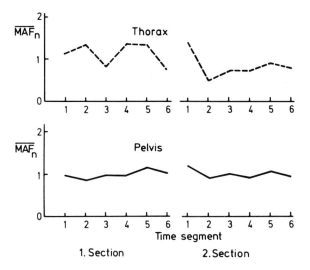

Fig. 11   Total average of the motion activity frequency MAF (level crossing) of all subjects; normalization with the individual average of each subject.

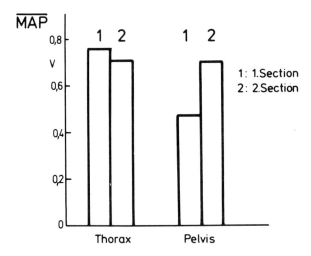

Fig. 12   Comparison of the average RMS-values of the motion activity for both test sections.

remained constant over the test duration, not only the myo-integral being a linear average, but also the RMS-values being quadratic averages. Spontaneous alterations in the driver-vehicle-environment control-circuit by external influences like traffic situations or internal influences like vigilance alterations were, however, clearly indicated.

The registration of the motion activity was also strongly influenced by artefacts as could be seen from the individual time histories. Often high motion activity was noticed at the beginning and at the end of each test section. This was explained by a habituation phase and a final activation increase ('Coming Home'-effect). A comparison between both test

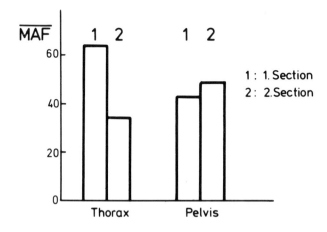

Fig. 13   Comparison of the average motion activity frequency (level crossing) for both test sections.

sections showed a transfer of motion activity from the upper part of the seat-back (thoracical area) to the lower part (pelvis area); this was found not only for the segmented continuous RMS-values (MAP) of the motion activity but also for the frequency of level crossing (MAF). The alteration of the RMS-values and the level crossing frequency between both test sections differed clearly in size; this was explained by a transfer of intentional motions: The frequency of intentionally started motions in the pelvis area with high amplitudes remained nearly constant while their RMS-values increased. This could only be explained by a higher number of motions with smaller amplitude. Conversely the RMS-values in the thoracical region remained nearly constant while motions with high amplitude occured less often. Therefore an increase of the motion frequency with small amplitudes at the expense of the motion activity power was concluded.

Summing up it may be said that a general increase of the motion frequency was found accompanied by a transfer of motions with high amplitude from the upper to the lower region of the seat-back (or from the thorax to the pelvis area) with proceeding test duration.

REFERENCES

Helander, M. 1974. Driver's Physiological reactions and control operations as influenced by traffic events. Z.f.Verkehrssicherheit, 20, 174-187.
Jonsson, S. and Jonsson, B. 1975. Function of the muscles of the upper limb in car driving. Ergonomics, 18, 375-388.
Küting, H.J. 1976. Belastung und Beanspruchung des Kraftfahrers. Technischer Überwachungsverein Hannover. Ed: Bundesanstalt für Straßenwesen.
Lemke, M. 1980. Monotoniebelastung von Kraftfahrern. Automobil-Industrie, 25, 3, 77-85.
Lemke, M. 1982. Correlation between EEG and driver's actions during prolonged driving under monotonous conditions. Accid. Anal. & Prev., 14, 1, 7-17.
Radl, G.W. 1969. Untersuchungen zur Quantifizierung der psychischen Beanspruchungen bei simulierten Fahrzeugführungsaufgaben. Aviation Psychological Research Reports of the 8. Conf.f. Aviation Psychology, Zürich.
Walz, L. 1976. Dynamisches Regelverhalten und hirnelektrische Vigilanzregulierung des Menschen bei der Durchführung von Regelaufgaben. Thesis, Berlin.

SYMBOLS

EMG	Electro-myogram
EMI	Electro-myo-integral
RMS	Root-mean-square
MAP	Motion activity 'power' (total RMS-value)
MAF	Motion activity frequency (level crossing)
U	EMG-voltage
W	Coefficient of concordance
k	Hardware-constant
n	Index for variables normalized with the average of the respective subject
t	Time variable
α	Statistical significance

HUMAN PERFORMANCE IN SEAFARING

A. Low, H. Goethe
Bernhard-Nocht-Institute for
Nautical and Tropical Diseases,
Department of Nautical Medicine,
2000 Hamburg, Federal Republic
of Germany

ABSTRACT

With a few exceptions, ships represent a combination of living quarters and working places. Many psychophysical difficulties and problems arise from this situation. In addition, ship machinery, environmental factors and social problems (separation from family etc.) have an effect on human performance on board. In this paper, these stressors and their consequences for mental and physical performance and efficiency are described in detail and verified with statistics from national and international investigations. Many negative stressors exist even today, e. g. noise and vibrations, whereby their levels are sometimes very much higher than accepted tolerance levels. Nautical medicine and naval architects should cooperate more closely with each other in order to improve the current situation.

INTRODUCTION

In contrast to the two transport systems (land and air transport) discussed at this Workshop, ships, with exception of short ferry-connections, are places of work and places of living at the same time for the crew as well as the officers. Members of the crew live together in small or cramped spaces for days, weeks, months and sometimes even for years, cut off from social connections on land. Short stopovers in ports can hardly be regarded as an interruption of the monotony. A multitude of psychophysical problems and difficulties stem from this particular situation. Many more factors, such as environmental stress (noise, vibrations, macroclimate, ship-movements, etc.), workloads / demands through the machinery on board, as well as human problems and interindividual problems must be included. Long sea-journeys and the resulting separation from family and friends, as well as other more subtle factors should be collected and taken into consideration if human performance on ships is to be correctly assessed, analysed and, where necessary, improved. Nautical Medicine is one of the special disciplines that deals intensively with this complex on a national and international level. In the following chapters, the particular aspects and important sections of human performance in sea transport operations will be dealt with in detail.

1. JOINT PROBLEMS
1.1. High risk load

Dangers for ships can occur through the malfunction of technical/mechanical equipment, through false conduct or behaviour of the crew and through environmental influences ("Flensburger Studie" etc.). These dangers can be averted through special protective technical installations on board, through procedures/activities of ship crews and through installations in the environment. If these should malfunction, then an accident will or can occur. When an accident or disaster occurs, it can be fought with rescue equipment on board the ship, by crew activities and through rescue equipment in the environment.

Environmental danger for ship transport systems can be caused by dense shipping traffic and the ensuing higher collision risk, e. g. in the British Channel, through severe weather conditions, treacherous waters, as for example around Cape Horn and in the North West Passage. The necessity for increased alertness on the part of the crew in these regions causes a significant increase of mental burden. Weather conditions that can lead to an increased risk are among others strong winds, bad vision, icy seas, low air and water temperatures, eventually also darkness, high relative air humidity and extremely low air pressure. The most important of these factors for seamen are strong winds and bad vision.

According to Flaatrud (1981) the seafarer is exposed to three types of risks:

Ship risks, work risks and social and health risks. He states that 29 % of the fatal accidents were caused by ship risks, 24 % by work risks such as falls and poisoning, and 47 % through social risks such as danger in leisure time, disappearing without a trace etc. in the Norwegian Merchant Marine Fleet between 1970 and 1976. This risk pattern was verified by statistics of the British Merchant Marine Fleet.

Risk to life and health on board the ship as a place of work is not primarily a problem of heavy loss of life, but more a problem of injuries from a great range of accidents, associated with the work situation. The seamens' profession is hard and arduous and brings with it many possible dangers in the form of accidents, injuries and diseases. According to Flaatrud, social risks on board can be regarded as a virtually endless list of human tragedies. This type of risk encompasses alcohol abuse, suicide and disappearing without a trace from the ship. In the Norwegian and British Merchant Marine Fleets, social risks are responsible for about 50 % of the fatal accidents. Nowadays this type of risk is receiving increasing attention and measures are being taken to counteract it.

1.2. Mental Load

Mental and physical loads on ships often occur combined and can only be separated from each other with difficulties. Psychophysical stress often differs from ship to ship. Control and monitoring activities, for example on the bridge, in the radio-rooms, in the engine-room and during loading and unloading of cargo, are primarily a mental stress.

Parameters used to register mental load are among others skin moisture, heart-frequency and breathing frequency as well as catecholamines in blood and urine. Mental load is used in connection with saturation, fatigue, monotony, great mental concentration etc. The expression "mental load" is very complex and, among other things, dependent on environmental factors, such as work climate, disturbance by noise, by vibrations, as well as personality structure and therefore motivation.

The "Flensburger Studie" has shown that mental load on modern ships is due to the following factors:
    a) Type of work:
    Monitoring, control, steering, command

b) Environmental stress factors (noise, vibration, micro- and macroclimate, watch-system),

c) social and psychological factors:

Separation, isolation, monotony, boredom etc.

Investigation methods for measuring mental load are quoted in the above mentioned study. These methods are:

EEG-measurements, EOG, heart-frequency-monitoring, skin-resistance measurements, skin-moisture measurements, questionnaires, biochemical tests as well as psychotechnical test procedures. Especially unusual or extraordinary circumstances or occurrences, such as sudden engine trouble or changes in the ship's course due to impending collision danger, i. e. occurrences that suddenly appear in the monotony of general activities on board, mean a very great mental load or stress.

Personnel with control or command functions as well as the rest of the nautical engineers and officers on deck and in the engine-room have to cope with medium mental stress. However, this mental load is felt very differently from person to person. In ports, deck personnel is especially prone to mental stress, whereas engine-room personnel experiences mental stress primarily out at sea. Mental stress becomes greater the more responsibility a person has, for example during control and monitoring functions.

Goethe et al. (1978) found that continuous mental load is low among nautical and technical ship officers. In difficult and dangerous situations, especially in very densely travelled water-ways and during approaches to port, high to extremely high mental stress levels occur for the captain and the engineer in charge. Nautical and technical officers are exposed to medium continuous mental stress, but this can become very high in risk situations etc. Mental stressors for ratings lie at a low level.

1.3. Environmental load

The climatic environment of the seafarer may be approximately subdivided into macro- and microclimate. Macroclimate is identical with the ship's environment, i. e. with the sea region. Here heat, cold, wet cold and wind are the primary factors. Microclimate is equivalent to the climate in the ship's interior. Optimal influence and control of this microclimate can be achieved through installation of air-conditioning systems and ventilating systems. On ships, further environmental stressors must be included, such as noise, vibrations, toxic gases and dusts from the cargo, insufficient noise insulation as well as stressors stemming from other persons on board (no possibility of fleeing for the individual as the ship consists of work areas and living quarters in one).

The climate is often a great stress for the crew, for example during physical work in a hot environment combined with high air humidity. The microclimate is most stressful in the engine-room ($\geq 30^{\circ}$ET). Besides on deck in the tropics, heat radiation is greatest in the engine-room. Even in the ship's galley, a high environmental load, especially radiation heat, can be found.

Ship movements caused by sea conditions lead among other things to considerable additional compensatory muscle activities and therefore to increased stress. The capability of rest and recuperation during the sleep phase is significantly decreased when ship movements (pitch, yaw and roll) increase.

On many ships, vibrations originating in the propulsion system are so strong that work activities for a part of the crew and especially recuperation during leisure time and sleep at night are significantly diminished.

Noise on board is a stress factor at the place of work and in the recreation area. On many ships, the noise level lies above the permissible levels. This leads to psychophysical stress and eventually to noise damage in the ear, especially in the engine-room.

And finally, the "Flensburger Studie" proved that illumination on board the 24 investigated ships was far from optimal, this being so in places of work as well as often in the living quarters.

1.4. Abnormal and irregular work/rest-cycles

As yet, watchkeeping, i. e. shift work cannot be omitted on board ships. Several watch schemes are in existence. Nowadays, the three watch systems, i. e. four hours watch/eight hours leisure time, is used internationally in ocean going shipping. In coastal and local shipping, the two watch system is found, i. e. six hours watch/six hours leisure time. Sometimes the six hour watch/four hours leisure time alternating with the four hours watch/six hours leisure time is also used. The watch times also stay the same for the watch keeper, at least with the nautical officers. On modern ships with reduced crew numbers and automatic engine control and monitoring, watch keeping is usually only performed by the three nautical officers and three sailors. Night watch, especially from twelve midnight till four in the morning, is most difficult for the human being to bear. Night work or night shift must always be regarded as a pathogenic factor as the human body does not adapt to night work. For example, there is no adaptation of physiological functions (= inversion), and therefore also no core temperature adaptation. Watch-keeping represents a great physiological stress, among other things, because the feeling for time remains the same and social connections are shifted.

Performance efficiency, especially mental performance, for example monitoring functions, decreases during the night and the number of mistakes rise. Decline in performance efficiency is not so evident among purely physically working men. On board ship, sleep during the daytime decreases in duration, in depth and in its recuperative value due to environmental acoustic disturbances, vibrations and noise. The stress becomes even greater in combination with time-zone shifts that can occur rapidly.

Rutenfranz also found no physiological adaptation to night work on board. He did, however, in some cases find a great reduction in sleep-duration as well as an increase in reaction times.

Shift-work on board as well as on land presents a psychophysical stress. The four hour watch scheme currently used on board, does not coincide with the normal biorhythm of human beings (Souchon). According to Goethe (1978), watchkeeping represents an essential stress factor which so far has been very difficult to assess. Time shift, noise, vibration and ship movement must also be mentioned in this connection as they have an effect as additional stressors.

Harrington (1978) points to some interesting connections between shift-work and health that are without doubt valid for conditions aboard ships. According to him 20 to 30 % of the working population do not like shift work, probably the majority of people tolerate shift —work and some, approximately 10 %, seem to like shift work. Those who cannot adapt to shift-work are more prone to get nervous gastro-intestinal diseases, these occurring quite frequently in seafaring.

Colquhoun also mentioned the watch system problem on board, especially the shift in the circadian rhythm in rapid transmeridional voyages.

However, on a whole, the question of watch-systems on board ships has so far been given little attention, completely in contrast to the problem of shift-work on land which has constantly and intensively been investigated by occupational physicians. This even though watch-keeping on board is the oldest type of shift-work in existence. Captain Cook was the first to introduce the three watch-system instead of the universal two watch-system in existence to then.

1.5. Constrained working-spaces

Monotonous body postures during work are unfavourable. Alternating between many possible body movements together with interspaced resting (sitting) possibilities is regarded as optimal. The officer on watch on the bridge usually stands and can walk to and fro. Occasional sitting down is possible, but in the majority of the cases the captain does not agree with this. The second watchkeeper (sailor) also stands and/or goes on control rounds. The remaining deck- and engine-personnel work in different body postures, often from time to time in unfavourable postures especially during maintenance and repair work. Workload in particular in the engine-room (oiler, greaser, mechanic), where spaces are predominantly very constrained, is in part very great. This large workload is for example the case during the taking apart and fitting together of heavy oil separators, when repairing engine pistons, during bilge pump control and cleansing, the screwing together of cylinder-lids and heavy oil-separators and when taking oil coolers apart and repairing them ("Flensburger Studie"). Constrained working spaces can also be found in ship galleys and in the cargo-holds. Apart from leading to an increase in physical stress these unphysiological conditions can cause cramps, over-extension of muscles, bursitis, myalgia "rheumatism", symptoms of cervical and lumbar vertebral disorders, lumbago, ischialgia and headaches. We very frequently diagnose these symptoms in our out-patient seamens' practice.

2. MAIN OBJECTIVES
2.1. Health

Most countries with merchant marine fleets have health authorities, departments or agencies for seafarers, as for example the "Seeberufsgenossenschaft" (SBG) in the Federal Republic of Germany. These authorities perform physical health examinations and issue health certificates for seamen. In general though, they do not have disease statistics (distribution and classification of diagnoses) of seafarers at their disposal. Only very few agencies or institutions publish these disease statistics. When accidents or diseases occur on board ships, medical treatment is performed by an officer as ship doctors no longer travel on cargo ships and are only to be found on passenger ships.

National studies as regards health of seafarers are often very difficult to compare with each other as many factors play a role for the health of seamen, for example very different crew sizes, mentality, education (i. e. Japanese versus, European training / education), religious denomination and following from the latter, consuming of alcohol yes or no, eating or diet habits. Therefore, a Japanese investigation can differ for instance from an American investigation and so forth. Because of different political, ethnological and living systems and conditions as well as different insurance systems it is also very difficult to compare morbidity and accident statistics of different countries (Goethe and Vuksanović 1975). Many countries keep no statistics at all regarding their disease and accident figures on board or they do not publish these data. Furthermore, the methods used for data acquisition are quite different. The review of diagnoses distribution from 30 publications about diseases of seafarers leads to the following sequence (Goethe and Vuksanović 1975):

gastro-intestinal 22.7 %
respiratory diseases 13.9 %
cardio-vascular diseases 9.8 %
musculo-skeletal system 9.2 %
urogenital system 8.4 %
skin diseases 8.9 %
neuro-psychiatric diseases 7.6 %
accidents 12.7 %
and further diseases 13.8 %.

In a comparative study of diagnoses distribution of diseases among seafarers from 7 seafaring countries Vuksanović and Goethe (1981) found that 70 % of all diagnoses belonged to the following main groups:

Accidents come first, followed by diseases of the digestive system, then by diseases of the respiratory system and fourthly by diseases of the musculo-skeletal system.

Only very little information as regards unfitness for duty on board can be found in the literature. An international comparison in this field also seems impossible. However, it looks as if accidents are primarily responsible for unfitness on board, followed by gastrointestinal and then respiratory diseases. As regards causes of death on board, drowning seems to come first.

Nakamura (1973) describes a twelve-year-study (1956 - 1968) on board Japanese oil tankers of 10,000 to 100,000 gross registered tons in comparison with the results of a one-year stay on board a 64,000 ton tanker. Within those twelve years body weight and height of a crew increased, the number of paces per day declined, blood pressures and their average values were lower than in the same age groups on land in Japan. Serum cholesterin was a bit lower and, as regards food intake, approximately the same qualitative and quantitative values were found as in other countries. Automation and rationalization lead to a drifting apart of interindividual relations, to disfunctions of the cardiovascular system and to psychosomatic disturbances.

Carter (1976) investigated unfitness for work or rather absence due to sickness in 1410 British deck- and engine-room-crew-members aboard oil tankers during a time period of 29 months. The average frequency of absence from work was 0.23 per man year with a mean duration of absence from work of 41 days. The absence frequency differed with rank and place of work of the seafarers. 23 % of the deck officers and 43 % of the engine-room personnel accounted for one or more absences during the 29 months. The occurrence of absence among officers was five times more frequent while they were on leave then when at sea. More than half of all cases of absence among younger officers that happened while they were on leave occurred at the end of the leave period. At sea the most frequent reasons for absence were accidents and psychiatric disorders. It is an astonishing fact, however, that fractures and other injuries occurred more often while on leave than out at sea.

Psychosomatic diseases on board German seagoing ships were investigated by Haeberle in 1979.

When comparing a collective of 43,429 active seamen with an AOK (Municipal Health Insurance) collective of 388,513 working non-seamen Haeberle found that mental disturbances, diseases of the gastrointestinal-, the cardio-vascular-, the skeletal system as well as of the muscles and the interstitial tissues showed significantly higher figures in the collective of active seamen. All analysed diagnoses within the main group "Mental disturbances" as well as alcoholism, drug addiction and ensuing diseases with exception of neurosis show substantial differences in distribution in favour of the seamens' collective. Diagnoses classified as genuinely psychosomatic occur in average twice as frequent in the seamen's collective as in the AOK-land-based collective and lead to a one and a half time longer absence from work period.

In 1979, the Norwegian Merchant Marine Fleet encompassed a total of approximately 24,000 seamen on 1000 different ships. Offer-Ohlsen who investigated unfitness for work in the Norwegian Merchant Marine Fleet between 1977 and 1979 states that on an average 900 seamen per year had to sign off because of diseases and injuries. He found that women had to sign off more frequently than men. The age group from 20 to 29 years of age were relatively more often signed off than the other age groups in the Merchant Marine Fleet. Engine-room personnel were especially liable to injuries and diseases which led to them signing off. As regards distribution of diagnoses, gastrointestinal diseases were more most frequent (16%), followed by diseases of the musculo-skeletal system (13 %), fractures (11 %) and mental disorders (9 %). Two thirds of the injuries occurred in the first two months after signing on board the ship. Many of the diseases and injuries that led to signing off do not appear to be very serious in nature.

Sugar (1981) reports about mental diseases leading to unfitness for sea-duty and repatriation for Norwegian seafarers. From the 1960ies onward the number of Norwegian seafarers who were repatriated because of mental diseases diminishes, however, the number of mental diseases gets larger until the 1970ies. It is astonishing that the diagnoses distribution of different types of mental diseases changes within these years. Oligophrenia, psychopathia and probably psychosis became less numerous whereas neurosis, alcoholism and narcomania increased in number. According to Sugar, the figures quoted in his paper represent only the "top of the iceberg" as only severe cases of neurosis lead to unfitness for work and only they qualify for repatriation at public expenses.

2.2. Performance

The multitude of professions or jobs required on board

ships demand physical as well as mental activity on the part of the seafarer. As the work processes on board get more and more automated the requirements for mental efficiency and thus responsibility increase. In the course of the last decade, we have experienced a complete shift from physical to mental work. Nowadays control and monitoring activities or duties must be looked upon as the main tasks on the ship. Less and less physical work and especially very arduous work is required on board. At present very heavy work is only required when repairing equipment and in emergency situations.

Mental performances or rather efficiency of people on board is influenced or diminished by a multitude of disturbing factors on most ships. These negative factors are often present for 24 hours of each day and have a subtle or manifest effect on the seamen, and can lead to an accelerated acute or chronic psycho-physical fatigue during sea-voyages of long duration.

2.3. Stress and fatigue

Time intervals occur on many ship journeys in which the total stress is reduced, for example during calm weather on the high seas with engaged autopilot. In situations like this, demands on performance and efficiency are more likely to decrease than increase due to monotony while traversing wide stretches of water. In sea lanes with dense traffic and especially in rivers in foggy conditions, however, stress can become very great for personnel in charge of the ship. Investigations by Erikssen et al. emphasize the great strain for the captain and the cardio-vascular dangers connected with it. Stress on board arising purely from work cannot be separated from environmental stress. Both stresses are interwoven with each other. Stress must furthermore be defined as a psycho-physical combination. Mental stress predominates among the officers in charge of the ship, whereas the other crew members are more affected by physical stress. Among the latter, above average stress can be found with engine-room and deck personnel during exceptional (unusual) work tasks.

Fatigue

Factors leading to fatigue on land must be substituted on board by further ship-specific factors. These being vibrations, ship-movements around several axes together with accompanying accelerations and amplitudes (reduced recuperation capability, physical stress through compensatory muscle activity, kinetosis), watch-keeping, time-zone shifts and monotony.

In the current four hour watch plus eight hour leisure time the individual gets maximally eight hours sleeping-time. This time shortens to approximately seven hours because of meals and food uptake. However, environmental factors lead to a reduction in depth and duration of sleep so that up to more than a third of the seafarers complain of disturbances in their sleep. Progressive recuperation deficiency and an ensuing decline in general mental efficiency with an increased excitability, nervousness and eventually sensory disturbances occur.

Jobs requiring frequent lengthy periods of standing on board such as on the bridge, lead to an increase in static muscle-load, blood pooling in the legs, a rise in heart frequency and to earlier fatigue. Performance while standing is $\leq$ 19 % lower than when sitting. Accelerations between one and three meters/sec^2, maximally seven to nine meters/sec^2, afford compensatory muscle work, lead to a decline in the recuperation value of leisure time and reduce the depth of sleep. The propeller and the engines on board are the main course of vibrations on a ship. These vibrations are frequently significantly higher than permissible (K=0.63 and 4.0 for living quarters and working areas) in the working areas and in the living quarters ("Flensburger Studie", Volume 5). Oscillations in the form of ship movements represent a significant stress- and also fatigue-factor. According to Goethe et al., we still do not know enough about the side-effects of vibrations on human being.

2.4. Safety and reliability

We still do not have a precise definition for ship safe-

ty. The term traffic safety is more universal. This term encompasses:
> Ship and crew, traffic/shipping lanes, i. e. ports, surrounding waterways, rivers, etc., coastal waters, seas and oceans.

Thus follows a new definition:
> Ship safety = safety of the shipping traffic system.

Reliability

Reliability is defined as the capability of a system to meet the requirements represented by the behaviour of its characteristics during a given time and stemming from the purpose of its use.

All measures concerning ship safety are prophylactic in nature and supposed to protect humans and material goods from danger. Safety is not equal to reliability. According to the "Flensburger Studie", Volume III, safety of the ship-traffic-system is greater the less dangers for humans and material goods ensue and the better unavoidable dangers can be warded off.

Additional safety equipment is available apart from the currently used equipment in order to increase ship safety, for example
> stabilizers, rescue and fire-fighting equipment,
> Decca-navigation, automatic pilots, sonar and radar.

National and international efforts and projects have been initiated to ameliorate ship safety and to decrease the accident rate. A current project (1977) of this type is, e. g. the Norwegian "3-S (system for a safe ship) project" with active participation of 5 research institutions and of active seamen. It has 6 main goals, among other things to obtain a general review of accidents and their consequences on board and a general review of dangerous work, places of work and equipment on board. Right from the beginning the question of safety as regards the ship as a place of work was a main aim with the target of reducing the accident rate.

The question of social factors on board the ship and at

home among the families on land was given priority as mental stress and problems for seafarers and their families can ensue from them. Fire-fighting and protection from fire is part of the "3 - S Project".

Flaatrud (1981) states that the main driving force behind the continuing changes and improvements of the technical system (ship, engine and equipment for propulsion and safety) is not safety, but economy alone.

According to the "Flensburger Study" the safety of the ship-traffic-system is substantially influenced by environmental factors which occur as "promoters" and "averters" of unfavourable conditions/happenings.

Factors acting as promoters are:
 Heavy shipping density, extreme weather conditions, difficult waters,

averters on the other hand are:
 navigation aids, tugs, shipyards and docks, search and rescue as well as service facilities.

3. SPECIFIC OBJECTIVES

3.1. Field studies

Almost all studies of human performance in ship transport operations (merchant shipping) are field studies. Several of these studies as well as the results derived from them have been mentioned elsewhere in this publication, so that we need not deal with them here.

Special mention must be made of the following studies conducted in the last few years and consisting in part of human performance:

- The "3 - S - Study" (Norway), the "Flensburger Study" (Federal Republic of Germany, Netherlands);
- "Sealife Programme" (United Kingdom)
- "Ship with reduced crew (18-men-crew) Study" (Japan).

We would like to bring to your attention two of the studies currently going on:

"The 18-Men Ship-Study" and "The Ship of the Future (12-men-crew) Studies" both being conducted in the Federal Republic of Germany.

3.2. Simulated Studies

Simulated studies of performance and stress of personnel on board have hardly been conducted in civilian merchant fleets. These studies can readily and easily be realized in the military, i. e. in the navies of different countries, as the project contracts there lie in the answering of different questions (Combat-Programme) and the financial support is more easily obtained.

Several countries now have simulators at their disposal which can copy many but not all functions of a completely equipped ship's bridge including some ship movements. These simulators are primarily used for the training of nautical officers and less for research of performance and stress tolerance of the human being. A research series regarding human performance of ships' officers under the influence of ethanol is at present going on in the ship control simulator of the Nautical Training School in Bremen in conjunction with the Institute of Forensic Medicine of the University of Hamburg.

A very sophisticated installation of this type, "Susan" (ship command and simulation equipment), was recently put into operation at the Nautical Training School's Department of Seafaring, Hamburg, Federal Republic of Germany. The computer section "Susan" can present up to 20 ships simultaneously on TV-screens and projection screens and the approximately 6 x 7 meters large shipbridge mock-up can perform pitching and rolling up to $\pm$ 5 degrees. In addition it is possible to simulate / replay engine and environmental sounds and noises. The "Susan"-Simulator therefore enables us to simulate genuine mental stress situations. Potential possibilities for future nautical medical stress studies with the "Susan" simulator can thus be envisioned.

3.3. Rest / Duty Regulations

Working time on board is subjected to very different regulations in seafaring countries as far as it is not predetermined by the watch-systems. ILO-Convention 197 concerning minimum standards on merchant ships requests that each signature country etc. must have laws or regulations laying down hours of work and manning on board ships. In the Federal Republic of Germany the "Seemannsgesetz" (Seamens' law) determines among other things the duration of work (hours of work per day etc.).

3.4. Selection and Training
3.4.1. Selection:

The seaman's profession must be regarded as strenuous (stressful) due to the multitude of factors from within and without the ship that come to bear on him. This stress depends on the type of activity or work on board. It is therefore only too understandable that health-selection criteria (fitness for duty at sea) must take these special conditions into consideration. Deck-personnel, for instance, works primarily in the open air and is thus in part subjected to extreme climatic conditions. Besides being physically strong, these men should not be prone to dizziness, as they are required not only to overcome great heights, like climbing up masts or down into or up from cargo-holds but also to handle heavy ropes. Deck officers, on the contrary, are more subjected to mental stress and their colour vision should be good. Hardly any physical work is required from the captain, but primarily mental work. He, however, carries a very great responsibility, must be able to make fast decisions and must cope with great inner loneliness from time to time. Engine-room personnel is exposed to noise, oily fumes, vibrations etc. From time to time they are required to perform heavy work in unfavourable or unphysiological body postures and they should have the capability of being able to tolerate oily smells which, among other things, can lead to nausea. Engine-crews must also work in dirty environments. Watchkeepers must have good hearing and colour-discriminating capabilities.

From the information quoted above it becomes evident that a thorough medical history must precede the "Fitness for duty at sea" medical examination. Mental disorders can become worse out at sea. Previous histories of stomach and duodenal ulcers as well as kidney stones should cast doubts as to fitness for duty at sea. Some skin-diseases as well as outer ear-infection (otitis externa) can re-occur or get worser in hot or tropical climate zones. Detailed minimal requirements as regarding vision, hearing and colour-discriminating capabilities exist.

When selecting seamen it should be remembered that medical care out on the open seas is only an emergency solution in comparison with a stay on land and that therefore a strict selection, medically speaking, represents the most optimal solution.

In Western Germany, medical fitness-for-duty-at-sea-examinations are compulsory for captains as well as for all crew members in accordance with the seaman's law (1948) (Seemannsgesetz). This medical examination is performed by the "Seeberufsgenossenschaft" (Seediensttauglichkeitsverordnung or regulations for fitness at sea).

The SBG-Rules and -Regulations mention fitness criteria in general in form of minimal requirements as to body weight and height and contain a list of diseases, disfunctions and underdeveloped or weakly developed parts of the body which can lead to absolute or relative unfitness for duty. However, this list is not a rigid scheme but gives the examining doctor a certain degree of freedom of choice.

3.4.2. Training:

A review of the literature shows that the education or training of seafarers, seen on a world-wide scale, still differs considerably. Although it is true that the highly developed countries have good to excellent training facilities and schools, training in a country of the Third World is at present partially substandard. For years, therefore, many underdeveloped countries have been sending their seafarers, in the majority future nautical officers, to receive training in the United States, Great Britain, France and the Federal Republic of Germany etc. This is also suggested by the ILO in its Re-

port IV: Vocational Training of Seafarers (1970).
Furthermore, national certificates of qualification for different special professions of seafarers are to be issued. According to ILO, emphasis should be put on theoretical and practical training during the education of seafarers. This of course includes the use of adequate demonstration models and also simulators, such as navigation-, bridge- and radar-simulators as well as video-equipment.

Instructors at these training centres should have a good general education, technical qualifications as well as sufficient practical experience in the Merchant Marine Fleet.

ILO (1970) states that training given to seafarers in most countries regarding first aid, surgical treatment, diagnoses and therapy of diseases was inadequate. Only a few countries have any compulsory education in such matters, although very few merchant ships have doctors on board. In 1976 the WHO published a book "International Medical Guide for Ships". However, there still remains a lot to do at national and international levels regarding medical training for a ship crew. This fact is also mentioned by the ILO in 1975 in its publication "An International Maritime Training Guide".

According to Goethe (1982), medical equipment on board as well as first aid training is still different from country to country. In most cases one of the nautical officers is put in charge of controlling health on board. Qualification for captains' or officers' licenses are also dependent on proof of basic training in first aid. In the highly developed countries such as the United States, Great Britain or the Federal Republic of Germany this training is either conducted in a form of Red Cross Courses, theoretical or theoretical-practical instruction courses at nautical schools and sometimes in hospitals. As in Germany, for instance, these courses can last one, two or four weeks, depending on whether they are refresher-courses or basic courses (Hamburg). Turnbull et al. gives a good review of the situation in different European Shipping Nations in form of a comparative study of nautical medical care and/or standards.

3.5. Accidents

Two categories of accidents occur in shipping, ship accidents or disasters and accidents involving people.

As this paper deals with human performance or rather psycho-physical stress of humans on board, we will not discuss ship accidents any further.

It is true that statistics about accidents of people on board are compiled in many countries, especially in the highly developed ones, but, according to Goethe and Vuksanović (1975) these statistics can often hardly be compared with each other as they often are based on different criteria. Diagnoses and treatment on board are often made and/or performed primarily by laymen so that a certain margin of error can be found here. As to type of injury, fractures and wounds come first in the publications currently at our disposal. In international literature comparison, deck personnel seems more prone to accidents than engine-room personnel.

A comparison of accident distribution among seafarers from 6 countries between 1954 and 1972 is as follows:

31.8 % fractures, 28.3 % wounds, 25.6 % contusions, 12.2 % burns, 9 % distorsions, 2.3 % luxations, 24.9 % other diseases, 37 % skin injuries and 1 % lethal accidents.

Carter (1976) investigated the frequency of absence from work among 1410 men of the British deck and engine personnel on ESSO Tankers during a time-period of 29 months. The most frequent reasons for absence from work at sea were injuries and mental disorders. However, fractures and other injuries occurred more frequently during leave than at sea. Accidents and/or injuries led to a total absence from work of 1139 days or 14 % of all the lost time. According to Carter, the relative scarcity of reported injuries while at sea as compared with those on shore, especially fractures where the diagnosis is almost invariably confirmed by X-rays, suggests that with good safety standards the frequency of accidents at sea can be reduced to well below those occurring in every day life on land.

The Norwegian Merchant Fleet consists of approximately 2500 ships and employs about 5000 people. 80 % of the accidents on Norwegian ships are connected with the work environment and social environment on board. The Norwegians found that between 1970 and 1976 the following distributions of accidental deaths in the Norwegian Merchant Marine Fleet occurred:

29 % through ship risks, 24 % through works risks, such as falling, poisoning etc., and 47 % through social risks such as risks in one's spare time, disappearances, etc. (Flaatrud 1981).

In the Norwegian Fleet the number of injuries and accidents per 100 seamen climbed continuously from 10 to 19 until 1975 and then fell from this time onward until it reached 12 in 1979. The number of people killed through accidents sank from 0.4 % (1973) until below 0.2 % (1979). According to Norwegian opinion there is approximately a one to five chance per year that a seaman is injured or has an accident. In the Norwegian Fleet, so called social risks are the cause for almost 50 % of all seafarers killed within the last five to ten years and it seems that the situation is similar within the British Merchant Fleet. Flaatrud found that the working climate on board as well as separation from family and social contacts on land are one of the main dangers to life for seamen.

Offer-Ohlsen (1981) states that 4 to 5 % of Norwegian seafarers sign off each year due to injuries and diseases. Thereby, engine-room personnel was especially susceptible for injuries and diseases and their percentage was 50 % higher than expected.

Apart from diseases (70 %) the most important injuries that led to signing off between 1977 and 1979 were: fractures (11 %), contusions (8%), wounds (3%), burns (1%), luxations (1%) and other injuries (2%). Half of the registered fractures concerned the upper extremities, especially the hands and fingers. Tibia and costal fractures were frequent. Contusions and sprains were most frequent in the lower extremities. Goethe and Vuksanović (1975) found that the number of injured seamen of those involved in accidents aboard West

German ships between 1954 and 1972 always lay between 12.2 and 20 % of the total Merchant Fleet personnel per year. The percentage of fatal accidents in the above-mentioned figures averages 1.06 % (0.74 to 2.29 %). Furthermore, 61.5 % of all fatal accidents in the West German Fleet between 1954 and 1972 were caused by drowning.

Vrcelj (1981) analysed 7810 accident reports of the West German Merchant Fleet that were made between 1974 and 1976. In these 3 years 109 fatal accidents through injuries occurred, whereby drowning came first with 56 % followed by closed fractures (17.3 %), for instance due to falls. 6.6 % of the male seafarers were involved in accidents, but only 4.7 % of the female seafarers. As regards the types of injuries among those 7810 accidents, contusions were most frequent (31.8 %), followed by closed bonefracture (23.6 %) and external wounds (16.6 %). Most accidents occurred between 8 and 11 o'clock in the morning (30.1 %) and 1 to 5 o'clock in the afternoon (32.2 %) with peak values at 10 in the morning and at 4 in the afternoon. A distribution of accidents at work according to ship type and place of work is shown in the following table (table 1):

Place of work	Cargo Ship	Tanker	Tug
Open deck	53.9 %	39.8 %	68.9 %
Engine room	17.1 %	24.9 %	14.8 %
Cargo hold	7.8 %	-.-	-.-
Galley, Mess-room, pantry	7.7 %	10.6 %	6.8 %
Repair shop	-.-	7.7 %	-.-
Living quarters	-.-	-.-	2.9 %

TABLE 1 Distribution of accidents at work according ship type and place of work.

The percentage of accidents on board cargo ships was higher than on tankers.

The Federal Republic of Germany with its relative accident frequency of 13 % lies in a favourable position in comparison with accident statistics from America, Italy, Sweden, Norway and Denmark. The death-rate of almost all above-mentioned countries lies between 0.1 and 0.2 %, whereby Germany's death rate of 0.1 % also lies in a relatively favourable position. The accident rates of men and women on board are higher than when working on land. Young seafarers (till 30 years of age) show a higher accident rate than elder seafarers. However, this accident rate again begins to climb after the age of 60 has been reached.

Subjective factors playing a role in the cause of accidents on board are very difficult to assess and analyse. Human errors, however, are the cause of more than 50 % of the accidents, e. g. overexertion or fatigue, emotional imbalance and excessive alcohol-drinking. But sea-movements also have an influence on the cause of accidents and disasters on German Merchant Ships (= 13.4 % of the accidents). The number of accidents due to subjective factors as well as alcohol still remain obscure but are probably very high. Lightheartedness and foolishness also play a role in this context.

Accident frequency on medium sized ships (500 to 10,000 gross registered tons) between 1974 and 1976 was almost two times as high as on smaller or larger ships. It was found that 30 to 40 % of all accidents occurred as so called "accidents in the working environment on board" (the most frequent cause of accidents). Collisons, fire or the sinking of the ship led only to 0.9 % of the accidents.

Ship accidents as well as accidents of seafarers during 1970 and 1971 were investigated in Volume III of the "Flensburger Study". 572 ship accidents and 1351 accidents concerning seafarers occurred in the West German Merchant Fleet during this time period. A total of 56 possible causes of accidents and/or reasons for injuries are listed in the above publication.

Accident frequency is highest between eight and twelve in the morning as well as from 4 to 8 in the evening, whereby younger seafarers (≤ 30 years of age) are more endangered than elder seamen. On board the accident rate was greater for deck personnel, engine room personnel and electricians than in the other speciality groups. It was highest for the first two groups mentioned. Nautical and radio officers were least endangered by accidents. Most of the injuries involved hands, feet, legs and arms, in this sequence. When grouped according to type of injury, skin abrasions and contusions came first followed by fractures and internal injuries. In the opinion of the SBG, accidents on board seldom have one but rather several causes. Even as early as back in 1970 and 1971 the accident rate on container ships was significantly lower than on dry cargo ships, bulk-carriers and tankers.

The "Flensburger Study" states that most accidents occurred through "falling and stumbling" and while working on deck. On deck they happened during the opening and closing of cargo holds and while handling cables and ropes. Finally, it was determined that accident rates in "ports, locks and docks" and in "regions near ports and canals" were the highest and not those out on the high seas.

In summing up, it can be said that viewed from a national as well as an international level, accidents on board are caused by many factors, whereby human errors as for instance disregarding safety regulations, light-heartedness/foolishness and alcohol-abuses (the exact figure is obscure but expected to be rather high) come first.

Social as well as interindividual and therefore psychological problem areas that can be stressful and thus lead to accidents, e. g. jumping overboard, suicide etc., must be added to the above-mentioned factors. It is true that statistics from different countries cannot be correlated with each other very well due to a variety of reasons and that the combination of accident types and the distribution of these accidents change depending on ship type, age, state of modernisation, crew number, etc., but the total sum of accidents, however, lies significantly higher than in professions on land. It thus becomes obvious that even nowadays the seaman is exposed to a dangerous and very stressful environment.

It is therefore necessary to approach, to work on and to improve this situation from several planes. We thus have a lot to do in this field, let's get started!

REFERENCES

Azzuqa, A.S. 1979. Die funkärztliche Beratung in der Seefahrt, ihre Probleme und die Möglichkeiten ihrer Verbesserung. Hamburg, Med.Diss.

Carter, J.T. 1976. Absence attributed to sickness in oil tanker crews. Brit.J.Industr.Med. 33, 9-12.

Colquhoun, W.P., Hamilton,P. and Edward,R.S. 1975. Effects of circadian rhythm, sleep deprivation, and fatigue on watchkeeping performance during the night hours. In: "Experimental Studies of Shiftwork." (Eds. P. Colquhoun, S. Folkard, P.Knauth et al. (Westdeutscher Verl., Opladen) pp 20-28.

Colquhoun, W.P., Paine,M.W.P.H. and Fort,A. 1975. Watchkeeping studies on a nuclear submarine. Final report. (Ed. Medical Research Council Royal Naval Personnel Research Committee. SMS 4/75) (Alverstoke).

Colquhoun, W.P. 1977. Watchkeeping and safety. In "Human Factors in the Design and Operation of Ships" (Ed. D. Andersen). (Ergonomi laboratoriet, Stockholm). pp 538-549.

Colquhoun, W.P., Paine,M.W.P.H. and Fort,A. 1978. Circadian rhythm of body temperature during prolonged undersea voyages. Aviat. Space Environ.Med. 49, 5, 671-678.

Colquhoun, W.P., Paine,M.W.P.H. and Fort, A. 1979. Changes in the temperature rhythm of submariners following a rapidly rotating watchkeeping system for a prolonged period. Int. Arch.Occup.Environ.Hlth 42, 185-190.

Erikssen, J., Johansen, A.H. and Rodahl, K. 1981. Coronary heart disease in Norwegian sea-pilots; part of the occupational hazard? Acta Med.Scand. (Suppl.0) 645, 79-84.

Flaatrud, A. 1981. System for a safe ship, presentation of the Norwegian research project "3 S". Norsk Bedriftshelsetjeneste 2, 3, 135-149.

"Flensburger Studie", Vol. 3 + Appendix 1974. Schiffssicherheit (Verkehrssicherheit). Forschungsauftrag "Schiffsbesetzungsordnung" (Ed. Forschungsstelle für Schiffsbetriebstechnik). (Flensburg).

"Flensburger Studie", Vol. 5 + Appendix 1974. Untersuchungen über die psychischen und physischen Belastungen von Schiffsbesatzungen. Forschungsauftrag "Schiffsbesetzungsordnung". (Ed. Forschungsstelle für Schiffsbetriebstechnik). (Flensburg).

Gesetz zum Übereinkommen Nr. 147 der Internationalen Arbeitsorganisation vom 29.10.1976 über Mindestnormen auf Handelsschiffen vom 28. April 1980.1980. BGBl. Teil II, 17, pp 606-620.

Goethe, H. and Vuksanović, P. 1975. Distribution of diagnoses, diseases, unfitness for duty and accidents among seamen and fishermen. Bull.Inst.Mar.Trop.Med. 26, 2,133-158.

Goethe, H., Zorn, E., Herrmann, R. et al. 1978. Die psycho-physische Belastung des Personals moderner Seeschiffe als aktuelles Problem der Schiffahrtsmedizin. Zbl.Bakt.I.Abt.Orig.B 166, 1, 1-36

Goethe, H. and Vuksanović, P. 1982. Diseases and accidents among seamen - an international comparison. In "The third European nautical medical officers meeting 1981. Panel discussion of day to day problems." (Ed. D.T. Jones) 1st-5th June 1981, Norway. (Oslo) pp 2/1-2/24.

Goethe, H. 1982. Sicknesses, injuries and health care on board - medical education of shipboard personnel. In "Int.Workshop on human relationship on board" (Ed. H. Böhm). Bremen, January 21-22, 1982. (Bremen) pp 119-130.

Haeberle, E.G. 1979. Psychosomatische Erkrankungen bei Seeleuten. (Hamburg, Med.Diss.)

Harrington, J.M. 1978. Shift work and health. A critical review of the literature. (Ed. Employment Medical Advisory Service of the Health and Safety Executive). (Her Majesty's Stationery Office, London).

An international maritime training guide prepared jointly by the International Labour Organisation and the Inter-Governmental Maritime Consultative Organisation 1975. (ILO,Geneva). =Document for guidance.

Internationales Übereinkommen von 1978 über Normen für die Ausbildung, die Erteilung von Befähigungszeugnissen und den Wachdienst von Seeleuten 1982. BGBl. Teil II, 14, 297-372.

Jacobs, W. 1980. Physische und psychische Belastungen an Bord (Hamburg, Abschlußarbeit, Fachhochschule Hamburg, FB Seefahrt).

Kleemann, A. 1976. Arbeitsmedizinische Untersuchungen über mögliche Einflüsse auf die Leistungsfähigkeit der Besatzung eines Containerschiffes am Beispiel des CTS Tokio Express. (Bonn).

Nakamura, F. 1973. Health status of ocean-going oil tanker seamen. Bull.Tokyo Med.Dent.Univ. 20, 3, 221-244.

Offer-Ohlsen, D. 1981. Seamen signed off owing to injuries and diseases in the Norwegian merchant fleet 1977-1979. Norsk Bedriftshesetjeneste 2, 3, 150-152.

Rutenfranz, J., Mann, H. and Aschoff, J. 1969. Circadianrhythmik physischer und psychischer Funktionen bei 4-stündigem Wachwechsel auf einem Schiff. Studia Laboris et Salutis 4, 1-11.

Sander, C. 1982. Schiffsführungs- und Simulationsanlage "Susan". Hansa 119, 5, 327-328.

Seemannsgesetz in der im BGBl. Teil III, Gliederungsnummer 9513-1, veröffentlichten bereinigten Fassung (1957). Original-Version BGBl. Teil II, 713.

Souchon, F. 1972. Der Einfluß von Umweltfaktoren auf die Leistungsfähigkeit von Schiffsbesatzungen. Wehrmed.Mschr. 16, 2, 33-39.

Sugar, E. 1981. Repatriation and unfitness for sea duty caused by mental disease. Norsk Bedriftshelsetjeneste 2, 3.

SUSAN: Informationsbroschüre Schiffsführungssimulator. (Ed. Krupp Atlas-Elektronik) (Bremen).

Turnbull, T.A. 1976. Medical standards, treatment and training in Western European Countries. (Ed. The Merchant Navy and Airline Officers'Ass.). (London).

Turnbull, T.A. 1979. A review of medical and training facilities in the U.K. and other countries. (Ed. Merchant Navy and Airline Officers'Ass.).(London).

Valentin, H., Klosterkötter, W., Lehnert, G. et al. 1971. Arbeitsmedizin. Ein kurzgefaßtes Lehrbuch für Ärzte und Studenten (Thieme, Stuttgart).
Vocational training of seafarers 1970, (Ed. ILO)(Geneva 1970) = ILO Report VI.
Vrcelj, J. 1981. Die Unfälle an Bord von deutschen Seeschiffen und ihre Ätiologie von 1974-76. Hamburg, Med.Diss.
Wojtczak-Jaroszowa, J. 1977. Physiological and psychological aspects of night and shift work. (Ed. US Dept. of Health, Education and Welfare) (U.S. Government Printing Office, Washington). =DHEW (NIOSH) Publ. No. 78-113.

STRESS FACTORS AND COUNTERMEASURES IN NAVIGATION

G. Athanassenas

3rd Psychiatr.Dept. State Mental Hospital

Daphni-Athens, Greece

ABSTRACT

In this study an effort is made to analyze in brief the performance of operators in sea-traffic, applicating the well known model of "sensory input - translation process - motor output", and to study the effect on it of the particular stress-factors connected with work and life on board. Measures intending to eliminate these factors, or at least to prevent their adverse effect are proposed. These measures are classified in four categories: Technical, social, medicolegal and organizational.

INTRODUCTION

The systematic study of the particular stress factors associated with work and life on board, and their influence on human performance is not only of theoretical importance. It deserves also an uppermost practical interest not only from the point of view of improving safety in sea-traffic, but as contributing also to the humanization of work conditions in navigation and to the enhancement of seamen's welfare.

HUMAN PERFORMANCE IN SEA-TRAFFIC

The sea occupying the major part of the surface of our planet is covered by a rich net of sea-roads. The traffic on these roads is done by ships and the act (art or science) to manage a ship is navigation.

The main operator in sea-traffic is the captain. Studies on human performance in this traffic are usually attracted by and concentrated upon the skilled performance of this man. However, it must not be forgotten that there are many other officers of different specialization on the ship, as well as the whole crew. All these men must also be seen as operators.

In psychological terms the operator represents the "translation process", i.e. the decision element intervening between the sensory input and the motor output of the sensorimotor activity, or the response to receptor and the stimulus to effector processes (Wright et al., 1972). The latter is expressed as spoken command or operation of control instruments. Furthermore a feedback loop exists coveying information about the results of effector processes and generating new action in order to correct probable discrepancies between aimed and obtained results.

The evolution of skilled operator from the primeval navigation till the modern one, was in general as follows: At the beginning it was a lonely man in a curved trunk using his sense perception to detect the changing outline of coasts, the position of the sun, the coming storm and the currents. He had only a paddle to move his boat. He was the first navigator praised by Horace: "..qui fragilem truci commisit pelago ratem primus, nec timuit...rabiem Noti...qui vidit mare turgidum et infames scopulos Acroceraunia..." (Carmina, II). Gradually his receptor capabilities were extended through a look-out on the top of the mast, a man sounding to measure the depth of the waters, as well as by different instruments: telescopes, sounding apparatuses, wireless, radiometers and recently radar and satellite navigation systems. His effector capabilities were also increased with the aid of rowers, steersmen, sailors (the original meaning of the word is men managing the sails), engines with their engineers, messengers, telephones, etc.

It is interesting to trace the origin of the world government as well as of the modern term cybernetic, both coming down from the ancient Greek word κυβερνήτης or κυβερνητήρ (Homer, Odyss.θ 557) meaning the steersman or captain. Plato (Politeia, 488 D) refers to cybernetics (κυβερνητική) as the art of governing a state corresponding to the art of steering and governing a ship.

Nowadays the control room of a modern ship is furnished with a display, like those of other transport media, conveying all information and corresponding to a logical extension of the sensorial input of the captain. On the other hand the control instruments can be considered as the logical extension of the effector part of his sensorimotor activity. A step further comes "dynamic positioning", a system that automatically controls vessel state: heading, course and position.

This evolution, beneficial in its major part, brought with it also some inconveniences, e.g. the sensory overloading of the operator from the abundance of information inflow and the great chain of persons intervening either in the receptor or in the effector processes. Studying the historical evolution of nautical hygiene, we have many times encountered this interplay between technical advances and worsening of conditions on board. For example, the development of sailing and the invention of compass led to the realization of long oceanic voyages. These brought with them malnutrition and the consecutive different forms of avitaminoses, scurvy being the most fatal between them with its well known devastating effects among mariners. Therefore, before the application of any new technical achievement in navigation, one must foresee the possible harmful effects on man. This is particularly true in this time of rapid technological progress. As we shall see in the next chapter, among the stress factors mentioned and especially those related to the internal environment of the ship, we can trace many of them as originating from imperfect or faulty constructions.

SPECIAL STRESS-FACTORS IN NAVIGATION

Work and life on board is full of adversities, some of them innate to the environment (climatic factors) and many other caused either by the construction of the ship or by social factors (special terms of work and life on board). All of them have a direct or an indirect inhibiting effect on performance. We can classify them into the following four categories, one of natural, one of structural and two of social origin.

1. Climatic factors:

With these we mean not only the need to perform under the uppermost contravening weather conditions (D'Ippolito, 1969), but also the frequent exposition to sharp meteorological changes (Tromp, 1968) and the sojourn in extreme climatic environments (arctic, tropic).

There are also the movements of the ship with the consequent motion-sickness.

2. Internal environment of the ship

Restricted space, noise, vibrations, air pollution, abnormal temperature.

3. Special conditions of work on board

The limited number of personnel resulting in the overloading of some crew members in case of absence of one of them.

The shift-work with its disruptive influence on circadian rhythms (Reinberg, 1981).

The alternation of periods of sensory overloading and fatigue with periods of sensory underloading.

4. Special conditions of life on board

The monotony of the internal environment combined with the lack of any change during long voyages (Castellani, 1972).

The poverty of opportunities for amusement during free-time because of the restricted space, persons and means (Giorda, 1972).

The lack of real free-time: In any moment a critical situation may arise, and the work of otherwise resting people may urgently be needed. Although this happens rather seldom, it is a factor having an anticipatory influence on the relaxation of crew-members and on their long-term well-being.

The lengthy separation from family, friends and familiar surrounding (all known under the term home) with the following nostalgia.

The irregular sexual life consisting of periods with compulsory abstinence alternating with periods of sexual abuse.

The difficulty in getting medical aid in time of emergency cases.

Last but not least the quality of meals.

All the above mentioned factors combined with the fear from the well known major dangers innate to navigation (Otterland, 1963) (accidents, fire, shipwreck) act as stress-factors contributing not only to the deterioration of performance but also to premature ageing, considered as a professional risk

(Benedetti et al., 1961).

In former times, when conditions on board were even worse, they led to the denomination of ships as "floating jails". It is a well known statement in court made by Dr. Johnson, the lawyer of a criminal who had the choice between serving in the eighteenth century navy or being sentenced to hard punishment and who prefered the punishment: "Why, Sir, no man will be a sailor who has the contrivance to get himself into a jail, for being in a ship is being in a jail with the chance of being drowned" (Roddis, 1950).

Dealing with performance we must also keep in mind the different degree and kind of reaction exerted by different personality types as a response to the same stimulus.

This is especially true in sea-traffic where abnormal reactions must be expected due to the higher vulnerability as the consequence of a major preponderance in psychopathology found among mariners in comparison to other professional groups (Petiziol and Rizzo, 1965). According to some authors, psychic normality is found only in 30%, subnormality (expressed as solitary and unsettled personalities) in 60% and psychopathy and psychosis in 10% (Brain, 1945) of the seaman population. (A rough estimation of the author among Greek seamen brings about more optimistic values: raising normality to 50%, subnormality to 45% and psychopathy and psychosis to 5% (Athanassenas, 1971).

The high range of abnormality has a twofold explanation: First, the fact that unstable personalities having difficulties in getting a job at land consider the merchant marine as an easy way to earn their livings (Petiziol et al., 1966). Second, the lack of readiness for entering the nautical profession, partly justified because of its difficulties and disadvantages, compels the employers to accept persons of doubtful mental capacities.

PROPOSED COUNTERMEASURES

There are many measures that can be proposed in order to ameliorate performance by eliminating the above mentioned stress factors, or at least by preventing their adverse effects. They can be classified in the following four categories: Technical, social, medicolegal and organizational.

1. Technical Provisions

The application of means and methods derived from anthropotechnic studies for the amelioration of performance on board by unfolding human capacities in their best (Mau et al., 1974). Indicatively one can mention:

The orthological design of displays in order to concentrate informational input, taking care at the same time to avoid sensorial overloading of the persons concerned (Hinsch, 1972).

The application of automatization in such a manner that human control can be possible in any moment by simple methods. In both cases sophisticated procedures must be excluded.

As to the deck of the ship (considered as the decision center) the number of persons needed must be exactly calculated for every kind and phase of navigation and for different conditions. Retrospective studies by analyzing shipping casualties can be of uppermost help in this direction (Hinsch, 1973).

Withal the widespread application of modern shipbuilding technology in order to ameliorate life conditions.

2. Social Provisions

A second step would be the improvement of conditions of life and terms of work on board in order to eliminate stress factors connected with their inconveniences. Among other measures, the duration of service on board and of the subsequent paid leave must be defined according to the kind of work and class of ships.

Another measure aiming also at the humanization of terms of work is to study the possibility to substitute part of the high wages by social benefits, such as long leaves and early pensioning. In this manner, voluntary overexhaustion motivated by profit, can be prevented, as well as premature ageing resulting from it. On the other hand measures of this kind will have a dispelling effect on abnormal personalities coming temporarily to the marine profession, attracted only by the possibility of an easy and quick enrichment.

3. Medicolegal Provisions

The codification of rules for the professional orientation and the psychotechnic selection of seamen in order to obtain the best quality of operators (Boganelli, 1959; Boganelli and Di Curzio, 1960).

4. Organizational Provisions

Although in many scientific institutions a great deal of work is done concerning performance and related subjects in sea-traffic, corresponding literature has not the spread as in the respective fields of road and air traffic. Furthermore the existing literature gives the impression that contact between research centers is loose.

The author believes that there exists a real need for a close cooperation between centers of research and relevant organizations dispersed in different countries in order to promote the study of these problems at an interdisciplinary level and the avoid unnecessary repetitions.

A solution of these problems can perhaps be offered by the creation of an international society or committee intending to promote closer contacts between all interested parties.

In this case, the European Undersea Biomedical Society and the European Diving Technology Committee, both concerned with problems of underwater activities, can serve as a model. A similar organ could aim to provide a place to discuss standards and techniques, to recommend harmonization so that common standards may be achieved and the safety improved; to provide expert advice to Governments, organizations and individuals.

REFERENCES

Athanassenas, G. 1971. Medical guide for Ships (in Greek), 2nd Edition. (N. Stavridakis & Son, Piraeus).
Benedetti, G., Mazza, F., Lazzeri, S., Lopane, F. 1961. Rilievi clinico-statistici sulla senescenza precoce dei marittimi. Atti della Sezione Studi. Centro Internazionale Radio Medico, Roma. 37-49.
Boganelli, E. 1959. Orientamento e selezione psicotechica dei marittimi. Atti della Sezione Studi. Centro Internazionale Radio Medico. Roma. 122-126.
Boganelli, E., DiCurzio. U. 1960. Profilo professionale presuntivo dell' Ufficiale di coperta della Marina Mercantile.

Atti della Sezione Studi. Centro Internazionale Radio Medico. Roma. 115-122.
Brain, D. 1945. Neuropsychiatric aspects and treatments of convoy and torpedo casualties. In "Manual of Military Neuropsychiatry" (Ed. H. Salomon & P. Yakovlev). (Saunders, Philadelphia and London). pp. 631-641.
Castellani, A. 1972. La solitudine dell' uomo di mare: aspetti psicologici e psicopatologici. Atti della Sezione Studi. Centro Internazionale Radio Medico. Roma. 31-37.
D'Ippolito, M.L. 1969. Recenti acquisizioni in tema di meteoropatie nei marittimi. Atti della Sezione Studi. Centro Internazionale Radio Medico. Roma. 47-51.
Giorda, R. 1972. Il "tempo libero" del marittimo. Atti della Sezione Studi. Centro Internazionale Radio Medico. Roma. 39-44.
Hinsch, W. 1972. Eine Brücke für Seeschiffe. Hansa 109, 25-29.
Hinsch, W. 1973. Abwendung von Schiffszusammenstößen. Hansa 110, 1-3.
Mai, G., Gabner, V. und Langhans, K. 1974. Erarbeitung von Kriterien für eine funktionsgerechte Besetzung deutscher Seeschiffe. Forschungsstelle für Schiffsbetriebstechnik. Flensburg.
Otterland, A. 1963. The mortality among seafarers. Atti della Sezione Studi. Centro Internazionale Radio Medico. Roma. 39-65.
Petiziol, A., Rizzo, N. 1965. Considerazioni per una psicopatologia dell marittimo. Atti della Sezione Studi. Centro Internazionale Radio Medico. Roma. 63-77.
Petiziol, A., Rizzo, N., e Bianchini, A. 1966. La condizione esistenziale dell marittimo. Atti della Sezione Studi. Centro Internazionale Radio Medico. Roma. 17-52.
Reinberg, A., Vieux, N., and Andlauer, P. 1981. Night and Shift Work, Biological and Social Aspects. (Pergamon Press, Oxford). pp. V-VII.
Roddis, L. 1950. James Lind. (Henry Schuman Inc., New York). p.27.
Tromp, S.W. 1968. Physiological changes in seamen as a result of rapid weather and climatic changes. Atti della Sezione Studi. Centro Internazionale Radio Medico. Roma. 255-259.
Wright, D., Taylor, A., Davies, D., Sluckin, W., Lee, S. and Reason, J. 1972. Skilled Performance. In "Introducing Psychology. An Experimental Approach". (Penguin Books Inc., England).

SHIP OF THE FUTURE: HUMAN PROBLEMS AND PERFORMANCE

H. Boehm
Bremen Polytechnic,
Department of Nautical Sciences
Werderstrasse 73
2800 Bremen, F.R.G.

ABSTRACT

The construction of the so-called "ship of the Future" will not only intensify most of the known specific conditions of seafaring, it will confront us with new human problems and influence the performance on board.
Models of this type of ship are part of national development research programmes in different countries.
The known "negative" factors
- long absence and separation from society and family, living together with others in a small (compulsory) group without a chance to alter the "organizational" role, the necessity of shiftwork, the environmental conditions etc. -
will be increased by new ones
- port of the future, centralization, isolation etc. -
which will change the self-realization of the seafarer.
The new conditions will not meet the motivational structure and thus lead to frustration and demotivation.
To help the seaman endure his situation without endangering his mental and psychical health, two measures can be taken:
improvements on board and compensations ashore.
Further necessary steps are
differentiated psychological accident analysis,
psychological selection and a
specific group training.

INTRODUCTION

This paper presents the state of the art not as a review but more as a sort of prevision of necessary future research. Reviews are very helpful if the development follows a continuous, unchanging and predictable path - but we need more if changes appear unexpected, "illogical" and surprisingly even to experts.

Seamanship and navigation have a very long tradition - and seafarers do like their tradition. According to this pattern of behaviour changes are reluctantly accepted. "We always have done it this way" is one of the barriers and touch-stones which innovations have to meet. There have been changes which altered the conditions of seafaring constantly - but now we are entering a dimension which whill shake the basis of the profession:

the construction of the so-called "Ship of the Future".

The realization of this ship will not only intensify most of the known specific conditions of seafaring but will confront us with new human problems and influence the performance on board.

It requires the interdisciplinary cooperation of experts with medical, psychological, physical, technical and nautical knowledge and experienced operators, masters and officers of ships to answer the following questions:

Which are the specific factors/parameters of the ship of the future?

How will these factors influence the conditions of working and living on board?

What can be done to preserve the seafarer in mental and psychical health?

The technical development has been always ahead of the human operator - what we can and must do now, is to demand that the construction of the future ship will not become a mere technical matter and that human sciences are not again reduced to an alibi- and fire-brigade-function. The seaman - different from the airman - never had the chance to start from the beginning when modern ships were constructed. Some parts of his behaviour set - very effective and useful in their time - may now hinder a quick and smooth adaptation.

IMO-RESOLUTION

The first step to improve human relationships on board internationally was taken in 1978 by the IMO-resolution (IMO = International Maritime Organization) 22:

"Human Relationships"

The Conference

Having adopted the International Convention on Standards of Training, Certification and Watchkeeping for Seafarers 1978,

Recognizing that only safe operation of the ship and its equipment but also good human relationships between the seafarers on board would greatly enhance the safety of life at sea,

Noting that the knowledge of personnel management, organization and training aboard ships is required for certification of supervisory personnel,

Recommends that this knowledge includes knowledge of basic principles of human relationships and social responsibility,

Invites all governments:

a) to establish or encourage the establishment of training programmes aimed at safeguarding good human relationships on board ships;

b) to take adequate measures to minimize any element of loneliness and isolation for crew members on board ships;

c) to ensure that crew members are sufficiently rested before commencing their duties.

MODELS OF THE "SHIP OF THE FUTURE"

Models and projects of the "Ship of the Future" are part of national development research programmes in different countries. Differentiated results or - better - tendencies can be recognized in the following projects:

1) "Ship 80" - The Netherlands
2) "Ship Operation of the Future" - Norway
3) "Ship of the Future" - West Germany
4) "Super Rationalized Ship" - Japan.

All models show the well-known similar and uniform factors: highly automated and sophisticated hard-ware and soft-ware of modern (transport) systems. They share the factor "reduced manning" too. And this to a degree which will change the human, the psychological and sociological aspects of shipboard life.

The Japanese researchers aim at a crew of 3 to 5 members (1993) and hope to eliminate all. The European conceptions tend to approximately 12-15 crew-members.

SPECIFIC PSYCHOLOGICAL AND SOCIOLOGICAL CONDITIONS

Similar for all transport systems is the fact that the operator (if he is still part of the vessel and not operating in a central-station far from it) is absent from his society, family and friends for some time. This time of absence is especially long for seafarers (sometimes six to nine months and even more). During this time, the connection with the groups they belong to is loose and this can lead to significant changes and effects.

1) The differentiated norms of modern societies are more or less regarded as unimportant and neglected.

2) The self-made standards and rules on board compel everyone to adjust his behaviour to this demand - otherwise his personal situation

will become difficult.

3) The separation from the family influences the emotional relations between the family-members and the bringing-up of the children.

4) The seaman has to live together with the other crew-members - independent from the fact whether he likes them or not. He misses - if he has no real friends aboard - the favours and privileges of friendship: confidence, veracity, relaxation, common interests, etc.

5) A ship is a sort of a "total institution" because of the unseparableness of working and living conditions. The working place, the social field and the individual sphere are close together and each part influences the others in a high degree. The consequence of this is that there is almost no chance for the single crew-member to forget his function within the working system and play another role during leisure time. His role is like an armour he cannot leave. Psychological theories agree on the importance of changing roles for the psychical hygiene. If a person is denied this chance, the consequences can be frustration and rigidity.

6) The seafarer as operator is - like all the other operators of highly automated systems - subject to the advantages and disadvantages of ergonomical changes of the working conditions. Until now psychological and ergonomical efforts aimed at the same target: to help operators. But now we have reached a point where - according to my opinion - both sciences start to fight one against the other in some situations. For example: The ergonomical science helped to develop working conditions which enable one man to do the work of many other men. This probably is a progress but it also leads to the isolation of this man - especially on board. If the operator is off duty he will find no man aboard who is willing and able to share the leisure time. This is no problem ashore - but a growing one aboard. It is the task of the psychological expert to help the operator - even against ergonomical experts.

Most of the above mentioned problems will become stronger aboard future ships - if we are not able to develop new models which are not more than elongated models of those now known.

SPECIFIC ENVIRONMENTAL CONDITIONS

The specific natural environmental conditions of ships, such as motion of the sea, climate and weather will not change even for the "Ship of the Future" - if this ship is still a surface-ship! There are ideas to construct cargo ships which "sail" beneath the surface to evade the influence of weather conditions.

An other group of environmental conditions will change on the future ship: the living quarters. Traditionally the living quarters of ships are situated according to the hierarchy of the organisational structure: those on top of the pyramid live above the others etc. Now there are plans (realized already on board of some ships) to alter this generally: all crewmembers will have their individual and social living quarters on the same deck of the superstructure. The so-called "village" - "atrium" or "tract-principle" proposes to build the individual cabins (designed like small appartments: corridor, shower/WC, sleeping- and living-room) around a central meeting place on one deck and recreation area (mess, table-tennis, cinema, smoke-room, sauna, hobby-room etc.) around a swimming pool on the next deck.

This alteration will change many of the deeply rooted patterns of behaviour and influence communication and even cooperation. Although this model seems to be reasonable and an optimum there are dissentient voices: some of the seafarers fear the loss of their individual privileges and independences - and this on both ends of the hierarchy.

One of the specific environmental conditions of seafarers is the harbour. Landfall with the prospect of a haven with all its changes and chances has been always an important part of the life at sea. The ship of the future will seldom belay its ropes in this shelter but in its counterpart: the port of the future. This port will be situated far from towns in the "desert". Even if there are (expensive) traffic-connections with the next human settlements the short time of work in the port will not allow to visit them. This will bring additional stress to the seafarer because of his inability to change his surroundings and find relaxation, entertainment and stimulations. According to this he will be confined to his isolated situation with its more or less rigid rules.

MOTIVATION AND IMAGE

The performance of an operator is directly influenced by his motivation.

Studies of the motive pattern of young seafarers regarding their vocational choice show a not so easy to define but nevertheless strong configuration which can be described with the following words: freedom, decision making, responsibility, independence, self-realization, self-assertion, adventure, knowledge of far countries and people etc.

The traditional life at sea was able to satisfy most of these motives - but it needs no prophet to predict that the ship of the future will not meet these pretensions. The consequences will be frustration and a still higher fluctuation.

When talking to young seamen about the future prospects of their profession I am always affected and startled by the strong and deep emotional rejection of the ship of the future. They do not belong to this part of the young generation which is highly interested in sophisticated automation. They do not want to be reduced to a - more or less important - part of a nearly perfect technical system. Their self-rating and evaluation is stamped in and reinforced by the "image" of the seafarer. Results of studies and analyses of self-representations of the "image" revealed constant factors:

The seaman is a man who is able to improvise. He is an improvisor because he had to become one. Very often nobody could help him and his comrades - they had to help themselves. He learnt that he was able to solve his problems and this gave self-confidence to him and the knowledge that he could find his way if needed. The privations of his profession taught him to develop a high grade of toleration of uncomfortable circumstances and to be proud of this trait.

I do not believe that improvements on board and compensations ashore will bind the "salts" - it is more probable that another group of men with other motivations will sail the future ships.

To find this group and to develop the "new" image we can compare - within certain limits - the situation of the ship-operator with that of the airplane-operator.

ORGANISATIONAL AND SOCIAL STRUCTURE

The traditional organisational structure of the crew is the hierarchical system. This principle determines nearly all working and living conditions on board. It is readily accepted in the working area but - nowadays - called in question regarding the living and private sphere. The hierarchical thinking has always strongly influenced all connections and relations between the crew-members, everyone had to accept his organisational role even during his leisure time.

There are some models and experiments to change this. One of the first steps was to transform the mess-system: from three messes (crew, petty-officers, officers) and one saloon (Captain, I. Officer, Chief

Engineer) to one mess and a so-called "dirty-mess". This alteration was readily accepted - with some restraints regarding privacy and group-cohesiveness. Another change was the introduction of the MPC, "Multi Purpose Crew": crew-members were trained to work alternately in the two departments deck and engine. The experiences seem to confirm this trial. The idea to "integrate" (responsible for the nautical and engine department) the officer is still in (sometimes passionate) discussion. Further models try to introduce common working in all departments if there is need and the specialists can't finish the task in time.

All experiments aim at a reduction of the manning - but simultaneously bring the chance to reduce the frictions and limitations which arise because of the stamping-in-effect of the organisational structure on the social structure in the rigid traditional system. But we have to take precautions: the advantages of better communication and more personal relations which arise from closer cooperation and common tasks are endangered by a reduction which follows solely the technical/ergonomical progress and possibilities. Instead of the "old" organizational structure the "new" functional structure could disturb and hinder the social life on board. We must never forget that men need more than a best constructed and equipped working place to preserve their mental and psychical health.

CENTRALIZATION

In former times ships have been relative autonomous units. Now - and more and more in the future - they are parts of a greater system: the transport-chain. And as a part each ship depends on informations coming from a central organization ashore which controls all parts.

This centralization changes the work and the image of the ship- officers: decision-making and taking the responsibility were always his main tasks. The new central system demands subordination in a high degree and with this it influences the self-realization and -evaluation and the job-satisfaction of the officer. He will assess this as a devaluation and look at himself as degraded to a mere "operator".

Already some officers left their job because of this development - they did not like their restricted and "monotonous" work with the now and then chance of subordinated decision-making. "We were not trained for this, we want to take responsibility and make our own decisions."

There is a tendency to bring back on board more "real" work - but if it is only a sort of "occupational therapy" it will demotivate still more of the highly trained officers.

Insight - obtained by job-rotation - could help tolerate some of the drawbacks of centralization.

SHIFT-WORK

The watch-system on board is one of the working conditions which did not change since the time of sailing-ships.

The "normal" (German) systems:
- a) Officer A: 0.00 - 4.00   12.00 - 16.00
     Officer B: 4.00 - 8.00   16.00 - 20.00
     Officer C: 8.00 -12.00   20.00 - 24.00
- b) Officer A: 0.00 - 6.00   12.00 - 18.00
     Officer B: 6.00 -12.00   18.00 - 24.00

It overcharges the psychical capabilities, especially if it does not alter during many months on board.

Ships work twenty-four hours and we can't stop them, but what we can do is to develop other watch-systems and minimize the negative effects by selection and compensations.

PSYCHOLOGICAL ACCIDENT ANALYSIS

Safe navigation - the prevention of collisions, strandings etc. - is no longer solely the task and responsibility of the navigator. Accidents of modern ships (with a rising share of chemical cargoes) threaten the environmental conditions of large areas. To help prevent accidents it is important to know "accident-prone" factors.

The analysis of accidents of ships has been always a difficult problem because of different reasons (no neutral witnesses, loss of documents, national and international considerations etc.). In spite of this we have to find methods and results to support the operator. The psychological analysis is one of these methods which is used successfully in accident prevention in different working conditions. It is necessary to lay stress on one point: the object of this analysis is not to find the culprit and solve insurance problems but to reveal the human factors.

If it is correct that 80 and more percent of accident causes are in the category "human factors", it is unjustifiable to do so little about it. International cooperation and exchange of methods and results are needed to develop a specific human concept of safety at sea.

ISOLATION

The reduction of manning automatically brings with it the dangers of isolation.

The single crew-member will be separated from the others not only during the long periods of the working-hours but also during the still longer periods of the non-working time.

There are some proposals to neutralize the well-known drawbacks of isolation:
- facilities to take the whole family on board,
- direct and permanent TV - communication with the family ashore,
- rebuilding of the living and recreation areas to obtain maximum communication, etc.

Probably no final satisfactory solution will be found on board. Isolation can be overcome by communication - if there is just a handful of people on board no reasonable "internal" communication is possible and "external"-communication (via TV) will have a "second-hand"- and substitute-satisfaction-quality for many seafarers.

Improvements aboard and compensations ashore can soften the effects of isolation - additionally it would be sensible to select seafarers according to their capability to withstand the consequences of isolation.

LEISURE-TIME

The useful organization of the leisure-time on board is an old problem. The specific circumstances (the ship as a relatively isolated unit, psychophysical stress of sea-motion and climate, hierarchical and racial barriers etc.) restrict the possibilities. There are a lot of proposals and even reasonable equipment (swimming-pool, table-tennis, hobby-room, library, etc.) to solve this problem, but the difficulties remain.

The ship of the future will add a new dimension: leisure-time will become solitary-time. This leads to some questions: How can one lonely man - or at the most two men - spend his time usefully over a long period? How can he find relaxation and pleasure without social contact? How will his motivation and stress-tolerances be affected if he is not able to talk to others in a relaxed atmosphere?

We can assume that the crew-members will be not overstrained neither bodily nor mentally during the working-time. Consequently it is necessary to reduce surplus energy and to compensate the stresses of the working conditions and the whole living-situation during the leisure-time. Experiences taught us that free-time activities without help soon flagged - moreover nobody cares for the equipment and the power of motivation of defective utensils is very low. Therefore it seems to be reasonable to train a so-called welfare-officer who is responsible for the facilities. We tried to introduce the training of welfare-officers in Germany, but until now withouth success - it is as far as I know no problem in Scandinavia.

Even if there is a welfare-officer on board the possibilities of solitary activities are limited. Additional to the known facilities: a fitness-room equipped with apparatusses which register the performance (as a sort of "supervisor" and "shadow-opponent"), a TV-room with direct communication with the home-programme and films as well as TV-games, a language-laboratory and a library with many books for entertainment and education.

The prospects are not encouraging! It would be reasonable to look at the self-concept and the ability of self-sufficiency as a part of the psychological selection.

COMPENSATION

One of the frequent arguments in discussions about the stress situation aboard future ships is - additional to the improvements on board - the question of compensations ashore.

There is a close connection between the time on board and that ashore. Two suggestions show the direction of possible solutions:

1) The time aboard should not be as extended as it is now. There should be a time of relief every one or two months.

2) The time aboard per year should be reduced to approximately 6 months. This does not necessarily mean 6 months holidays - the model of job-rotation should be discussed.

If the privations aboard will become half as many as we fear and can give reasons for, the costs of travelling will be justified. It is known that the performance of an operator depends on the time and chance of his recreation and regeneration.

This solution will bring with it another fundamental change for the "salts": the loss of the emotional bond to "my ship", where he felt at home because he was able to furnish his cabin according to his taste and where he knew all details and specific traits of the ship and his comrades.

The emotional involvement did help the seafarers a lot to compensate the tensions and stress of their situation.

SELECTION

One of the until now neglected necessities is the psychological selection of seafarers in most countries. Regarding the factors of the ship of the future it is without question that selection according to defined standards of professional knowledge and body health is not sufficient but must be completed by an examination of the psychical structure.

To find out which capacities are needed an exact psychological analysis of the working and living conditions - requiring inter-disciplinary cooperation - is necessary.

To minimize tension, friction and resulting frustration it would be reasonable to select the members of working groups - of crews - according to the standards of each group. All members should get the chance to help select the other members of the crew. This would strengthen the group cohesiveness and solidarity.

The importance of psychological tests is known since long in other transport systems where selection is obligatory.

TRAINING

It is without question that training with (in) simulators is a successful method to help the operator. This method predominates in the training of the performance in the man-machine-system: ship.

The same method, i.e. simulating average and exceptional cases, can be used in the man-man-system on board.

If the number of crew-members of the ship of the future is very small, it becomes a vital factor to train those people who live together in an isolated, limited situation for a relatively long time in techniques of communication, co-existence etc.

Example:

Department of Nautical Sciences, Bremen: In the optional course "Group dynamic processes - shown by the example of a group on board ship", the ability to recognize own effects in groups and to perceive conscious and unconscious occurrences within the group more clearly through greater perceptiveness is trained. The applied methods range from observation exercises to discover perceptual distortions and from models of communicating to process analyses and playing roles. The simulation of specific situations on board brings the student closer to his future professional scenery, while the experience is continuously remoulded by theoretical considerations and analyses.

This method should be developed and specified according to the conditions of future ships.

ACCIDENTS ON BOARD MERCHANT SHIPS

N. Rizzo and F. Amenta

Centro Internazionale Radiomedico ,
C.I.R.M.,Via dell'Architettura,n.41
00144 Roma , Italy

ABSTRACT

   The frequency of accidents on board merchant ships during the period 1962-1981 was studied according to data supplied by the International Radiomedical Centre (C.I.R.M.).
   In comparison with the results published in an earlier paper fewer accidents occurred among certain categories of personnel concerned with machinery (engine-room staff,mechanics , etc.). It is pointed out that there has been a particular marked increase in their number during the above mentioned period (from 13.7 to 15.6 %).
   Two principal causes are involved in the occurrence of accidents on board merchant ships. The nature of seaman's environment on board ship - characterized by the instability of the vessel,by vibrations and noise,by various conditions of microclimate,by the consequent difficulty of movement:all negative elements that cannot easily be changed even in modern ship design. The prominence of the human component of the accidents in its various aspects (care,experience,adaptation to environment , psychophysical aptitude) seems to be confirmed by the statistics compiled regarding the lenght of period of embarcation,the age of seaman and his qualification.
   It is hoped that,in the near future,crews may be recruited after having undergone not only a careful generical medical examination,but also a psychotechnical examination.

INTRODUCTION

   The study of accidents on board merchant ships is one of the principal interests of the Research Section of the International Radiomedical Centre (C.I.R.M.). This is an organization founded in 1935 and which gives medical assistance via radio to ships without a doctor on board in all the seas of the world.

   Previous publications have shown that the accidents on board merchant ships which have requested C.I.R.M. for medical assistance were 515 out 4,043 cases assisted in the five year period 1954-1958 (12.7%); and 833 out of 6,433 in the seven year period 1959-1965 (13.7 %) (see Rizzo,1959;1967).

   In the present study we wish to analyse the frequency and the type of accidents on board merchant ships which have occurred in the twenty year period 1962 - 1981.

METHODOLOGY

To carry out this research a total number of 20,135 files were examined. These related to the medical assistance by the C.I.R.M. in the period 1962-1981.

RESULTS

The results of an examination of files was that in the 20 year period C.I.R.M. was called on for a total of 3,141 cases of accidents on board. These accidents, therefore, represented 15.7% of the total number of cases treated. The highest percentage of increase was in the last four year period with an average of 19% of accidents out of the total number of cases treated (see Table 1).

TABLE 1   Frequency of accidents on board merchant ships during the period 1962 - 1981.

Year	Total cases	Accidents	%
1962	1,036	139	13.6
1963	927	119	12.8
1964	800	136	17.0
1965	916	157	17.1
1966	835	145	17.4
1967	1,007	170	16.9
1968	1,073	156	13.7
1969	1,275	164	13.9
1970	1,193	177	14.8
1971	1,191	162	13.6
1972	1,326	212	16.0
1973	1,098	156	14.2
1974	1,033	184	17.8
1975	988	166	16.8
1976	1,031	166	16.1
1977	806	142	17.6
1978	939	163	17.4
1979	846	158	18.7
1980	875	177	20.2
1981	825	163	19.8

To explain this progressive increase of accidents we need to keep constantly in the mind the dual concept man-machine (or, more generally, man-work environment) since external factors (like metereological ones) do not seem to play a decisive role in the cause of accidents. As a matter of fact, the accidents show a certain uniformity during the whole course of the year as is indicated in Fig. 1.

Having excluded external factors as decisive, we must now concentrate attention on the work environment and on the human element which must adapt itself to that environment. The particular structural characteristics and the instability of the ship, the fact that the ship is at the same time a living quarters and a work environment for the whole duration of the crossing, night watches and the need to carry out certain emergency operations, expose sailors more to accidents than workers on land. This greater risk is aggravated by the fact that crews are hardly ever homogeneous, they often change from voyage to voyage and so, for the most part, are never able to become a well organized group and to reach maximum efficiency in individual and collective safety.

In the present study particular attention is paid to the parameters which will be indicated later. These are examined especially with a view to the possibility of putting into pratice in the future measures which would be capable of reducing or preventing accidents among sailors. We have studied, in particular, the type of accidents that most frequently happen on board merchant ships, the age of those involved and their professional qualification.

The most common accidents on board are listed in Table 2. The age of sailors involved in accidents is shown in the Fig. 2 ;from this figure it can be observed that the greater part of the accidents happened to people between the ages of 21 and 50.

With regard to their qualification on board, the deck personnel and the engine room personnel were victims of accidents in almost equal measure (Table 3). This datum is easily obtained if we consider that engine room personnel, numerically inferior to the deck personnel, gave a greater contribution to the survey in question.

DISCUSSION

TABLE 2 Most common accidents on board merchant ships.

Type of accident	%
Wounds	28
Contusions	25
Fractures	12
Ocular lesions	11
Burns	9
Intoxications	6
Amputations of fingers or falanxes	5
Dislocations and sprains	3
Heat strokes	1

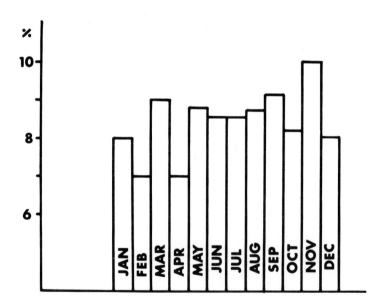

Fig. 1 Incidence of accidents on board merchant ships during the whole course of the year (1962-1981).

TABLE 3 Percentage of accidents on board merchant ships with regard to the qualification of crew's members (1962-1981).

Qualification	%
Sailors	30
Engineers	15
Officers	7
Mechanics	7
Ship boys	7
Kitchen personnel	6
Boat swains	5
Electricians	3
Stewards	2
Others	18

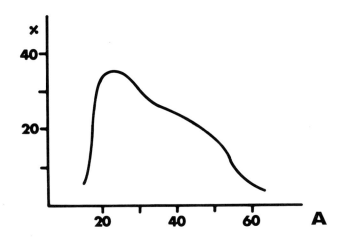

Fig. 2 Age (A) of sailors involved in accidents on board merchant ships (1962 - 1981).

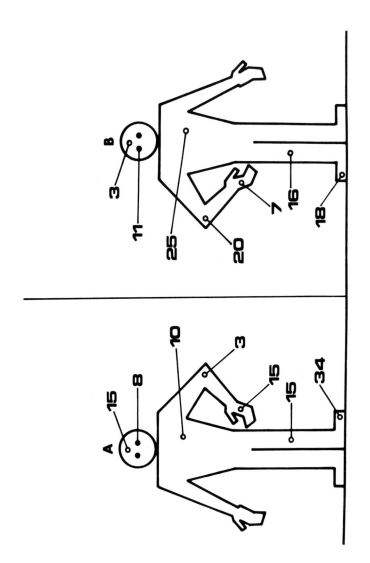

Fig. 3 Percentage distributions of injuries by part of body affected. A-Land workers (data from Accident Prevention, 1978) B-Sailors

To analyze better the incidence of certain types of accidents among certain categories of personnel on board, we need to keep in mind the principal factors of accident risk on board.

The agents of professional risk on board are various: a) they are physical (high temperatures in the engine rooms); b) they are chemical (paint vapours in enclosed areas, gasses from loading, almost always in the case of tankers, of crude oil); c) they are mechanical (parts of engines in movement, winches, cranes, capstans, steel cables and ropes). But as well as these specific danger factors, we must also consider the general ones connected with the instability and especially with the particular structural characteristics of the ship: the slope of the decks, the narrowness of the rooms, the steepness of the stairs, obstacles littered everywhere above and below deck, floors often slippery because they are wet or dirty with grease. So, accidents which are typical of life on board ship happen (heat strokes, intoxications with gasses or vapours of crude oil) as well as others which are similar to those met with in industry and in workshops on land even though they are conditioned in an important way by the characteristic of work environment (see Fig. 3).

In relation to what has been stated above it is easy to see how the pathology encountered has relationship to the different activities which the various members of the crew engage in and the relative differences in the environments in which they have to work. The deck personnel, in fact, give a relatively higher contribution to bruises, wounds and occular lesions caused by foreign bodies, while among the engine room personnel burns, heat strokes and accidental amputations of terminal phalanxes are more common.

Comparing the results obtained from an examination of the cases in the present study with those published elsewhere (Rizzo, 1959; 1967), it is possible to observe how the number of wounds, contusions and occular lesions have increased in the course of recent years, while the amputation of fingers and terminal phalanxes have reasonably diminished, and, in a consistent way, heat strokes and occular lesions caused by foreign bodies. The remaining pathology remains at more or less constant levels.

While it is difficult to interpret the increase of wounds

and contusions - which are the consequences of accidents of the most varied kind - one can consider, with regard to the greater number of occular burns caused essentially by acids and alkalis, the increased use of these chemical substances by personnel who clean kitchens, latrines and floors. Shipboys and cabin-boys in particular have suffered from accidents of this kind.

With regard to the steep decrease of heat strokes, which happen especially in engine rooms, one must consider both the improved environmental conditions and the presence of younger and more physically sound people. In fact, it is favoured by organic causes (heart, liver and kidney diseases) especially if they are associated with a non-rational diet and the use of unsuitable clothing.

For the lesser incidence of occular lesions caused by foreign bodies as well as accidental amputations of terminal phalanxes, the employment of non-qualified personnel in services different from the traditional ones (like, for example, scaling) and the availability of expert and specialized workers for the operation and maintenance of the modern machinery is not to be excluded. Finally, a more prudent work control by the Captain and his Officers, who are nowadays more sensitive to the problems of the physical well-being of their subordinates, should not be underestimated as well as the putting into practice of the relevant safety measures.

From the study we have carried out on the frequency of accidents on board, direct and indirect causes of these events seem to emerge.

Among the first, let us recall, the duties carried out on board, the particular environmental conditions, the instability of the ship, the age and somatic constitution of the individuals.

Let us analyse the second in a more detailed way and deal with the nature of each single lesion.

Lacerated and contused wounds, caused by a cut or a point, affected the upper limbs, the head, the lower limbs and the trunk. They were caused by activity in the engine room and in the workshop; by steel cables and winches; by work on deck; by falls in the engine rooms, on the upper bridges, in the holds and on stairs between the various levels; by nails, knives, broken glass and the fall of weights.

Contusions - more numerous in the trunk and, successively, in the head, the lower limbs and the upper limbs - are the result of falls or collisions with metallic structures both on deck and in the engine room; by slipping on stairs; and by the fall of weights.

Fractures - in the first place, those of the lower limbs followed by the upper limbs, the rib cage and finally the head can be attributed to falls in the holds or hatchways (from a height of, generally, more than 3 metres), on decks, in engine rooms and from stairs; to parts of machines in movement; to steel cables and chains during mooring and unmooring operations and to the fall of weights.

Burns - almost exclusively concern the uncovered parts, the upper limbs and thorax - are caused especially by jets of water or boiling steam (because of a fracture of the level indicator tubes or the explosion of boilers or relative valves or oversleeves) and, in lesser measure, by chemical substances (acids and alkalis); by flame returns of the carter motor during the lighting of the boilers; by boiling liquids such as soup or oil.

Ocular lesions by foreign bodies were caused by iron splinters (during work at the electric drill, at the lathe, at the grindstone, at the rotary-brush polisher) or by rust splinters (while scaling) or by coal particles (in the boiler rooms).

Accidental amputations of fingers and terminal phalanxes took place in the engine rooms (during work around the pistons of the servomotors of the rudder, at lubrification belts, at valves of steam turbines, at motor pumps); in the workshop (with lathes, saws, vertical drills and circular wheels). The crushing of hands took place as a result of the violent closing of tank doors, of gratings and of boiler, fridge and metal-locker doors.

Acute gas intoxications - involving in a particular way the crews of oil tankers - were caused by crude oil vapours (because of damage to the principal ventilation collectors of the tanks); by petrol (while the tanks were being cleaned); by paratoluidine, tetrachloride and trichloroethane; by ammonia (for work on refrigeration plants).

Occular burns originated from splashes of caustic potash (while the work of checking the battery mixture was being carried out); of tetrachloride (used for cleaning the motors); of muriatic acid and of other chemical products like paint solvents, insecticides, sulphuric acid and mercuric oxide.

80% of the cases of heat stroke happened in engine rooms and the rest in the kitchen.

Dislocations and sprains which affected in particular the shoulder and thereafter the elbow, the ankle and the knee were all caused by slipping on various levels or down stairs.

From an examination of the nature of the lesions we can understand how some of these are closely connected with the duties carried out by members of the crew and with the environment in wich they must work. On board, in fact, the importance of the interdependence man-machine (including the surrounding structures too in that dual concept) must be emphasized more than ever before in the dynamics of accidents.

Thus, many accidents which happen on deck (wounds, contusions, fractures) can be attributed to lack of foresight in avoiding the obstacles that accumulate there; to a poor knowledge of how winches and cranes work; and to mistaken manoeuvres in mooring and unmooring.

Not using the hand-rail and the presence of grease or oil on the metal steps favour falls from stairs which themselves are sometimes not easily accessible or poorly lit. Often the hatch-openings - for the movement of persons and goods in the vertical sense between the various levels of the ship - are not closed by gratings (wooden or metal grates). The same can be said for the hatches - openings in the lower decks - which should be surrounded by iron fences of a height of more than 5 - 10 cm from the ground.

In the engine room - where the necessity to concentrate the greatest power in the least possible space, the complexity of the equipment and the high temperatures contribute to creating particular conditions of work risk - the personnel do not always try to prevent even small drippings of fuel on the floor or to avoid oil patches on the stairs or on the protective grilles or to put aside oily rags or woolwaste so as to limit the danger of slipping or of frequent falls.

The accidental amputation of fingers and terminal phalanxes often comes about as a consequence of crushing by the violent and sudden closure of metal wings and doors - either hinged or sliding - without having secured them with the proper stop system or during manoeuvres at the hatches without making use of the handles with which they are provided. In small mechanical jobs, failure to use certain instruments or solid means of support cause this kind of accident because of the imprecision of the blow or the instability of the piece, held manually, upon which work is being done. This holds good for engine room personnel too since they prefer to grease the various machines by hand rather than use automatic lubricators.

Ocular lesions caused by foreign bodies are to be attributed to carelessness in particular. As a matter of fact, with the objection that glasses are awkward (those of mica, for example, are simple and handy) the scalers or those in the workshop (beaters and welders) avoid using these efficacious means of protection because they don't want to undergo an initial period of discomfort.

Burns, as had already been said, concern kitchen personnel and especially engine room personnel. In boiler rooms explosions or bursts of tubes, valves and cylinders often occur because of the inadequate checking of the safety equipment (pressure gauges which indicate the level), because of the lack of general maintenance or the failure to check particular electromechanical devices.

In order to avoid heat strokes - which are favoured by individual factors such as age, constitution, not being used to particular environmental conditions as well as by dietetic excesses - it would be necessary to maintain the ventilation system of the places concerned (kitchens and engine rooms) fully efficient especially when sailing in tropical or sub-tropical climates.

Acute gas intoxications take place aboard oil tankers or on other merchant ships among sailors engaged in jobs which have to be done in closed and confined spaces, in the holds or in the double bottoms, as well as among refrigerator personnel.

On tankers, if the supervision of the tight closure of the tanks is scrupulous and the control of the refrigeration plants and the artificial ventilation - to take away the gas by aspiration systems - is continuous, then there shouldn't be any accidents which depend on the cargo. People have complained about cases of acute intoxication during the work of cleaning the tanks of petroleum residues or diesel oil caused by insufficient natural or artificial ventilation (sometimes because of the wrong disposition of the collector tubes or wind sleeves) or by the non-use of special filtre masks. The complete degassing of the areas - which is carried out every day by powerful rotating automatic irrigation machines and completed by the cleaning of the tanks with non-caustic chemical solvents which require the minimum use of workmen - should be ascertained not empirically but by suitable exposure meters. The danger of asphyxiation- either because of the lack of oxygen or because of the presence of unbreatheable or poisonous vapours or gasses - exists in the ballast holds, holds and double bottoms. They are caused by poor ventilation and inadequate cleaning of the places in question.

The intoxications which affect refrigerator personnel are sometimes to be attributed to the inaccurate checking of the state of the tubes and valves through which the gasses used for the operation of the machines pass (ammonia, methyl chloride, sulphur dioxide), especially where recourse is not had to leak detectors (with chemical reaction, with mixture combustion and with infra-red rays).

To cope with the unfavourable environmental factors, much has been done on more modern ships about the size of the rooms and of the access to them, systems of lighting and artificial ventilation; the conditioning of the various environments; the segregation of the moving mechanical parts. But despite the progress of naval engineering, certain work risks connected with particular construction requirements cannot, unfortunately, be completely eliminated. Thus, for example, the partitions of the tight closure doors and of the doors above and below deck constitute an obstacle at floor level; the steepness of the stairs; the hindrance created by the various pieces of equipment on deck; the narrowness of some rooms; the noise and the protruding metallic structures in the engine room.

So, it is necessary to consider at the same time in all its importance the human component of accidents under its various aspects (attention, experience, apprenticeship, specific ability, psychophysical attitude). We cannot, in fact, forget that even if the conditions of life and work aboard merchant ships have undergone remarkable improvements, the danger of accidents has not diminished. Rather, the new equipment and modern machinery require the personnel to have considerable psycho-intellective capacity and precise specialist notions. This requirement has been partly met by the recruiting of more qualified workers which has, in addition, favoured that rejuvenation of crews which was mentioned above. However, besides the preferred age and technological knowledge, sailors need certain psycho-sensorial and physical qualities. Many dangerous situations, in fact, require a speed of reflexes and a mastery of movement which only individuals used to fatigue and in normal psychological condition are able to neutralize.

Finally, experience plays a not unimportant part. It permits the individual to weigh up adequately the exact terms of the work risk. In fact, among sailors under 25 accidents like ocular burns and amputations of the phalanxes predominate. These are probably to be attributed to scarse attention in carrying out a task and neglect of the elementary norms of personal safety. Among older sailors, however, fractures, dislocations and heat strokes are more common. These are presumably to be placed in relation to a poorer capacity for adaptation of the organism and to the slowing down of reflexes. These phenomena are particularly evident with advancing age.

But in the light of the data explained and discussed above, how is it possible to lessen the number of accidents on board, a number which in spite of ever greater technological progress, has shown a particulary marked increase in recent years?

Much has already been done to prevent an intrinsic situation of particular risk by recruiting younger and more qualified personnel and by the entrance into service of new ships constructed according to the latest discoveries of naval engineering. However, it is inconceiveable that the environmental conditions on board can be radically modified. These conditions are the instability of the ship, vibrations, different microclimates (deck and engine room), difficulty of movement, rapid climatic variations. All of these elements have a negative influence on the physical efficiency of the crew.

It is, therefore, on the human factors of accidents that attention and study must be concentrated. So, for example, besides professional capacity, the sailors should be required to know the safety regulations; both the general ones (connected, that is, with the structure of the ship) and the specific ones (concerned with the operation and maintenance of the various machines and equipment).

In the light of the new requirements different countries have already set up commissions with the task of examining and revising the recruitment system of sailors and of laying down severe medico-scientific criteria for it. If technical progress has freed man, in large part, from physical fatigue and from the danger of handling work-tools, it has also obliged him to follow the perfect rhythm of the machine which requires of those who use it precise intellectual and sensorial capacities.

It is the sum total, therefore, of measures of a technical, medical and educative order which will be able to determine the success or the failure of any action against accidents. For this reason, a closer collaboration between designers, builders, hygienists and doctors would be very desireable in order to study the possibility of adapting the requirements of naval engineering to the improvement of the living and working conditions on board. Just as useful would be a more confident relationship between the competent authorities, bodies and specialized Institutes, employers and employees, in the effort to overcome the fatalistic idea of the inevitability of accidents on board merchant ships.

REFERENCES

International Labour Office.1978. Accident Prevention.9th
    impression. (Geneva).
Rizzo,N.1959. Gli infortuni e loro prevenzione a bordo delle
    navi mercantili. Atti Centro Studi C.I.R.M. (Roma).
Rizzo,N.1967. Studio clinico-statistico sugli infortuni a bordo di navi mercantili. Atti della Sezione Studi del C.I.
    R.M. (Roma).

SLEEP-DATA SAMPLED FROM THE CREW OF A MERCHANT MARINE SHIP

P. Knauth[+], R. Condon[++], F. Klimmer[+],
W.P. Colquhon[++], H. Hermann[+++] and J. Rutenfranz[+]

[+]   Institut für Arbeitsphysiologie an der Universität Dortmund, Ardeystr. 67, D-4600 Dortmund 1, FRG
[++] MRC Perceptual and Cognitive Performance Unit, Laboratory of Experimental Psychology, University of Sussex, Brighton, Susses, BN1 9QG, U.K.
[+++] Esso AG, Kapstadtring 2, D-2000 Hamburg 60, FRG

ABSTRACT

The effects on sleep of a permanent watchkeeping schedule on board a merchant ship were investigated by a time-budget study.

MATERIALS AND METHODS

The investigation took place on board a 250,000 tons oil tanker, the "Esso Europa", transporting crude oil from the Persian Gulf to Rotterdam and Milford Haven. The research team joined the ship at Tenerife when the tanker slowed down to exchange crew members.

The crew comprised 24 members with a further officer taken on at Rotterdam. 12 of the crew volunteered to take part in the study (Tab. 1). The study required the subjects to complete performance tests, make subjective ratings, and note their oral temperatures at 4-hourly intervals. They were also asked to keep an account of the hours they spent working and sleeping. In addition to this, 6 subjects (A-F) agreed to deliver 4-hourly urine samples on specific days during the study, and 3 officers (A, B and C) co-operated further by wearing a heart rate recorder on these same days, allowing their watch periods to be monitored for events, and permitting their movements to be recorded.

TABLE 1 Details of subjects

Subject	Position	Shift	Age
A	Officer	00-04/12-16	39
B	Officer	04-08/16-20	33
C	Captain	days*	40
D	AB	08-12/20-24	41
E	Engineer	days**, 00-04/12-16	44
F	Asst. Engineer	08-12/20-24	28
G	Officer	08-12/20-24	29
H	Steward	days	25
I	AB	04-08/16-20	36
K	AB	days	33
L	AB	days, 04-08/16-20	37
M	Electrician	days	32

* 'days' is used throughout to indicate that the subject concerned generally worked an 8-hour day and slept during the night as would be typical of normal daytime workers on land

** Where two sets of work schedule are mentioned, the subject in question changed from one to the other during the course of the study

Table 2 illustrates the watchkeeping schedule in use on board the Esso Europa; this schedule is also known as the "4 on - 8 off schedule". The off-duty period from 0000 to 0800 approximates the sleeping period of a normal dayworking population. One of the three watchkeeping "crews" can sleep during the whole of these 8 hours every night, but the other two "crews" can only ever sleep for 4 of these particular hours.

TABLE 2   Watchkeeping schedule on board the Esso Europa

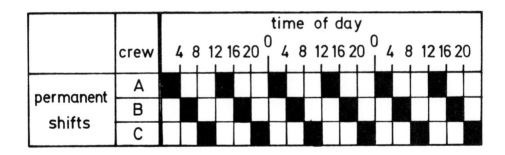

We studied the sleeping behaviour, i. e. the duration of all sleeps, and the positioning of those sleeps within the 24-hour period, of the three watchkeeping crews and of a control group without watchkeeping duties, by means of a "time-budget" method. We have previously used this method in non-marine studies to investigate the behaviour of about 1800 shiftworkers (e. g. Knauth and Rutenfranz, 1972; Rutenfranz et al., 1974; Knauth et al., 1975; Landau et al., 1978; Rutenfranz et al., 1980; Knauth et al., 1980).

The 12 volunteers on board the Esso Europa were asked to fill out special diaries during the voyage. The diary had one page for each day. On time axes the volunteers had to mark the start and the end of each of the following "activities": "watchkeeping", "other duties" and "sleep".

After each sleep the volunteers were also asked to rate its quality on a 10 cm line with the extremes "very bad" and "very good". A total of 130 such completed diaries days were obtained.

For testing differences of sleep length between the groups the t-test was used. Differences between the frequency distributions of subjective ratings of sleep quality were tested by a χ^2-test for contingency tables (Sachs 1978, pp. 366-368).

RESULTS

In Fig. 1 sleep fractions of both the dayworkers and the three watchkeeping crews on board the Esso Europa are shown. The data of the dayworkers also include days off on board. The position of the night sleep in the two lower graphs in Fig. 1 seems to be more similar to that of the dayworkers sleep than does the night sleep of the crew with watchkeeping from 0000 to 0400 (and 1200 to 1600). This latter crew had the shortest average duration of total sleep per 24 hours (6.5 hours) compared with 7.4 hours (crew: 4-8, 16-20), 7.2 hours (crew: 8-12, 20-24) and 7.0 (dayworkers on board). However, the differences in sleep duration between the 4 groups were not significant.

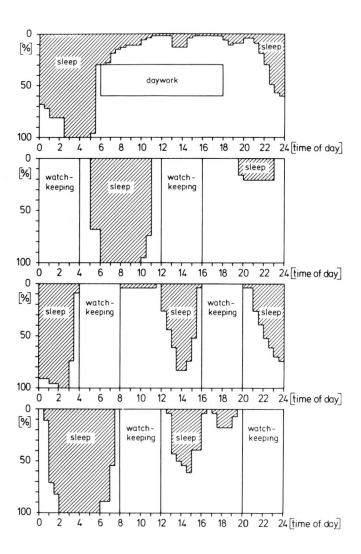

Fig. 1 Sleep fractions over the 24-h period sampled from the dayworkers and the three watchkeeping crews.

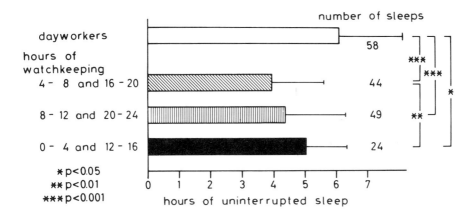

Fig. 2 Average duration of single sleep episodes of the dayworkers and the three watchkeeping crews.

In contrast to this finding, the average duration of <u>single sleep episodes</u> for the three watchkeeping groups (see Fig. 2) indicates that sleep in these groups was more fragmented than in the group of dayworkers on board. Furthermore, the subjective ratings of the sleep quality of the two watchkeeping crews with the shortes average durations of uninterrupted sleep (crew: 4-8, 16-20; and crew: 8-12, 20-24) were not as good as the ratings of the two other groups, as can be seen in Fig. 3.

Fig. 4 shows the frequency distribution of the hours of uninterrupted sleep of the three watchkeeping crews considered as a single group. Though there is a peak at 6 - 6.5 hours the majority of sleeps were shorter than this. Fig. 5 shows the distribution of intersleep intervals in the watchkeepers. If a person on the 4 on - 8 off - watchkeeping system is awake for 12 hours he has to stay awake for another four hours (next watch). Therefore the frequency distribution of intersleep intervals has a gap between 12 and 16 hours. In a normal dayworking population the sleep intervals center around 16 hours.

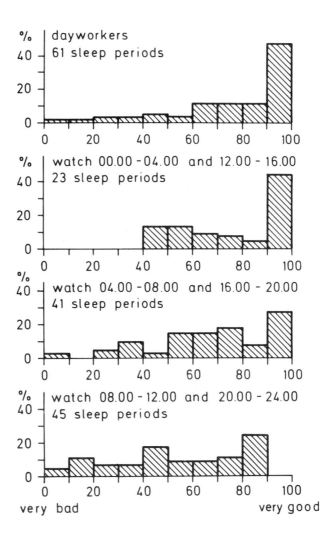

Fig. 3 Subjective ratings of sleep quality of the dayworkers and the three watchkeeping crews ($\chi^2 = 28.64$, df = 3, $p < 0.001$).

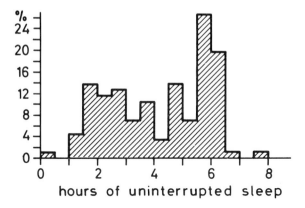

Fig. 4 Frequency distribution of the durations of sleep episodes of the watchkeeping crews.

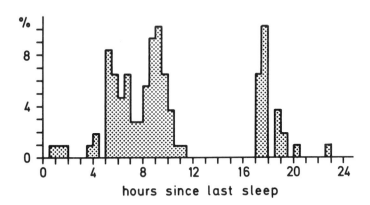

Fig. 5 Frequency distribution of intersleep intervals of the watchkeeping crews.

SUMMARY

On board a merchant marine ship, the sleeping behaviour of the watchkeeping crews with a "4 on - 8 off" schedule was found to differ from that of a dayworking group on board the same ship.

The fixed watchkeeping system allowed only a part of the total sleep to occur between the "normal" hours of 23.00 and 07.00, and was shown to reduce the duration of single sleep episodes. The sleep quality ratings of the two crews with the shortest average sleep episodes were lower than those both of the crew with the relatively longest average sleep episode, and of the dayworking group.

This study demonstrates that the relatively simple time-budget method provides a successful means of quantifying the effects of organisation of working time on sleep on board ships, as it has already been shown to do in the case of shiftworking in land-based industries.

REFERENCES

Knauth, P. und Rutenfranz, J. 1972. Untersuchungen über die Beziehungen zwischen Schichtform und Tagesaufteilung. Int. Arch. Arbeitsmed. $\underline{30}$, 173-191

Knauth, P., Landau, K., Dröge, C., Schwitteck, M., Widynski, M. and Rutenfranz, J. 1980. Duration of sleep depending on the type of shift work. Int. Arch. Occup. Environ. Health $\underline{46}$, 167-177.

Knauth, P., Romahn, R., Kuhlmann, W., Klimmer, F. und Rutenfranz, J. 1975. Analyse der Verteilung verschiedener Tageselemente bei kontinuierlicher Arbeitsweise mit Hilfe von "time-budget-studies". In: Nachreiner et. al.: Schichtarbeit bei kontinuierlicher Produktion. Arbeitssoziologische, sozialpsychologische, arbeitspsychologische und arbeitsmedizinische Aspekte. pp. 17-82, Forschungsberichte der Bundesanstalt für Arbeitsschutz und Unfallforschung, Dortmund, Nr. 141, Wirtschaftsverlag Nordwest, Wilhelmshaven.

Landau, K., Knauth, P., Rohmert, W. und Rutenfranz, J. 1978. Untersuchungen zur Tagesaufteilung und Schichtform in Rechenzentren. In: H. Loskant (Ed.): Möglichkeiten und Grenzen des Biological Monitoring. Arbeitsmedizinische Probleme des Dienstleistungsgewerbes. Bericht über die 18. Jahrestagung der Deutschen Gesellschaft für Arbeitsmedizin e.V. in Frankfurt/M.-Höchst, 24.-27.5.1978, pp. 299-310, Gentner, Stuttgart.

Rutenfranz, J., Knauth, P., Hildebrandt, G. und Rohmert, W. 1974. Nacht- und Schichtarbeit von Triebfahrzeugführern. 1. Mitteilung: Untersuchungen über die tägliche Arbeitszeit und die übrige Tagesaufteilung. Int. Arch. Arbeitsmed. $\underline{32}$, 243-259.

Rutenfranz, J., Knauth, P., Küpper, R., Romahn, R. and Ernst, G. 1980. Pilot project on physiological and psychological consequence of shift work in some branches of the services sector. In: European Foundation for the Improvement of Living and Working Conditions (Ed.): Effects of shiftwork on health, social life and family life of workers. Dublin.

Sachs, L. 1978. Angewandte Statistik, Springer-Verlag Berlin, Heidelberg, New York.

TRANSPORT OPERATORS AS RESPONSIBLE PERSONS IN STRESSFUL SITUATIONS

Paul Branton

22 Kings Gardens, London NW6 4PU, U.K.

ABSTRACT

The reliability of a transport operator depends on his ability to control not only the vehicle but also his own states of awareness and alertness. As stress is frequently reported in monotonous situations when mental work load appears to be low, doubts about the appropriateness of current theories arise. Usually, the operator is modelled as an 'engine', as a physiological organism, as information processor or as basically differentiated by personality type. Each of these models conflicts with the reality of transport, where practical necessity treats the human as a responsible person. In field studies, brief episodes of absent-mindedness have been found which normally recur cyclically at about 100 min. intervals, akin to the dream stages of REM sleep. When suddenly emerging from these lapses, 'minipanics' have been reported. It is thought that the sudden consciousness of having lost control over oneself, whilst being fully responsible for the lives of others, may be the major source of stress in transport. A model of the operator as a goal-directed controller, autonomous, yet liable to natural fluctuations in consciousness, is presented. Some counter-measures to maintain the arousal level are suggested.

INTRODUCTION

Reliability is often regarded as the inverse of stress: the less stress a person experiences, the more reliable he is thought to become. This relationship does not hold true in those transport situations where operators are eminently reliable and yet are under very real stress as they are necessarily responsible for the lives of others. No one - and no mechanical or other contrivance - can relieve them entirely of that responsibility. The charge of responsibility tacitly assumes an autonomous, selfcontrolled person, which is right in principle, but which raises problems in the real-life situation because no one can be certain that the operator is always under full, effective self-control. The uncertainty arises from the common knowledge that those body and mental functions which affect awareness and alertness vary and are not wholly controllable.

The thesis of this paper is that any improvement toward

The complexity of the problem is demonstrated when explanations are sought from a simple analogy taken from the 'engine' model: mental 'load' as a source of stress. Many studies have confirmed the task related differences between the response required of pilots and, say, train drivers. The pilot's task is characterised by relatively short but intense periods of heavy mental load during take-off and touch-down with long periods of relative inaction in between; whereas the train driver must constantly monitor his progress closely and frequently intervene actively throughout the journey. The pilot has too much to do in too little time; the train driver has relatively little to do, other than watch, most of the time. If there is stress, he suffers from underload, where the pilot does so from overload. The difference has received little attention from researchers, probably because of the more dramatic situations of high speed travel. Public transport long distance coach drivers and train drivers have been rather neglected.

Compared with research in other fields, the traindriver's task is akin to the less glamorous process controllers in refineries (Bainbridge, 1978) or anaesthetists at work (Branton and Oborne, 1979). As long as all is going well, a low activity level combines with long lag between action and feedback of result. The problem common to these operators is one of knowing whether all is indeed well or not.

THE ENGINE MODEL

Even on the engine model alone, the difference between train driver and pilot demands different methods of fact finding, different measurements and different remedies. Direct measurements of energy expenditure by the usual metabolic parameters are not likely to be sensitive enough to bring out that difference. Heart rate measurement is indeed shown to reflect the intuitively judged high mental load of the pilot (Nicholson, 1970) and the racing car driver. However, in the low activity, long lag, underload case, heart rate does not reveal directly any response which could be related to stress and the 'mental exhaustion' widely and totally convincingly reported by opera-

self-determined behaviour which can be wrested from the interplay between man and environment will help to increase overall safety and reliability. The operator as a self-activated being in a mixed man-machine system is, however, not yet well understood. The usual theoretical models of him are not exact enough to bring about significantly greater confidence. They are often internally inconsistent and not wide enough to grasp the generality of real life. Four typical models may be characterized as influencing recent thinking on stress:

1) the 'engine' model - energetic - mechanistic - input/output
2) the 'organic' model - physiological - reflex/response - hormone balance
3) the 'personality differences' model - attitudinal - life style
4) the 'information processing' model - consciously cognitive decisions

These models will be considered critically in general and then with particular regard to the methods used to study them. Certain practical conclusions are drawn and an alternative model is briefly presented of the operator as a purposive vehicle controller, essentially autonomous but always subject to naturally fluctuating levels of awareness and adaptation.

THE CONCEPT OF MENTAL LOAD

In transport operations man is said to pit himself against physical nature. But engineers and designers 'adapt' that nature to their purposes and so create an artifical environment which stretches the operator to the limits of his own natural, adaptive capacities. Especially in modern vehicles, the operator as driver and pilot neither produces nor expends much physical energy, yet he controls large energy resources very accurately. If he is to be treated as a reliable controller, it is appropriate to view him as pitting himself against the limitations within his own nature. The struggle is now mainly a mental one - between keeping calm and keeping awake - a matter of the limits of reasonable self-control.

tors. Could that be because the heart is in the first place a pump and so an energy supplier and only indirectly an index of mental events? There is no need to enter here into a critique of the 'energetic' side of the engine model since this has been done recently and very effectively by Sanders (1981) among others. Unfortunately that does not go far enough. Perhaps the inadequacy of language to express subtle distinctions is at fault; one easily overlooks that any model is at best only an analogy and not the object itself or concept to be desribed or explained.

Similarly, 'mechanistic' models fail to satisfy most of those who deal with the actual operators day by day Naturalistic description of events by practitioners, as distinct from theoretical approaches to vehicle guidance, persistently reveals a dynamic mental control activity of such flexibility and immediate adaptability to suddenly changing circumstance that any automatisms and mechanisms become mere figures of speech. In mental functions, nowadays generally called 'cognitive', it is arguable that the term 'mechanism' denotes a mythical entity. There is nothing to be found in the pilot or the driver which functions strictly automatically as in a machine or in a computer. A case can be made that the term 'faculty' would be less misleading.

This is not to say that models have no use at all, just that researchers must not 'adapt' to them and confound them with the real thing. Such is also the case with the cybernetic model, a subset of the engine model. Ever since Ross Ashby, their explanatory power has been used to illustrate, for instance, the generality of the biological tendency towards homeostasis, a concept resorted to in explaining human stress response. The evidence is used to show the ever available capacity to compensate for 'adverse' environmental conditions. When and how the system decides what is adverse and what is not, still remains mythical, or is shifted to a higher level of abstraction. At the level of most frequent usage, to discard these kinds of mechanisms would be a gain in scientific rigour.

Nevertheless, the cybernetic model has the great historical merit that it overcame the static view of the person as just sitting there passively to wait for a stimulus before it responded. The model favours the dynamic view of continual internal process, where a stimulus must 'fight its way into the system'. Furthermore, its considerable heuristic calue is that it demonstrated how a weak David of a man can control a disproportionately larger Goliath of a system with an infinitesimally small amount of energy.

Even the sophisticated cybernetic models break down at the practical point at which it becomes obvious to all that the human operator has a mind of his own and is essentially and indubitably an open system. If the system of the human operator is open in real life outside the laboratory, that open-endedness may turn out to be a strength when compared with the rigidity of an automatic, mechanistic device.

THE ORGANIC MODEL

The physiological phenomena of stress need hardly be elaborated here; they are far better known than the mental events antecedent or concurrent. This assumes that at least one primary source of stress is 'in the mind' and, if true, takes the present discussion deliberately to a stage before that of the more descriptive than explanatory models of Selye, Levi or Frankenhaeuser. The facts established by the physiologists are not made any less true by critique, only their interpretation will be affected. For instance, homeostasis, as expressed in physiological balance is, in fact, often overridden by the kind of social responsibility of pilots and drivers, mentioned at the outset. Really effective remedies for this kind of stress have still to be found. Maybe some effort should now be spent on seeking solutions at a higher level of abstraction in metatheory. As far as the transport operator's case goes, Cannon's 'fight or flight' schema should now be given a decent burial.

On closer analysis physiological models are often a strange mixture. They still present basically a mechanistic, stimulus-response approach, looking for probable external 'causes' of stress where there are obviously multiple causes, some of them internal in origin. Not merely do they look for causal explanation, but tacitly assume a deterministic succession of events directly observable by outsiders. At the same time, physiologists happily accept the possibility of voluntary movements and distinguish them form involuntary ones. They thus rush in where psychological angels fear to tread. The possibility of there being an indeterminate or self-determined 'system' at work is thus just as tenable - even if difficult to contemplate because it is unfamiliar. The assumption of indeterminism in human behaviour introduces, no doubt, an element of uncertainty into the discussion, but if physicists and philosophers are able to live with this uncertainty so will human factors researchers.

PERSONALITY DIFFERENCES

The evidence of physiology almost inevitably leads to the third typical set of theoretical models: stress as the effect of differences in personality. Whether or not physiological stress syndromes present themselves after prolonged exposure to transport situations would then depend on whether a person is, say, of Type A or B (Friedman et al., 1974). In this case, a presumed primary causal factor becomes active only when stress becomes 'harmful'. How the body (in its wisdom) 'knows' when that is the case and communicates it to the mind remains an important question for the practitioner.

In the last analysis, one may distinguish a strong and a weak version of the influence of personality upon stress. The strong regards the transport operator's whole personality so entrenched as to be strictly unchangeable. Any attempt at modifying his behaviour would then not only be in vain but also dangerous, as in frequent cases stress appears to result in reversion to basic type. Insights into these basic differences, if true, would only be of practical use if individuals were definitely identifiable, by means of generally acceptable tests, and then only if job selection is feasible on scientific as well as on social grounds. Remembering the controversies over selection by intelligence tests, the prognosis for this approach is not bright.

The weak version of the personality approach to operator stress is supported by the common experience that there are no clear-cut types but an infinitely graded variety of personalities. Moreover, behaviour under stress is not totally invariant within the person from one time to another. If even slight modification of behaviour is possible, it must be attempted for the sake of greater safety and reliability.

One recent development in this area is work on 'locus of control'. It is worthy of special mention as it may lead to a promising growing point. Although recent authors (Rotter, 1976; Lefcourt, 1966) treat the stress problem as one of attitudes or beliefs in the social context of work, they all share the tacit assumption that control over an operator's fate could possibly be in his own hands. The idea of 'perceiving the locus of control within the person' necessarily implies the kind of autonomy pursued here, albeit at a level of actual individual (mental) function, whether or not it is merely imagined. Whether perceived or not, whether believed to be within the person or not, the ultimate control over stress effects on behaviour must be presumed to rest with the person. Who else moves his hand on the joystick or his foot on the brake? Control of an operator's action, not least in the transport context, cannot mean anything other than that a discrete action must be decided and carried out by the individuum and no one else. That is the common meaning of individual responsibility in practice. It is applied here explicitly to the initiation of even the smallest muscular movement to control a vehicle. Outsiders, such as engineers, designers and managers, may be able to help or hinder this goal.

MAN AS INFORMATION PROCESSOR

Because human presence makes systems inherently open-ended, situational uncertainty has been identified as a source of stress. Quantitative measures of information, derived originally by communications engineers, have given good service - in the controllable 'closed-loop' laboratory. Yet the same experiments also showed up the almost infinitely extendable capacity of man to process successfully ever larger 'chunks' of 'bits' (Crossman, 1955). One aspect of the model is that it assumes man to have only a 'limited channel capacity'. To be practically useful, reliable measurement would be essential because interfaces need to be designed so that they deliver just the right amount and range of information as will reduce the operator's uncertainty to levels at which he can cope safely. Unfortunately, real-life cases of intuitively obvious overload, e.g. air traffic controllers, have so far defied accurate measurement.

In fact, the transport operator uses not only whatever information the design engineers care to present to him. Branton (1978) showed in the relatively simple case of the train driver that he greatly depends on extraneous information, both currently perceived and recalled from the past. The real process is genuine guesswork, intelligent anticipation, forecasting of system states, all actions in the future tense. Unless one is able to specify this information, the value of conceptualizing the transport operator as information processor is limited. This does not detract from the virtues of that model; it broke away from behaviourism altogether and highlighted the role of uncertainty in the control of behaviour.

CONCEPTUAL LIMITATIONS TO ADAPTATION

The phenomena under observation are boredom, monotony and sensory deprivation (cf. O'Hanlon, 1981). It is noteworthy that all these are closely related to those fluctuations of consciousness or awareness which are manifestations of the general function of adaptation, the familiar and ubiquitous faculty of all sense organs. Since organs do not directly perceive objects, but immediately merely report changes in their surroundings, they 'adapt to' invariances in the rate of such change. From the present viewpoint, therefore, regularity and familiarity of stimulation are the enemies of the senses.

ADAPTATION IMPLIES VALUE JUDGMENTS

Adaptation is not an all-or-nothing function but varies infinitely in degree; though this is virtually impossible to measure, the slightest change evokes response. However, unless there is a clear response, the determination of a point at which adaptation 'breaks down' must be arbitrary, just as it must be a matter of judgment when a state of arousal turns from 'sufficiently high' into 'harmful stress'. Equally, 'adverse' environmental conditions were said earlier to be compensated by homeostatic mechanisms. Such statements must be recognized for what they are, namely value judgments. There is no need to ban value judgments from discussion. They will anyhow only re-emerge in hidden form. As long as they are acknowledged as such, they help to clarify matters. The danger comes when they, in their turn, are 'adapted to' and so taken for granted that they may be unknowingly misapplied.

UNCERTAINTY AND SELF-GENERATED INFORMATION

How then does a person in a compulsively restricted situation, say, a driving cab or cockpit, respond to uncertainty? Analysis of field study protocols shows that his 'adaptive response' is to create his own information. More or less consciously, he takes the risk that his quasi-information may not conform to the real state of affairs. To cover that gap between his 'imagination' and the future 'real world outside', he continuously seeks for further information from the environment. The term 'adaptation' now changes its meaning from passive compliance to active endeavour. He actively scans the surrounds with all perceptual modes at his disposal and when it is poor, say in fog, he strains his perceptual apparatus to the highest degree. If the environment is impoverished by imposition of artefacts, as in the uniformity of motorways or railway lineside equipment, or even by the well-intended soundproofing of his work station, he takes greater risks and guesses where he does not know for certain.

It may seem rather irrelevant or strange to introduce the distinction between 'imaginary' and 'real' into a discussion of the practicalities of transport operation. But the relevance is justified as it arises particularly from common experience in studies in which mental underload is reported. The phrases repeatedly used are 'mind-wandering', 'day-dreaming', (Oswald; Kripke), 'absent-mindedness' (Reason), 'neglect of surroundings' (Broadbent), 'shift of attention from environment to internal events' 'relaxation of vigilance' and 'under-arousal'. If the

senses are lulled, interests and controls are diminished and the
ever ongoing mental activity seeks its material from fantasy by
a random walk among 'internal representations'. Control over
thought processes necessarily involves some monitoring of the
surroundings. The more intense and exact the control over the
exteroceptive senses, the greater the influence of the 'real
world' on the person. At certain times of day - and night - a
natural struggle occurs between a 'realm of fantasy' and one of
'reality'. In this struggle, all will depend for the operator's
reliability on whether mental activity is more or less controlled.

'MINI-PANICS'

Field observations of repetitive manual work and boring,
vigilance-type inactivity have directed this author's attention
to the occurrence of spasmodic episodes of hyperactivity, anxiety and stress, tending to arise at intervals of 90-100 minutes. Branton and Oborne (1979) characterize these episodes of
day-dreaming followed by sudden actions as 'mini-panics'. Self-reports by articulate professional operators make it clear that
they are aware after the event of their minds having wandered
from the task. They know that there was a time gap during which
their mental activity had slipped out of control but are quite
uncertain about the duration of that gap. Depending on the retrievability from any mishaps during that moment, the uncertainty will be more or less stressful. Even though these are experienced operators, they do not seem to be able to anticipate or
prevent these lapses.

ULTRADIAN RHYTHMS AND POSTURAL CONTROL

While the stress aspect has been dealt with by these authors, the reliability of alertness needs further discussion.
Once it was realized how similar these day-dreaming episodes
were to the sleep stage REM cycles of dreaming, the obvious
question arose whether such cyclical events did not generally
and naturally extend into daytime. In the literature of sleep
research and on biorhythms some empirical and clinical evidence
was found. It can be subsumed under the heading of 'ultradian
rhythms' in physiological changes throughout wakefulness as well

as sleep (Oswald, 1970; Othmer, 1969; Kripke, 1974; Lavie, 1979; and others). The findings are that, in the right circumstances of enforced inactivity throughout the waking day, almost everyone is subject to such changes in arousal at about 90-100 minute intervals. They are reported to be regularly accompanied by brief spells of daydreaming. Of course, body movements are largely suppressed during sleep. But even if one is not asleep, yet in a well-supported posture, whether seated or horizontal, the anti-gravity muscle action control is 'switched off'. If in addition, the eyelids close and visual perception is suppressed, the person easily enters into twilight states in which the balance between attention to the environment and self-awareness becomes unstable. These transitional (hypnagogic and hypnapompic) states are frequently associated with hallucinations and with highly original and creative thoughts, in other words, with fantasies. A connection between secure suspension of the body and reduced information intake on the one hand and enhanced abstract mental activity on the other is suggested. Conversely, it is arguable that the two factors which normally prevent one from being conscious of these states of diminished external awareness, both in daytime and during the transitions to sleep, are the active search for stimulation, particularly visual, and antigravity muscular work. In short, day dreams are less likely to occur if one keeps oneself upright and actively seeks to occupy eyes and ears.

To sum up this section: It was said at the start that in transport, man pits himself against nature, but that this alone is probably not the direct source of stress. By various considerations, these sources have now been narrowed down to the need for knowing what is 'real'. In a boring task situation, the operator should be abel to control his mental process so as to be as certain as possible that he can distinguish between what is real and what he imagines at any given moment, so that he can act accordingly. The uncertainty uncovered in this case refers not to the overcoming of (inanimate) natural forces, but to the inevitable interdependence with other persons. Because others rely on the operator's constancy of awareness of them and their interests, the critical factor in this type of stress is social.

AN EXPLANATORY MODEL

While the foregoing conjectures about natural fluctuations in awareness explain some of the observed behaviour, the self-reports of stress can best be explained by the assumption that operators actually feel responsible for their human charges. They have made it their purpose to conduct themselves safely. A philosophical, meta-psychological case for a purposive explanation of a broader spectrum of human behaviour has been set out briefly elsewhere (Branton, 1982). For the present, a subsidiary model is proposed which takes the form of a strictly autonomously controlled person, necessarily possessing value standards and interests in social relations, perpetually seeking and evaluating information from the surroundings. The search varies in intensity, depending on arousal level. This level normally fluctuates in cycles of about 100 minutes. The model concerns the form of mental operations, their contents being material either taken immediately from the surrounding world or from stored, primarily emotive experiences. The explanation is purposive, rather than causal, as it is argued that the thoughts which determine behaviour are forecasts of future states of affairs and their consequences for the person, rather than past experiences in themselves of speculative origin. Needless to say, this statement can only be the briefest of telegram-style sketches in the space available.

METHODS APPROPRIATE TO NEW PARAMETERS

To find ways out of this conceptual maze, one guiding thread may be offered to the researcher. It is to adopt an 'epistemic' strategy: to ask himself in the first place at each stage, "How do I come to know whether this or that statement by the operator is true?" "What is the source of my knowledge?" "How direct or indirect is my perception of the measurement?" "How far is my conjecture based on analogy and how far does it penetrate to the 'real thing'?" Having thus become conscious of their own inevitable bias, observers (and their readers) are better able to speculate on their Subjects' knowledge, values and actual purposes.

Even though it is necessary to trust in self-reports elicited post hoc that will not be enough; the endeavour must be to obtain convergent information from a number of disciplines and angles. The matter under investigation is serious enough to demand in addition a search for the precursors of states of low arousal and of loss of control.

If reliability depends on the operator's control over himself as well as over the system as a whole, parameters will be more informative the more they are sensitive to changes in the control of his behaviour. This is why energy consumption measures were rejected earlier on as unrepresentative of mental work. Control can be exerted either over task related 'cognitive' and potentially cognitive elements of actions, e.g. hand movements, or over behaviour of which the Subject is not normally immediately aware at all. The latter category includes such control functions as affect perception (e.g. eye movements, eye blinks, width of the papebral fissure, increases in latency of overt 'orienting responses') and postural maintenance (e.g. body sway, standing or sitting, and tonic head/neck muscle activity). In consonance with the autonomy view, records of spontaneous behaviour and physiological activity should be particularly informative - provided their meaning and function can be hypothesized. Examples are pupillary movement and limb tremor, but there are likely to be others like transients in EEG, ECG and EMG, which still await meaningful interpretation.

An important implication of the insight into the continuously floating threshold af awareness of the person about to be observed is the requirement that recording and measurement of behaviour and functions should be unobtrusive and non-invasive. Here the ever present capacity to adapt may actually help the researcher to overcome the Subject's self-consciousness. In this author's experience, Subjects quickly adapt to the observer's presence in quite close proximity provided the latter studiously avoids attracting attention. The slightest movement, let alone an overt intrusion into the Subject's world, say, to administer a reaction-time test or a questionnaire, are bound to awaken him. This would crucially defeat the purpose of the whole exercise, namely to capture and record a reality of internal events expressed in unprovoked behaviour.

Other methodological requirements will depend on specific circumstances and the attitudes and co-operation of managements and organisations involved. It can, however, be said in conclusion that, given the premise of this paper - the individual responsibility of the operator - managements and organisations must make an increased effort to allow operators to develop their potential by design for enhancing self-awareness and confidence in their own conscious judgments. Social support, if it is given sincerely by a management committed to safety and reliability, will enhance the confidence of operators in their own conscious judgments. To prove this proposition, a specific effort must be directed towards a fresh attempt at purposive explanation of operator behaviour. The practical usefulness of the attempt will be to enhance understanding of one's own body functions, which is likely to arouse one to act before drowsiness or stress become overwhelming.

SUMMARY OF PRACTICAL RECOMMENDATIONS

A) Organisational:
  1. Active programme of stress/fatigue prevention, with periodical review built into system.
  2. Environmental Design
     a) avoiding rhythmical 'lulling' stimulation in sight or sound
     b) introducing mild climatic variations, including frequent air changes and humidity/ionisation balance
  3. Positive provision of secondary activities not competing with main task. Operations designed for some walking about and standing as well as sitting
  4. Job enrichment, alternation and rotation

B) Individual:
  1. Consciously self-induced changes of posture (abt. one per minute) especially when low on the arousal curve
  2. Deep breathing for transport of oxygen to brain
  3. Occasional check of time on task - rather than mere clock time - to establish position on arousal curve in relation to 100 minute cycle
  4. Intake of glucose/fructose or caffeine

5. Re-creation of positive interest in the job
6. Equipment design and standardisation through representative experimental validation.

5. Cooling of head, ears, wrists, ankles to counteract too warm or stale environmental conditions.

REFERENCES

Bainbridge, L. 1978. The Process Controller. In "The Study of Real Skills I: Analysis of practical skills" (Ed. W.T. Singleton). (MTP Press, Lancaster).

Branton, P. 1978. The Train Driver. In "The Study of Real Skills I: Analysis of Practical Skills" (Ed. W.T. Singleton). (MTP Press, Lancaster).

Branton, P. 1982. The Use od Critique in Meta-Psychology. Ratio, 24, 1-13

Branton, P. and Oborne, D. 1979. A behavioural study of anaesthetists at work. In "Psychology and Medicine" (Eds. Oborne, D., Gruneberg, M. and Eisler, J.R.). (Academic Press, London).

Broadbent, D.E. 1953. Neglect of the Surroundings in Relation to Fatigue Decrements in Output. In "Fatigue: A Symposium" (Eds. Floyd, W.R. and Welford, A.T.) (H.K. Lewis, London).

Crossman, E.R.F.W. 1955. The Measurement of Discriminability. Quart.J.Exp.Psychol. 7, 176-195.

Friedman, M. and Rosenman, R.H. 1974. Type A Behaviour and your Heart. (Knopf, New York).

Kripke, D.F. 1974. Ultradian Rhythms in Sleep and Wakefulness. In "Advances in Sleep Research" Vol.I. (Ed. E.D. Weitzman). (J.Wiley, London).

Lavie, P. 1979. Ultradian Rhythms in Alertness - A pupillometric study. Biological Psychology, 9, 46-62.

Lefcourt, H.M. 1966. Internal versus External Control of Reinforcement. Psychol.Bull., 65, 206-220.

Nicholson, A.N., Hill. L.E., Borland, R.G. and Ferres, H. 1970. Activity of the Nervous System during the Let-down, Approach and Landing: A Study of Short Duration High Workload. Aerospace Medicine, 41, 436-446.

O'Hanlon, J.F. 1981. Boredom: practical consequences and a theory. Acta Psychologica, 49, 53-82.

Oswald, I., Merrington, J. and Lewis, H. 1970. Cyclical 'On Demand' Oral Intake by Adults. Nature, 225, 959-960.

Othmer, E., Hayden, M.P. and Segelbaum, R. 1969. Encephalic cycles during sleep and wakefulness in humans. Science, 164, 447-449.

Rotter, J.B. 1976. Generalized expectancies of internal versus external control of reinforcements. Psychol.Monogr. 80, (1.whole number 609).

Sanders, A.F. 1981. Stress and Human Performance: a working model and some applications. In "Machine Pacing and Occupational Stress" (Eds. G. Salvendy and M.J. Smith). (Taylor & Francis, London).

STRESS RESPONSE AS A FUNCTION OF AGE AND SEX

J. Vernikos

Biomedical Research Division
Ames Research Center, NASA
Moffett Field, CA 94035 U.S.A.

ABSTRACT

The data in the literature on the response to stress in males and females is reviewed and discussed. The effects of age on this response are also considered. Factors that may affect the response to stress in field studies and applied situations and thus bear on the validity or interpretation of the results are also discussed. The stress response as a function of age and sex is considered in terms of the tolerance of individuals to physical stresses such as exercise, heat, acceleration, and in terms of the responses of the pituitary-adrenal and the sympathetico-adrenal systems.

INTRODUCTION

Every body reaction to the external or internal environment is in some way related to the neuroendocrine and the autonomic nervous systems. Parameters have therefore been sought in these systems for the discrete quantitation of the stress response. It has become important to find ways of detecting and quantitating stress responses for two reasons: first, as a measure of the magnitude of the stress response, and secondly as a possible indicator of the subsequent physiological cost, particularly where exposure to repeated or continued stressful situations is involved. Heart rate, cortisol excretion, galvanic skin response, and catecholamine excretion usually increase in response to a stress stimulus or a stressful situation but not always. Such increases while measurable in well controlled experimental conditions, where subjects are naive volunteers, are often not seen in real life situations. Lack of correlation between physiological response and real life stressful conditions have been generally attributed to "individual variation" or to the subject's "experience" or "adaptation".

Unlike experimental animals where the variables can be controlled to a great extent, humans, particularly in field research, bring with them a host of complex psychological and physiological variables which can influence their stress response. Therefore, in trying to evaluate field research data, we should know as much as possible about the exact conditions in which the individual will find himself, as well as be aware of any alterations in his physiological base line. For instance, we should know something about whether the conditions are familiar or unfamiliar, expected or unexpected, do they represent an inescapable situation, can the individual cope with these conditions either because of prior learning or experience or because of his emotional and personality make-up, his rank and social environment, is the situation controllable and predictable? In addition, information should include the individual's age, sex, and clinical history, the time of day when he is expected to operate, his state of hydration, meal composition, fluid intake and sleep habits, medication and motivation. Has his physiological base line been altered by such situational factors as the work shift, the light intensity, environmental gas, humidity, temperature or pressure conditions, degree of confinement, activity or posture, the social structure of his environment (marital, job, family and community status, psychosocial stresses), and degree and recovery from a previous task or stressful experience? Finally the time of sampling and particularly the reference value or "control" that is used for the response can particularly influence the interpretation of the data. It is therefore obvious that the best kind of program for the investigation of stress responses in field operations would involve not only field research, but also simulation studies (preferably using the same test subjects to enable the investigation of individual variables) and where necessary, supporting animal research.

SEX

Sex differences have been known to exist in the responsiveness of the pituitary-adrenal and sympathetic-medullary systems in both animals and humans. These can be generally characterized by lower sympathetico-medullary and greater pituitary-adrenal reactivity to stress in females than in males (Vernikos-Danellis and Marks, 1962; Frankenhaeuser et al., 1976; Aslan et al., 1981). Lower catecholamine responsiveness in females is evident in laboratory studies using excerpts of action or erotic films (Levi, 1967), a 6-hr examination stress (Frankenhaeuser et al., 1978), a cognitive conflict test (Collins and Frankenhaeuser, 1978) or a color-word conflict test as the stressor (Aslan et al., 1981). This is particularly true for epinephrine responses. Similar differences have been reported in work stress conditions (Frankenhaeuser et al., 1976). This lower responsiveness in females cannot be attributed to the menstrual cycle. No differences in catecholamine excretion have been found during the different phases of the menstrual cycle (Couche et al., 1975) and pre-menstrual mood variations are not accompanied by increased epinephrine excretion (Silbergield et al., 1971; Patkai et al., 1974).

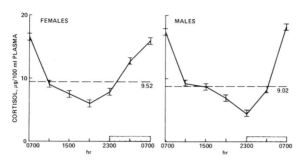

Figure 1 Diurnal Rhythm in Plasma Cortisol of Male and Female Subjects Aged 35-45 (16L:8D, lights on 0700)

Although the diurnal rhythm in plasma cortisol is not significantly different between the sexes (Figure 1), adrenals are larger in size and responsiveness of the pituitary-adrenal system to various stressful procedures is generally greater in females (Vernikos-Danellis, 1964). For instance, increase in plasma ACTH following epinephrine infusion is considerably greater in females than in males (Vernikos-Danellis and Marks, 1962) and surgical stress under comparable conditions provokes a markedly greater cortisol response in females (Brandt et al., 1976; McQuay et al., 1979; Moore et al., 1981). Prolactin secretion under these conditions is also increased and that response is also greater in females. It is of interest to note that post-operative fatigue has been attributed to an excessive catabolic response to the stress of anaesthesia and surgery (Rose and King, 1978). Various attempts have therefore been made to block the endocrine stress response by a variety of means including morphine and other analgesics (George et al., 1974; Hall et al., 1978; Kehlet, 1979). Moore et al., (1981) report a sex difference in the ability of the opiate buprenorphine to inhibit cortisol secretion. Although an abrupt and significant fall in cortisol levels was seen in men following buprenorphine administration, no such inhibition was seen in any of the women. Such sex differences in the suppressibility of the pituitary-adrenal system have not been reported previously.

Tolerance to various physical stresses requiring muscular work has been generally lower in females. For instance, in a series of studies on groups of males and females ranging in age from 25-65 (Table 1), Sandler and his associates (Sandler and Winter, 1978; Goldwater and Sandler, 1982) have shown that oxygen uptake during maximum exercise levels (VO_2 max) was considerably less in the females than in males, although the heart rate

SEX		AGE	+3G$_z$ TOLERANCE, sec	LBNP -50 mm Hgx 15 min	MAXIMAL EXERCISE	
					VO$_2$ MAX, ml/kg/mm	HR, bpm
FEMALE	(N = 8)	25-35	283 ± 20		35.5 ± 2.04	181 ± 5
FEMALE	(N = 10)	35-45	266 ± 26	13.77 ± 0.5	31.6 ± 1.54	175 ± 3
FEMALE	(N = 8)	45-55	557 ± 93	15	29.2 ± 1.90	163 ± 4
FEMALE	(N = 7)	55-65	541 ± 111	12.88 ± 1.2		
MALE	(N = 8)	25-35	396 ± 29	15	47.24 ± 0.96	185 ± 2
MALE	(N = 6)	35-45	749 ± 110	14.57 ± 0.43		
MALE	(N = 7)	45-55	536 ± 120	15	33.9 ± 2.8	160 ± 6
MALE	(N = 8)	55-65	724 ± 83	14.38 ± 0.63	37.1 ± 3.0	157 ± 6

Table 1: Tolerance levels to acceleration, LBNP, and exercise in Male and Female Subjects

attained for that effort was the same. With increasing age VO$_2$ max as well as the maximum heart rate were lower in both the males and the females. Similarly, females appear to tolerate thermal stress less well. When males and females must perform the same work in heat, females do poorly (Wyndham et al., 1965). When the work load is relative to their VO$_2$ max, metabolic and cardiovascular responses to the combined stress of exercise and heat seem to be similar for both sexes (Brouhas et al., 1960). However, with maximal work to exhaustion and heat, the duration of this effort is reported to be shorter in women than in men. In contrast, tolerance to cold reflects individual subcutaneous fat thickness and women may have an advantage, as has been suggested by the study of the diving women of Korea. Rennie et al., (1963) speculated that women rather than men originally took up this work because of greater adaptability to cold.

Nunneley (1978) has concluded that the differences recorded are probably related to women's relatively low level of physical fitness, their lack of heat acclimatization, anthropometric factors and body composition. Menstrual cycles also appear to have no significant effect on heat tolerance in women (Fein et al., 1975; Kamon and Avellini, 1976). More recently, Horvath and Drinkwater (1982), studying the effects of the menstrual cycle on thermoregulation, found that resting core temperature and oxygen uptake in females is higher in the luteal phase than any other part of the cycle but these differences disappear under the combined influence of exercise and heat stress. Exercise heart rate, core temperature and sweat rate were similar in all cycle phases. The only significant difference during the female cycle occurred in the greater relative decrease in plasma volume during the luteal phase in subjects exercising in 48°C as compared to flow days.

AGE

Studies of hemodynamic performance in the cardiovascular response to physical stress in man have been well documented over the past 50 years (Gerstenblith et al., 1976). Maximum cardiac output and oxygen consumption during dynamic exercise decrease with advancing age. The heart rate response in particular, declines with age during isometric exercise (Kino et al., 1977) and hypoxia (Kronenberg and Drage, 1973). Strandell (1964) has reported that whereas the increase in heart rate on standing was nearly 20 beats/min. in young adults it was only 10-15 beats/min. in middle and old age. This postural hypotension with age is accompanied by a reduced vasoconstrictor response to cooling and decreased baroreflex sensitivity during LBNP (Collins et al., 1980) and phenylephrine injections (Gribbin et al., 1971), and has been attributed to changes in the structure of the vasculature (MacLennan et al., 1980). Most of these studies have involved subjects that are older than 65 years of age. Thus although postural hypotension in those aged 66 or older is a well recognized phenomenon (Johnson et al., 1965; Wollner, 1978), in the series of studies conducted by Sandler and his associates (Sandler and Winter, 1978; Goldwater and Sandler, 1982) tolerance to +G_z centrifugation and to LBNP were found to increase or were at least maintained with age in both sexes through age 65 (Table 1) as has also been reported for blood pressure responses to stress (Palmer et al., 1978).

To what extent this decreased responsiveness in the elderly is due to aging per se or the deconditioning associated with the relative inactivity of aging remains to be determined. It certainly does not appear to be due to decreased noradrenergic release in response to sympathetic stimulation. Lake et al., (1976) in a study on individuals aged from 10 to 70 found increasing basal plasma norepinephrine (NE) levels. Similarly, we found progressively increased excretion of NE in normal healthy males between 45 and 65 years of age (Table 2). However, epinephrine excretion decreased in the same age groups and females showed no such changes with age. Measurement of the stress-induced changes in circulating catecholamines provide an even better estimate of the secretion of catecholamines in response to

Sex	Age (years)	Urine Cortisol (μg/24 Hr)	Urine Epinephrine (μg/24 Hr)	Urine Norepinephrine (μg/24 Hr)
Female	35 - 45	56.1 ± 5.5	24.5 ± 5.4	57.9 ± 5.1
Female	45 - 55	47.0 ± 2.2*	29.5 ± 4.1	59.0 ± 4.3
Female	55 - 65	37.2 ± 4.3**	23.5 ± 2.1	58.9 ± 5.2
Male	35 - 45	44.2 ± 5.6	35.0 ± 3.4	57.5 ± 4.3
Male	45 - 55	43.2 ± 2.2	21.1 ± 2.6**	78.1 ± 7.6**
Male	55 - 65	52.2 ± 3.7	22.2 ± 3.6**	89.2 ± 5.4**

*Significantly different from 35-45 age group at P<0.05
**Significantly different from 35-45 age group at P<0.01

Table 2: Twenty-four Hour Cortisol, Epinephrine and Norepinephrine Excretion in Male and Female Subjects of Different Age Groups

stress. It has been shown that on standing from the supine position or during isometric exercise (Ziegler et al., 1976; Palmer et al., 1978; Rowe and Troen, 1980) or in response to emotional stimuli (Aslan et al., 1976), the increase in circulating catecholamines in the elderly is at least equivalent and usually greater than that in the young adult. This suggests that if a diminished sympathetic response does occur with age, it cannot result from a failure to release NE in response to the stress. This conclusion has been supported by in vitro evidence of decreased post-synaptic sympathetic responsiveness in both cardiac and vascular tissue (LaKatta, 1980).

Reduced target organ responsiveness with increasing age is not limited to the sympathetico-adrenal system. Decreased gonadal sensitivity to gonadotrophins during the menopause is the classic example (Meites et al., 1980). The pituitary-adrenal system seems to show similar changes which are not limited to the "very old". We have found significant alterations in the function of various components of this system as early as 45-55. The diurnal rhythm in circulating cortisol (Figure 1) which is not very different between males and females till age 45, shows a marked downward shift to lower levels in females aged 45-55 (Figure 2), with the greatest decreases in circulating cortisol at 0300 and 1100 hours so that except for the early morning rise in plasma cortisol most of the 24 hour period is spent in the trough of the rhythm. This downward shift and consequent reduced daily mean in circulating cortisol is maintained at least till age 65. A similar change

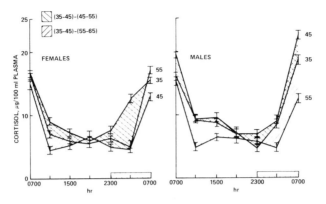

Figure 2 Shift in Diurnal Rhythm of Plasma Cortisol with Age in Male and Female Subjects

in this rhythm occurs in males, but only after 55 years of age.

A decline in the plasma cortisol response to the stress of centrifugation ($+3G_z$) is also evident in both sexes as early as 35 years of age. This is not due to reduced duration of the stress since tolerance and hence duration of exposure at the $+3G_z$ level was in fact longer in the older groups of both sexes (Table 1). Reduced cortisol responses with age have also been reported after fasting or surgery in animals and man (Jensen and Blichert-Toft, 1971; Sartin et al., 1980). The reduced cortisol response to stress might suggest decreased pituitary ACTH responsiveness. We have found this not to be the case, and in fact the ACTH response to centrifugation was <u>at least</u> the same or greater in the older subjects (Vernikos et al., unpublished).

These findings suggest that as with the gonads, adrenal sensitivity to the trophic hormone declines with age but there is no good, direct evidence in the literature for such a change. On the other hand, there is some evidence that the system becomes less sensitive to suppression by glucocorticoids. There is a reduction in glucocorticoid receptors in aging rat brain, a resistance to the inhibition by dexamethasone of the response to surgical stress (L. Johnson, personal communication) and the negative feedback increase in ACTH after metyrapone in humans is greater in the elderly (Jensen and Blichert-Toft, 1971). A change in metabolic or renal clearance with age is also possible since cortisol excretion was unchanged in the males of any age, but decreased in females after 45 (Table 2).

In assessing the physiological responses to a particular stress, it is also important to be aware of what factors other than sex or age might be contributing to an altered response. For instance, ample evidence exists to suggest that prior exposure to stressful conditions creates a susceptibility to infection, drugs, or a subsequent stress (Sakellaris and Vernikos-Danellis, 1974). Previous exposure to chronic cold (Vernikos et al., 1982) or intermittent or repeated immobilization stress in animals (Mikulaj et al., 1976) initially potentiates the pituitary-adrenal response to another stress. However, after a period of time when the system is considered to have "adapted", adrenal corticosteroid responses are the same or less than in animals exposed to that same stress for the first time, though the response to a different stress may be unimpaired. This "adaptation" to the same stress is also apparent at the adrenal level as evidenced by decreased responsiveness of the adrenal to ACTH in animals exposed to chronic exercise (Tharp and Buuck, 1974). In contrast, E and NE responses to the same intermittent stress (immobilization) increase and remain elevated even after the 49th exposure (Mikulaj et al., 1976). Similar findings have been reported in a study of parachutists where with repeated jumps cortisol responses declined after the first or second jump but catecholamine responses persisted (Levine, 1978). The increased secretion of E from the adrenal medulla and of NE mainly from the adrenergic vasomotor system serve to prevent metabolic and cardiovascular disturbances. The variation observed in the secretion of E may be attributable to hypoglycemia and mental stress. As to NE secretion, factors such as training may play an important role in altering the activation of the vasomotor system.

In a recent review of endocrine responses to stressful psychological events, Rose (1980) pointed out that adaptation of the adrenocortical system to initially provocative events often occurs rather rapidly. The ability to cope with a particular stressful situation certainly affects the magnitude of the pituitary-adrenal response (Conner et al., 1971). This may account for a great deal of data, especially in field research where no cortisol response is evident under conditions that would generally be considered highly stressful. For instance, Cullen et al., (1979) studying experienced truck drivers during days of particularly difficult driving tasks and days off, found no increase in plasma cortisol levels. Dutton et al., (1978) found no difference between days worked and days off in paramedics. Similarly, Bourne et al., (1967) were perhaps the first to show that helicopter medics during the Vietnam War, showed no increase in urinary corticosteroids on days they flew as compared to non-flying days.

Similar observations have been made more recently by other investigators. Hale et al., (1971) and Rose and his associates (1978; 1982a; 1982b) could find no increases in urinary steroids in air traffic controllers, except during particularly busy parts of the work shift.

In a study of three U-2 pilots on routine 2 hour and 15 minute flights, urinary cortisols were found to be unchanged or indeed suppressed in flight and only showed post-flight increases in two cases of unexpected headwinds during landing (Vernikos-Danellis et al., 1975). What was routine for these pilots would be considered highly stressful for any other individual. The data from these studies suggest that in experienced pilots during highly complex flight conditions which nevertheless are controllable, the pituitary-adrenal system may actually be suppressed.

The data reviewed here indicate that small but interesting differences exist in the neurohumoral and neuroendocrine mechanisms underlying the male and female response to stress. Although insufficient evidence exists, the data are suggestive that males may respond to stress primarily via their sympathetico-adrenal system while females may use primarily pituitary-adrenal mechanisms. The significance of decreased responsivity of the sympathetico-adrenal system to stress in younger females may be related to the lower incidence of coronary artery disease. Estrogen is believed to be directly related to triglyceride levels and to high density lipoproteins (HDL) (Bradley et al., 1978). Epidemiologic evidence (Berg et al., 1976; Gordon et al., 1977) points to a protective role of higher HDL levels decreasing the risk of heart disease. It is of interest in this respect, that unlike males where exercise training increases HDL levels, (Wood and Haskel, 1979), it does not alter HDL levels in young (19-30) females (Wynne et al., 1980) except possibly in extreme cases such as marathon runners (Wood et al., 1977) With the reduction of ovarian function in older females, there is a greater tendency to show coronary artery disease than during the reproductive life. Since catecholamines appear to contribute to metabolic changes which may lead to atherosclerotic disease than (Carlson et al., 1968), the greater noradrenergic responsivity to stress in older females may well be related to this particular pathology.

There is no doubt that age alters the stress response, but more, and more systematic studies are needed to determine the causes and mechanisms of these changes. It is obvious that age-related changes in the neuroendocrine response to stress occur much earlier than is usually considered.

Finally, in interpreting data from human studies, particularly in the work place or in field research, an awareness of sampling time and other factors that may be altering the physiological baseline of the individual is essential.

REFERENCES

Aslan, S., Nelson, L., Carruthers, M. and Lader, M. 1981. Stress and age effects on catecholamines in normal subjects. J. Psychosom. Res. 25, 33-41.

Berg, K., Borresen, A. L., Dahlen, G. 1976. Serum high density lipoprotein and atherosclerotic heart disease, Lancet i 499-501.

Bourne, P. G., Rose, R. M. and Mason, J. W. 1967. Urinary 17-OHCS levels: data on seven helicopter ambulance medics in combat. Archiv. Gen. Psychiat. 17, 104-110.

Bradley, D. D. Wingerd, J., Petitti, D. B., Krauss, R. M. and Ramcharan, S. 1978. Serum high density lipoprotein cholesterol in women using oral contraceptives, estrogens and progestins. N. Engl. J. Med. 299, 17-20.

Brandt, M., Kehlet, H., Binder, C., Hagen, C. and McNeilly, A. S. 1976. Effects of epidural analgesia on the glycoregulatory endocrine response to surgery. Clin. Endocrinol. 25, 7-14.

Brouha, L., Smith, P. E., Lanne, R. D. E., and Maxfield, M. E. 1960. Physiological reactions of men and women during muscular activity and recovery in various environments. J. Applied Physiol. 16, 133-140.

Carlson, L. A., Levi, L. and Oro, L. 1968. Plasma lipid and urinary excretion of catecholamines in man during experimentally induced emotional stress and their modification by nicotinic acid. J. Clin. Invest. 47, 1795-1805.

Collins, K. J., Exton-Smith, A. N. James, M. H. and Oliver, D. J. 1980. Functional changes in autonomic nervous responses with ageing. Age and Ageing. 9, 17-24.

Collins, M. and Frankenhaeuser, R. W. 1978. Stress responses in male and female engineering students. J. Hum. Stress 4, 43-52.

Conner, R. L., Vernikos-Danellis, J., and Levine, S. 1971. Stress, fighting and neuroendocrine function. Nature (Lond.) 234, 564-566.

Couche, J. L., Kuchel, O., Barbeau, A. and Genest, J. 1975. Sex differences in urinary catecholamines. Endocr. Res. Commun. 2, 549-550.

Cullen, J. H., Fuller, R. and Dolphin, C. 1979. Endocrine stress responses of drivers in a 'real-life' heavy goods vehicle driving task. Psychoneuroendocrinol. 4, 107-115.

Dutton, L. M., Smolensky, M. H., Leach, C. S., Lorimor, R. and Hsi, B. P. 1978. Stress levels of ambulance paramedics and fire fighters. J. Occup. Med. 20, 111-115.

Fein, J. T., Haymes, E. M. and Buskirk, E. R. 1975. Effects of daily and intermittent exposures on heat acclimation of women. Int. Journal Biometeor. 19, 41-52.

Frankenhaeuser, M., Dunne, E. and Lundberg, U. 1976. Sex difference in sympathetic-adrenal medullary reactions induced by different stressors. Psychopharmacology. 47, 1-14.

Frankenhaeuser, M., Wright, M. R. V., Collins, A., Wright, J. V., Sedvall, G. and Swahn, C. G. 1978. Sex difference in psychoneuroendocrine reactions to examination stress. Psychosom. Med. 40, 334-340.

George, J. M., Reier, C. E., Larese, R. R. and Rower, J. M. 1974. Morphine anaesthesia blocks cortisol and growth hormone response to surgical stress in humans. J. Clin. Endocrinol. Metab. 38, 736-741.

Gerstenblith, G., Lakatta, E. G. and Weisfeldt, M. L. 1976. Age changes in myocardial function and exercise response. Prog. Cardiovasc. Dis. 19, 1-21.

Goldwater, D. J. and Sandler, H. 1982. Orthostatic and acceleration tolerance in 55 to 65 year old men and women after weightlessness simulation. Aerosp. Med. Assoc. Preprints (Miami, Florida) pp.202-203.

Gordon, T., Castelli, W. P., Hjortland, M. C., Kannel, W. B., and Dawber, T. R. 1977. High density lipoprotein as a protective factor against coronary heart disease. The Framingham study. Am. J. Med. 62, 707-714.

Gribbin, B., Pickering, T. G., Sleight, P. and Peto, R. 1971. Effect of age and high blood pressure on baroreflex sensitivity in man. Circulation Research. 29, 424-431.

Hale, H. B., Williams, E. W., Smith, B. N. and Melton, C. E. 1971. Excretion patterns of air traffic controllers. Aerosp. Med. 42, 127-128.

Hall, G. M., Young, C., Holdcroft, A., Alaghbandzadeh, J. 1978. Substrate mobilization during surgery: A comparison between halothane and fentanyl anaesthesia. Anaesthesia. 33, 924-930.

Horvath, S. M. and Drinkwater, B. L. 1982. Thermal regulation and the menstrual cycle. Aviat. Space and Environ. Med. 53, 790-794.

Jensen, H. K. and Blichert-Toft, M. 1971. Investigation of pituitary-adrenocortical function in the elderly during standardized operations and post-operative intravenous metyrapone test assessed by plasma cortisol, plasma compound S and eosinophil cell determinations. Acta. Endocrinologica. 67, 495-507.

Johnson, R. H., Smith, A. C., Spalding, J. M. K. and Wollner, L. 1965. Effect of posture on blood pressure in elderly patients. Lancet i, 731-733.

Kalbfleisch, J. H., Reinke, J. A., Porth, C. J., Ebert, T. J. and Smith, J. J. 1977. Effect of age on circulatory response to postural and Valsalva Tests. Proc. Soc. Exp. Biol. Med. 156, 100-103.

Kamon, E. and Avellini, B. 1976. Physiological limits to work in the heat and evaporative coefficient for women. J. Applied Physiol. 41, 71-76.

Kehlet, H. 1979. Stress free anaesthesia and surgery. Acta Anaesthesiol. Scand. 23, 503-504.

Kino, M., Lance, V. Q., Shahamatpour, A., and Spodick, D. H. 1975. Effect of age on responses to isometric exercise. Am. Heart J. 90, 575-581.

Kronenberg, R. S. and Drage, C. W. 1973. Attenuation of the ventilatory and heart rate responses to hypoxia and hypercapnia with aging in normal men. J. Clin. Invest. 52, 1812-1819.

Lakatta, E. G. 1980. Age-related alterations in the cardiovascular response to adrenergic mediated stress. Fed. Proc. 39, 3173-3177.

Levi, L. 1967. Stressors, stress tolerance, emotions and performance in relation to catecholamine excretion. Forsvarsmedicin 3, 192-211.

Levine, S. 1978. Cortisol changes following repeated experiences with parachute training. In "Psychobiology of Stress: A Study of Coping Man" (Eds. H. Ursin, E. Baade and S. Levine). (Academic Press, New York). pp. 51-56.

MacLennan, W. J., Hall, M. R. P. and Timothy, J. I. 1980. Postural hypotension in old age, is it a disorder of the nervous system or of blood vessels? Age and Ageing 9, 25-32.

McQuay, J. J., Moore, R. A., Patterson, G. M. C. and Adams, A. P. 1979. Plasma fentanyl concentrations and clinical observations during and

after operation. Br. J. Anaesthesia. 51, 543-550.
Meites, J., Steger, R. W. and Huang, H. H. H. 1980. Relation of the neuroendocrine system to the reproductive decline in aging rats and human subjects. Fed. Proc. 39, 3168-3171.
Mikulaj, L., Kvetnansky, R., Murgas, K., Parizkova, J. and Vencel, P. 1976. Catecholamines and corticosteroids in acute and repeated stress. In "Catecholamines and Stress". (Ed. E. Usdin) (Oxford, Pergamon Press) pp. 445-455.
Moore, R. A., Smith, R. F., McQuay, H. J. and Bullingham, R. E. S. 1981. Sex and surgical stress. Anaesthesia. 36, 263-267.
Nunneley, S. A. 1978. Physiological responses of women to thermal stress: A review. Medicine and Science in Sports. 10, 250-255.
Palmer, G. J., Ziegler, M. G. and Lake, C. R. 1978. Response of norepinephrine and blood pressure to stress increases with age. J. Gerontol. 33, 482-487.
Patkai, P., Johansson, G. and Post, B. 1974. Mood, alertness, and sympathetic adrenal medullary activity during menstrual cycle. Psychosom. Med. 36, 503-512.
Rennie, D. W., Covino, B. G., Howell, B. J., Song, S. H., Kang, B. S. and Hong, S. K. 1963. Physical insulation of Korean diving women. J. Appl. Physiol. 17, 961-966.
Rose, E. A. and King, T. C. 1978. Understanding post-operative fatigue. Surgery, Gynecology & Obstetr. 147, 97-102.
Rose, R. M. 1980. Endocrine responses to stressful psychological events. Psychiat. Clin. N. Am. 3, 251-276.
Rose, R. M., Jenkins, C. D., and Hurst, M. W. 1978. "Air Traffic Controller Health Change Study" (Ed. M.A. Levin) (University of Texas Medical Branch, Galveston, Texas).
Rose, R. M., Jenkins, C. D., Hurst, M., Livingston, L. and Hall R. P. 1982a. Endocrine activity in air traffic controllers at work. I. Characterization of cortisol and growth hormone levels during the day. Psychoneuroendocrinol. 7, 101-111.
Rose, R. M., Jenkins, C. D., Hurst, M., Herd, J. A. and Hall, R. P. 1982b. Endocrine activity in air traffic controllers at work. II. Biological, psychological and work correlates. Psychoneuroendocrinol. 7, 113-123.
Rowe, J. W. and Troen, B. R. 1980. Sympathetic nervous system and aging in man. Endocr. Rev. 1, 167-179.
Sakellaris, P. C. and Vernikos-Danellis, J. 1974. Alteration of pituitary-adrenal dynamics induced by a water deprivation regimen. Physiol. Behav. 12, 1067-1070.
Sandler, H. and Winter, D. L. 1978. "Physiological Responses of Women to Simulated Weightlessness: A review of significant findings of the first female bed rest study". (NASA SP-430, Ames Research Center, Moffett Field, CA).
Sartin, J., Chaudhuri, M., Obenrader, M. and Adelman, R. C. 1980. The role of hormones in changing adaptive mechanisms during aging. Fed. Proc. 39, 3163-3167.
Silbergield, S., Brast, N. and Noble, E. B. 1971. The menstrual cycle: A double-blind study of symptom, mood, and behavior and biochemical variables using envoid and placebo. Psychosom. Med. 33, 411-419.
Strandell, T. 1964. Circulatory studies in healthy old men. Acta. Med. Scand. 175, Suppl. 414, 1-44.
Tharp, G. D. and Buuck, R. J. 1974. Adrenal adaptation to chronic exercise. J. Appl. Physiol. 37, 720-722.

Vernikos-Danellis, J. 1964. The regulation of the synthesis and release of ACTH. Vitamins and Hormones. 23, 97-152.

Vernikos-Danellis, J., Dallman, M. F., Bonner, C., Katzen, A., and Shinsako, J. 1982. Pituitary-adrenal function in rats chronically exposed to cold. Endocrinology. 110, 413-420.

Vernikos-Danellis, J., Goldenrath, W. L. and Dolkas, C. B. 1975. The physiological cost of flight stress and flight fatigue. U.S. Navy Med. 66, 12-16.

Vernikos-Danellis, J. and Marks, B. H. 1962. Epinephrine-induced release of ACTH in normal human subjects: A test of pituitary function. Endocrinology. 70, 525-532.

Wollner, L. 1978. Postural hypotension in the elderly. Age and Ageing. Suppl. 7, 112-118.

Wood, P. D. and Haskell, W. L. 1979. The effect of exercise on plasma high density lipoproteins. Lipids. 14, 417-427.

Wood, P. D., Haskell, W. L., Stern, M. P., Lewis, S. and Perry, C. 1977. Plasma lipoprotein distributions in male and female runners. Ann. NY Acad. Sci. 301, 748-763.

Wyndham, C. H., Morrison, J. F. and Williams, C. G. 1965. Heat reactions of male and female caucasians. J. Applied Physiol. 20, 357-364.

Wynne, T. P., Bassett Frey, M. A., Laubach, L. L. and Glueck, C. J. 1980. Effect of a controlled exercise program on serum lipoproteins in women on oral contraceptives. Metabolism. 29, 1267-1271.

Ziegler, M. G., Lake, C. R. and Kopin, I. J. 1976. Plasma noradrenaline increase with age. Nature. 261, 333-335.

DRUGS AND TRANSPORT OPERATIONS

Anthony N. Nicholson
Royal Air Force Institute of Aviation Medicine
Farnborough, Hampshire, United Kingdom

ABSTRACT

Personnel involved in transport operations may use drugs for a specific medical condition or for purposes which arise from the nature of their work. For instance drugs may be used in allergic states and in the management of mild hypertension, or may be used for the prophylaxis of malaria or in the management of disturbed sleep. Unfortunately many drugs have central effects and so careful attention must be given to the balance between the advantages of a drug and its possible adverse effects on performance. In this paper the experimental approach to two problems - hypnotics for sleep disturbance and antihistamines for allergic states - is discussed and tentative recommendations for the use of these drugs are given.

INTRODUCTION

Personnel engaged in transport operations may use drugs for a specific medical condition or for purposes which are related to the nature of their occupation. Examples of the former are drugs in the management of hypertension and antihistamines in the treatment of allergic states, while examples of the latter are drugs in the prophylaxis of malaria and hypnotics in the management of disturbances of sleep. Unfortunately many drugs have central effects and centrally acting drugs may have persistent activity, and so careful attention must be given to the advantages of a drug and its possible adverse effects on performance.

If drugs are used, then the correct choice is important. There may be little to choose between similar drugs as far as their therapeutic effect is concerned, but their potential for impaired performance may differ considerably. For instance, in the treatment of mild hypertension β-adrenoceptor antagonists are often used and with these it has been proposed that low lipophilicity may be less likely to give rise to central effects. With antihistamines it has been suggested that some may preferentially block peripheral receptors while others may cross the blood-brain barrier so slowly that tolerance may develop quite easily, while with hypnotics the duration of action may be short in those with fast elimination.

Essentially, the factors which influence the use of a drug in the context of occupational medicine vary. We have examined the use of hypnotics and antihistamines as they raise two very different issues. In the management of sleep disturbance the challenge is to provide an effective hypnotic free of adverse effects on sleep and of residual effects on performance, whereas with the allergic states the individual must remain alert while the drug is acting peripherally.

SLEEP DISTURBANCE AND HYPNOTICS

The management of sleep difficulties which arise from irregularity of work must be related to the cause, but in occupational medicine we are frequently faced with the use of hypnotics - particularly in the middle aged. There are no simple guidelines, but an understanding of the various issues involved help toward the wise use of these drugs in those who have to cope with irregular rest and carry out skilled activity. Some knowledge of pharmacokinetics is appropriate as persistence of effect and accumulation with repeated ingestion are important factors (Breimer, 1979; Curry and Whelpton, 1979).

PHARMACOKINETICS

Distribution and elimination

The way in which an hypnotic is distributed in the body involves a central compartment of blood and highly vascular tissues such as the heart, lung and liver, and a peripheral compartment of lesser vascularity such as voluntary muscle. The brain is also a highly vascular organ, and as hypnotics usually cross the blood-brain barrier with ease it is also considered part of the central compartment. When a drug is given intravenously there is for all practical purposes instantaneous mixing, and the changes in plasma concentration, which are usually assessed by measurement of venous plasma, may indicate the parts played by the central and peripheral compartments if the decay follows two clearly defined phases - both of which are exponential. However, when it is given orally or intramuscularly plasma concentrations are influenced by absorption, and so there is also a growth phase. It is absorption rather than penetration of the blood-brain barrier which tends to limit the rate of transfer of an hypnotic to its site of action.

The first part of the decay of the plasma concentration relates primarily to distribution and so to penetration of the peripheral compartment, while the second part relates to elimination by metabolism and/or by excretion from the central compartment (Fig. 1). Tissue penetration and elimination occur together, and the curve reflects the relative dominance

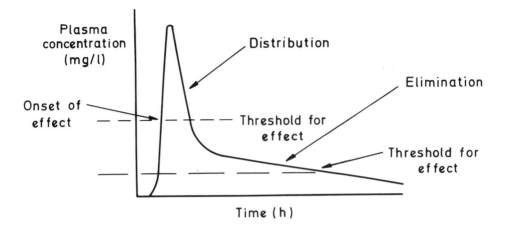

Fig. 1. Plasma concentration profile (semi-log plot) for a drug with a biexponential decay after oral dosing. There is an effect as long as the concentration remains above a certain level. The threshold may be related to the distribution or elimination phases. If it is above the concentration at which the inflexion of the distribution and elimination half-lives occurs then it will be related to the distribution phase and any effect will be of short duration.

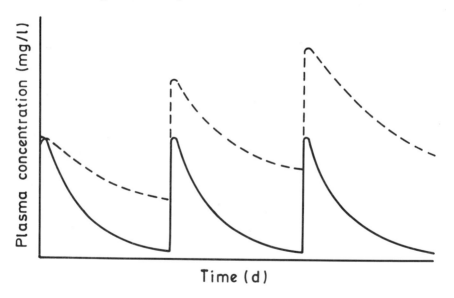

Fig. 2. Plasma concentration profiles of two hypnotics with different elimination half-lives (lower curves - 8 h: upper curves - 24 h) given every 24 hours. If the compound with a half-life of 8 h is given daily there will be no accumulation and there will be an intermittent type of drug action. However, the drug with an elimination half-life of 24 h will accumulate with daily ingestion.

of these events at different times. A drug has an effect as long as its
concentration remains above a certain plasma level. If the threshold is
reached during the phase which predominantly represents distribution - and
this part of the decay is usually rapid - then the duration of action is
likely to be short, but if the threshold is related to the elimination
phase then it will be much longer - particularly if this phase is prolonged.

It is evident that both distribution and elimination influence duration
of activity, though a relatively short duration of action may be attained
even though elimination is relatively slow. In principle two conditions are
required. The drug must become available in the blood in relatively high
concentrations soon after administration, otherwise the concentration profile will level out and no distinction will be found between the faster
distribution and slower elimination phases. High concentrations such as
those seen after intravenous injection can be achieved by drugs which are
rapidly absorbed. Secondly, the drug has to cross the blood-brain barrier
rapidly, and with hypnotics brain concentrations follow closely those of
the plasma.

Intermittent therapy and accumulation

These various pharmacokinetic aspects are important in the appropriate
use of hypnotics. With oral ingestion the major determinant of the onset
of action of a single dose is absorption. Rapid absorption is associated
with a quick onset of action, whereas with slow absorption activity may be
attenuated or even eliminated. When a drug is taken as an hypnotic rapid
absorption is desirable, whereas a slower absorption rate may be preferred
if a sustained anxiolytic effect without immediate drowsiness is sought.

In the management of sleep difficulties an intermittent type of drug
action may be desirable even when the drug is given daily. The theoretical
plasma profile levels of two hypnotics with very different half-lives are
illustrated in Fig. 2. When given every 24 hours a drug with an elimination
half-life of 24 hours or longer will accumulate, but one with a half-life of
8 hours will not, and this ensures the intermittent type of action. However,
clinical effects may not relate directly to plasma concentration.
Tolerance may develop particularly as far as the central nervous system is
concerned. Indeed, unwanted drowsiness may be experienced only early on in
the course of chronic therapy, though tolerance to the persistent effects
of an hypnotic is not a point in favour of its use when other
suitable hypnotics free of such effects are available.

The metabolism of long-acting benzodiazepines is likely to be influenced by the age of the patient. Individuals over 60, even when healthy, may have impaired ability to complete the usual biotransformations, and so the half-lives of long-acting benzodiazepines tend to be prolonged in the elderly as opposed to young individuals. Age related decrements can vary from slight to a very marked effect depending on the drug and on the gender of the patient, and metabolic impairment in the elderly males may be greater than that in elderly females.

Another question is what happens if the drug is taken daily. The rate of accumulation of a drug varies inversely with the elimination half-life, and so the longer the half-life, the slower the rate of accumulation. In general, steady state conditions are reached after an interval of about four times the half-life. Drugs with long half-lives accumulate slowly but extensively, whereas accumulation, though completed more rapidly, may not be a factor of clinical significance with drugs with shorter half-lives.

PHARMACODYNAMICS

The activity of a drug relates closely to absorption, distribution and elimination as well as to metabolism, and it is useful to follow these events with diazepam as an example. Diazepam is rapidly absorbed. A single dose, say 10 mg, has an immediate effect, which persists for a few hours (Fig. 3) even though its elimination half-life ranges from 14-90 hours (Clarke and Nicholson, 1978). This is due to rapid and extensive distribution into the tissues. On the other hand daily ingestion will lead to accumulation both of the parent compound and of its long-acting metabolite, nordiazepam, which also has a long elimination half-life (30-60 hours). Residual sequelae with a single dose of 10 mg diazepam are unlikely, but sustained daytime anxiolytic activity will follow daily ingestion. The pharmacokinetics of lorazepam are different from those of diazepam. With lorazepam the distribution phase is less extensive, and so concentrations above threshold are more likely to be related to the elimination phase with its half-life of 10-20 hours. Though this is much shorter than that of diazepam, lorazepam may have a more prolonged effect due to the absence of a sustained distribution phase.

The repeated use of benzodiazepines with half-lives of 16-24 hours or longer, of either the parent compound or of an active metabolite, usually lead to persistent activity and this property is relevant to anxiolytic therapy. Nordiazepam, which has a long elimination half-life, is a metabolite of several benzodiazepines and in one, potassium chlorazepate, the parent drug may not even reach the systemic circulation, and serves as a pro-drug or precursor of nordiazepam. However, pro-drugs or precursors with a common metabolite may not have an identical action as the rate at which the metabolite reaches the blood after a single dose may not be the same, and this could give rise to differences in immediate effects. Another drug which is essentially a precursor and has a long-active metabolite is flurazepam.

CLINICAL CONSIDERATIONS

In those whose insomnia is associated with anxiety, the use of benzodiazepines with more persistent activity is appropriate. A single daily dose of diazepam (5-10 mg), potassium chlorazepate (15 mg) or flurazepam hydrochloride (15 mg) at bedtime may be sufficient. Furthermore, the occasional omissions of doses will not lead to an acute reappearance of symptomatology. However, accumulation may give rise to unwanted daytime drowsiness and impairment of psychomotor performance. With low doses impaired performance the next day is likely to be minimal, and this may be offset, at least partially, by tolerance, and, in any case, performance in some patients with anxiety may even be improved by therapy.

Though benzodiazepines with persistent effects are useful in anxiety, drugs in which daily ingestion leads to an intermittent type of action are appropriate as hypnotics for those in which psychopathology is not a relevant feature and particularly so when residual sequelae the next day must be avoided. Active metabolites are less common, and their clinical effect is usually determined by the parent compound. With daily ingestion steady state conditions are reached relatively rapidly, and the washout of the drug after termination of treatment is quicker. Further, the metabolic transformation of some drugs with short half-lives - those which are conjugated with glucuronic acid - do not appear to be influenced by age as much as the longer acting compounds.

RESIDUAL SEQUELAE

Obviously, hypnotics with limited duration of activity are appropriate for those involved in skilled activity and in whom the predominant problem is that of disturbed sleep. However, residual impairment of performance with overnight ingestion may still arise, and it may extend well into the next day even with doses which are within the generally accepted therapeutic range. A variety of tasks has been used to investigate this problem, and there is broad agreement on the relative persistence of the various drugs available. Studies with the barbiturate, heptabarbitone, which has a limited duration of action, have shown that decrements in psychomotor performance may persist for 10 hours after 200 mg, 13 hours after 300 mg and 19 hours after 400 mg (Fig. 4)(Borland and Nicholson, 1974). It is evident that sequelae are likely to be related to dose both in the extent of the decrement at any given time and in the persistence of the impairment. Impaired performance is more severe and persists far longer with higher doses.

1,4-Benzodiazepines

In the context of hypnotics which may be used safely by those involved in skilled activity, attention has been largely directed toward the benzodiazepines. Impaired performance with a single dose of diazepam is limited - due to its sustained distribution phase - and residual impairment is highly unlikely with a single dose taken overnight as long as it does not exceed 10 mg. Unfortunately, daily ingestion of diazepam leads to accumulation of the parent compound and of its metabolite, N-desmethyl-diazepam, and so residual sequelae are only absent when used occasionally. Nevertheless, it can be suitable for those who carry out skilled activity, but it should not be taken more frequently than once every 48 hours and more than twice in 7 days.

The long elimination half-life of diazepam and of its metabolite, nor-diazepam, are clear disadvantages when intended for use in those involved in skilled activity, but the relatively short duration of action of a single dose due to its sustained distribution phase suggested that closely related compounds without long-acting metabolites might have a better profile. Developments have been along two broad lines. The benzodiazepine molecule has been modified by demethylation (nitrazepam and oxazepam) or substitution of the 1-position (flurazepam hydrochloride), by hydroxylation of the 3-position (temazepam and oxazepam), by substitution of a nitro group in the 7-position of the benzene ring (nitrazepam and flunitrazepam), and by addition of various halogens in the phenyl ring (flurazepam and flunitrazepam).

The hydroxylated metabolites of diazepam, temazepam and oxazepam, which are free of long-acting metabolites, have been studied in detail (Fig. 5). Temazepam has a well defined distribution phase and the elimination

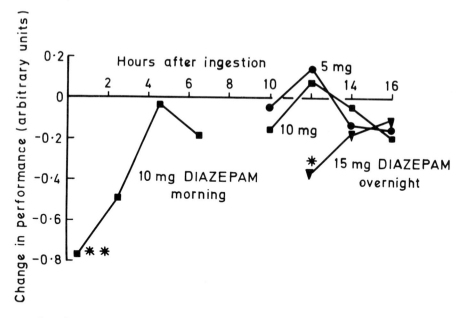

Fig. 3. Changes in performance on a visuo-motor coordination task (arbitrary units) after diazepam. Stars refer to the level of significance (5%, 1% or 0.1%) of the difference between performance with diazepam and with placebo. 5-10 mg diazepam overnight is without a residual effect, though it can be seen with 15 mg. Ingestion of 10 mg diazepam in the morning shows that the duration of impaired performance is limited to a few hours, and so likely to be contained by a sleep period. The limited duration of action of a single dose is related to its sustained distribution phase.

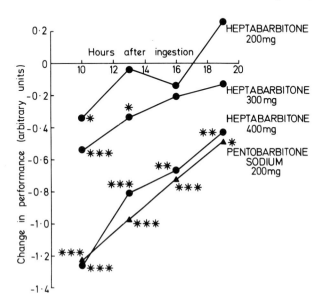

Fig. 4. Changes in performance on a visuo-motor coordination task (arbitrary units) after the barbiturates, heptabarbitone and pentobarbitone. Stars refer to the level of significance (5%, 1% & 0.1%) of the difference between performance with drug and with placebo. It can be seen that the severity and persistence of impaired performance is related to dose.

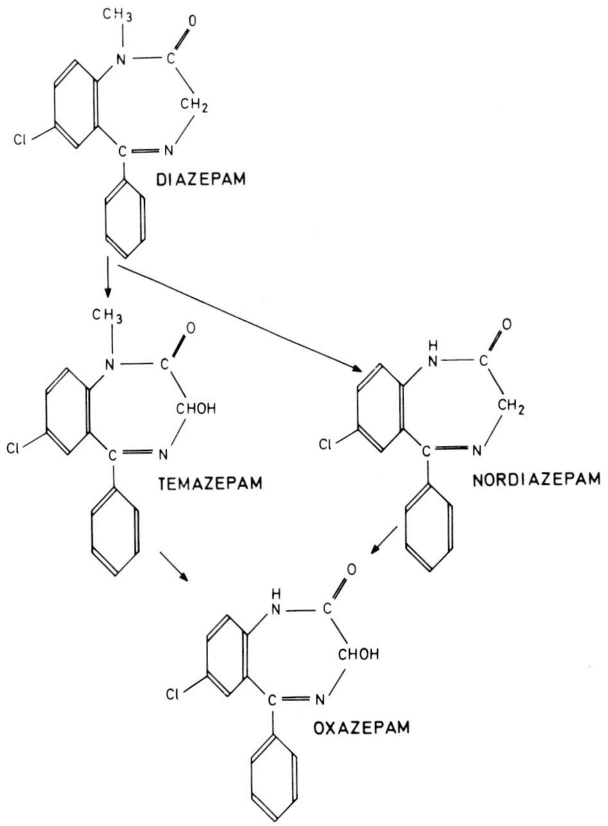

Fig. 5. Diazepam and its hydroxylated metabolites, temazepam (3-hydroxy-diazepam) and oxazepam (3-hydroxy,N-desmethyl-diazepam). Nordiazepam (N-desmethyldiazepam) is also a primary metabolite of diazepam, but has a very long elimination half-life (30-60 h) compared with those of temazepam (5-15 h) and oxazepam (5-20 h).

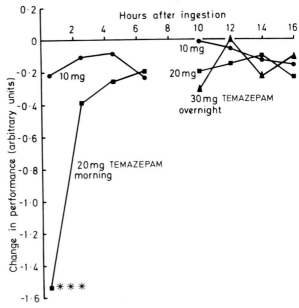

Fig. 6. Change in performance on a visuo-motor coordination task (arbitrary units) after temazepam. Stars refer to the level of significance (5%, 1% & 0.1%) of the difference between performance with the drug and with placebo. 10-30 mg temazepam overnight are without residual sequelae, though there is more likelihood of an effect as the dose increases. With 20 mg temazepam the duration of impaired performance is limited to a few hours, and so likely to be contained by a sleep period.

phase has a half-life of around 10 hours, while oxazepam has a single exponential decay with a somewhat longer half-life. There are no residual sequelae after overnight ingestion of 20 mg temazepam (Fig. 6), though with a higher dose (30 mg) they may appear. With oxazepam, the dose range 15-30 mg is also free of residual effects, though 45 mg leads to an obvious decrement (Fig. 7) (Clarke and Nicholson, 1978). Oxazepam (15-30 mg) has the disadvantage that in the formulations at present available it is slowly absorbed and so is unlikely to be effective when a reduction of sleep onset latency is necessary. Otherwise it is a useful hypnotic.

Triazolo- and imidazodiazepines

More recently attention has turned toward the triazolodiazepines and the imidazobenzodiazepines. These compounds have heterocyclic ring structures across the 1,2-position, and include triazolam and brotizolam of the triazolodiazepine group, and midazolam of the imidazobenzo-diazepine group. They have very fast elimination and appear to be without long-acting metabolites. Drugs with fast elimination may nevertheless have a slower fall in plasma concentration than the very fast distribution phase of drugs with a clear biexponential decay, and so provide a more sustained activity during the sleep period. Further, the rapid fall in plasma concentration around awakening is quicker than that of the elimination phase of the biexponential compounds, and so residual sequelae, if any, disappear rapidly. In this way drugs with fast elimination may provide a particularly favourable balance between efficacy and residual sequelae.

The plasma decay half-life of triazolam is around 3 hours, and brotizolam which has the same heterocyclic moiety across the 1,2-position but with the benzene ring replaced by a thieno complex, has a slightly longer half-life. Both are without residual effects on performance in doses up to 0.25 mg (Figs. 8 & 9) (Nicholson and Stone, 1980; Nicholson, Stone and Pascoe, 1980a) and accumulation with daily ingestion is highly unlikely. The half-life of midazolam is around 2 hours, and in doses up to at least 20 mg do not lead to performance decrements (Nicholson and Stone, 1983). Midazolam may prove to be particularly useful for the short periods of rest of shiftworkers.

SLEEP

Hypnotics with dose ranges free of residual sequelae must, nevertheless, preserve normal sleep patterns. A drug which can shorten the sleep onset

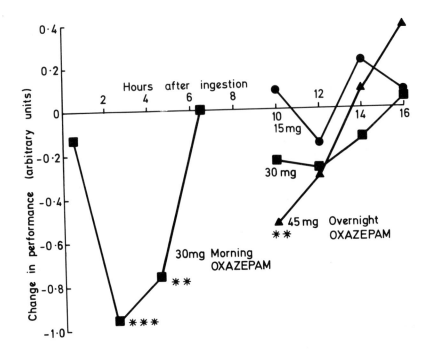

Fig. 7. Change in performance on a visuo-motor coordination task (arbitrary units) after oxazepam. Stars refer to the level of significance (5%, 1% & 0.1%) of the difference between performance with the drug and with placebo. 15-30 mg oxazepam overnight are without a residual effect, though there is a clear decrement with 45 mg. With 30 mg oxazepam in the morning onset of impaired performance is delayed, but the duration is limited to a few hours and so still likely to be contained by a normal sleep period. The delay in the onset of impaired performance indicates poor absorption of the drug in the formulation used. In sleep studies oxazepam has little, if any, effect on sleep onset latency.

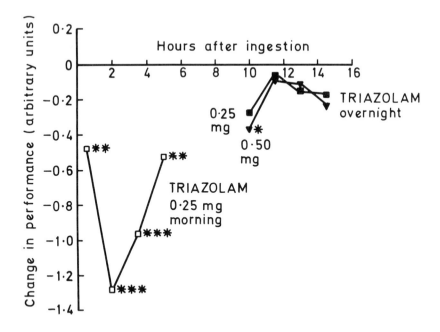

Fig. 8. Change in performance on a visuo-motor coordination task (arbitrary units) after triazolam. Stars refer to the level of significance (5%, 1% & 0.1%) of the difference between performance with the drug and with placebo. 0.25 mg triazolam overnight is without a residual sequel, though there may be an effect with the higher dose of 0.5 mg. Ingestion of 0.25 mg in the morning shows that the duration of impaired performance is limited to a few hours, and this is related to its high rate of elimination.

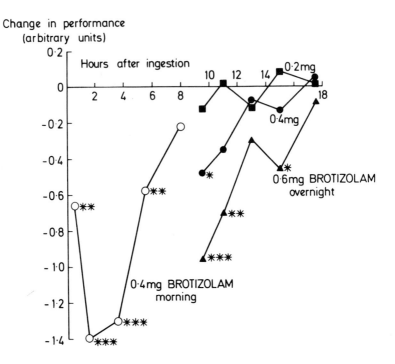

Fig. 9. Change in performance on a visuo-motor coordination task (arbitrary units) after brotizolam. Stars refer to the level of significance (5%, 1% & 0.1%) of the difference between performance with the drug and with placebo. 0.2 mg brotizolam ingested overnight is free of residual effects the next morning, but they appear with 0.4 mg, though limited to 10 h after ingestion. With ingestion of 0.4 mg brotizolam in the morning, impairment of performance is limited to a few hours. With brotizolam the overnight ingestion of 0.25 mg is likely to be free of sequelae the next day, and this is related to its high rate of elimination.

latency and reduce awake activity and possibly drowsy sleep without an
adverse effect on sleep architecture is required. Increased total sleep
time may not be essential. Diazepam (5-10 mg) and temazepam (10-20 mg)
fulfil these requirements (Nicholson, Stone and Clarke, 1976; Nicholson and
Stone, 1976). Oxazepam (15-30 mg) is poorly absorbed in the formulation at
present available and does not shorten sleep onset latency (Nicholson and
Stone, 1978). A good profile is also provided by triazolam (0.125-0.25 mg),
brotizolam (0.125-0.25 mg) and midazolam (10-20 mg), and these drugs are
without significant adverse effects on sleep (Nicholson and Stone, 1980;
Nicholson, Stone and Pascoe, 1980b; Nicholson and Stone, 1983). However
with many drugs the first period of REM sleep may be delayed, but though
REM sleep may be reduced during the early part of the night, duration is
not changed over the whole night.

Middle age

In the use of hypnotics in occupational medicine it is important that
they should be effective over the span of working life. The sleep of middle
age is more disturbed than that of the young adult. The sleep period contains more awake and drowsy activity, and so it may be expected than hypnotics would easily improve the sleep of the middle aged. However, diazepam
and temazepam are less effective than would be expected from studies with
the same drugs in young adults. Essentially, diazepam and temazepam reduce
awake activity in middle age. The possibility that some hypnotics may be
less useful in middle age is also supported by studies with other benzodiazepines. Somewhat higher doses may be required by this age group.
However, with diazepam doses above 10 mg should not be used, but with
temazepam the dose may be extended to 30 mg (Nicholson and Stone, 1979a),
and with triazolam 0.375 mg may well be appropriate. These doses are
unlikely to lead to performance decrements, and in any case, as they will be
used by those over 45 years, the question may not be quite so critical.
With brotizolam the dose range, 0.125-0.25 mg, is likely to be sufficient in
middle age (Nicholson, Stone and Pascoe, 1982).

Hypnotics and shiftwork

Although sleep disturbance associated with shiftwork has been well
documented, less attention has been given to the use of hypnotics at
unusual times, when the circadian desire for sleep is less - particularly
during the morning. It is true that an effect on sleep onset latency may

not be essential as many shiftworkers experience less difficulty in falling asleep than in maintaining their sleep, but daytime rest tends to be shorter than night time sleep and so a limited duration of action is essential.

Diazepam reduces awake activity and improves the efficiency of day time sleep, and in doses up to 10 mg is unlikely to impair performance beyond the sleep period. Temazepam has also proved to be effective for sleep during the day. Oxazepam, despite its slow absorption, is useful if there are no difficulties with sleep onset (Nicholson and Stone, 1979b). Interest also centres on the triazolo- and imidazo-diazepines for the management of sleep disturbances in shiftworkers. Doses of 0.125 mg triazolam and brotizolam are indicated, at least initially, for the shorter periods of sleep which are seen in shiftworkers, and would certainly be free of residual sequelae (Nicholson & Stone, 1980; Nicholson, Stone & Pascoe, 1980). Midazolam (10-20 mg) is also promising, but more information is needed on this drug.

SUMMARY

It is evident that much progress has been made over the past few years in the development of hypnotics which can be used for those involved in skilled activity. However, because of the critical nature of the requirements in occupational medicine, more information is needed, not only on the use of some of these drugs in the management of normal patterns of rest and activity, but also the management of irregular patterns of rest associated with shiftwork. There is a wide choice, though the disadvantages of some drugs must be borne in mind. Diazepam (5-10 mg) is an excellent hypnotic when used occasionally for both night time and day time sleep, but it should not be ingested more than once every 48 hours and more than twice in 7 days. Temazepam and oxazepam are also useful, but oxazepam (15-30 mg) is unlikely to shorten sleep onset. Triazolam (0.125-0.25 mg) can be used daily - if this is essential - and this is so with the other triazolodiazepine, brotizolam (0.125-0.25 mg). Midazolam (10-20 mg) with its very short elimination half-life is a promising hypnotic in the context of shiftwork.

Finally, it must be emphasised that if hypnotics are to be used in the management of sleep difficulties in those who carry out skilled work a drug with which the individual is familiar must be used. It should be given at the lowest dose, and as infrequently as possible. There should be an interval of 24 h between ingestion and commencement of duty unless their use is supervised. Under these circumstances the interval may be reduced to 12 hours. Perhaps it it unnecessary to emphasise that, if hypnotics are to be used in the management of sleep disturbances, the concomitant use of alcohol it to be avoided.

ALLERGY AND ANTIHISTAMINES

The use of antihistamines is usually associated with impairment of central nervous function, but it is unclear whether unacceptable decrements in performance are an inevitable sequel of the use of these drugs. Drowsiness has been attributed to various mechanisms such as inhibition of histamine N-methyltransferase and blockade of central histaminergic receptors, and it has been postulated that greater affinity for peripheral rather than central receptors could diminish their sedative effect. Serotonergic antagonism, anticholinergic activity and blockade of central alpha adrenoreceptors may also be involved, although central alpha adrenoceptor blockade and sedation have not correlated well for this group of drugs.

Central effects are dependent on the ability of a drug to cross the blood-brain barrier and most H_1-antihistamines (unlike histamine, the H_1 and H_2-receptor agonists and the H_2-antihistamines) are highly liposoluble and pass the blood-brain barrier with ease. Nevertheless, it has been shown in animals that high doses of some H_1-antihistamines cause little, if any, central effects. Such compounds may cross the blood-brain barrier only slowly and so tolerance of the central nervous system may develop gradually without immediate effects on performance.

Antihistamines without adverse effects on performance would be useful, and so over the past few years we have studied this group of drugs in man with particular reference to their effects on performance.

INITIAL STUDIES

Initial studies were concerned with chlorpheniramine (4 mg), clemastine (1 mg), promethazine (10 mg) and terfenadine (60 mg). Effects on visuomotor coordination after morning ingestion were compared with placebo in healthy females. Performance was impaired around 1.5 h after chlorpheniramine, from 3.0 to 5.0 h after clemastine and from 3.0 to 5.0 h after promethazine. However, it was not possible to establish performance decrements after ingestion of terfenadine. Indeed, the subjects reported improved alertness and wakefulness 0.5 to 3.5 h after ingestion of this drug (Fig. 10).

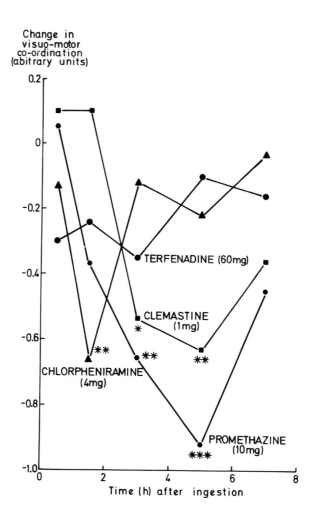

Fig. 10. Change in performance on visuo-motor coordination (arbitrary units) after 4 mg chlorpheniramine, 10 mg promethazine, 1 mg clemastine and 60 mg terfenadine.

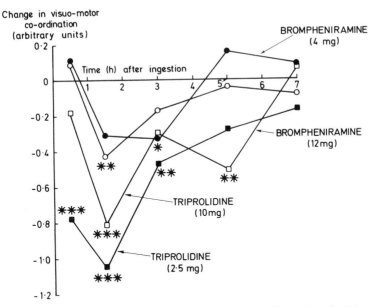

Fig. 11. Change in performance on visuo-motor coordination (arbitrary units) after 2.5 mg (filled square) and 10.0 mg (open square) triprolidine hydrochloride, and after 4.0 mg (closed circle) and 12.0 mg (open circle) brompheniramine maleate. Filled symbols - immediate release and open symbols - modified release preparations.

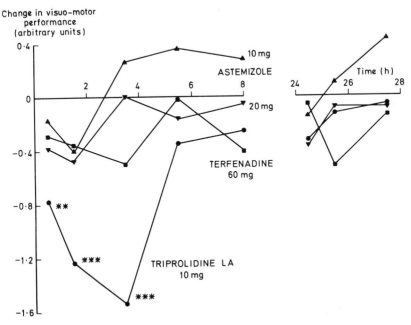

Fig. 12. Change in performance on visuo-motor coordination (arbitrary units) after 10 mg triprolidine in modified release form, 60 mg terfenadine and 10 & 20 mg astemizole.

This study clearly demonstrated that the appearance of central effects with antihistamines may vary between drugs, and that, though some decrease in performance may be likely, persistence and severity may differ. With chlorpheniramine impaired performance was limited and appeared shortly after ingestion, whereas with clemastine and promethazine performance decrements were more serious, though delayed. On the other hand terfenadine may be without effects on performance, and subjectively, at least, alertness and wakefulness may be increased.

Studies on three of these antihistamines had been reported previously. The appearance of impaired performance with promethazine (25 & 30 mg) was similar to that observed in our studies, though we had studied a much lower dose. The delay we had observed with clemastine had also been reported and was consistent with pharmacokinetic data. Peck, Fowle and Bye (1975) had observed impaired performance using prolonged and monotonous tasks, and they suggested that such tasks may be more sensitive to the effects of antihistamines than brief, interesting tasks. However, our studies, which are in close agreement with Peck et al. (1975), suggested that impaired performance may be observed with tasks of limited duration.

Impaired performance may not always be detected by subjects. With chlorpheniramine, clemastine and promethazine, there was little subjective evidence of impaired performance at the time when the drugs were having their maximum effect. The reliability of subjective assessments in studies with antihistamines is therefore questionable. Even with objective measures of performance there is much variation between the response of subjects. However, this may suggest that the appropriate choice of antihistamine may reduce the possibility of impaired performance in the individual, and even avoid it in some subjects.

Overnight ingestion of long-acting antihistamines may avoid daytime sedation and at the same time provide useful antihistaminic activity. This possibility has been explored using drugs available in sustained release form. Brompheniramine maleate and triprolidine hydrochloride were each studied in females. Though triprolidine hydrochloride (2.5 mg) had an immediate effect on performance which persisted to 7.0 h, the sustained release form (10 mg) only impaired performance from 1.5-5.0 h. Brompheniramine maleate (4 mg) impaired performance from 1.5-3.0 h, and with this drug the sustained release preparation (12 mg) only impaired performance around 1.5 h (Fig. 11).

The observations with triprolidine (2.5 mg) were in close agreement with the previous studies of Bye, Dewsbury and Peck (1974) and Peck, Fowle and Bye (1975). Essentially, the onset of impaired performance was rapid and recovery occurred around 7 h and, although the decrease in performance appeared later with the sustained release form, recovery occurred within a similar period of time. A similar pattern was observed with brompheniramine as with both the immediate and sustained release preparations recovery occurred within a few hours. These observations would suggest that sustained release preparations could be used overnight if their antihistaminic activity persisted during the next day. With triprolidine both weal and flare measurements show an antihistaminic effect up to 24 h after ingestion (Fowle, Hughes and Knight, 1971), and with brompheniramine maleate there is pharmacokinetic evidence of maintained plasma levels.

The question also arises whether tolerance of the central nervous system may develop during continued exposure to some antihistamines. There is some evidence to support this suggestion in the study of Bye, Claridge, Peck and Plowman (1977) with repeated 12 hourly ingestion of triprolidine, and with our studies with the sustained release preparations. Antihistaminic activity may lead to impaired performance for only the initial part of its peripheral activity, and so it would appear that the different durations of impaired performance and antihistaminic activity of sustained release preparations may have clinical importance.

ASTEMIZOLE AND TERFENADINE

The possibility that terfenadine may be free of central effects when ingested orally prompted further studies with this drug and with two other antihistamines, astemizole and mequitazine. The effects of terfenadine (60 mg) and astemizole (10 & 20 mg) on visuo-motor coordination, arithmetical ability and digit symbol substitution and on mood were again studied in healthy females. The study included triprolidine (10 mg) in a sustained release form as a control with known central effects. Triprolidine (10 mg) caused a decrement in visuo-motor coordination from 0.5 to 3.5 h, but there were no changes in performance after terfenadine (60 mg) and astemizole (10 & 20 mg) (Fig. 12). The subjects assessed their performance as impaired with triprolidine (10 mg), and their mood assessments were also altered.

This study confirmed our previous observations with triprolidine. Peck et al. (1975) had reported that triprolidine (1.25, 2.5 & 5.0 mg) impaired auditory vigilance and prolonged reaction times, and that the highest dose impaired digit symbol substitution. In our previous study we found that the persistence of effects with 2.5 mg triprolidine in the standard formulation and with the 10 mg long-acting preparation were similar, and so the absence in the study with 10 mg triprolidine of an effect on digit symbol substitution was to be expected.

ANTIHISTAMINES AND VISUAL PERFORMANCE

Our studies on astemizole and terfenadine have been extended to visual function. The effects of triprolidine (10 mg) in a sustained release form, astemizole (10 mg) and terfenadine (60 mg) on dynamic visual acuity (Fig. 13), pupillary response to light (Fig. 14), critical flicker fusion (Fig. 15), digit symbol substitution and letter cancellation, as well as subjective assessments of mood were measured. Triprolidine impaired dynamic visual acuity and reduced the frequency at which a flickering light appeared steady up to 2.5 h after ingestion, but there were no changes with astemizole or terfenadine. The diameter of the pupil and its response to light were not changed by the drugs, and performance on digit symbol substitution and cancellation were also not altered.

Dynamic visual acuity is related to such variables as age and gender (Burg, 1966; Reading, 1972a & b) and fatigue (Behar, Kimball and Anderson, 1972). Changes in this measure with drugs may be due to a number of factors. An increase in reaction time will lead to greater displacement of the target image before any corrective movement is made, while displacement of the image from the fovea will persist even after ocular movement if a decrease in the size of corrective saccades or in saccadic velocity occurs. Further, a decrease in the velocity of the smooth pursuit phase may lead to a continued displacement, and even to movement of the image on the retina.

Impairment of dynamic visual acuity by triprolidine also varied with target velocity. At low velocities there was a reduction in acuity 1.0 & 2.5 h after ingestion, and this pattern was similar to that observed with visuo-motor coordination (Clarke and Nicholson, 1978; Nicholson and Stone, 1982) in which the stimulus had an average velocity of 15°/sec. Slowed smooth pursuit may be involved both in the impairments of dynamic acuity at these velocities as well as in visuo-motor coordination.

With the highest velocity (83°/sec) an improvement in acuity after placebo was observed during the same day, and this may have been due to learning or to an improvement in performance related to circadian rhythmicity. 1.0 h after ingestion, performance was so low under placebo that impairments were not possible. At the later testing times the improved

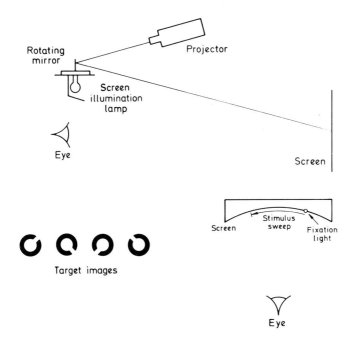

Fig. 13. Measurement of dynamic visual acuity. The target images were four Landolt C rings with gaps at the oblique meridians, which were projected on to a curved screen by a rotating mirror. A fixation light was provided, and the subject was required to determine the orientation of the gap.

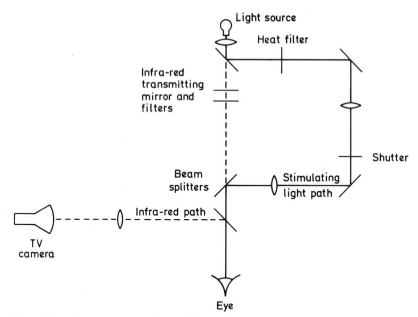

Fig. 14. Measurement of pupillary diameter and the response to light. Light from a source was split by an infra-red transmitting mirror into visible and infra-red beams. The visible beam stimulated the retina, and the change in pupil size was detected by the infra-red sensitive camera.

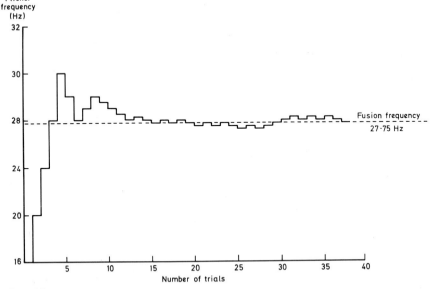

Fig. 15. Measurement of critical flicker fusion. Initially, the subject indicated that a light flickering at 16 Hz was not fused. In response the frequency was altered stepwise. Fusion was defined as the frequency at which 50% or more of the last 25 responses was considered to be fused.

acuity under placebo was decreased by triprolidine. These decrements were more likely to be related to slowing of saccadic eye movements than to slowing of smooth pursuit, which occupied very little time at this target velocity.

The studies on triprolidine suggested that antihistamines may slow both saccadic eye movements and smooth pursuit, and this conclusion is similar to that drawn from studies with barbiturates (Rashbass, 1961; Tedeschi et al., 1981) and with benzodiazepines (Bittencourt and Richens, 1981; Bittencourt et al., 1981). Peak saccadic velocity and smooth pursuit are impaired after ingestion of amylobarbitone and after diazepam and temazepam, and so with barbiturates, benzodiazepines and antihistamines impaired performance of complex tasks could involve ocular mechanisms.

MEQUITAZINE

Performance studies have also been carried out with the H_1-histamine receptor antagonist, mequitazine, which is used widely in the management of allergic states. Clinical studies indicate that at the normal therapeutic dose of 5 mg, mequitazine may be free of sedation (Gervais et al., 1975; Halpern et al., 1977; Muler and Blum, 1977; Orlando, 1977; Caille, 1979).

Visuo-motor coordination, digit symbol substitution, critical flicker fusion and dynamic visual acuity were measured. The subjects also assessed their mood and performance. As previously, triprolidine (10 mg) used as an active control impaired visuo-motor coordination from 1.5-5.5 h after ingestion. In these subjects it also reduced the number of substitutions in the digit symbol substitution test from 0.5-5.5 h, lowered the threshold at which a flickering light appeared to fuse (0.5-3.5 h after ingestion) and impaired the visual acuity of a moving target around 0.5 h. There were no changes with terfenadine.

Effects of mequitazine at its therapeutic dose of 5 mg were minimal. With 5 mg mequitazine visuo-motor coordination was impaired 7.5 h after ingestion (Fig. 13), but there were no changes in co-ordination at other times, and there were no effects on digit symbol substitution test, critical flicker fusion or dynamic visual acuity. Examination of the data suggested that the unexpected change in visuo-motor coordination around 7.5 h was due to one subject whose performance had been low throughout the day. With the higher dose of 10 mg (which is double the recommended dose) visuo-motor performance was decreased from 3.5-5.5 h after ingestion and digit symbol substitutions were reduced around 5.5 h. However, 10 mg mequitazine did not alter the threshold of fusion of a flickering light or impair visual dynamic acuity.

CONCLUSIONS

It is clear that decreased performance is not an inevitable accompaniment of the use of antihistamines. Studies with terfenadine (Kulshrestha et al., 1978; Clarke and Nicholson, 1978; Moser et al., 1978; Fink and Irwin, 1979) have failed to establish central effects, and in this way terfenadine is a promising antihistamine for those involved in skilled activity (Nicholson, 1982). This could be related to absence of anticholinergic activity in the doses used or to its affinity for peripheral receptors. It would appear to cross the blood-brain barrier with difficulty and so tolerance may be able to develop without obvious central effects on initial exposure.

Astemizole is also a promising antihistamine. It is rapidly absorbed and distributed with peak plasma levels around 1 h. However, it is exceptionally slowly eliminated with a half-life of around 20 days and so more information is needed on the possible significance of this property, and on the appropriate dose regime. Activity is due to direct antagonism of H_1-receptors. It has no affinity for acetylcholine receptors and is devoid of β-adrenergic receptor activity, and there is no evidence of H_2-receptor blockade. There is some evidence of 5-HT antagonism in high doses as well as affinity for alpha-adrenergic receptors, but as it also crosses the blood-brain barrier with difficulty, interactions with central receptors are unlikely to lead to immediate effects.

Mequitazine possesses a greater affinity for peripheral H_1-receptors than for central H_1-receptors and this has been proposed as an explanation for the absence of sedative effects at doses which may nevertheless be optimum for its peripheral activity. However, at higher doses blockade of central alpha-adrenoceptors could lead to a sedative effect: both mequitazine and promethazine produce such a blockade, though chlorpheniramine is practically ineffective. It is therefore unlikely that such a mechanism applies generally to the sedative activity of the H_1-antagonists. It is more likely that sedative effects are due to blockade of central H_1-receptors (Uzan et al., 1979) and so lack of sedative properties may be attributed to poor or slow access to cerebral receptors (Quach et al., 1980). Mequitazine would appear to have a therapeutic window around 5 mg with useful antiallergic activity free of central effects

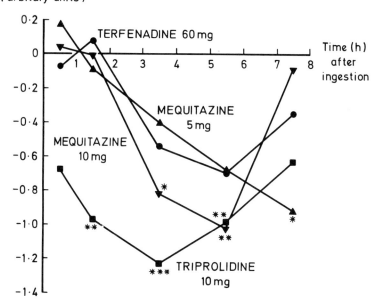

Fig. 16. Change in performance on a visuo-motor coordination task (arbitrary units) after mequitazine. Stars refer to the level of significance (5%, 1% & 0.1%) of the difference between performance with the drug and with placebo. The effect with mequitazine (5 & 10 mg) is compared with terfenadine (60 mg) and with triprolidine (10 mg).

Though we have used recommended or anticipated doses of these antihistamines it must be emphasised that they may not necessarily have equivalent histamine antagonism. In particular triprolidine is a potent inhibitor. Nevertheless, present clinical evidence suggests that terfenadine (Huther et al., 1977; Cheng et al., 1977; Dubuis et al., 1978; Reinberg et al., 1978; Brandon and Viner, 1980), astemizole (De Volder and Van de Heyning, 1980; Chapman and Rawlins, 1982; Callier et al., 1980) and mequitazine (Gervais et al., 1975; Halpern et al., 1977; Muler and Blum, 1977) also provide useful antihistaminic activity free of effects on the central nervous system.

Indeed, the careful use of antihistamines may lead to greater safety as well as to an adequate anti-allergic effect. Triprolidine and brompheniramine may be ingested overnight if their persistent anti-allergic activity during the next day is sufficient, while mequitazine (5 mg), terfenadine (60 mg) and astemizole (10 mg) may be ingested in the doses cited during the day. However, an important question remains concerning their relative efficacy. Brompheniramine, triprolidine and mequitazine are well established antihistamines, whereas the effectiveness of astemizole and terfenadine in day-to-day use in clinical practice remain to be explored in detail, and more information is needed on the appropriate dose and dosing interval for astemizole.

REFERENCES

Sleep Disturbance and Hypnotics

Breimer, D.D. 1979. Pharmacokinetics and metabolism of various benzodiazepines used as hypnotics. Br. J. clin. Pharmac., $\underline{8}$, 9-13S.

Borland, R.G. and Nicholson, A.N. 1974. Human performance after a barbiturate (heptabarbitone). Br. J. clin. Pharmac., $\underline{1}$, 209-215.

Clarke, C.H. and Nicholson, A.N. 1978. Immediate and residual effects in man of the metabolites of diazepam. Br. J. clin. Pharmac., $\underline{6}$, 325-331.

Curry, S.H. and Whelpton, R. 1979. Pharmacokinetics of closely related benzodiazepines. Br. J. clin. Pharmac., $\underline{8}$, 15-21S.

Nicholson, A.N. and Stone, B.M. 1976. Effect of a metabolite of diazepam, 3-hydroxydiazepam (temazepam) on sleep in man. Br. J. clin. Pharmac., $\underline{3}$, 543-550.

Nicholson, A.N. and Stone, B.M. 1978. Hypnotic activity of 3-hydroxy,N-desmethyldiazepam (oxazepam). Br. J. clin. Pharmac., $\underline{5}$, 469-472.

Nicholson, A.N. and Stone, B.M. 1979a. Diazepam and 3-hydroxydiazepam (temazepam) and sleep of middle age. Br. J. clin. Pharmac., $\underline{7}$, 463-468.

Nicholson, A.N. and Stone, B.M. 1979b. Hypnotic activity during the day of diazepam and its hydroxylated metabolites, 3-hydroxydiazepam (temazepam) and 3-hydroxy,N-desmethyldiazepam (oxazepam). In "Chronopharmacology" (Ed. A. Reinberg and F. Halberg). (Pergamon Press,

Oxford). pp. 159-169.
Nicholson, A.N. and Stone, B.M. 1980. Activity of the hypnotics flunitrazepam and triazolam in man. Br. J. clin. Pharmac., 9, 187-194.
Nicholson, A.N. and Stone, B.M. 1983. Imidazo-benzodiazepines: Sleep and performance studies in man. J. clin. Psychopharmac. In Press.
Nicholson, A.N., Stone, B.M. and Clarke, C.H. 1976. Effect of diazepam and fosazepam (a soluble derivative of diazepam) on sleep in man. Br. J. clin. Pharmac., 3, 533-541.
Nicholson, A.N., Stone, B.M. and Pascoe, P.A. 1980a. Studies on sleep and performance with a triazolo-1,4-thienodiazepine (brotizolam). Br. J. clin. Pharmac., 10, 75-81.
Nicholson, A.N., Stone, B.M. and Pascoe, P.A. 1980b. Efficacy of some benzodiazepines for day time sleep. Br. J. clin. Pharmac., 10, 459-463.
Nicholson, A.N., Stone, B.M. and Pascoe, P.A. 1982. Hypnotic efficacy in middle age. J. clin. Psychopharmac., 2, 118-121.

Allergy and Antihistamines

Behar, I., Kimball, K.A. and Anderson, D.A. 1976. Dynamic visual acuity in fatigued pilots. US Army Aeromedical Research Laboratory, Fort Rucker, Alabama, Report ADA 027663.
Bittencourt, P.R.M. and Richens, A. 1981. Smooth pursuit eye movements and benzodiazepines. Electroenceph. clin. Neurophysiol. In Press.
Bittencourt, P.R.M., Wade, P., Smith, A.T. and Richens, A. 1981. The relationship between peak velocity of saccadic eye movements and serum benzodiazepine concentration. Br. J. clin. Pharmac., 12, 523-533.
Borland, R.G. and Nicholson, A.N. 1974. Human performance after a barbiturate (heptabarbitone). Br. J. clin. Pharmac., 1, 209-215.
Brandon, M.L. and Weiner, M. 1980. Clinical investigation of terfenadine, a non-sedative antihistamine. Annals of Allergy, 44, 71-75.
Brown, B., Adams, A.J., Haegerstrom-Portnoy, G., Jones, R.T. and Flom, M.C. 1975. Effects of alcohol and marijuana on dynamic visual acuity: I. Threshold measurements. Perception and Psychophysics, 18, 441-446.
Burg, A. 1966. Visual acuity as measured by dynamic and static tests. J. appl. Psychol., 50, 460-466.
Bye, C., Dewsbury, D. and Peck, A.W. 1974. Effects on the human central nerve system of two isomers of ephedrine and triprolidine, and their interaction. Br. J. clin. Pharmac., 1, 71-78.
Bye, C.E., Claridge, R., Peck, A.W. and Plowman, F. 1977. Evidence for tolerance to the central nervous system effects of the histamine antagonist, triprolidine, in man. Eur. J. clin. Pharmac., 12, 181-186.
Caille, E.-J. 1979. Étude comparative du Primalan et de la Polaramine sur l'eeg, la vigilance et la thymie. Psychologie Médicale, 11, 1541-1555.
Callier, J., Engelen, R.F., Ianniello, I., Olzem, R., Zeisner, M. and Amery, W.K. 1980. Astemizole (R 43512) in the treatment of hay fever. An international double blind study comparing a weekly treatment (10 mg & 25 mg) with a placebo. Curr. therap. Res., 29, 24-35.
Chapman, P.H. and Rawlins, M.D. 1982. A randomized single blind study of astemizole and chlorpheniramine in normal volunteers. Br. J. clin. Pharmac. In Press.
Cheng, H.C., Reavis, O.K., Munro, N.L. and Woodward, J.K. 1977. Antihistaminic effect of terfenadine. Pharmacologist, 19, 187.
Clarke, C.H. and Nicholson, A.N. 1978. Performance studies with antihistamines. Br. J. clin. Pharmac., 6, 31-36.

De Volder, M. and Van de Heyning, J. 1980. Concurrent use of histamine skin tests and histamine nose provocation tests for the study of a new antihistamine. Proceedings of the International Alpine Surgical Society Congress, Astria (Janssen Pharmaceutica Clinical Research Report R 43512/2).

Dubuis, P., Charpin, J., Orlando, J.P. and Gervais, P. 1978. Le traitement antihistaminique des pollinosis aigues a l'epreuve du double insu. Thérapeutique, 2, 137-140.

Fink, M. and Irwin, P. 1979. CNS effects of the antihistamines diphenhydramine and terfenadine (RMI 9918). Pharmakopsychiat., 12, 35-44.

Fowle, A.S.E., Hughes, D.T.D. and Knight, G.J. 1971. The evaluation of histamine antagonists in man. Eur. J. clin. Pharmac., 3, 215-220.

Gervais, P., Gervais, A., de Beule, R. and Van de Bijl, W. 1975. Essai compare d'un nouvel antihistamine, la mequitazine et d'un placebo. Acta Allergologica, 30, 286-297.

Halpern, G., Sabouraud, D. and Garcelon, M. 1977. Recherche par la technique a double insu d'une diminution de vigilance induite par deux antihistaminiques chez des patients traites en ambulatoire. Allergie et Immunologie, 9, 15-18.

Hedges, A., Hills, M., Mackay, W.P., Newman-Taylor, A.J. and Turner, P. 1971. Some central and peripheral effects of meclastine, a new antihistaminic drug, in man. J. clin. Pharmac., 11, 112-119.

Huther, K.J., Renftle, G., Barraud, N., Burke, J.T. and Koch-Weser, J. 1977. Inhibitory activity of terfenadine on histamine-induced skin wheals in man. Eur. J. clin. Pharmac., 12, 195-199.

Krnjevic, K.J. and Phyllis, J.W. 1963. Actions of certain amines on cerebral cortical neurones. Br. J. Pharmac., 20, 471-490.

Kulshrestha, V.K., Gupta, P.P., Turner, P. and Wadsworth, J. 1978. Some clinical pharmacological studies with terfenadine, a new antihistaminic drug. Br. J. clin. Pharmac., 6, 25-30.

Large, A.T.W., Wayte, G. and Turner, P. 1971. Promethazine on hand-eye coordination and visual function. J. Pharm. Pharmac., 23, 134-135.

Molson, G.R., Mackay, J.A., Smart, J.V. and Turner, P. 1966. Effect of promethazine hydrochloride on hand-eye co-ordination. Nature, 209, 516.

Moser, L., Huther, K.J., Koch-Weser, J. and Lundt, P.V. 1978. Effects of terfenadine and diphenhydramine alone or in combination with diazepam or alcohol on psychomotor performance and subjective feelings. Eur. J. clin. Pharmac., 14, 417-423.

Muler, H. and Blum, F. 1977. Essai compare de deux antihistaminiques: mequitazine et dexchlorpheniramine. Gazette Médicale de France, 84, 3-7.

Netter. K.J. and Bodenschatz, K. 1967. Inhibition of histamine-N-methylation by some antihistamines. Biochem. Pharmac., 16, 1627-1631.

Nicholson, A.N. 1979. Effect of the antihistamines, brompheniramine maleate and triprolidine hydrochloride, on performance in man. Br. J. clin. Pharmac., 8, 321-324.

Nicholson, A.N. 1982. Antihistaminic activity and central effects of terfenadine. A review of European studies. Arzneimittel Forschung, 32, 1191-1193.

Nicholson, A.N. and Stone, B.M. 1982. Performance studies with the H_1-histamine receptor antagonists, astemizole and terfenadine. Br. J. clin. Pharmac., 13, 199-202.

Nicholson, A.N., Smith, P.A. and Spencer, M.B. 1982. Antihistamines and visual function: Studies on dynamic visual acuity and the pupillary response to light. Br. J. clin. Pharmac., 14, 683-690

Orlando, J.-P. 1977. Utilisation du primalan dans une consultation d'allergologie respiratoire. Le Journal des Agreges, 10, 119-120.

Peck, A.W., Fowle, A.S.E. and Bye, C. 1975. A comparison of triprolidine and clemastine on histamine antagonism and performance tests in man: Implications for the mechanism of drug induced drowsiness. Eur. J. clin. Pharmac., 8, 455-463.

Quach, T.T., Duchemin, A.M., Rose, C. and Schwartz, J.C. 1979. In vivo occupation of cerebral histamine H_1-receptors evaluated with 3H-mepyramine may predict sedative properties of psychotropic drugs. Eur. J. Pharmac., 60, 391-392.

Quach, T.T., Duchemin, A.M., Rose, C. and Schwartz, J.C. 1980. Labeling of histamine H_1-receptors in the brain of the living mouse. Neuroscience Letters, 17, 49-54.

Rashbass, C. 1961. The relationship between saccadic and smooth tracking eye movements. J. Physiol., 159, 326-338.

Reading, V.M. 1972a. Visual resolution as measured by dynamic and static tests. Pflugers Arch. ges. Physiol., 333, 17-26.

Reading, V.M. 1972b. Analysis of eye movement responses and dynamic visual acuity. Pflugers Arch. ges. Physiol., 333, 27-34.

Reinberg, A., Levi, F., Guillet, P., Burke, J.T. and Nicolai, A. 1978. Chronopharmacological study of antihistamines in man with special reference to terfenadine. Eur. J. clin. Pharmac., 14, 245-252.

Tedeschi, G., Bittencourt, P.R.M., Smith, A.T. and Richens, A. 1981. Specific oculomotor defects after amylobarbitone. In Press.

Turner, P. 1968. Critical flicker frequency and centrally-acting drugs. Br. J. Ophth., 52, 245-250.

Uzan, A., Le Fur, G. and Malgouris, C. 1979. Are antihistamines sedative via a blockade of brain H_1 receptors? J. Pharm. Pharmac., 31, 701-702.

MECHANICAL VIBRATION IN TRANSPORT OPERATIONS

L. Vogt

DFVLR Institut für Flugmedizin
Linder Höhe
5000 Köln 90, F.R.G.

ABSTRACT

The response of humans to vertical (z-axis) vibrations encountered in transport operation is reviewed. The human body is a resilient system with resonances in the frequency range between 2 and 100 Hz. Vibration effects on various body systems (intestinal, cardiovascular, pulmonary, visual) are described. Although national and international standards are available, vibration evaluation still poses problems, when complex motions are encountered. Health hazards and chronic vibration effects have been described in the past, however, it is not yet possible to establish a dose-response relationship.

INTRODUCTION

Mechanical vibration is an environmental factor which is closely related to transport operations. Harmonic motions are produced by man during natural locomotion (Simic, 1974) and are in some cases even considered as pleasant. However if vibration amplitude and frequency exceed certain limits discomfort is produced and performance may be impaired. Under severe conditions mechanical vibrations may even threaten the health of the operator.

Through the centuries great efforts have been made to isolate man from the vibrations of coaches and other vehicles. But it was until the thirties of this century that scientists realized that the human body could be regarded as a resilient mechanical system and that its dynamic properties have to be known if its response shall be accessed (Bekesy v., 1939), (Reiher, et al., 1931). The purpose of this paper is to point out the basic facts of human vibration response and to elucidate some open questions.

THE REACTION OF THE WHOLE BODY TO MECHANICAL FORCES

Although there is an indefinite number of possibilities how mechanical vibrations are transmitted to the human body, the following have operational importance:
a) Vibrations transmitted simultaneously to the whole body surface or substantial parts of it. This occurs when the body is immersed in a vibrating medium.
b) Vibrations transmitted to the body as a whole through the supporting surface like the feet of a standing, the buttocks of a seated, or the supporting area of a reclining person.
c) Vibrations applied to particular parts of the body such as the head or limbs.

This presentation will mainly deal with vibrations transmitted along the long body axis (Z-axis) via the buttocks or the feet.

Vibrations are defined by amplitude, phase and frequency. According to the physical response of the human body four frequency bands between 1 Hz and 30 KHz can be defined:

Below 1 Hz the body moves like a pure mass and no significant relative displacements of body parts take place. In this band the vestibular organ has its greatest sensitivity and motion sickness occurs (ISO, 1982b), (Reason, et al., 1975), (Shoenberger, 1975).

Between 1 Hz and 20 Hz the main body resonances are located. One possibility to gain an insight into the dynamic behavior of the human body is to regard its mechanical impedance (Coermann, 1963), (Vogt, et al., 1968). It is the complex ratio of transmitted force to velocity and reflects the sum of the relative displacements within the body. Figure 1 gives the standard impedance of sitting humans (ISO, 1981). It exhibits a main resonance at 5 Hz and a second peak around 11 Hz. Frequencies between 20 Hz and 100 Hz are damped progressively by the human body and only a fraction of the input forces reach the upper body parts and the head.

However, some important body resonances are located in this frequency band. They affect the dynamics of the cervical part of the vertebral column and the visual system. In this frequency range there are also a number of non-specific responses of the neuromuscular system which influence autonomous reflexes (Loeckle, 1941). Vibrations of this kind can therefore alter motor actions of the operator.

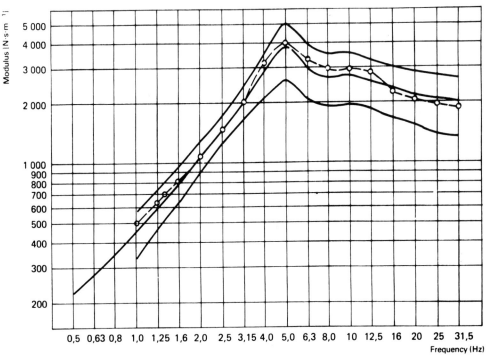

Fig. 1   Driving point impedance of humans, sitting position (ISO 1981).

Vibrations of 100 Hz and more have an effect on some bony structures like the skull. It has its main resonance in the frequency range around 400 Hz. These data have an important meaning for the evaluation of the impact sensitivity of the head. Also hand-arm vibrations have to be assessed up to 1000 Hz. White finger disease, bone and joint affections of the upper limbs result from exessive exposure in this frequency range (Wasserman, et al., 1982).

In summary it can be stated, that for whole-body vibration exposure, the frequency range from 2 Hz to 100 Hz will contain the maximum response.

EFFECTS ON THE INNER ORGANS

At resonance the abdominal and thoracic organs exhibit a piston-like up and down-motion. This leads to disturbances in respiration. A typical example is shown in Figure 2. It indicates that large amounts of air are moved into and out of the respiratory duct during resonance around 4 Hz (Coermann, et al., 1960). But if the arterial CO_2 partial pressure is determined simultaneously, it becomes obvious that this air volume is effective mainly in ventilating the functional dead space, however in some cases signs of hypercapnia were observed (Hood, et al., 1965).

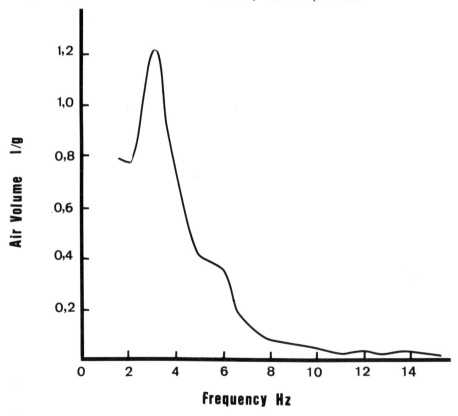

Fig. 2 Pulmonary ventilation during vibration (after Coermann 1960).

The motion of the viscera has effects on speach which can be heard when listening to radio communications with helicopters. The voice is modulated by vibrations and acquires a tremolo quality with an increase in pitch. Both effects lead to a decreased intelligibility of single words. However no major decrements in communications occur if whole scentences and standard phrases are used (Nixon, 1962), (Nixon, et al., 1963).

Relative displacements and associated pressure changes in the gastrointestinal system have been recorded by various authors (Clark, et al., 1963), (Nickerson, et al., 1962). Others have reported an increase in affections of the stomach of the operators of various vehicles like buses, trucks, tractors and construction machines (Gruber, et al., 1974). It has been proved in the past that mechanical vibrations lead to an increased and prolonged production of gastric juice compared to a resting condition (Dupuis, et al., 1966). Although there is a significant increase in ulcera, chronic gastritis and "nervous stomach" the aethiology remains complex. Traffic conditions, noise, mental strain, and eating habits may contribute considerably to an overall stress situation of which mechanical vibration is only another component. The difficulty to obtain significant findings from statistical surveys is a typical problem in vibration research.

THE CARDIOVASCULAR SYSTEM

Changes in cardiovascular functions are similar during vibration in all axis. The maximum response is again located in the 6 to 10 Hz range. Figure 3 shows some representative reactions of the cardiovascular system (Hood, et al., 1965), (Hood, et al., 1966).

They are: mean arterial blood pressure, heart rate, cardiac index, and oxygen consumption index plotted against frequency. The solid lines represent control values which were taken during a rest period before a new vibration frequency was applied. The experiment was made with 0.6 and

1.2 g acceleration amplitudes. These values exceed the published ISO 16 minutes exposure limits considerably and would therefore not be allowed in a transport system. In spite of this the reactions of the circulatory system resemble only the response of a human to moderate exercise. The increased effort in bracing against the vibration and possibly the increased work of breathing may account for some but not all of these responses. The same trends are also seen in anesthetized animals. Mechanical stimulation of various mechanical receptors may also contribute to the observed changes (Gierke v., et al., 1971).

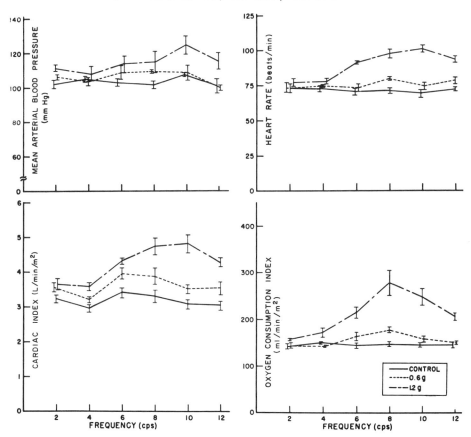

Fig. 3   Vibration effects of the cardiovascular system (after Hood, 1966).

Other authors reported an increase in heart rate during the first minutes of vibration exposure, which disappears after some time. An effect, which is typical for an altered state of arausal during the initial stress phase.

This leads to the question of how task performance is affected by vibrations: Any given task must involve input and output as well as central processing functions to a greater or lesser degree (Shoenberger, et al., 1971).

EFFECTS ON VISION

In vehicles the main input to the operator is via the visual channel. Visual decrements in the frequency range of less than 2 Hz are minimal since the body moves in phase with the vehicle and there is little relative movement of the eye with respect to the instruments. Furthermore the eye is able to track targets even at higher frequencies by means of the vestibulo-ocular-reflex which stabilizes the image on the retina. This mechanism, however, only functions if the target is stationary and the operator moving. In this case, reading tasks are unaffected by mechanical vibrations up to 10 Hz. Beyond this value performance drops off rapidly. The first peak in decrement of visual acuity appears around 12 Hz (Lewis, et al., 1980).

Two additional frequency ranges were identified where peaks in visual decrement occur:
One is in the range of 25-40 Hz corresponding to the resonance of the head with respect to the torso (Coermann, 1939) and another between 60 and 90 Hz corresponding to resonance of the eyeballs (Dupuis, et al., 1979). These higher resonances are of limited practical significance because the body attenuates these motions considerably if they are not transmitted directly to the upper body where they may easily be damped.

If the observer is stationary and the display moves the pursuit or fixation reflex of the eye is used to track the moving object in order to maintain a foveal image. Under these conditions already 0.5 Hz produce a significant impairment in reading speed and accuracy. The cause is that the retinal image becomes smeared and visual acuity decreases. Reading accuracy under these conditions is highly dependent on image size, luminance and contrast. An image velocity of 30°/s seems to be a limiting value (Barnes, 1980).

MOTOR RESPONSE

On the output side vibrations induce involuntary movements of the body and the extremities. This leads to control problems especially in the low frequency range. Performance drops when the frequency increases form 1.5 to 2.5 Hz. Obviously, with very low frequency vibration man tries to correct for each individual peak in the oscillation while at higher frequencies he is not able to respond rapidly enough to do so (Hornick, 1962).

At frequencies above 2 Hz the large amplitude motions of the torso and the limbs impose involuntary movements of the extremities. They may seriously decrease operator functions such as pushing buttons, writing, and adjusting knobs. Proper human factors engineering can overcome these drawbacks (Gierke v., et al., 1971).

CHRONIC VIBRATION EFFECTS

Long term vibration effects have already been mentioned in connection with the gastrointestinal system. In the sitting and standing position vibrations are transmitted via the bony structures of the body and the vertebral column plays the major role. It is not so much the internal structure of the bones which is affected by the mechanical energy which has to be dissipated by the spine, but rather the interface structure of the vertebra and the intervertebral disks. Mechanical vibrations affect the pathways along which nutrition normally diffuses into these tissues. Whether the boundary between bone and cartilage of the vertebral joints is also affected is still an open question (Junghanns, 1979).

In order to correlate vibration exposure and affections of the spine many other factors, which also lead to back problems, have to be considered. Even in the normal population there is a certain amount of spinal disorders with increasing age, although these individuals have never had any excessive vibration exposure. In some cases poor seat design and misplacement of controls has caused back pain which was originally attributed to excessive vibration. An example is given in Figure 4. It shows the seating posture in an older helicopter (a) compared with the published comfort angles for sitting humans (b). When the configuration of the seat and the controls were changed the complaints related to the vertebral column decreased considerably (Delahaye, et al., 1970). The statistical surveys investigating the affections of the spine of operators are contradictory (Heide, 1977), (Junghanns, 1979), (Dupuis, 1969). Only for drivers of agricultural tractors and for earth moving machinery a significantly higher incidence of back problems could be established (Junghanns, 1979).

Fig. 4 Unfavourable posture in an older helicopter (a) and "comfort position" (b) (after Delahaye et al., 1970).

VIBRATION EXPOSURE CRITERIA

Man has no sensory organ for the perception of mechanical vibrations. They are sensed by the joint efforts of the vestibular organ, the pressure receptors in the skin and the proprioceptors in the joints and muscles. However, curves of equal sensation have been established. They are produced by the comparison method (Dieckmann, 1957) which is well known from accustics: The subject is given a vibratory stimulus of a certain amplitude and frequency. Then the frequency is changed by the investigator and the subject has to adjust the vibration amplitude to a value which corresponds to his former sensation. The results have been outlined and form the basis of national and international standards for the evaluation of human exposure to whole body vibration. (ISO 2631), (ISO, 1978).

The recommended exposure criteria are contained in two sets of curves. The first is a weighting curve for evaluating vibration response as a function of frequency. (Figure 5). Curves of equal sensation are plotted for various exposure times with frequency on the abszissa

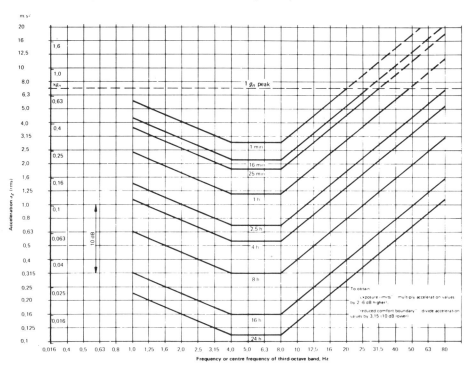

Fig. 5 Acceleration limits (az) as a function of frequency and exposure time (ISO, 1978).

and rms acceleration on the ordinate. The basic criterion is "fatigue decreased proficiency". Exceeding the exposure specified by these boundaries will in most situations cause noticable fatigue and decreased proficiency on most tasks.

The degree of task interference depends on the complexity of the task. The criteria are derived from studies on aircraft pilots and vehicle operators.

The upper limits of the exposure where health hazards are expected is indicated by levels twice as high (+ 6 dB). The reduced comfort boundary as derived from studies by the transportation industries is assumed to be at one-third of the levels shown in figure 5 (- 10 dB) and follows the same time and frequency dependence.

It is not difficult to apply these curves under operational conditions as long as the acting vibrations are sinusoidal. However, in a transport environment sinusoidal motion is the exception. The crucial point is how to produce an rms value for random motion and which integration time constant is used. There are international proposals for a value of 125 msec (Bobbert, 1982).

In everyday transport operations shocks are often superimposed on harmonic and random vibrations. If the rms value of such acceleration-time histories is formed and evaluated by the ISO-Standard, the load imposed on the operator is underestimated. Therefore a definition of shocks which are superimposed on vibration has to be established. ISO 2631 has introduced the "crest factor". It is defined as the ratio of the peak value to the rms value of a shock superimposed on vibration. At present a crest factor of 6 is allowed. If this is exceeded other evaluation procedures have to be applied. There are several proposals how to characterize shocks and how they are sensed by the operator. All the proposed procedures increase the apparent energy content of the impact by mathematical manipulations.

The effect of mechanical vibration on humans is not only a function of amplitude and frequency but also of exposure time. In Figure 6 vibration exposure criteria are given as a function of total daily exposure time. In the meantime these curves have been surpassed by a new proposal on the international scene. It says that exposure times between 1 and 10 minutes represent equal loads, that means, the curve will be horizontal in this time range. For longer exposure times vibration stress is considered equal to the weighted energy which results in a straight, downwardsloping line (ISO, 1982a). The recommendation does not take into account any interruption in vibration exposure. This is due

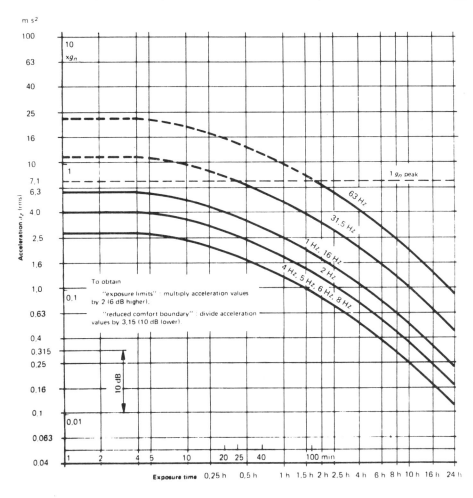

Fig. 6 Acceleration limits (az) as a function of exposure time and frequency (ISO, 1978).

to the fact that no sufficient knowledge is available on the effects of intermittend exposure. Here is a serious gap in our knowledge. Another drawback of this recommendation is, that it is not clear, up to which exposure times the curves may be extrapolated. The question is especially urging in occupational situations where exposure is not limited by an eight hours work-shift but where vibrations are present for very long times as is the case on seagoing ships.

Finally national and international standards do not and should not state exposure limits. This is up to government authorities. However, the scientists have to supply the data and their expert opinion. Although there is an extended literature and good knowledge about most vibration effects on man some problems are still unsolved. They concern the occupational safety and health of the operator. National and international standards and recommendations indicate that if the stated tolerance curves are exceeded the morbidity may increase. However a firm dose-response relationship has not yet been established. Longterm studies are required during which the vibration exposure and health status of operators is examined regularly over many years.

REFERENCES

Barnes, G.R. 1980. The effects of aircraft vibration on vision. In "High-speed, low-level flight: Aircrew factors", AGARD-CP-267. (AGARD, Neuilly-sur-Seine, France).

Bekesy v., G. 1939. Über die Empfindlichkeit des stehenden und sitzenden Menschen gegen sinusförmige Erschütterungen. Akust.Z., 4, 360-369.

Bobbert, G. 1982. Schwingungseinwirkung auf den Menschen, Wahrnehmung, Beanspruchung, Bewertung, Beurteilung. VDI-Berichte Nr. 456, (VDI-Verlag, Düsseldorf).

Clark, W.S., Lange, K.O., Coermann, R.R. 1963. Deformation of the human body due to uni-directional forced sinusoidal vibration. In "Vibration Research" (Ed. S. Lippert). (Pergamon Press, London). pp. 29-48.

Coermann, R.R. 1939. Untersuchungen über die Einwirkung von Schwingungen auf den menschlichen Organismus. Luftfahrtmed., 4, 73-117.

Coermann, R.R., Ziegenruecker, G.H., Wittwer, A.L., v. Gierke, H.E. 1960. The passive dynamic mechanical properties of the human thorax-abdomen system and of the whole body system. Aerospace Med., 31, 443-455.

Coermann, R.R. 1963. The mechanical impedance of the human body in sitting and standing position at low frequencies. In "Vibration Research" (Ed. S. Lippert). (Pergamon Press, London). pp. 1-27.

Delahaye, R.-P., Pannier, R., Seris, H., Auffret,R., Carre, R., Mangin, H., Teyssandier, M.-J. 1970. Physiopathology and pathology of affections of the spine in aerospace medicine. AGARD-CP-140. (AGARD, Neuilly-sur-Seine, France).

Dieckmann, D. 1957. Einfluß vertikaler mechanischer Schwingungen auf den Menschen. Int.Z.angew.Physiol.einschl.Arbeitsphysiol., 16, 519-564.

Dupuis, H., Christ, W. 1966. Über das Schwingungsverhalten des Magens unter dem Einfluß sinusförmiger und stochastischer Schwingungen. Int.Z.angew.Physiol., 22, 149-166.

Dupuis, H. 1969. Zur physiologischen Beanspruchung des Menschen durch mechanische Schwingungen. VDI-Z Reihe 11, Nr. 7, (VDI-Verlag, Düsseldorf).

Dupuis, H., Hartung, E. 1979. Tierexperimente zur Ermittlung des biomechanischen Schwingungsverhaltens des Bulbus. Albrecht von Graefes Arch.klin.exp.Ophthal., 210, 167-174.

Gierke v., H., Clarke, N.P. 1971. Effects of vibration and buffeting on man. In "Aerospace Medicine" (Ed. H.W. Randel). (The Williams & Wilkins Co., Baltimore, M.D.).

Gruber, G.J., Haskell Ziperman, H. 1974. Relationship between whole-body vibration and morbidity patterns among motor coach operators. NIOSH-75-104. National Institute for Occupational Safety and Health, Cincinati,OH.

Heide, R. 1977. Zur Wirkung langzeitiger beruflicher Ganzkörpervibrationsexposition. Dissertation Zentralinstitut für Arbeitsmedizin der DDR, Berlin-Ost.

Hood, W.B.Jr., Higgins, L.S. 1965. Circulatory and respiratory effects of whole-body vibration in anesthetized dogs. J.Appl.Physiol., 20, 1157-1162.

Hood, W.B. Jr., Murray, R.H., Urschel, Ch. W., Bowers, J.A., Clark, J.G. 1966. Cardiopulmonary effects of whole-body vibration in man. J.Appl.Physiol., 21, 1725-1731.

Hornick, R.J. 1962. Effects of whole body vibration in three directions upon human performance. J.Engineering Psychol., 1, 93-101.

ISO 1978. Guide for the evaluation of human exposure to whole-body vibration. 2nd edition. Ref. No. ISO 2631-1978 (E) International Organization for Standardization, Geneva.

ISO 1981. Vibration and shock - Mechanical driving point impedance of the human body. 1st edition. Ref. No. ISO 5982-1981 (E) International Organization for Standardization, Geneva.

ISO 1982a. Guide for the evaluation of human exposure to whole-body vibration. Amendment 1. Ref. No. ISO 2631-1978/A 1-1982 (E) International Organization for Standardization, Geneva.

ISO 1982b. Guide for the evaluation of human exposure to whole-body vibration. Addendum 2: Evaluation of exposure to whole-body z-axis vertical vibration in the frequency range 0.1 to 0.63 Hz. Ref. No. ISO 2631-1978/Add. 2-1982 (E) International Organization for Standardization, Geneva.

Junghanns, H. 1979. Die Wirbelsäule in der Arbeitsmedizin, Teil II: Einflüsse der Berufsarbeit auf die Wirbelsäule. (Hippokrates Verlag, Stuttgart).

Lewis, Ch. H., Griffin, M.J. 1980. Predicting the effects of vibration frequency and axis, and seating conditions on the reading of numeric displays. Ergonomics, $\underline{23}$, 485-501.

Loeckle, W.E. 1941. Über die Wirkung von Schwingungen auf das vegetative Nervensystem und die Sehnenreflexe. Dissertation Friedrich-Wilhelm-Universität Berlin.

Nickerson, J.L., Coermann, R.R. 1962. Internal body movements resulting from externally applied sinusoidal forces. AMRL-TDR-62-81, Wright-Patterson AFB, OH.

Nixon, Ch. W. 1962. Influence of selected vibratons upon speech. I. Range of 10 CPS to 50 CPS. The Journal of Auditory Research, $\underline{2}$, 247-266.

Nixon, Ch. W., Sommer, H.C. 1963. Influence of selected vibrations upon speech. III. Range of 6 CPS to 20 CPS for semi-supine talkers. Aerospace Med., $\underline{34}$, 1012-1017.

Reason, J.T., Brand, J.J. 1975. Motion Sickness (Academic Press, London).

Reiher, H., Meister, F.J. 1931. Die Empfindlichkeit des Menschen gegen Erschütterungen. Forschung auf dem Gebiete des Ingenieurwesens, Z. Techn. Mechanik und Thermodynamik, $\underline{2}$, 381-385.

Shoenberger, R.W. 1975. Subjective response to very low-frequency vibration. Aviat.Space,Environ.Med., $\underline{46}$, 785-790.

Shoenberger, R.W., Harris, C.S. 1971. Psychophysical assessment of whole body vibration. Human Factors, $\underline{13}$, 41-50.

Simic, D. 1974. Contribution to the optimisation of the oscillatory properties of a vehicle: Physiological foundations of comfort during oscillations. Royal Aircraft Establisment Library Translation 1707.

Vogt, H.L., Coermann, R.R., Fust, H.D. 1968. Mechanical impedance of the sitting human under sustained acceleration. Aerospace Med., $\underline{39}$, 675-679.

Wasserman, D., Taylor, W., Behrens, V., Samueloff, S., Reynolds, D. 1982. Vibration white finger disease in U.S. workers using pneumatic chipping and grinding hand tools - I. Epidemiology. DHHS (NIOSH) Publication No. 82-118, U.S. Department of Health and Human Services, Cincinnati, OH.

CONTINUOUS ELECTROPHYSIOLOGICAL RECORDING

T. Åkerstedt and L. Torsvall

Laboratory for Clinical Stress Research
National Institute for Psychosocial Factors and Health
10401 Stockholm, Sweden

ABSTRACT

This paper reviews the applicability of electrophysiological registration to problems of human functioning and breakdown in adaptation. It is suggested that physiological measures will be most useful when they allow evaluation of data in absolute (acceptable/non-acceptable) rather than relative terms. It is suggested that sleep and extreme cardiac activation are states which allow such judgements. The variables pertaining to these states are reviewed and it is concluded that both states may be successively monitored even under rather severe field conditions. A combination of the EEG and EOG may be used to indicate sleep intrusions, recognizeable as eye closure with increased power (spectral analysis) in the alpha and/or theta bands. Cardiac activation may be indicated by conventional heart rate measures and different kinds of cardiac arrhythmias. It is suggested that recordings are extended outside work hours and that sleep latency tests are used as complements.

INTRODUCTION

The rationale for using physiological parameters in research into health and functioning is, of course, that the physiological state of the individual reveals something about his performance capacity, his health, his environmental load, or about the mechanisms behind changes in performance and health. There is a number of physiological systems which are of interest in this context. One is e.g. endocrine and other measures obtained from body fluids. In this chapter we have, however, elected to treat parameters that allow continuous recording. The reason for this is that the environmental demands fluctuate rapidly, that the functional state of the individual may fluctuate equally rapidly, and that a transient mismatch between them may have very serious consequences in transport operations. Falling asleep when driving does, for example, make up a substantial part of single vehicle road accidents (Harris, 1977).

Using the arousal continuum (Duffy, 1962) for descriptive purposes, we are not as interested in arousal per se, but rather in the extremes where it becomes too low to sustain efficient performance or too high to be compatible with health requirements. Changes in the intermediate section are seldom possible to interprete in absolute terms and may be of less interest in applied studies.

At the lower end of the arousal continuum our major interest is sleep since this state, by definition, involves non-responsiveness to environmental stimuli as well as cessation of purposeful behavior (Kleitman, 1963), both of which are clearly undesirable during attention-demanding tasks. It should also be borne in mind that transport operations frequently involve irregular work hours and monotonous stimulation, which both lead to reduced alertness (O´Hanlon, 1981; Åkerstedt et al., 1982). Sleep-like states, fatigue and similar concepts have been measured in many different ways (Grandjean, 1979; Wierwille, 1979) but sleep proper is physiologically defined by electroencephalographic (EEG) and electrooculographic (EOG) measurements (Rechtschaffen and Kales, 1968).

At the upper end of the arousal continuum there is not as obvious a choice as the sleep concept and there is a large amount of measures which are of interest (cf. Ursin and Ursin, 1979; Sharit and Salvendy, 1982). Our own preference is electrocardiographic (ECG) parameters since there exist criteria of acceptability both with respect to heart rate (Åstrand and Rodahl, 1970) and heart irregularity (Hinkle et al., 1974) and since cardiac-behavioral relationships have received intensive research attention (Obrist, 1982).

In the following sections we will discuss transport operations and arousal in relation to continuous measurement of the EEG, the EOG, and the ECG.

EEG AND EOG MEASURES

The EEG registers the changes of electrical potentials on the scalp which are considered to represent brain electrical activity (cf. Creutzfeldt, 1976). For clinical EEG:s the international 10-20 system for placing (19) electrodes is frequently used (Jasper, 1958). Early studies and those concerned with sleep (Rechtschaffen and Kales, 1968) may use fewer electrodes. The EEG pattern is often described in terms of its frequency content; 0-3 Hz = delta activity, 3-8 Hz = theta, 8-12 Hz = alpha, 12-30 Hz = beta. This may be evaluated visually but is nowadays also automatically analyzed via e.g. bandpass filtering and subsequent integration of power, zero crossing analysis, autocorrelation, spectral analysis, etc. (cf. Rémond, 1977).

Basic studies of arousal and sleep

In their pioneering work on sleep Loomis et al. (1935; 1936; 1937) showed that relaxed subjects lying with closed eyes showed a prominent alpha activity and responded to environmental stimuli. When alpha started to break up, however, the subjects ceased to respond. Further progression of sleep showed that the EEG frequency decreased into the theta and delta range. In other studies it has been shown that the disappearance of alpha ("downwards") coincides with the loss of muscle tonus (Blake et al., 1939) and with a feeling of "drifting away" (Davis et al., 1938). Similar results have been obtained more recently by Simon and Emmons (1956), Kamiya (1961), Kuhlo and Lehmann (1964), Liberson and Liberson (1966).

In a similar vein it has been studied how sleep-like states intrude on waking behavior. Bjerner (1949), in an early study of sleep loss, showed that exceptionally long reaction times were associated with alpha blocking (downwards) and interpreted as "transient phenomena of the same nature as sleep". A similar connection between alpha disappearance and vigilance omissions has been demonstrated by Williams et al. (1959), Mirsky and Cardon (1962), and Davis and Krkovic (1965). Similarly, Guilleminault et al. (1975a; b) have demonstrated in narcoleptics that inadequate performance was associated with intrusion of theta activity.

Most laboratory studies, it should be emphasized, have used relaxed and closed eyes conditions in which case alpha signals alertness. If the eyes are open, however, this is reversed (Daniel, 1967). In a rather detailed study of simulated radar watching O´Hanlon and Beatty (1977) showed a high correlation between the performance measure ("sweeps" to detect the target) and the % theta ($\bar{r} = 0.44$), % alpha ($\bar{r} = 0.68$), and % beta ($\bar{r} = -0.67$). The authors particularly emphasize the need to include several EEG bands since alpha activity may change either upwards or downwards, indicating increased and decreased arousal, respectively.

Several studies of sleep loss have used scheduled sessions with eyes closed and found that the time alpha remains after eye closure decreases with increasing sleep loss, fatigue, and performance deterioration (Blake and Gerard, 1937; Armington and Mitnick, 1959; Rodin et al., 1962; Naitoh et al., 1969). In studies where controlled periods with eyes open have been used as well, rather little alpha has been found (Armington and Mitnick, 1959; Rodin et al., 1962). Some alpha may occur with eyes open, however, (Oswald, 1960; Rodin et al., 1962; Legewie et al., 1969).

Not only closure of the eyelid but also the movement of the eye is related to EEG changes. Usually, slow (0.1 - 0.6 Hz) eye movements (SEM;s) starts to appear when the alpha activity starts to break up and the subject starts to "drift off" (Aserinsky and Kleitman, 1959; Liberson and Liberson, 1966; Kuhlo and Lehmann, 1964; Hori, 1982). As sleep is firmer established the slow eye movements begin to disappear. The SEM:s also appear when a person experiences short transient sleep episodes during waking activity (Guilleminault et al., 1975a; b).

It should be emphasized that the earlier mentioned subjective experience of "drifting away" at sleep onset is not synonymous with the subjective experience of sleep. The latter seems to be delayed far into stage 2 sleep (Foulkes and Vogel, 1965; Agnew and Webb, 1972; Campbell and Webb, 1981; Bonnet and Moore, 1982), implying that subjective estimates of the occurrence of sleep will be too low.

Field studies in transport operations

Field studies with EEG measures present a more difficult recording situation than that in the laboratory since the task and the environment may interfere with the recording and since the recording, usually, must not interfere with performance of the task. The solution is often telemetric or ambulatory recording with amplification close to the signal source and fewer leads recorded.

One of the early studies of the EEG in transport work was carried out by Adey et al. (1967) on an astronaut on the Gemini flight GT-7. The data were mainly obtained from a vertex-occipital bipolar derivation and tape recorded. The data were analyzed with autospectral and cross spectral density distributions and coherence functions. With increasing flight time the power in the theta band increased and it was concluded that the EEG patterns "followed in excellent detail the fine aspects of transiently focussed attention or brief episodes of drowsiness or light sleep".

Lecret and Pottier (1971) recorded EEG from eight car drivers. They used a bioccipital electrode derivation recording via head-worn amplifiers and a tape recorder. The EEG was band pass filtered for the alpha frequency and yielded an alpha index (time with alpha). On the whole, the amounts of alpha increased with decreasing levels of stimulation, the early part of the drive showing less alpha than later parts, as did city driving as compared rural driving. Eyelid closure as monitored by a vertical EOG were fewest during city driving and early in driving.

Caille and Bassano (1977) studied 4 subjects during night driving with different amounts of carbon monoxide. The EEG was recorded telemetrically via fronto-parietal and perieto-occipital derivations (also EOG was recorded). The data were analyzed with spectral analysis as well as other methods. Towards the end of driving alpha bursts frequently appeared, followed by theta and sometimes sigma waves (apparently micro-sleep episodes).

In another study of night driving O´Hanlon and Kelley (1977) recorded a parieto-occipital derivation with amplification close to the electrodes. The EEG was bandpass filtered into the usual frequency bands. The results showed that mainly poor drivers increased their power in the alpha band over the duration of the drive and for monotonous segments, in both cases covarying with performance deterioration (lane-drifting). The authors also observed several cases of subjects´ apparently falling asleep while driving, forcing the observer to take over controls. Sleep intruded while the driver still had his eyes open, staring absently straight ahead and it was accompanied by theta waves, sleep spindles, and k-complexes. Interestingly, the subjects denied awareness of the fact that they had been driving the car some distance while asleep. It was also concluded that the alpha band is sensitive to changes in medium alertness while the theta and delta bands are necessary for distinguishing in the lower ranges of arousal.

Fruhstorfer (1977) applied spectral analysis to various EEG derivations from three subjects driving at night. The results (preliminary) showed a clear increase in alpha during monotonous segments, as well as occasional theta bursts and a 14-18 Hz activity which "was not sleep spindles". Interestingly, blink duration increased and EOG velocity decreased in connection with the increase of alpha. No correlations were presented, however.

In our own studies we have applied spectral analysis to the parieto-occipital EEG of train engineers driving during both day and night (Torsvall and Åkerstedt, 1982). The EEG was amplified by miniature head-worn amplifiers a few centimeters from the electrodes and recorded on a small four-channel tape recorder worn by the subject (Medilog - Oxford Instruments). This set-up allows continuous 24-hour recording of ambulatory subjects with only moderate amounts of artefacts.

Figure 1 illustrates the qualitative aspects of the EEG data of one driver, presented as a compressed spectral array (Bickford, 1977). During the day trip (left) essentially no frequency changes were seen except for the four "snack artefacts" in the lower frequencies. The night trip (middle), with fewer stops and slower speed, in contrast, exhibited repeated and long bursts of alpha activity. This tendency increased with increasing time but was interrupted when the driver entered stations or populated areas. The third spectral array was obtained from an observer beside the driver. In this case alpha started to appear after 100 min (midnight) and continued throughout the drive except for two 30 min bursts of theta activity between 190 and 240 min. The observer complained of extreme fatigue during the drive and was conscious of having fallen asleep.

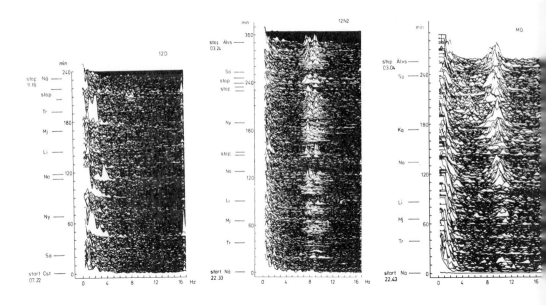

Fig. 1 Compressed spectral array from a train engineer during day driving (left), night driving (middle), and from an observer beside the driver (right). On the y-axis are indicated time from start and abbreviations of station names. Also given are clock hours for the beginning and the end of the drive.

A detailed analysis of the raw data may also be used to obtain information about electrophysiologically indicated alertness and performance deterioration. Figure 2 shows the physiological state involving a signal miss. When passing a red "pre-signal" (channel 2) the driver exhibits a slow heart rate (ch.1), closed eyelids and slow eye movements (ch.3), and alpha (ch.4). Normally, the driver should start breaking at the pre-signal. Some seconds later the driver opens his eyes and starts blinking, alpha is blocked, and the heart rate is trebled (presumably when he perceives the red main signal). At that point the driver starts breaking but fails to bring the train to a stop before the main signal is passed. No accident occurred, however, since the line by that time had been cleared.

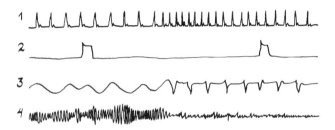

Fig. 2 Original recording (filled in with India ink for better reproduction) of EKG, signal occurrence (pre-signal and main signal), EOG (vertical), and EEG (O_2-P_4). The time marker did not reproduce but the whole segment is 20 sec. long.

To obtain group results the EEG power in a certain band was, for each individual, averaged for each line segment and expressed as percent deviation from the first segment. These data were then averaged across individuals. Figure 3 shows that there was no change in alpha during day driving (broken line). During night driving (full line), however, there was a 60% increase towards the end. The pattern is similar for theta activity. The same figure also shows that the number of 10 second periods per minute without eyeblinks increased during night driving and remained the same during day driving. Slow eye movements (not in the figure) were visually scored per segment as the proportion of four-minute intervals in which SEM:s were observed. During day driving the number of such intervals was between 0 and 0.2 % per individual, without any trend over time. During night driving it varied between 0.1 and 1.4 with a strong increase towards the end of the trip. On the whole, these results were parallelled by ratings of sleepiness.

It should also be mentioned that in a recent study Haslam (1982) has shown that it is quite feasible to record EEG and other bioelectric signals also from soldiers during actual battle exercise, and for one to two weeks continuously. Also in this study there was found an increase of alpha (visually scored) accompanying increased fatigue during sleep loss.

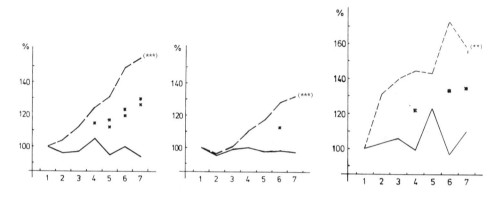

Fig. 3   Amount of power in the alpha (left) and theta (middle) bands as well as the number of 10 sec periods per minute without eye blinks. All values are expressed as percent change from the initial value and plotted against the seven segments of the trip (between major stations). Night driving is represented by the broken line and day driving by the solid line. Within parenthesis is given the level of significance for change over time (F-test) and between curves are indicated the level of significance for differences between the curves (t-test).
* = p<.05, ** = p<.01, *** = p<.001.

Considering the data available it must be concluded that EEG recordings can be made under field conditions and with very few leads reflect the state of alertness of an individual. Whereas SEM:s and theta activity may indicate intrusions of sleep, the occurrence of alpha activity may tentatively be interpreted as sleepiness, at least if the latter is defined as a drive to fall asleep as has been proposed by Dement and Carskadon (1982). The EEG will, however, also reflect other processes than arousal, such as e.g. information processing (Gale, 1977). Very little is known about how to separate the effects of such processes.

Sleep latency tests

Before closing the discussion on sleep and alertness we would like to bring attention to a non-continuous complement to continuous recording. This is the so called (Multiple) Sleep Latency Test (MSLT), based on the previously discussed tendency to fall asleep. The latter may be defined operationally (Carskadon and Dement, 1982a) as the time between "lights out" and the first appearance of sleep signs according to conventional EEG criteria (Rechtschaffen and Kales, 1968). In a series of studies Carskadon and Dement have measured sleep latency in recumbent subjects and shown that the measure reflects the sleepiness due to various manipulations of sleep (Carskadon and Dement, 1982b).

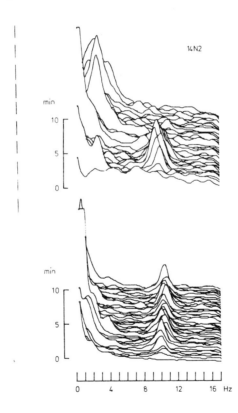

Fig. 4 Spectral analysis (O_2-P_4) of a 10 min sleep attempt before (below) and after (above) a night journey.

In our own study on irregular work hours and train engineers we used a 10 min version of the MSLT. This was done with the ambulatory recording equipment in a quiet room at the station before and after the trip. The results showed that no subjects fell asleep during the test before work. When tested after work, day driving was associated with a 13% incidence of subjects falling asleep while 65% did so after night driving. With spectral analysis the latency may be illustrated as in fig. 4. Sleep starts after night driving when the "ridge" in the alpha regions turns into the low frequency region and becomes a delta "ridge" (Torsvall and Åkerstedt, 1982). Before night driving the alpha "ridge" is continuous, indicating continuous wakefulness.

ECG AND OTHER VARIABLES

The ECG

Heart rate is probably the most well researched parameter in psychophysiological contexts. Its role as an indicator of work load through oxygen uptake is of course well established (Åstrand and Rodahl, 1970). Laboratory studies have shown a direct relation between mental load and HR (cf. Rutenfranz, 1960; Ettema and Zielhuis, 1971) and the number of applied studies abound. In pilots, for example, landing, take off, and other critical manoeuvres are associated with considerable increases of HR (Smith, 1976; Bateman et al., 1970; Opmeer and Krool, 1973). Similar responses may be observed in e.g. ships´ pilots (Cook and Cashman, 1982) or automobile drivers (Burns et al., 1966; Simonson et al., 1968).

The opposite reaction, i.e. a decrease in HR, is frequently found in prolonged driving, e.g. in automobile drivers (Caille and Bassano, 1977; Riemersma et al., 1977; Fagerström and Lisper, 1977) and train engineers (Hashimoto, 1964; Kogi and Saito, 1971; Endo and Kogi, 1975). Although the decrease in HR during prolonged driving would seem to indicate lowered arousal/alertness, there is not always a correlation between HR decrease and e.g. performance indicators (Fagerström and Lisper, 1977). Neither does sleep loss affect HR in the driving situation (Riemersma et al., 1977).

Using the variation between heartbeats a number of measures of heart rate variability (HRV) may be constructed (cf. Luczak and Laurig, 1973; Sayers, 1973).

On the whole, it seems that HRV decreases with increased work load/ stress (Obrist et al., 1964; Kalsbeek, 1971; Ettema and Zielhuis, 1971;

Thackray et al., 1974). In the field, Opmeer and Krol (1973) showed that difficult flight phases were accompanied by a decreased HRV whereas O´Hanlon (1971) has shown an increased HRV during prolonged automobile driving and suggested that it is a measure of driver fatigue. Some workers have failed, however, to find a connection between work load/stress and HRV (Sayers, 1973; Hicks and Wierwille, 1979). A rather recent development in this area is the spectrum of heartbeat intervals which Mulder (1979), Egelund (1982) and Mulders et al. (1982) have shown to be sensitive in certain components to operator loading and fatigue.

On the whole, it seems that both HR and HRV may be used to indicate changes in the arousal state of an operator. Both variables are, however, affected by a number of other factors such as physical activity, temperature (Åstrand and Rodahl, 1970), and information processing (Mulder and Mulder, 1981; Obrist, 1982). Thus great care must be taken in interpreting results, and it is clearly an advantage if these parameters can be integrated with other parameters such as the EEG, EOG, activity measures etc.

When the ECG variability extends to the stage where the ECG complexes start to interfere with one another, it becomes of interest to the cardiologist. It appears that such arrhythmias, especially ventricular ectopic beats (VEB:s), may be precursors of both ventricular fibrillation and sudden death (Hinkle et al., 1974; Lown et al., 1976). There are suggestions that the work environment may play a direct role in the genesis of such arrhythmias as shown in car drivers (Taggart et al., 1969), auto workers (Baxter et al., 1978) or ships´ pilots (Cook and Cashman, 1982). In our own work with train engineers we found that arrhythmias occurred particularly in connection with difficulties along the line (Torsvall and Åkerstedt, 1982).

Some other measures

Certainly, there are many more variables of interest than those described above. One such of clinical as well as performance interest is blood pressure. This may be recorded through the usual inflatable cuff with repeated inflation as e.g. demonstrated during car driving by Schneider and Costiloe (1975). No effects differing from normal daily activities were found in that study. Registration may also be done through direct arterial measurement as demonstrated by e.g. Littler et al. (1973). Also in this study systolic and diastolic pressures remained stable except for small transient increases when e.g. overtaking.

Another measure of interest is electrodermal activity (GSR) which seems to reflect transient arousal during e.g. driving (Hulbert, 1957; Taylor, 1964; Brown and Huffman, 1972; O'Hanlon and Kelley, 1974, Helander, 1975). Since electrodes are usually placed in the palm of the hand there is, however, risk of interference from manoeuvres even if the non-preferred hand is used.

The electromyogram may also be useful indicating arousal or activity (Kennedy, 1953; Lecret and Pottier, 1971; Helander, 1975).

Body temperature may be continuously recorded through a rectal termistor. This may be of particular importance in transport operations with irregular work hours, since rectal temperature reliably predicts sleep length and the occurrence of dream sleep (REM) (Czeisler et al., 1980; Zulley et al., 1981; Gillberg and Åkerstedt, 1982).

Finally, a variable of potential future importance is motor activity, By attaching an accelerometer or a similar device to the wrist or leg and recording all movements exceeding some minimum acceleration it is possible to obtain a measure of gross motor activity (Webster et al., 1982). As yet, the application has been in clinical settings, showing that motor activity closely describes the sleep/wake pattern and degree of depression in psychiatric patients (Wehr et al., 1982).

CONCLUSIONS AND COMMENTS

There is no doubt that continuous physiological recording from non-stationary subjects in their natural occupational setting is quite feasible. With a battery of variables it is usually possible to form an impression of the variation of the physiological state of the person recorded. As yet, however, most information pertains to the arousal dimension and very little is known about effects of information processing.

REFERENCES

Adey, W.R., Kado, R.T., and Walter, D.O. 1967. Computer analysis of EEG data from Gemini flight GT-7. Aerospace Med., 38, 345-359.
Agnew, H. and Webb, W.B. 1972. Measurement of sleep onset by EEG criteria. Am. J. EEG Technol., 12, 127-134.
Åkerstedt, T. and Gillberg, M. 1982. Displacement of the sleep period and sleep deprivation. Hum. Neurobiol., 1, 163-171.
Armington, J.C. and Mitnick, L.L. 1959. Electroencephalogram and sleep deprivation. J. Appl. Physiol., 14, 247-250.
Aserinsky, E. and Kleitman, N. 1959. Regularly occurring periods of eye mobility and concomitant phenomena during sleep. Science, 118, 273-274.
Åstrand, P.-O. and Rodahl, K. 1970. Textbook of Work Physiology. (McGraw-Hill, New York).

Bateman, S.C., Goldsmith, R., Jackson, K.F., Ruffel Smith, H.P., and Mattocks, V.S. 1970. Heart rate of training captains engaged in different activities. Aerospace Med., 41, 425-429.

Baxter, P.J., White, W.G., Barnes, G.M., and Cashman, P.M.M. 1978. Ambulatory electrocardiography in car workers. Br. J. Industr. Med., 35, 99-106.

Bickford, R.G. 1977. Computer analysis of background activity. In "EEG Informatics. A Didactic Review of Methods and Applications of EEG Data Processing" (Ed. A. Rémond). (Elsevier, Amsterdam). pp. 215-232.

Bjerner, B. 1949. Alpha depression and lowered pulse rate during delayed actions in a serial reaction test. Acta Physiol. Scand., 19, suppl. 65.

Blake, H, and Gerard, R. 1937. Brain potentials during sleep. Am. J. Physiol., 119, 697-703.

Bonnet, M.H. and Moore, S. 1982. The threshold of sleep: Perception of sleep as a function of time asleep and the auditory threshold. Sleep, 5, 267-276.

Brown, J.D. and Huffman, W.J. 1972. Psychophysiological measures of drivers under actual driving conditions. J. Safety Res., 4, 172-178.

Broadbent, D.E. 1979. Is a fatigue test now possible? Ergonomics, 22, 1277-1290.

Burns, N.M., Baker, C.A., Simonson, E., and Keiper, C. 1966. Electrocardiogram changes in prolonged driving. Perc. Mot. Skills, 23, 210.

Caille, E.J. and Bassano, J.L. 1977. Validation of a behavior analysis methodology: variation of vigilance in night driving as a function of the rate of carboxyhemoglobin. In "Vigilance" (Ed. R.R. Mackie). (Plenum Press, New York). pp. 59-72.

Campbell, S.S. and Webb, W.B. 1981. The perception of wakefulness within sleep. Sleep, 4, 177-183.

Carskadon, M.A. and Dement, W.C. 1982a. The multiple sleep latency test: What does it measure? Sleep, 5, suppl. 2, 67-72.

Carskadon, M.A. and Dement, W.C. 1982b. Nocturnal determinants of daytime sleepiness. Sleep, 5, suppl. 2, 73-81.

Cook, T.C. and Cashman, P.M.M, 1982. Stress and ectopic beats in ships' pilots. J. Psychosom. Res., 26, 559-569.

Creutzfeldt, O. (Ed.). 1976. Morphological basis of EEG mechanisms. In "Handbook of Electroencephalography and Clinical Neurophysiology" (Ed. A. Rémond). (Elsevier, New York). part A.

Czeisler, C.A., Weitzman, E.D., Moore-Ede, M.C., Zimmerman, J.C., Knauer, R.S. 1980. Human sleep: its duration and organization depend on its circadian phase. Science, 210, 1264-1267.

Daniel, R.S. 1967. Alpha and theta EEG in vigilance. Perc. Mot. Skills, 25, 697-703.

Davies, D.R. and Krkovic, A. 1965. Skin conductance, alpha-activity, and vigilance. Am. J. Psychol., 78, 304-306.

Davis, H., Davis, P.A., Loomis, A.L., Harvey, E.N., and Hobart, G. 1938. Human brain potentials during the onset of sleep. J. Neurophysiol., 1, 24-38.

Dement, W.C. and Carskadon, M.A. 1982. Current perspectives on daytime sleepiness: the issues. Sleep, 5, suppl. 2, 56-66.

Duffy, E. 1962. Activation and Behavior. (Wiley and Sons, London).

Egelund, N. 1982. Spectral analysis of heart rate variability as an indicator of driver fatigue. Ergonomics, 25, 663-672.

Endo, T. and Kogi, K. 1975. Monotony effects of the work of motormen during high speed train operation. J. Hum. Ergol., 4, 129-140.

Ettema, J.H. and Zielhuis, R.L. 1971. Physiological parameters of mental load. Ergonomics, 14, 137-144.

Fagerström, K.-O. and Lisper, H.-O. 1977. Effects of listening to car radio, experience, and personality of the driver on subsidiary reaction time and heart rate in a long-term driving task. In "Vigilance" (Ed. R.R. Mackie). (Plenum Press, New York). pp. 73-86.

Foulkes, D. and Vogel, G. 1965. Mental activity at sleep onset. J. Abnorm. Psychol., 70, 231-243.

Fruhstorfer, H., Langanke, P., Meinzer, K., Peter, J.H., and Pfaff, U. 1977. Neurophysiological vigilance indicators and operational analysis of a train vigilance monitoring device: A laboratory and field study. In "Vigilance" (Ed. R.R. Mackie). (Plenum Press, New York). pp. 147-162.

Gale, A. 1977. Some EEG correlates of sustained attention. In "Vigilance" (Ed. R.R. Mackie). (Plenum Press, New York). pp. 263-285.

Gillberg, M. and Åkerstedt, T. 1982. Body temperature and sleep at different times of day. Sleep, 5, 378-388.

Grandjean, E. 1979. Fatigue in industry. Br.J. Industr. Med., 36, 175-186.

Guilleminault, C., Billiard, M., Montplaisir, J., and Dement, W. 1975a. Altered states of consciousness in disorders of daytime sleepiness. J. Neurol. Sci., 26, 377-393.

Guilleminault, C., Phillips, R., and Dement, W. 1975b. A syndrome of hypersomnia with automatic behavior. Electroencephalogr. Clin. Neurophysiol., 38, 403-413.

Harris, W. 1977. Fatigue, circadian rhythm, and truck accidents. In "Vigilance" (Ed. R.R. Mackie). (Plenum Press, New York). pp. 133-146.

Hashimoto, K. 1964. Estimations of the drivers' workload in high speed car operation on the new Tokaido Line in Japan. Ergonomics, 7, 463-469.

Haslam, D.R. 1982. Sleep loss, recovery sleep, and military performance. Ergonomics, 25, 163-178.

Helander, M.G. 1975. Physiological reactions of drivers as indicators of road traffic demand. Transportation Research Record, 530, Washington, D.C.

Hicks, T.G. and Wierwille, W.W. 1979. Comparison of five mental workload assessment procedures in a moving-base driving simulator. Hum. Factors, 21 129-143.

Hinkle, L.E., Carver, S.T., and Argyros, D.C. 1974. The prognostic significance of ventricular premature contractions in healthy people and in people with coronary heart disease. Acta Cardiol., 18, suppl. 5.

Hori, T. 1982. Electrodermal and electro-oculographic activity in a hypnagogic state. Psychophysiology, 19, 668-672.

Hulbert, S.F. 1957. Driver's GSRs in traffic. Perc. Mot. Skills, 7, 305-315.

Jasper, H.H. 1958. Report of the committee on methods of clinical examination in electroencephalography. Electroencephalogr. Clin. Neurophysiol., 10, 371-375.

Kalsbeek, J.W.H. 1971. Sinus arrhythmia and the dual task method in measuring mental load. In "Measurement of Man at Work" (Ed. W.T. Singleton, J.G. Fox, and D. Whitfield). (Taylor & Francis, London). pp. 103-113.

Kamiya, J. 1961. Behavioral, subjective, and physiological aspects of sleep and drowsiness. In "Functions of varied Experience" (Ed. D.W, Fiske and S.R, Maddi). (Dorsey, Homewood, Ill). pp. 145-174.

Kennedy, J.L. 1953. Some practical problems of the alertness indicator. In "Symposium on Fatigue" (Ed. W.F, Floyd and A.T. Welford). (Lewis, London). pp. 144-153.

Kogi, K. and Saito, Y. 1971. Psychophysiological reactions of electric locomotive drivers with or without an assistant driver. Jap. J. Industr. Health, 13, 89-99.

Kuhlo, W. and Lehmann, D. 1964. Das Einschlaferleben und seine neurophysiologischen Korrelate. Arch. Psychiat. Z. ges. Neurol., 205, 687-716.
Lecret, F. and Pottier, M. 1971. La vigilance, facteur de sécurité dans la conduite automobile. Trav. Human., 34, 51-68.
Legewie, H., Simonova, O., and Creutzfeldt, O.D. 1969. EEG changes during performance of various tasks under open- and closed-eyed conditions. Electroencephalogr. Clin. Neurophysiol., 27, 470-479.
Liberson, W.T. and Liberson, C.W. 1966. EEG records, reaction times, eye movements, respiration, and mental content during drowsiness. Proc. Soc. Biol. Psychiat., 19, 295-302.
Littler, W.A., Honour, A.J., and Sleight, P. 1973. Direct arterial pressure and electrocardiogram during motor car driving. Br. Med. J., 2, 273-277.
Loomis, A.L., Harvey, E.N., and Hobart, G. 1935. Potential rhythms of the cerebral cortex during sleep. Science, 81, 597-598.
Loomis, A.L., Harvey, E.N,, and Hobart, G. 1936. Electrical potentials of the human brain. J. Exp. Psychol., 19, 249-279.
Loomis, A.L., Harvey, E.N., and Hobart, G. 1937. Cerebral states during sleep as studied by human brain potentials. J. Exp. Psychol., 21, 127-144.
Lown, B., Temte, J.V., Reich, P., Gaughan, C., Regestein, Q., and Hai, H. 1976. Basis for recurring ventricular fibrillation in the absence of coronary heart disease and its management. N. Engl. J. Med., 294, 623-630.
Luczak, H. and Lauring, W. 1973. Analysis of heart rate variability. Ergonomics, 16, 85-97.
Mirsky, A.F. and Cardon, P.V. 1962. A comparison of the behavioral and physiological changes accompanying sleep deprivation and chlorpromazine administration in man. Electroencephalogr. Clin. Neurophysiol., 14, 1-10.
Mulder, G. 1979. Sinus arrhythmia and mental workload. In "Mental Workload, its Theory and Measurement" (Ed. N. Moray) (Plenum Press, New York). pp. 327-344.
Mulder, G. and Mulder, L.J.M. 1981. Information processing and cardiovascular control. Psychophysiology, 18, 392-402.
Mulders, H.P.G., Meijman, T.F., O´Hanlon, J.F., and Mulder, G. 1982. Differential psychophysiological reactivity of city bus drivers. Ergonomics, 25, 1003-1011.
Naitoh, P., Kales, A., Kollar, E.J., Smith, J.C., and Jacobson, A. 1969. Electroencephalographic activity after prolonged sleep loss. Electroencephalogr. Clin. Neurophysiol., 27, 2-11.
Obrist, P.A. 1982. Cardiac-behavioral interaction: A critical appraisal. In "Perspectives in Cardiovascular Psychophysiology" (Ed. J.T. Cacioppo and Petty, R.E.). (Guilford Press, New York). pp. 265-295.
Obrist, P.A., Hallman, S.I., and Wood, D.M. 1964. Autonomic levels and lability and performance time on a perceptual task and a sensory-motor task. Perc. Mot. Skills, 18, 753-762.
O´Hanlon, J.F. 1971. Heart rate variability: A new index of driver alertness/fatigue. (Hum. Factors Res. Inc., Goleta, Calif.). Tech. Rep. 1712-1.
O´Hanlon, J.F. 1981. Boredom: practical consequences and a theory. Acta Psychol., 49, 53-82.
O´Hanlon, J.F. and Beatty, J. 1977. Concurrence of electroencephalographic and performance changes during a simulated radar watch and some implications for the arousal theory of vigilance. In "Vigilance" (Ed. R.R. Mackie). (Plenum Press, New York). pp. 189-202.

O´Hanlon, J.F. and Kelley, G.R. 1974. A psychophysiological evaluation of devices for preventing lane drift and run-off-road accidents. (Hum. Factors Res. Inc., Goleta, Calif.). Tech. Rep. 1736-F.

O´Hanlon, J.F, and Kelley, G.R. 1977. Comparison of performance and physiological changes between drivers who perform well and poorly during prolonged vehicular operation. In "Vigilance" (Ed. R.R. Mackie). (Plenum Press, New York).

Opmeer, C.H.J.M. and Krol, J.P., 1973. Towards an objective assessment of cockpit workload: 1. Physiological variables during different flight phases. Aerospace Med., 44, 527-532.

Oswald, I. 1960. Falling asleep open-eyed during intense rhythmic stimulation. Br. Med. J., 1, 1450-1455.

Rechtschaffen, A. and Kales, A. (Ed.). 1968. A manual of standardized terminology, techniques, and scoring system for sleep stages of human subjects. Brain Information Service/Brain Research Institute, University of California.

Rémond, A. (Ed.). 1977. EEG informatics. A Didactic Review of Methods and Applications of EEG Data Processing. (Elsevier, Amsterdam).

Riemersma, J.B.J., Sanders, A.F., Wildervanck, C., and Gaillard, A.W. 1977. Performance decrement during prolonged night driving. In "Vigilance" (Ed. R.R. Mackie). (Plenum Press, New York). pp. 41-58.

Rodin, R.A., Luby, E.D., and Gottlieb, J.S. 1962. The electroencephalogram during prolonged experimental sleep deprivation. Electroencephalogr. Clin. Neurophysiol., 14, 544-551.

Rutenfranz, J. 1960. Über das Verhalten der Pulzfrequenz bei Arbeit unter Zeitdruck. Int. Z. angew. Physiol. einschl. Arbeitsphysiol., 18, 264-279.

Sayers, B.M. 1973. Analysis of heart rate variability. Ergonomics, 16, 17-32.

Schneider, R.A. and Costiloe, J.P. 1975. Twenty-four hour automatic monitoring of blood pressure and heart rate at work and at home. Am. Heart J., 90, 695-702.

Sharit, J. and Salvendy, G. 1982. Occupational stress: Review and reappraisal. Hum. Factors, 24, 129-162.

Simon, C.W. and Emmons, W.H. 1956. Responses to material presented during various levels of sleep. J. Exp. Psychol., 51, 89-97.

Simonson, E., Baker, C., Burns, N., Keiper, C., Schmitt, O.H., and Stockhouse, S. 1968. Cardiovascular stress (electrocardiographic changes) produced by driving an automobile. Am. Heart J., 75, 125-135.

Smith Ruffel, H.P. 1976. Heart-rate of pilots flying scheduled airline routes. Aerospace Med., 38, 1117-1119.

Taggart, P., Gibbons, D., Somerville, W. 1969. Some effects of motorcar driving on the normal and abnormal heart. Br. Med. J., 4, 130-134.

Taylor, D.H. 1964. Driver´s galvanic skin response and the risk of accident. Ergonomics, 7, 439-451.

Thackray, R.I., Jones, K.N., and Tochstone, R.M. 1974. Personality and physiological correlates of performance decrement on a monotonous task requiring sustained attention. Br. J. Psychol., 65, 351-358.

Torsvall, L. and Åkerstedt, T. 1982. Trötthet under dag- och nattarbete; en fältundersökning av lokförare. Stressforskningsrapporter. Lab. Clin. Stress Res., Karolinska Institute, Stockholm. No 156.

Ursin, H. and Ursin, R. 1979. Physiological indicators of mental load. In "Mental Workload, its theory and Measurement" (Ed. N. Moray). (Plenum Press, New York). pp. 349-366.

Webster, J.B., Kripke, D.F., Messin, S., Mullaney, D.J., and Wyborney, G. 1982. An activity-based sleep monitor system for ambulatory use. Sleep, 5, 389-399.

Wehr, T.A., Goodwin, F.K., Wirtz-Justice, A., Breitmaier, J. and Craig, C. 1982. 48-hour sleep/wake cycles in manic depressive illness. Arch. Gen. Psychiat., 39, 559-565.
Wierwille, W.W. 1979. Physiological measures of aircrew mental workload. Hum. Factors, 21, 575-593.
Williams, H.L., Lubin, A., and Goodnow, J.J. 1959. Impaired performance with acute sleep loss. Psychol. Monogr., 73.
Zulley, J., Wever, R., and Aschoff, J. 1981. The dependence of onset and duration of sleep on the circadian rhythm of rectal temperature. Pflügers Arch., 391, 314-318.

DIMENSIONS OF FLIGHT CREW PERFORMANCE DECREMENTS:
METHODOLOGICAL IMPLICATIONS FOR FIELD RESEARCH

R.C. Graeber, H.C. Foushee, J.K. Lauber

Man-Vehicle Systems Research Division
NASA Ames Research Center
Moffett Field, California, U.S.A. 94035

ABSTRACT

Operational flight crew research into the etiology of performance breakdown in the cockpit requires an integrated methodology sensitive to both individual and group performance factors. This paper provides an introduction to some of the relevant dimensions and an overview of previous research. The discussion focuses upon the classic concern with individual human performance parameters, shortcomings of this approach, the need for research dealing with the performance characteristics of the group, and the implications of advanced technology for future aircrew performance issues. By way of example, a current NASA study is described which seeks to address some of these issues.

INTRODUCTION

Despite the dramatic technological advances in aviation over the past three decades and a steady decrease in the overall accident rate, the persistence of flight crew performance decrements continues to impress all those who monitor the operational transport environment. Each new aircraft promises better aeronautical performance and reduced pilot workload, yet mishaps primarily attributable to flight crew error persist. In fact, about 65 percent of all accidents continue to fall into this category. It would be reasonable to expect an improvement in flight safety if a comprehensive approach were undertaken to understand thoroughly the factors responsible for these statistics. This paper examines some of the variables of interest as well as the research design requirements dictated by this challenge. Both are discussed in relation to ongoing and past research concerning the effects of fatigue and circadian dysrhythmia on flight crew performance.

SCOPE OF PREVIOUS RESEARCH

Laboratory Investigations

Aside from classic human engineering research focusing on cockpit design, there have been relatively few human factors studies carried out in the operational environment of commercial aircraft. Perhaps by necessity, traditional investigations have focused upon the effects of various biomedical factors, such as hypoxia or alcohol consumption, under controlled laboratory conditions where flight safety could not be jeopardized when factor levels were raised beyond prescribed limits. Others have examined the influence of operational factors at levels commonly encountered by flight crews, but they have again chosen the laboratory because of the inherent advantages it provides in controlling the effects of other variables. In both cases, laboratory results must be extrapolated in order to provide tangible benefits and predict actual operational implications. It is this process of extrapolation that often produces a credibility gap among airline flight operations personnel who are urged to address the implications of these laboratory findings. This gap often widens when the implications deal with subtle behavioral phenomena instead of straightforward biomedical effects and when increased costs are involved. Consequently, the scientific advantage gained by laboratory control is often ignored because of an historic failure to address the issue of generalizability.

Flight Related Studies

An alternative research strategy is to collect data on physiological and behavioral responses associated with an actual flight, thus obviating the criticism that the data are unrealistic. With regard to the influence of fatigue and circadian factors, these studies take several forms. The majority are limited to data derived from sleep logs and self-ratings of fatigue by crews flying transmeridian routes (Hartman and Cantrell, 1967; Nicholson, 1970; Preston and Bateman, 1970; Nicholson, 1972; Preston, 1973; Mohler, 1976; Aviation Human Engineering Research Team, 1976). Some investigators have also collected supplemental biochemical measurements or performance data based on laboratory tasks (Buck, 1976; Storm et al., 1977). Unfortunately, none of these studies have collected fatigue data during actual flight operations, and only some have included baseline and recovery data obtained beyond the days immediately surrounding the trip (Nicholson, 1970; Preston and Bateman, 1970; Nicholson, 1972; Preston, 1973). This latter procedure is particularly important when trying to interpret the results as a function of different trip schedules. The widely recognized variability in individual sleep habits makes it difficult to draw meaningful conclusions about trip-induced fatigue and sleep loss, especially given the limited number of subjects included in such studies. Moreover, partly because of difficulties in data collection, there is no associated flight performance data with which to address the important question of the findings' operational significance.

While these reports have served a very useful purpose in alerting the aviation industry to the potential problems associated with transmeridian flight operations, their scientific usefulness has suffered from the subjective nature of the data. There have been several attempts to remedy this problem by carrying out EEG sleep recordings on individuals before and after transmeridian flights (Evans, 1970; Evans et al., 1972; Klein and Wegmann, 1980; Athanassenas and Wolters, 1981; Endo et al., 1981; Wegmann and Klein, 1981). As a result, we now know that a single transmeridian flight can alter the structure of sleep in addition to the length and timing of sleep and that these effects can be partially predicted by concomitant alterations in the circadian rhythm of body temperature. However, from an operational standpoint, the generalizability of these studies is greatly restricted by the fact that only one experiment (Endo et al., 1981) has examined an actual flight crew after their arrival in a new time zone. All other subjects have been non-pilots who rode as passengers and had not been exposed to the repeated transmeridian flying that international pilots experience. Consequently, it may be reasonable to expect that their results may overstate the case for disrupted sleep and that seasoned crews develop strategies to cope more effectively with the consequences of time zone changes. Unfortunately, there are insufficient data available to address this issue.

In addition to the emphasis on sleep loss and fatigue, flight related research has also concentrated on the circadian rhythm effects of transmeridian flight. Much of this effort has been directed at measuring the rates of resynchronization for various physiological rhythms and determining how these rates are affected by different trip related variables. This work has been primarily carried out by Klein and Wegmann and recently reviewed by them (1979a, b, 1980). They and others have also investigated the effects of transmeridian flight on performance, with the aim of better predicting post-flight performance efficiency. These studies were recently reviewed by Graeber (1982). In general, this behavioral research has demonstrated the need to measure post-flight performance at different timepoints throughout the circadian cycle and not just once or twice per day. Unfortunately, the generalizability of other specific findings more pertinent to flight crew performance proficiency is seriously limited by some of the same design factors already discussed. Almost all of the subjects have been non-pilots traveling as passengers on a single transmeridian flight or on a round trip involving an extensive stay before returning. While this procedure is necessitated by the scientific requirement to document the circadian rhythmicity of time zone induced changes in performance, it does not provide operational guidance regarding typical international flight crew schedules with short layovers and repeated bidirectional time zone crossings within the same trip. Furthermore, the emphasis in these studies has been on the use of laboratory tasks to measure psychomotor and simple cognitive performance.

Only one experiment has measured performance in flight simulators (F-104) using experienced jet pilots (Klein et al., 1970). Twelve subjects were tested every two hours for several days before and after being transported on successive westward and eastward flights across eight time zones. Performance, as measured by percent deviation from preset flight parameters, exhibited statistically significant deficits depending on the time of testing, direction of flight, and number of days in the new time zone. While this study is the first to clearly demonstrate an operationally significant effect of multiple time zone shifts and time of day, it leaves many questions unanswered. The subjects were military fighter pilots who were transported as passengers and were not accustomed to multiple time zone crossings. Because of these factors, their performance data represent more an estimate of the effects of time shifts on raw flying skill than an estimate of the effects on flying in line operations. This conclusion is further supported by the realization that the simulator tests lasted only twelve minutes, were limited to manual control skills, and did not tap any of the higher level cognitive skills required on the flight deck.

In Flight Research

Perhaps no greater credibility can be gained among pilots and flight managers than that based on data collected from crew members flying an aircraft. While many researchers would relish the opportunity to gather such data, accessibility to the cockpit, for obvious reasons, is highly restricted; and the amount of reported research is still quite limited.

Aside from military research, most in-flight efforts have concentrated on examining the physiological effects of acute stress on the autonomic nervous system during high workload situations (e.g., Bateman et al., 1970; Nicholson et al., 1970). Fewer studies have been designed to examine the impact of fatigue on the flight deck; nevertheless, their results reveal the importance of this type of work as well as its potential pitfalls.

Carruthers and his colleagues (1976) carried out ECG and EEG recordings on a B-707 crew flying a trip from Buenos Aires to London. Their data, collected during a nine-hour segment from Rio to Madrid, represent the only available objective evidence of microsleep and superficial sleep patterns in pilots functioning on the flight deck. During the early morning hours (0400-0600 GMT) while the airplane was cruising on autopilot, all crew members displayed various EEG patterns characteristic of drowsiness including sleep spindles and increased 12-14 Hz activity with mass EEG synchronization. While such fatigue may not be uncommon at that time of the night during long overwater flights, the authors caution that some sleepiness may have been due to the cumulative sleep deprivation caused by unusually warm weather in Buenos Aires for several days before departure. The lack of any sleep recordings or self-reports before or after the eastward trip make it difficult to assess the contribution of this extraneous factor. Although follow-up studies were apparently initiated (Carruthers et al., 1978), no subsequent reports have been forthcoming.

Howitt et al. (1978) also examined the effects of fatigue on EEG and ECG during flight but focused on changes induced by carefully controlled variations in workload. A single pilot flew repetitions of the same three-part flight plan which included an instrument approach flown with and without use of the autopilot and flight director, a coupled ILS approach followed either by an overshoot or visual landing, and a simulated engine failure upon takeoff. All flights originated and terminated at the same airport. The fatigued condition included those flights which followed either 30 hrs. sleep deprivation or prior iterations of the same flight earlier in the day. As in the previous study, the physiological recordings substantiated the subjective feelings of fatigue, i.e., increased workload produced a significantly smaller increase in EEG activity during fatigued flights than during fresh ones. However, because of the combined statistical treatment of both fatigue conditions, it is not possible to determine whether fatigue generated by sleep loss or continuous activity affected the EEG differently. Some suggestion of a possible difference is provided by the behavioral observations of the pilot and the Training Captain occupying the right seat. Both reported that only the sleep deprived condition produced a marked narrowing of attention and a tendency to commit gross errors due to short-term memory losses whenever attention was diverted. Fatigue produced by repeated flights on the same day was characterized more by boredom and a lack of concern about maintaining precise accuracy on the instruments. While these behavioral observations are less compelling than quantitative data, they represent the only behavioral data reported in the open literature on actual flying proficiency as a function of fatigue.

So far, only one study has examined the psychophysiological reaction of commercial flight crews to flying typical short-haul trips involving daily multiple takeoffs and landings for several days. Klein et al. (1972) monitored eleven different B-737 crews across two different three-day trips involving either a 0600-1400h schedule or a 1200-2300h schedule. Focusing on standard physiological indices of acute stress, these investigators found flight related increases in pulse and respiration rate comparable to those reported for other moderate workload activities (e.g., intense administrative work or driving a car a long distance). Since the amount of increase (15-20 percent over rest values) did not change with successive days on the trip, the authors concluded that there was no accumulation of stress. Supplementary data on urinary catecholamines and 17-OCHS generally supported this claim. Concentrations of both hormones exhibited substantial increases during flight and across each trip day in comparison to control values. While these values increased further after day 1, subjective pilot estimates indicated that workload was also lowest on the first day for both schedules and that subsequent increases in workload most likely accounted for the further increase in hormone levels. Despite the reported lack of any cumulative effect, the finding of elevated stress hormones during each nighttime sleep period suggests that the psychophysiological impact of flying short-haul trips is substantial enough so as not to dissipate overnight. It should also be noted that the generalizability of these findings may be limited by the relatively young age of the flight crew sample (mean age for pilots = 32.2 yrs, co-pilot = 26.8 yrs).

Ruffell Smith (1967) reported similar percent increases in heart rate over resting levels in Captains flying commercial Trident aircraft on selected short-haul flights. Although no comparable data are available on the cumulative effect of these flights, the author emphasized that the highly variable heart rates observed during approach and landing warranted the inclusion of the number of daily flight segments as a stress factor in designing flight crew schedules.

OPERATIONAL APPROACH: THE TWO DIMENSIONAL COCKPIT

The future success of transport flight crew research rests on developing a multidimensional strategy for examining the operational significance of performance factors under study. We have attempted to organize such a comprehensive effort around the concept of the two-dimensional cockpit. The first and more traditional dimension is that of the individual flight crew member. The second encompasses all those variables related to the performance of the crew as a whole. While it is obvious that these two dimensions are not orthogonal, they are better discussed separately in order to clarify the methodological requirements that they each entail.

The Pilot as an Individual

If there is one theme which characterizes all of the past research reviewed above, it is the remarkable unity of emphasis on the individual pilot. Ignoring for the moment the obvious importance of integrated crew performance factors, there are several methodological points which need to be made concerning the individual dimension of the two dimensional cockpit. Whether the area of interest has been physiology or behavior, the intended focus has been on the individual and how he reacts to the demands of flight operations. In many cases, however, the findings are of limited usefulness due to the restricted nature of the data collection process. Often no information is available regarding preflight activities or physiological state. Nor is there usually any data collected during the recovery period other than immediately following exposure to the flight variables under investigation. Both types of extended data are needed particularly when the sample size is so limited and the dependent variables are often subtle and subject to wide individual differences. Frankly, this situation is often the result of limited subject and aircraft availability and not necessarily the fault of the investigator.

Another reason for collecting more extensive background data on the individual is to facilitate the transition from scientific findings to operational application. Numerous industrial shiftwork studies have already shown that individuals differ remarkably in their ability to tolerate shiftwork and to adjust to repeated changes of work schedule (Colquhoun and Rutenfranz, 1981). It is the ability to predict these individual differences, not only describe them, that is needed to convince the airline industry that a reasonable effort can be undertaken to incorporate fatigue related research findings into operational policy. Some success has been achieved already in shiftwork studies which have demonstrated the predictive value of personality dimensions, lifestyle indices, and circadian rhythm characteristics (Colquhoun and Rutenfranz, 1980). Similar efforts need to be made in air crew research especially regarding sleep habits both at home and during trips.

The two other requirements for improving the scientific-operational interface are: (1) the increased use of line pilots as subjects and (2) measuring realistic performance on flight deck tasks. While Klein and his colleagues (1970) have taken an initial step in this direction, there needs to be more effort directed at examining the cognitive dimension of pilot performance. The large percentage of aircraft incidents and accidents attributed to "human error" has intensified the focus of research upon the performance characteristics of the individual pilot. This effort has traditionally emphasized the importance of manual control skills. Besides being of direct consequence in aircraft handling, these skills are easily measured and therefore easy grist for the human factor laboratory. Nevertheless, the vast majority of human factor air transport accidents are not related to manual control skills, but result rather from inadequate knowledge, communication errors, poor pilot judgement, or failure to utilize available resources. It is in this arena that the interactive nature of the two dimensional cockpit becomes rapidly apparent. While most of the discussion regarding these topics is better suited to the role of pilot as a group member, there is still a need for individually oriented research examining the influence of flight related factors on short-term memory, attention, cognitive flexibility, and other personal cognitive skills required on the flight deck and which contribute to the crew's overall ability to function effectively.

The Pilot as a Group Member

As a direct result of the physiological, perceptual, and information processing limitations of individual human operators, multi-pilot aircraft cockpits were designed to ensure an acceptable level of redundancy. Yet, this system of redundancy has failed in many cases. It has failed too often because captains have not heeded the warnings of other crew members. It has failed because crew members who possessed adequate information have, for some reason, not provided it to others. Very few accidents or incidents are the result of a single gross error by one individual (Cooper et al., 1979). The individual pilot in air transport operations does not (or should not) perform in isolation; and, while research focusing upon the individual is of critical importance, it is equally critical that comprehensive investigations of flight crew behavior deal with the entire performance unit.

As a small group, the flight crew is susceptible to the laws of group dynamics. Designing research which includes this macroscopic perspective should improve our understanding about how human factors interact to affect crew performance. For instance, it is well known that the psychophysiological state of a person is directly tied to his or her emotional state. Likewise, it should be expected that the relative emotional states of group members, such as irritability and tiredness, can affect many of the dimensions related to group function. These include leader-subordinate relationships, leadership styles, the personality structures of group members, and communication patterns within the group. It is becoming increasingly apparent that these factors sometimes interact to produce breakdowns in the crew coordination process.

The incident and accident record strongly suggests the occurrence of these types of problems. An examination of transcripts of the cockpit voice recordings contained in U. S. National Transportation Safety Board accident reports has revealed a number of cases where, for some reason, subordinate crew members did not communicate their concerns assertively enough to prevent a mishap. Similarly, in the course of our research, subordinate crew members have complained that insensitive and intimidating captains often make them hesitant to verbalize their discomfort with certain situations. Foushee (1982) described reports from the NASA/FAA Aviation Safety reporting system which characterize this phenomenon:

> I was the first officer on an airlines flight into Chicago O'Hare. The captain was flying, we were on approach to 4R getting radar vectors and moving along at 250 knots. On our

> approach, Approach Control told us to slow to 180 knots. I acknowledged and waited for the captain to slow down. He did nothing, so I figured he didn't hear the clearance. So I repeated, "Approach" said slow to 180," and his reply was something to the effect of, "I'll do what I want." I told him at least twice more and received the same kind of answer. Approach Control asked us why we had not slowed yet. I told them we were doing the best job we could and their reply was, "You almost hit another aircraft." They then asked us to turn east. I told them we would rather not because of the weather and we were given present heading and to maintain 3000 ft. The captain descended to 3000 ft. and kept going to 2500 ft. even though I told him our altitude was 3000 ft. His comment was, "You just look out the damn window."

Although the preceeding is a rather flagrant example of poor leadership and ineffective group performance, the role structure inherent in any leader-subordinate relationship can easily foster more subtle occurrences of this type. While there are no reliable data available on the extent of this type of problem in the air transport environment, anecdotal evidence suggests that the situation occurs frequently enough to "condition" this hesitancy into the behavior of some subordinate crew members. Thus, the problem can manifest itself even in an environment managed by captains with more functional leadership styles:

> I was the copilot on a flight from JFK to BOS. The captain was flying. Departure turned us over to center and we were given FL210 which was our flight plan altitude. I noted that we had reached FL210 and were continuing through it, but was reluctant to say anything. As we climbed through 21,300 ft., I mentioned it to the captain, but not forcefully enough, and he did not hear me. I mentioned it again and pointed to the altimeter. We were at 21,600 ft. when the climb was stopped, and we descended back to 21,000. As we started our descent, center called and told us to maintain FL210. The captain said he had misread his altimeter and thought he was 1000 ft. lower than he was. I believe the main factor involved here was my reluctance to correct the captain. This captain is very "approachable" and I had no real reason to hold back. It is just a bad habit that I think a lot of copilots have of double-checking everything before we say anything to the captain (Foushee, 1982).

Crew Coordination Research Strategies

Unfortunately, very few studies exist to document such problems; however, the preceding observations suggest a methodological approach for studying group performance deficiencies. While the measurement of individual crew member variables provides only limited insight into the crew coordination process, the interactions of group members provide a rich source of data which can yield valuable insights. In fact, pilot-to-pilot communication is the only outward manifestation of the crew coordination process. While only a limited amount of research attention has centered on cockpit communication patterns, the measurement of relational communication has been utilized by a number of researchers over the years in other social settings (e.g., Bales, 1950; Mark, 1970).

The rapid advancement of simulator technology has finally provided an ideal laboratory for the study of these aircrew operating problems. It is now feasible to realistically simulate every aspect of line operations to the point where actual trips can be flown in a simulator which are almost indistinguishable from those in the airplane. Because of this high degree of realism, it is possible to study individual and group parameters with almost complete confidence that results generated in the simulator are strongly representative of the real world (e.g., Lauber and Foushee, 1981). Moreover, the simulator affords a high degree of experimental control and allows the study of operational problems too dangerous to examine in an actual aircraft.

The first example of this use of simulation was Ruffell Smith's (1979) classic simulator study of crew performance, which provided strong evidence for the importance of the group performance dimension. In his study, B-747 crews were asked to fly a highly realistic simulated flight from New York to London. Because of an oil-pressure problem, the crew was forced to shut down an engine. This created a diversion problem which was further compounded by a hydraulic system failure, poor weather, less than ideal air traffic control, and a cabin crew member who always chose inopportune times to request assistance from the cockpit crew. Despite the very controlled simulator setting, flight crews exhibited marked variations in their performance. Perhaps the most salient finding was that the majority of performance deficiencies were related to breakdowns in crew coordination. "High error" crews experienced difficulties in the areas of communication, crew interaction, and integration. For example, some of the more serious errors occurred when the performance of an individual crew member was interrupted by demands from other crew members. Other performance deficiencies were associated with poor leadership and the failure of the flight crew to exchange information in a timely fashion.

Recently, Foushee and Manos (1981) analyzed the cockpit voice recordings from the Ruffell Smith study utilizing a technique adapted from

Bales' (1950) interaction process analysis. Each statement or phrase was coded into predetermined communication categories and correlated with the errors made in the simulated flights. Several interesting relationships emerged which were germane to crew coordination. Overall, there was a tendency for crews who did not perform as well as others to communicate less, but the type or quality of communication played a more important role. There was a negative correlation ($r=-0.51$) between crew member observations about flight status and errors related to the operation of aircraft systems. In a similar fashion, a negative relationship ($r=-0.61$) was evident between aircraft systems errors and acknowledgements to information which had been provided. Acknowlegements were also related to fewer errors overall ($r=-0.68$). It would appear from these data that acknowledgements serve the important function of validating that a certain piece of information has, in fact, been transferred. These types of communication also serve as reinforcements to encourage the input of other crew members.

Commands were associated with a lower incidence of flying errors ($r=-.64$): errors related to power settings, neglect of speed limits, altitude errors, and the lack of formal transfer of control between captain and first officer. Often communications of this type seem to assure the proper delegation of cockpit duties, but their overuse may have negative consequences depending on the interpersonal styles involved. An identical piece of information can be related to other crew members in one of several different ways. For instance, a communication such as, "Check the plates for that profile descent procedure," which would constitute a command, could also be relayed, "I think we should check the plates for that profile descent procedure," an observation; or "Why don't we check the plates for that profile descent procedure?"-- an inquiry. Each style of communication may produce a different cumulative effect on the attitudes and behavior of other crew members during a trip.

Finally, the communication analysis revealed higher rates of response uncertainty, anger, and embarrassment, and lower rates of agreement in crews who tended to make more errors. Despite the fact that these correlational data do not allow inferences of causation, it would be safe to infer that discord related to the commission of errors, whether cause or effect, may be related to crew coordination deficiencies downstream. These findings also suggest that more methodological attention be paid to the crew coordination process and the role of such variables as communication. It is entirely possible, given the interpersonal effects of fatigue or sleep loss and mood, that communication difficulties are even more prevalent under these conditions, although insufficient data exist to directly support this notion.

Another approach to the study of group dynamics involves the examination of personality factors and their relationships to various group phenomena. While there are undoubtedly many personality variables which merit further exploration, two have received particular attention in the group performance context (Blake and Mouton, 1978; Spence and Helmreich, 1978). The first is referred to as "instrumentality" or "goal orientation." Persons high on this dimension tend to be very performance oriented, decisive, capable of getting the job done, etc. The second dimension is usually referred to as "expressiveness" or "group orientation." Highly expressive people tend to be sensitive to the feelings of others, warm in interpersonal relationships, and communicative.

The most obvious prediction with respect to these attributes is that instrumentality would play a pivotal role in pilot performance while expressiveness would be generally irrelevant. However, Helmreich (1982) has presented data which would seem to indicate otherwise. In a study of air carrier pilots, both instrumentality and expressiveness were positively related to check airman ratings of flight crew performance. The best crew performance was associated with the joint contribution of these attributes.

These data conform well with the communication and crew coordination findings. Effective group performance is heavily reliant upon the ability of group members to coordinate their individual activities with those of other group members. This in turn is related to one's ability to work effectively with others, a skill which involves sensitivity and effective communication. It should also be apparent that expressive tendencies may well counteract some of the negative consequences of the increased irritability associated with fatigue. Most importantly, these observations have serious implications for research methodology, suggesting a multidimensional approach for studying crew fatigue and different criteria for measurement selection depending on whether the focus is upon individual or group performance.

The Impact of Advanced Technology

State-of-the-art microprocessor technology is creating a revolution on the flight deck. Modern aircraft such as the B-767, B-757, DC-9-80, and the A-310 incorporate systems which automatically perform many of the functions which used to be performed manually. With respect to pilot performance, there are many advantages associated with this new technology, such as reduced workload and presumably lower levels of stress and fatigue. However, this revolution also raises many areas of concern.

Wiener and Curry (1980) have suggested that automation can potentially produce many undesirable effects which may magnify the importance of personality factors among flight crew members. Advanced technology aircraft are almost by definition changing the pilot's role into that of systems monitor. They speculate about a new class of pilot performance problems such as errors in equipment setup, responses to false alarms, failure to heed automatic alarms, failure to monitor, loss of flying proficiency and the implications for coping with situations where systems fail, etc. There are also serious motivational and psychosocial implications. It has been suggested that while automation certainly reduces workload, it may cause serious problems with boredom and complacency.

Largely as a result of self-selection processes, pilots have tended to be individuals with relatively high levels of achievement motivation compared to the population as a whole. Highly achievement oriented individuals tend to derive satisfaction from the successful performance of difficult tasks, to perceive a high degree of control over their environment, and to attribute their successes to their own superior abilities and efforts. Thus, pilots are accorded prestige, tend to have a positive self-concept, and experience a high degree of job satisfaction.

One inherent danger of the increasing automation of human operator tasks is the potential for creating work situations devoid of many factors which contribute to the motivation of highly achievement oriented individuals. For example, it might be very difficult for a pilot to derive much satisfaction from monitoring an autoland in adverse weather conditions, but the completion of a well-flown manual approach under similar conditions often provides great satisfaction to most pilots. In many cases, automated tasks have very little to do with the operator's own ability or level of effort except in those rare cases where the operator must exercise manual control because of a system malfunction. Past research has shown that when people attribute success to internal factors such as ability and effort (e.g., Weiner et al., 1971) there is usually a high degree of motivation for future performance. Conversely, when individuals attribute the cause of successful performance to external factors or perceive a lack of control over their environment (e.g., the ease of the automated task), motivation for subsequent performance is decreased.

These motivational factors have substantial implications for pilot performance: the resultant boredom could lead to lower levels of vigilance. There is considerable irony in this observation--the lower levels of workload associated with automated tasks should presumably allow more time for pilots to monitor system performance and detect the presence of abnormalities. However, the nature of the task could sometimes remove the pilot too far "out of the loop" thus inducing a psychological sense of loss of control or diffusion of responsibility (Darley and Latane, 1968). This line of research suggests that under such conditions, humans are not as quick to take action or tend to define potentially threatening situations as non-threatening.

Problems associated with a lack of vigilance, extreme boredom, or complacency could easily be exacerbated by fatigue. The level of stimulation associated with flying an approach in severe weather might well overcome the fatigue associated with moderate sleep loss, but autoland approaches (as they become more and more routine) may not provide the same level of stimulation.

A COMPREHENSIVE APPROACH

In summary, previous studies of pilot performance in general, and studies of fatigue in particular, have suffered from methodological problems related to the lack of adequate generalization from the laboratory to the "real world"; the difficulties of collecting fatigue data during actual flight operations; the lack of adequate performance data with which to address the question of operational significance; a limited focus on manual control skills; and a failure to consider factors related to overall crew performance.

To address the issues discussed above and to overcome some of the limitations and shortcomings of previous studies of pilot fatigue and circadian dysrhythmia, NASA has undertaken a long-term study of these factors in air transport flight operations. The program consists of a field study designed to document the physiological and psychological effects of demanding flight schedules and a simulator study designed to address pilot and flight crew performance issues, including the issue of operational significance.

Field Studies of Air Transport Flight Operations

The objectives of the field study are to: (1) determine the psychological and physiological responses of individual crew members to various flight schedules, with an emphasis on documenting circadian physiology, sleep quantity and quality, and fatigue and mood states; (2) to identify relevant individual attributes which may determine, or help to predict, the responses of pilots to fatigue and circadian factors; (3) to identify any personal adaptive strategies already being used by flight crewmembers; and (4) to identify significant operational factors which affect individual responses to fatigue and circadian factors.

To accomplish these objectives, extensive use is made of biomedical monitoring equipment, logs, rating scales, observer comments, and interviews with flight crew members. Monthly "lines of flying" (crew flight schedules) are scanned to identify target schedules based on considerations such as early morning departures, late evening departures, long duty days with a large number of takeoffs and landings, multiple time zone crossings or some combination of these factors. Once target schedules have been selected, the flight crew members assigned to these schedules are contacted and asked to volunteer as participants.

Subjects wear a solid state, digital "Vitalog" physiological recorder continuously (except for bathing) during the entire trip sequence, which is usually three to four days for a typical short-haul airline, and can be up to ten to twelve days for a long-haul operation. The recorder automatically samples and stores heart rate, rectal temperature, and non-dominant wrist activity every two minutes. In addition, to obtain baseline and recovery data, volunteers are asked to wear the recorder for at least two days before the trip begins and for up to one week following their return home. They also maintain a daily log, which seeks information such as sleep and awakening times, meal times, caffeine intake, and exercise, also includes mood rating scales which are administered every two hours throughout the waking day. For some subjects, saliva samples are also taken every two hours. These are analyzed to estimate plasma cortisol and melatonin levels (Riad-Fahmy et al., 1982).

A background questionnaire is used to obtain extensive information on lifestyle variables, sleep habits, nutritional information, and personality traits which may be related to pilots' ability to cope with various types of duty cycles. These data are correlated with physiological and mood data in an effort to determine whether certain profiles are associated with better adaptivity.

A NASA observer accompanies the flight crew volunteers for the entire trip, riding in the cockpit jumpseat, recording various operational data, including weather, Air Traffic Control, equipment problems, delays, etc.

Simulator Studies of Flight Crew Performance

A second major thrust of this comprehensive approach is the study of flight crew performance in the highly controlled and realistic environment provided by advanced aircraft simulators. While the field studies are primarily concerned with the impact on the individual, the simulator studies are focused on the operational impact upon the crew. Thus, the objectives of these studies are: (1) to determine any behavioral and crew performance changes which may be associated with certain types of duty cycles; (2) to determine the operational significance of these changes with regard to flight safety and operational efficiency; (3) to identify possible adaptive strategies that well-coordinated crews may use on the flight deck to cope with the impact of various flight schedules; and (4) to determine whether certain individual attributes are linked to crew coordination and good performance.

Fully qualified crews will be asked to fly a full-mission simulation scenario which will include all aspects of an actual line flight. Half the crews will fly this segment immediately after completing one of the schedules targeted for evaluation in the field study. The other half will fly the scenarios during an extended off-duty period. As in the Ruffell Smith (1979) study, the scenario will include complex operational abnormalities designed to test the crew coordination process.

Peformance in the simulator will be assessed within two broad categories: (1) aircraft handling data, and (2) crew coordination and cockpit resource management. The aircraft handling data will originate from two sources. First, the data from all major aircraft systems will be recorded, stored, and coded by elapsed time since the beginning of the scenario. These data will provide an objective basis for comparing performance between conditions. More importantly, they will be particularly useful in examining performance as function of various group performance variables and during critical periods associated with operational anomalies. The second source of aircraft handling data will be obtained from trained airline cockpit observers who will be unfamiliar with the condition of the volunteer crew. These individuals will record, in detail, all errors which occur during the simulated flights, and thus provide the primary means for assessing operational significance.

Crew coordination and resource management data will also be obtained in two ways. All simulated flights will be videotaped for later use in data analysis, and the cockpit voice recordings will be content coded for communication categories, shown in previous work (Foushee and Manos, 1981) to be related to flight crew performance. These data will be analyzed as a function of experimental condition and also correlated with the various individual personality attributes as a means of determining the relationship of these variables to dimensions of group performance. Second, the trained observers will rate each crew on a number of dimensions relevant to crew coordination. The videotapes will also be utilized as a subsequent means of rating crew performance by observers blind to the experimental manipulations.

To the extent possible, crews recruited for participation in the simulator study will have also participated in the field study. Thus, an entire range of parameters will be available for the comprehensive analysis of individual and group performance dimensions.

Improving Flight Safety

The goal of human factors research in the aviation environment is to reduce the contribution of human errors to aircraft accidents. The comprehensive research plan just described was designed to do so in a manner somewhat different from that implied by previous research efforts in the areas of crew fatigue and circadian dysrhythmia. Earlier studies focused on describing the extent of sleep loss, fatigue, or dysrhythmia associated with different flight schedules. Our current effort is tailored not only to enhance our understanding about how these factors affect flight crew performance but also to determine how the negative impact of these factors may be reduced. While schedule changes are an often suggested solution, the economies of today's air transport environment make this option unlikely to gain serious consideration except in extreme cases. An alternative solution is to develop flight crew countermeasures which can be shown to successfully reduce the performance deficits associated with these factors.

The anticipated countermeasures will most likely fall into two categories: (1) those oriented toward minimizing individual crew member disturbance, and (2) those designed to maintain crew performance despite individual crewmember deficits. Thus, the first category represents more of a biomedical approach, while the second represents more of a training and operational approach to the problem. Included in the individual countermeasures might be items such as when to nap, eat, or exercise, what to eat or drink, and how to optimize the sleep environment at layover sites. Also pertinent may be those individual traits that can not be manipulated easily, such as the lability of a pilot's circadian timing system, his usual sleep patterns at home, his degree of physical fitness, and his personality. The countermeasure value of these traits is derived from their potential as crewmember selection factors for particular types of flight schedules. To develop both types of individual countermeasures, data obtained from the background questionnaires and daily logs will be compared with the corresponding physiological and psychological data gathered during the field study. Crew countermeasures will most likely include those types of behavior commonly referred to as resource management skills, but tailored to the performance requirements of crews likely to be tired and in a dyschronic circadian state. These measures may include certain communication techniques, rotating work-rest patterns, workload pacing, standard operating procedures and other management techniques which are used by those captains and crews who best cope with the operational demands observed in the field as well as in the simulator. The emphasis will be on the whole crew, including the cabin crew who can often make a crucial difference especially on long-haul flights.

Hopefully, this comprehensive and positive research approach will not only enhance our understanding of crew performance factors in long-haul and short-haul flight operations, but also will lead to the adoption of effective strategies for dealing with the human limitations of all flight crews. There is little doubt that among flight crews there is a wide range of individual differences in the ability to cope with demanding flight schedules. The challenge is to discover why certain pilots and crews perform better than others and why certain flight schedules exceed the performance capabilities of even the best flight personnel.

REFERENCES

Athanassenas, G. and Wolters, C.L. 1981. Sleep after transmeridian flights. In "Night and Shift Work: Biological and Social Aspects" (Ed. A. Reinberg, N. Vieux and P. Andlauer). (Pergamon, New York). pp. 139-147.
Aviation Human Engineering Research Team. 1976. Survey Report of the Sleep Time on the Moscow Route. Aviation and Space Laboratory. Tokyo, Japan. Transl. NASA TT F-17,530. 1977. (NASA, Washington, D.C.).
Bales, R.F. 1950. Interaction Process Analysis: Theory, Research, and Application. (Addison Wesley, Reading, Mass.).
Bateman, S.C., Goldsmith, R., Jackson, K.F., Ruffell Smith, H.P. and Mattocks, V.S. 1970. Heart rate of training captains engaged in different activities. Aerospace Med., 41, 425-429.
Blake, R.R. and Mouton, J.S. 1978. The New Managerial Grid. (Gulf, Houston).
Buck, L. 1976. Psychomotor test performance and sleep patterns of aircrew flying transmeridional routes. Aviat. Space Envir. Med., 47, 979-986.
Carruthers, M., Arguelles, A.E. and Mosovich, A. 1976. Man in transit: biochemical and physiological changes during intercontinental flights. Lancet, 1, 977-980.
Carruthers, M., Cooke, E. and Frewin, P. 1978. Ambulatory monitoring of aircrew. In "Proceedings of the Second International Symposium on Ambulatory Monitoring" (Ed. F.D. Stott, E.B. Raftery, P. Sleight, and L. Goulding). (Academic, London). pp. 23-27.
Colquhoun, W.P. and Rutenfranz, J. 1980. Studies of Shiftwork. (Taylor & Francis, London).
Cooper, G.E., White, M.D. and Lauber, J.K. 1979. Resource Management on the Flight Deck. NASA CP 2120. (NASA, Washington, D.C.).
Darley, J. and Latane, B. 1968. Bystander intervention in emergencies: Diffusion of responsibility. J. Pers. Soc. Psych., 8, 377-383.
Endo, S. and Yamamoto, T. 1981. Effects of time zone changes on sleep. In "Variations in Work-Sleep Schedules: Effects on Health and Performance. Advances in Sleep Research" (Ed. L.C. Johnson, D.I. Tepas, W.P. Colquhoun, and M.J. Colligan.). (Spectrum, New York). pp. 415-434.
Evans, J.I. 1970. The effect on sleep of travel across time zones. Clinical Trials Journal, 7, 64-75.
Evans, J.I., Christie, G.A., Lewis, S.A., Daly, J. and Moore-Robinson, M. 1972. Sleep and time zone changes. Arch. Neurol., 26, 36-48.
Foushee, H.C. 1982. The role of communications, sociopsychological, and personality factors in the maintenance of crew coordination. Aviat. Space Envir. Med., 53, 1062-1066.
Foushee, H.C. and Manos, K.L. 1981. Information transfer within the cockpit: Problems in intra-cockpit communication. In "Information Transfer Problems in the Aviation System" (Ed. C.E. Billings and E.S. Cheaney). NASA TP 1875. (NASA, Washington, D.C.). pp. 63-71.
Graeber, R.C. 1982. Alterations in performance following rapid transmeridian flight. In "Rhythmic Aspects of Behavior" (Ed. F.M. Brown and R.C. Graeber). (Erlbaum, Hillsdale, N.J.). pp. 173-212.
Hartman, B.O. and Cantrell, G.K. 1967. Sustained pilot performance requires more than skill. Aerospace Med., 38, 801-803.
Helmreich, R. L. 1982. Pilot Selection and Training. Paper presented at the annual meeting of the Amer. Psych. Assoc. (Washington, D. C.).
Howitt, J.S., Hay, A.E., Shergold, G.R. and Ferres, H.M. 1978. Workload and

fatigue: In-flight EEG changes. Aviat. Space Envir. Med., 49, 1197-1202.

Klein, K.E., Bruner, H., Holtmann, H., Rehme, H., Stolze, J., Steinhoff, W.D. and Wegmann, H.M. 1970. Circadian rhythm of pilots' efficiency and effects of multiple time zone travel. Aerospace Med., 41, 126-132.

Klein, K.E., Bruner, H., Kuklinski, P., Ruff, S. and Wegmann, H.M. 1972. The Evaluation of Studies of Flight Personnel of the German Lufthansa on the Question of the Stress During Flights on the Short European Routes. DFVLR Inst. Flt. Med. Report DLR 355-74/2. Transl. NASA TM 76660, 1982. (NASA, Washington, D.C.).

Klein, K.E. and Wegmann, H.M. 1979a. Circadian rhythms in air operations. In "AGARD Lecture Series 105. Sleep, Wakefulness and Circadian Rhythm". (NATO-AGARD, Neuilly-Sur-Seine, France). pp. 10/1-10/25.

Klein, K.E. and Wegmann, H.M. 1979b. Circadian rhythms of human performance and resistance: Operational aspects. In "AGARD Lecture Series 105. Sleep, Wakefulness and Circadian Rhythm". (NATO-AGARD, Neuilly-Sur-Seine, France). pp. 2/1-2/17.

Klein, K.E. and Wegmann, H.M. 1980. Significance of Circadian Rhythms in Aerospace Operations. AGARDograph No. 247. (NATO-AGARD, Neuilly-Sur-Seine, France).

Klein, K.E., Wegmann, H.M., Athanassenas, G., Hohlweck, H. and Kuklinski, P. 1976. Air operations and circadian performance rhythms. Aviat. Space and Envir. Med., 47, 221-230.

Lauber, J.K. and Foushee, H.C. 1981. Guidelines for Line-Oriented Flight Training. NASA CP 2184. (NASA, Washington, D.C.).

Mark, R.A. 1970. Parameters of Normal Family Communication in the Dyad. Unpubl. Doctoral Diss. (Michigan State University, Lansing, Mich.).

Mohler, S.R. 1976. Physiological index as an aid in developing airline pilot scheduling patterns. Aviat. Space Envir. Med., 47, 238-247.

Nicholson, A.N. 1970. Sleep patterns of a airline pilot operating world-wide east-west routes. Aerospace Med., 41, 626-632.

Nicholson, A.N. 1972. Duty hours and sleep patterns in aircrew operating world-wide routes. Aerospace Med., 43, 138-141.

Nicholson, A.N., Hill, L.E., Borland, R.G, and Ferres, H.M. 1970. Activity of the nervous system during the let-down, approach and landing: A study of short duration high workload. Aerospace Med., 41, 436-446.

Preston, F.S. 1973. Further sleep problems in airline pilots on world-wide schedules. Aerospace Med. 44, 775-782.

Preston, F.S. and Bateman, S.C. 1970. Effect of time zone changes on the sleep patterns of BOAC B-707 crews on world-wide schedules. Aerospace Med., 41, 1409-1415.

Riad-Fahmy, D., Read, G.F., Walker, R.F., and Griffiths, K. 1982. Steroids in saliva for assessing endocrine function. Endocrine Reviews, 3, 367-395.

Ruffell Smith, H.P. 1967. Heart rate of pilots flying aircraft on scheduled airline routes. Aerospace Med., 38, 1117-1119.

Ruffell Smith, H.P. 1979. A Simulator Study of the Interaction of Pilot Workload with Errors, Vigilance, and Decisions. NASA TM 78482. (NASA, Washington, D.C.).

Spence, J.T. and Helmreich, R.L. 1978. Masculinity and Femininity: Their Psychological Dimensions, Correlates, and Antecedents. (Texas, Austin).

Storm, W.F., Hartman, B.O. and Makalous, D.L. 1977. Aircrew fatigue in nonstop, transoceanic tactical deployments. In AGARD Conf. Proc. No. 217, "Studies on Pilot Workload" (Ed. D. Wheatley). (NATO-AGARD, Neuilly-Sur-Seine, France). pp. B7/1-B7/7.

Wegmann, H.M. and Klein, K.E. 1981. Sleep and air travel. In "Psychopharmacology of Sleep" (Ed. D. Wheatley). (Raven, New York). pp. 95-116.

Weiner, B., Frieze, I., Kukla, A., Reed, L., Rest, S. and Rosenbaum, R. M. 1971. Perceiving the Causes of Success and Failure. (General Learning, Morristown, N.J.).

Wiener, E.L. and Curry, R.E. 1980. Flight-Deck Automation: Promises and Problems. NASA TM 81206. (NASA, Washington, D.C.).

METHODOLOGY IN WORKSTRESS STUDIES

G.C. Cesana, A. Grieco

Istituto di Medicina del Lavoro
"Clinica del Lavoro L. Devoto" dell'Università
via S. Barnaba 8, 20122 Milano, Italy

ABSTRACT

A critical review has been carried out on the methodology of work stress studies. Current theory and measurement have been taken into consideration. The latter has been classified as subjective-psychological, physio-psychological and epidemiological. No single measurement has shown a satisfactory reliability and validity in order to assess the adaptive reactions of man to his work and life environment. Multiple measurement is more satisfactory, but does not exceed the basic biological mechanicism of stress theory. A hypothesis is here proposed to bring stress research within the utilitarian scope of medicine as regards the prevention, treatment and rehabilitation of diseases which may be related to breakdowns in adaptation. To this purpose the necessity of more accurate exploration of the socioeconomical costs of stress, of standardization, of methods, above all in psychology and epidemiology, and of intensification of biological research emphasized.

INTRODUCTION

As Yuwiler ironically stated in an excellent article, which will be dealt with subsequently,"the term stress itself is one of the more stressful aspects of stress research" (Yuwiler, 1976). Selye, on the other hand summarizing 4o years of research, says that, since his first letter to "Nature", entitled "A syndrome produced by diverse nocuous agents", more than 110.000 articles have been published about what ist known as "stress concept". He himself compiled an encyclopaedia in which 7518 fundamental references are reported (Selye, 1976). This is the extent of a problem which will surely not have a rapid solution, but nevertheless must urgently find a socially useful approach. That is the thesis which will be proposed in this paper, with particular reference to the safeguard of occupational health.

In recent years the organization of work has undergone, and is still undergoing very great changes, that require new adaptation patterns as well as new models of biomedical approach - this also holds true in the field of transports.

The development and organization of transport reflect most typically our age in cultural, economic and technological terms: they are related to the increase of mobility, more widespread exchanges, fundamental discoveries and their application in the modern world. All these, however, come at no little cost from the environmental (physical and chemical pollution), energy supply (decrease in fossil fuels) and, above all, human points of view, for the two previously mentioned reasons as well as for the risk to the workers. Work in transport, in particular on the roads, is referred to by Mc Donald as "dangerous" (Mc Donald, 1980). In the USA, haulage work ranks among the first ten industries (of several hundred categories) as far as sick-leave and accident rate resulting from work are concerned. In the U.K., although comparable data are not available, it can be estimated that among HGV drivers there is a rate of fatal accidents which is 2,5 times as high as in all the manufacturing industries taken together. These data are comparable to those of the 4-5 highest risk industries.

In the "Proceedings of the International Conference on Ergomics and Transport" (Oborne and Levis, 1980), from which the previously mentioned work has been taken, all the risks to which transport workers are exposed are clearly evident. They range from the physical - i.e., noise, vibration, visibility, posture, physical effort - to the more important psychosocial and organizational - i.e., shift work, overwork, irregular hours, boredom associated with the need to maintain a continuous state of alertness. The psychophysiological consequences are disorders of arousal, sleep, diet as well as increase in anxiety, aggression, consumption of alcohol, tobacco, drugs. In fact it has been suggested that 35-50 % of fatal accidents on the main roads could be attributed to fatigue (Mc Kenna, 1982).

Perhaps in no other job are there present, to such a typical and uniform extent, risk factors capable of producing marked break-downs in adaptation as well as the connected behavioral, biochemical and pathological disorders. To this, one must add important cultural changes in transport work, above all on the roads (Lille, 1976). Transport work has been subjected to a heavy industrialization process, that is eroding its traditional autonomy in favour of a greater control and routine segmentation of the job. Also this breakdown in "cultural canon" produces results which cannot be neglected in a sociobiological approach to medicine (Henry and Stephens, 1977). Therefore one could say that transport work is a singular opportunity to further the study of health problems related to acute and chronic stress as well as to the general problem of work and social maladjustment. This is due to the previously mentioned features as well as to the fact that the transport world is, in a sense, most representative of our society in its positive and negative aspects. Therefore it is worth going back to the initial premises to enlarge upon the theory and measurement of stress.

THE THEORY OF STRESS

The word "stress" in the anglosaxon language has a technological origin: it describes the action of a force deforming a body. Beyond the possible semantic discussions, this concept has been introduced into biology almost independently by behavioral sciences (psychiatry in particular) and physiology (Yuwiler, 1976).

According to the psychiatric sciences the researches on stress could be referred to the study about maladaptative consequences of trauma, especially in childhood, in the genesis of mental disease, above all neurosis.

In physiology, stress research stems from the ground prepared by the studies of C. Bernhard, W. Cannon and above all Selye. While the first two authors respectively pointed out the vital importance of homeostasis and the emergency role of adrenaline, the third discovered the ubiquitous adrenocortical activation in response to a wide range of stimuli. In the physiologically oriented research the stress reaction can be operationally defined as the activation of neurovegetative sympathetic and adrenocortical systems.

In their origins the two scientific approaches are quite separate. In the first approach the stress or trauma has mostly a negative connotation, subjectively defined. In the second approach, the stress reaction basically follows a genetic determinism: it is aspecific, may be apetitive or aversive and has an adaptative-restorative function, at least initially

Many steps have been taken in an attempt to formulate an unified theory, satisfactory both from the point of view of interpretation of the stress phenomenon and its measurement. With reference to a psychosomatic approach, the most widely followed hypothetical model is the one described by Henry and Ely (1976), also witnessed in the studies of Mac Lean (1975) and Lazarus (1975). The information collected by the peripheral senses runs through the midbrain and thalamus to the primary sensory cortical projection areas. The temporal, parietal and occipital associative areas integrate and elaborate it in order to form the symbolic-abstract substrate of the sociocultural experience. The frontal pole has a central role in connecting this social brain to the storehouse centers of the "ancestrally learned behaviors" (lymbic striatal system), to the hypothalamus and the brainstem which controls the basic vital instincts. The coordinating action, on which depends wether or not adequate psychological defenses are set up, takes place through this connection. The confrontation between the individual and the stimulus occurs also according to the type and the in-stimulus may be associated with different injuries in different organs.

The way the researchers have tried to deal with the difficulties in measurement, previously mentioned, is extremely complex and various. Here folllows a hypothesis of classification. A few comments will be given with particular reference to field studies on work stress.

Subjective-psychological measurement

Subjective-psychological measurement has been greatly developed in recent years, on account of both the criticism of scientific positivism (Kasl, 1978) and the necessity of evaluating all the subjective and psychological factors which may be important in producing stress reaction (Jenkins, 1979; Davidson and Cooper, 1981). We can distinguish:

a) general descriptive questionnaires about health, sleep, diet, life style and work environment. These questionnaires are becoming more and more important because the subjective evaluation of health and life conditions (by the individual and/or the group) has been proved to be an indispensable aid to the traditional parameters used in medicine (Hunt, et al., 1980). In addition life style - i.e. the habitual behavior freely adopted - has been stressed as the factor that ultimately accounts for more than half of the annual deaths in the USA (Lambert et al., 1982). With reference to occupational medicine, the "General Health and Adjustment Questionnaire", produced by the NIOSH (Tasto et al., 1978) for the study of shift work, is particularly interesting. Although the questionnaire works on a simple descriptive level, it provides an originally methodological contribution, that seems to deserve a satisfactory reliability and validity also in transcultural studies (Cesana, 1983).

b) subjective scales - they attempt to quantify self-reports with reference to the single topics already investigated by the descriptive questionnaires. They vary and are generally used to estabilish possible causal relationships, between subjective perception and biological disorders, presumed to be produced by acute or chronic stress phenomena. Two scales are particularly interesting on account of their standardization and widespread usage in recent years. The first is the "Social Readjustment Rating Questionnaire" set up by Holmes and Rahe (1967). It searches for possible temporal associations between events, requiring adaptative reactions from the individual, and disease. The weight given to single events has been shown to be stable enough in transcultural studies (Rahe, 1975). However, the results obtained are poor, and have been widely criticized owing to the scarce control of data that can be collected only retrospectively (Goldberg, 1977). The second scale is the questionnaire and/or interview set up in order to identify aggressive-competitive type A behaviors (Dembroski et al., 1979). Such a personality has been correlated to an incidence of coronary heart disease which is twice as great as the opposite type B. In addition it seems to be the only psychosocial factor clearly associated with this kind of pathology (Jenkins, 1976). The data are without doubt very interesting, in particular if one bears in mind the possibility that type A could be a cultural product of the technologically advanced society (Cohen and Cohen, 1978). In summary the results obtained from the above measurement are neither uniform nor final. Besides, the continuous proliferation of new scales for the evaluation of self-reports has caused confusion, rather than precision both in methods and results of research (Chadwick et al.,1979).

c) psychometric tests. Their aim is to measure the various personality aspects, which may be related both to the subjective perception of stress and to the consequent biological disorders. The most widely used tests are: EPI, MPI, IPAT 16 PF, MMPI (Davidson and Cooper, 1981). The last one is the most comprehensive, good for individual as well as for group diagnosis. At any rate, especially on account of its completeness (about 500 items),it is not very suitable for studying large groups, which may often be at low cultural level (e.g. manual workers). The third one merits particular consideration because, though short and easily understandable, looks into different personality aspects (just 16), that can be analyzed separately as well as, in our experience,through a total sore (Cesana, 1983). The first two, very short and easy to fill in, are interesting because formulated by Eysenck with reference to the neurophysiological interpretation of stress in term of arousal (Eysenck, 1975). According to this interpretation (Welford, 1974), the central nervous system in order to work adequately requires a constant influx of stimuli from the external environment. Both the lack and the excess of stimulation challenge the homeostatic mechanism, by which the organism maintains an adequate level of activation, according to a curvilinear relationship, in the form of a "U" or an inverted "U" if stress perception or performance respectively are considered. Eysenck's tests point out two important and independent dimensions of personality regarding the"arousal" of emotions: "extroversion" and "neuroticism". The introverts tend to have a higher level of arousal and to amplify the stimulus

tensity of conditioning. If the outcome of such a confrontation is negative, the hypothalamus together with the aspecific reticular activating system will no longer coordinate the body economics stemming from the interplay of varied pituitary, vagal and sympathetic reactions.

The described model well suits the exigences of modern psychophysiology. Its succes may be attributed to the following reasons: a) it is supported, or at least not contradicted, by experiments, carried out mostly on animals but also on man (Yuwiler, 1976; Henry and Stephens, 1977, Henry and Ely, 1976); b) it shows a good heuristic value. It allows to unify within a coherent scheme a large number of observations about the interrelationship between psychosocial factors, the symbolic activity of the individual and health disorders (Lipowsky, 1977); c) it is open to a deeper evaluation of "subjectivity", i.e. of the necessity to study the psychological concomitants of the events, as emphasized by several authors (Mason, 1975; Burchfield, 1979). The model, however, right in its final important point shows, in a sense, the great limits charaterizing it. In fact, the evaluation of subjectivity is a big question with regards to the biological mechanism rather than a hypothesis fixed in its methods and contents. Biological mechanism, widely criticized (Schmale and Ader, 1980), surely cannot be corrected by the heroic efforts to complete the theory by hypertrophyzing it with enumerable variables, which keep interplaying within a physico-chemical system. Man cannot be reduced to this. Nowadays the challenge to medicine, above all in the field of diagnosis and therefore measurement is: "to maintain and develop further the emerging unity of man health care at a conceptual level and to devise ways to permit this unified concept to impact favorouably on health care practice" (Swisher,1980).

MEASUREMENT OF STRESS

Methods and instruments of measurement, used in stress research, in particular with regard to occupational maladjustment, cannot avoid any of the questions previously mentioned.

As strongly stressed by Selye ('76), in spite of the aspecificity of the stress reaction, the different receptiveness of the individual determines: a) qualitatively different stimuli may cause the same stress reaction; b) equal power stimuli may cause equal intensity reactions in different individuals; c) the etiology of the stress reaction is never monofactorial but always multifactorial implying at least the interaction between stimulus and conditioning; d) the same level of stress, produced by the same

intensity, the extroverts the opposite, in consequence of the cortical activation produced by different thresholds in the "ascending reticular activating system". This situation is also interpretable in the more psychodynamic terms of "trait anxiety" (N) and "state anxiety" (E).

Psychophysiological measurement

Psychophysiological measurement is based on the arousal theory previously mentioned. It is worth noting that, on account of this approach, this measurement has been applied especially to the study of shift work (Colquhoun and Rutenfranz, 1980), which is an evident challenge to the vigilance mechanism. Here too, three kinds of measurement can be distinguished:

a) measurement of performance. Although in the arousal theory, this measurement refers to many interpretative models of the mental load. As Moray (1979) well synthesized, there are two empirical starting points for this measurement. The first, which refers to the structural models of the "Information" and "Signal Detection" theory, evaluates the scores derived from the increase in errors and latency times. The second, which stems from the concept of "effort", with its energetic and subjective features (Sander, 1979), assesses the performance more in dynamic terms of "processing capacity" than in structural terms of "channel capacity". The former approach has led to "a priori" measurement of discrete tasks (discrimination, attention, language and reaction-times) as well as of continuous tasks (bandwidth, amplitude, order of control, etc.). The latter approach has promoted the empirical reference of such a measurement to the operator as well as the consideration of physiological variables expressing subjectivity (heart rate, ventilation, catecholamines, etc.). With regard to this workshop, it is worth mentioning that measurement of performance has been extensively applied to transport operations, as reported in the experimental part of the volume, from which the two previous publications were taken (Moray, 1979). However, such a measurement is frequently criticized because it does not show in the real field the same efficacy as the one observed in the laboratory.

b) electrophysiological measurement. It refers to the recording of electrical activity of the organism as well as of non-electrical activity that can be transformed into electrical signals. One can distinguish between measurement of the central nervous activity -electroencephalogram and evoked potentials (Shagass, 1975)- and that of the peripheral neurovegetative activity - heart rate and galvanic skin responses. Other ways of measurement of the peripheral neurovegetativ activity (salivation, pupillometry, electrogastrogram) are far less important (Lader, 1975). Although all these methods of measurement can be a documented expression of psychological activity, their discriminative power is very low, between emotive and cognitive factors as well as between psychological and physical factors. Heart rate is the clearest example from this point of view, on account of its widespread use (Monod, 1967; Vogt et al., 1973; Kalsbeek, 1973). Though Shagass hypothesized that emotional processes may be better expressed by central measurement, there are no data supporting this thesis (Shagass, 1975). Then, these methods of measurement can hardly be applied to field studies because of their sophystication and low discriminative power, which tolerates badly the interplay of numerous and scarcely controlled intervening variables. Surely telemetry will help to deal with some of these problems. In summary, it can be said that electrophysiological measurement is useful for the assessment of conditions of marked occupational overload in quite well controlled situation, as shown by the recent publications about repetitive tasks (Weber et al., 1980) and irregular workhours (Torsvall et al., 1981).

c) biochemical measurement. It regards the ever growing knowledge in the field of psychoendocrinology. The most widely used neuroendocrinological parameters in stress research are in order of importance (Rose and Sachar, 1981): catecholamines, cortisol, testosterone, prolactin, growth hormone, β-endorphins. Many other cerebral hormones and peptides undergo changes in psychiatric diseases or have well known effects on behavior, but practically they have no importance for the purpose of research into environmental stress. With reference to the above neuroendocrinological parameters, thanks especially to Frankenhaeuser's studies (1980), catecholamines (adrenaline and noradrenaline in particular) come out as the most reliable stress indicators. They are increased by stimuli that also increase cortisol secretion. However, while the latter is influenced more by the novelty than the intensity of the stimuli and has a faster adaptation rate, the former remain high whenever a consistent level of vigilance and effort is required. The excretion of catecholamines seems to be related more to the intensity than the direction of the reaction. Most of the stimuli increase adrenaline as well as noradrenaline. However, adrenaline like cortisol seems to be mostly influenced by conditions of "uncertainty" and consistent arousal. On the other hand, the increase of noradrenaline is mostly related to a high level of physical stimulation. The other hormones mentioned are less important in field studies: however, a brief consideration of testosterone and ß-endorphins is recommended. The former is the only hormone that decreases in stress situation because of a still unknown mechanism (negative feedback from corticoids?). Such an observation was made also on HGV drivers (Cullen et al., 1979). The latter, whose function seems to be connected with a greater endurance to pain during stress, are particular interesting because they bring into the researches on maladjustment the latest and very important findings on the neuroregulatory role of cerebral peptides (Barchas et al., 1978). The biochemical measurement of stress is exposed to the same criticism addressed to the physiological one. However, regarding catecholamines in particular, to their advantage there is a more widespread experience in their use (Levi, 1972; Jenner et al., 1980) and greater ductility of measurement arising from the possibility of urinary dosing. Further advantage will come through wider application of more reliable and valid laboratory methods, such as high pressure liquid chromatography (Cesana et al., 1982).

Epidemiological measurement

Epidemiological measurement arises from the observed differences in morbidity and mortality rate associated with psychosocial factors. The matter has been frequently suggested in the previous paragraphs. According to studies carried out on very large communities in Australia, the 20 % of physical diseases and the 37 % psychiatric diseases might be related to psychosocial factors (Andrews et al., 1978). A significant and controlled association has been reported between longevity and work satisfaction (Haynes and Feinleib, 1980). A synthetic review of the matter, considering psychological, social, cultural and economic factors has been recently publishe by Cohen and Brody (1981). The most widely studied diseases in relation to work stress are surely mental and cardiovascular illnesses (Cooper and Marshall, 1976). Unfortunately the surveys on mental illnesses suffer from a very uncertain nosographic classification. However, they point to a wider prevalence of illnesses among the blue collar workers and a wider prevalence of symptomatology among white collar workers and professional people (Caplan et al., 1975; Cherry, 1978). Cardiovascular diseases are very widely studied on account of almost "epidemic" proportion of their diffusion in our society, of their seriousness and of the huge costs involved, also in occupational terms (Chadwick et al.,

1979; Harlan, 1981), in spite of the apparent decrease in these diseases in recent years (Hampton, 1982). Moroever there is a series of pathogenetic hypotheses supporting the relationship between psychosocial-behavioral factors and CHD-hypertension. Nowadays the pathogenesis of CHD and hypertension refers to a complex theory which considers, in addition to the status of vessels, the variation of flux and pressure on the vessels themselves and the reset of aortic and carotid baroceptors, as a consequence of hormonal and neurovegetative influences (Buell and Eliot, 1980). The more comprehensive data regarding the relationship between occupational-socio-economic status and morbidity-mortality caused by CHD can be found in Jenkins (1976) and Cooper and Marshall (1976). According to the former, the social status mostly affected by CHD risk tends to be different in relation to the different development stages of a country, from the rural to the industrialized civilization. In the first stage of industrialization the CHD is greater for higher social classes. In the advanced stage the lower classes are at a greater risk, the less specialized workers in particular. Occupation may influence the coronary risk by selecting life style (see Type A) and life habits connected to it (diet, smoking, etc.). Among the twenty important articles mentioned by Cooper and Marshall, five concern job features defined as quantitative overload, fifteen concern job features defined as qualitative overload. In the former category are listed factors such as overwork, second job, shiftwork; in the latter, work dissatisfaction and tension, low self esteem, role conflicts, occupational level, responsibility, career, status incongruity. All the quoted factors showed positive correlation to CHD as well as, above all, to the worsening of traditional risk factors (smoking and cholesterol in particular), according to coefficients between 0.1 to 0.4. Therefore there are data supporting the possible relationship between job, the induced life style and the most important chronic-degenerative disease in our countries. However, as rightly observed by Chadwick et al. (1979), though they are meaningful, they are, in a sense, "too many" and presented with such a disordered fashion as to obscure what is really going on. A consistent incapacity to replicate the results can be noted both within the same laboratory and among different research groups. We are again faced by a still unsatisfactory measurement.

CONCLUDING REMARKS

As underlined by what has been said so far and as stressed by many authors, none of the single described methods of measurement can evaluate with satisfactory reliability the consequences of the adaptative reactions of man to his work and life environment (Hopkins et al., 1979). Nor is an easy optimism allowed about the results produced by multiple measurement coming from the different disciplinary areas, as rightly requested by some researchers (Cooper et al., 1978; Levi, 1979). In fact, the use of multiple measurement makes more and more difficult the assessment of stress phenomena as well as it does not overcome the weakness of the underlying theory. On the contrary, from this point of view, the weakness of the theory is, in a sense, reflected to a greater extent. Finally an "umbrella" or "total" theory of stress will be hardly, if ever, produced (Hopkins et al., 1979). The development of this research would be therefore hopeless if it were not considered within the field of medicine. As said in the first chapter of "Harrison's Principles of Internal Medicine" (Isselbacher et al., 1980), "medicine is an art also in the sense that physicians can never be content with the sole aim of endeavoring to clarify the laws of nature; they cannot proceed to their labors with the cool detachment of the scientist whose aim is the winning of the truth, and who, in doing so, conducts a controlled experiment ... their primary and traditional objectives are utilitarian: the prevention and cure of diseases and the relief of suffering, wether of body or mind". That means that the aim of stress research is not the achievement of the outdate positivistic idea of a scientific model of man and his interactions with the environment. Were it so, the position of those asking for a moratorium on stress research would be quite justified (Burchfield, 1979). The primary task of stress research is to introduce into the basic aim of medicine the consideration of breakdowns in adaptation and the application of the knowledge so far obtained about it. It would be very dangerous not to achieve this scope, above all in occupational medicine: it would mean falling again into an individualistic-mechanicistic notion of disease which had already absolved, in the XIX century, an oppressive industrial world from any responsibility about illnesses (Cohen and Cohen, 1978).

Today health problems related to work are changing in a way that can by empirically summarized through the following three points (Cesana et al., 1981): a) psychological and organizational factors are becoming more and more important in work structure, with particular reference to the introduction of advanced technology; b) traditional occupational diseases (silicosis, asbestosis, lead intoxication, etc.) are decreasing in incidence; c) subjective complaints and aspecific pathology are increasing, as deduced from the growing level of sick-leave, which is almost unanimously considered to be the most reliable indicator of the discomfort of the working groups. Such a phenomenon, which a recent E.C. report (Laurence et al., 1981) states to be corresponding to 2-3 % of the gross national product in western Europe, takes place paradoxically while the average life expectancy, height and weight of population as well as the public health services are improving (Parmeggiani, unpublished manuscript). In addition it is worth taking notice of the prevalent trends in morbidity and mortality. In the previously mentioned E.C. report, they observe a very consistent increase in mental pathology as well as in organic pathology with strong psychosomatic component (CHD, hypertension, gastrointestinal and immunoallergic disorders, also in the form of mass psychogenic illnesses). This pathology is differently distributed in relation to job, being more widespread among jobs with lower cultural content. The causes of the phenomenon cannot be totally attributed to work, but there is no doubt in the fact that work and its changes are a peculiar feature of the highly industrialized societies. Work and the possible breakdowns in adaptation consequent to it must be seriously considered in the study, prevention, treatment and rehabilitation of chronic degenerative pathology with multifactorial etiology. (For this kind of pathology the term "job related disease" has even been proposed, Rutenfranz, 1982). All this is necessary in order to confront efficently the connected sociocultural phenomena: the previously mentioned high levels of sick-leave, the process of stressed medicalization of the social troubles, the abuse of drugs and sanitary services, the frequent failure of primary and secondary prevention interventions, the difficulties in the social and work rehabilitation of the handicapped. At this stage, it is not only a question of theory, but of a sound methodology for the application of the knowledge so far available. The research procedures which might be taken are:

   a) socio-economic - The cost/benefit relationship of the current technological changes has to be evaluated from a sanitary point of view, with particular reference to those groups which are more exposed to risk (e.g. transport workers). The extent of sick leave, the incidence of mental ill health and "psychosomatic" chronic degenerative diseases, the consumption of drugs and sanitary services will have to be carefully considered. The surveys will possibly have to be longitudinal with reference to the following point.

b) epidemiological - Jenkins (unpublished manuscript) in his hypothesis on a behavioral epidemiology suggests a series of questions, some of which are very interesting: are health conditions effectively improving? are the modern health problems qualitatively the same as in the past? what are the inequalities? which are the vulnerable groups? what are the preventive hypotheses which can be deduced from future projections? A systematic epidemiological research on the pathology possibly related to maladjustment should answer such questions. The existing knowledge does not justify the expenditure of money and time required by follow up studies (e.g. on workstress and CHD). However, it has sufficient elements to allow incidence studies through the introduction into the factories of morbidity and mortality registers, which could be at a relatively low cost and useful in order to investigate the possible relationship between workstress and the most prominent diseases in our society (i.e. CHD and mental ill health).

c) psychological - All research into stress must substantially rely on self-reports. In spite of this incontestable fact and the current reappraisal of subjectivity we are still seriously deprived of standardized methodology. This is the first point which must be dealt with regarding the debate among the several psychological schools about the reliability and validity of quantitative procedures handling qualitative data. A basic contribution can come from the appraisal of traditional methods of psychological diagnosis both of behavioristic origin and above all of psychoanalytical origin, owing to its particular consideration of descriptive procedures (Rapaport, 1968). These in fact seem most useful, as shown by the experience of questionnaires.

d) psychophysiological - It is perhaps the most complex procedure. In fact, to define in psychophysiological terms the risk of stress means to answer the questions about where, when and how a normal, aspecific and general biologic mechanism, for the adaptation to the environment, can become harmful and lead to disease. These questions which undoubtedly contain philisophical as well medical implications are by no means simple and have been subjected to many attempts at reduction according to statements like "everything is stress" or "it is not important what happens, but what one thinks has happened" (Yuwiler, 1976). Such a relativism, while not devoid of reasons which must be considered, cannot be allowed to ignore the essentials of the problem: how does a given assembly of environmental variables (physical, chemical, psychosocial) promote or not promote health; how do biological solutions, which are adaptative today, become malad aptative tomorrow; how finally can the wide diffusion of psychosomatic chronic diseases be related to psychophysiological disorders coming from the attempt to confront with the current work organization. The studies on shiftwork and the new pathogenetic hypotheses on CHD and hypertension may show a way which can be usefully followed by applying more systematically the definite knowledge so far acquired.

In spite of very big problems, the studies on stress can provide medicine with more hope than it appears at first. For this purpose, the present report wants to stress that, in a sense, the systematic organization of research is today more important than the theory supporting it. Such a choice is medical as well as cultural and political.

REFERENCES

Andrews, G. et al. 1978. The relation of social factors to physical and psychiatric illness.Am.J.Epid.,108,27-35.
Barchas, J.D. et al. 1978. Behavioral neurochemistry: neuroregulators and behavioral states. Science, 200,964-973.
Buell, J.C. and Eliot, R.S. 1980. Psychological and behavioral influences in the pathogenesis of acquired cardiovascular disease. Am.Heart J., 100,723-740.
Burchfield, S.R. 1979. The stress response: a new perspective. Psychosom. Med.,41,661-672.
Caplan, R.D. et al. 1975. Job demands and workers health(DHEW (NIOSH) publication No. 75-160).
Cesana, G.C. et al. 1981. Elementi di fisiologia del lavoro: la fatica industriale. In "Trattato di Medicina del Lavoro" (Ed. E. Sartorelli) (Piccin, Padova), pp.125-137.
Cesana, G.C. et al. 1982. Work stress and urinary catecholamine excretion in shift workers exposed to noise. Med.Lav., 2,99-109.
Cesana, G.C., Zanettini, R. et al. 1983. Work stress and CHD risk in a group of shift workers. Med.Lav. (in press).
Chadwick, J.H. et al. 1979. Psychological job stress and coronary heart disease(SRI International, Menlo Park (CA).
Cherry, N.1978. Stress, anxiety and work. J.Occup.Psychol.,51,259-270.

Cohen, C.I. and Cohen, E.J. 1978. Health education: panacea, pernicious or pointless. New Engl.J.Med.,299,718-720.
Cohen, J.B. and Brody, J.A.1981.The epidemiologic importance of psychosocial factors in longevity. Am.J.Epid.,114,451-461.
Colquhoun,W.P. and Rutenfranz, J.(Eds.)1980. Studies of shiftwork. (Taylor and Francis Ltd, London).
Cooper, C.L. and Marshall, J.1976. Occupational sources of stress:a review of the literature relating to coronary heart disease and mental ill health. J.Occup.Psychol.,49,11-28.
Cooper, C.L. et al. 1978. Prevention and coping with occupational stress. JOM,20,420-426.
Cullen et al., 1979. Endocrine stress responses of drivers in a "real life" heavy goods vehicle driving task. Psychoneuroend.,4,107-115.
Davidson, M.J. and Cooper, C.L. 1981. A model of occupational stress. JOM, 23,564-574.
Dembroski, T.M. et al. (Eds.) 1978. Coronary prone behavior. (Springer Verlag, New York).
Eysenck, H.I.1975. The measurement of emotion:psychological parameters and methods. In "Emotions their parameters and measurement". (Ed. L.Levi). (Raven Press, New York).
Frankenauser, M. 1980. Psychobiology effects of life stress. In "Coping and health" (Eds. S.Levine and H. Ursin). (Plenum Press, New York). pp.203-233.
Goldberg, L. and Comstock, C.N. 1976. Life events and subsequent illness. Am.J.Epid.104,146-158.
Hampton, J.R. 1982. Falling mortality in coronary heart disease. Brit.Med. J.,284,1505-6.
Harlan, W.N.1981. Physical and psychosocial stress and the cardiovascular system. Circulation, 63,266A-271A.
Haynes, S.G. and Feinleib, M. (Eds.) 1980. Second Conference on the epidemiology of aging.(DHEW (NIOSH) publications No. 80-969).
Henry, I.P. and Ely, D. 1976. Biological correlates of psychosomatic illness. In "Biological foundations of psychiatry" (Eds. R.G. Grenell and S.Gabay). (Raven Press, New York).pp.945-985.
Henry, I.P. and Stephens, P.M.1977. Stress, health and social environment: a sociobiological approach to medicine. (Springer-Verlag, New York).
Holmes, T.H. and Rahe, R.H. 1967. The social radjustment rating scale. J. Psychosom.Res.,16,213-218.
Hopkins, V.D. et al. 1979. Final report of the application group. in "Mental Workload" (Ed.N.Moray). (Plenum Press, New York). pp.469-495.
Hunt, S.M. et al. 1980. A quantitative approach to perceived health status: a validation study. J.Epid.Comm.Health,34,281-286.
Isselbacher, K.I. et al. 1980. Introduction to clinical medicine. in "Harrison's principles of internal medicine". (Ed.K.I. Isselbacher et al.) (McGraw Hill Book Company, New York).pp.1-7.
Jenkins, C.D. 1976. Recent evidence supporting psychological and social risk factors for coronary disease (First and second parts). New Engl. J.Med.,294,987-994,1033-1040.
Jenkins, C.D. 1979. Psychosocial modifiers of response to stress. In "Stress and mental disorders" (Ed. Barret J.E.). (Raven Press, New York). pp. 265-278.
Jenkins, C.D. (unpublished manuscript). Behavioral epidemiology as a specialized field.
Jenner, D.A. et al. 1980. Catecholamine excretion rates and occupation.Erg., 23,237-246.
Kasl, S.V. 1978. Epidemiological contribution to the study of work stress.

In "Stress at work". (Ed. C.L.Cooper and Payne R.L.). (John Wiley and Sons, New York).pp.3-48.
Kaslbeek, J.N.H. 1973. Do you believe in sinus arrhytmia? Erg.,16,99-104.
Lader, M. 1975. Psychophysiological parameters and methods. In "Emotions their parameters and measurement" (Ed. L.Levi). (Raven Press, New York). pp.341-368.
Lambert, C.A. et al. 1982. Risk factors and life style: a statewide health interview. N.Engl.J.Med.,306,1048-1051.
Lawrence, M.G. et al. 1981. Stress psicofisico sul lavoro. (Fondazione Europea per il Miglioramento delle Condizioni di Vita e di Lavoro). (Contratto No. 80-3030-4-ST-MK).
Lazarus, R.S. 1975. The self regulation of emotions. In "Emotions their parameters and measurement" (Ed. L.Levi). (Raven Press, New York). pp. 47-68.
Levi, L. 1972. Stress and distress in response to psychosocial stimuli. Acta Med.Scand. (Suppl. 528).
Levi, L. 1979. Occupational mental health: its monitoring, protection and promotion. JOM, 21, 26-32.
Lille, F. 1976. Etude psycho-sociologique de la profession de conductor rutiere. (Rapport No. 1: Synthese). (Secretariat d'Etat au Transport, Paris).
Lipowsky, Z.J. 1977. Psychosomatic medicine in the seventies. An overview. Am.J.Psychiatr., 134,3-11.
Mac Lean, P.O. 1975. Sensory and perceptive factors in emotional functions of the triune brain. In "Emotion their Parameters and measurement" (Ed. L. Levi). (Raven Press, New York).pp.71-92.
Mason, J.W.1975. Emotion as reflected pattern of endocrine integration. In "Emotions their parameters and measurement". (Ed. L.Levi) (Raven Press, New York). pp.143-182.
McDonald, N.1980. Fatigue, safety and the industrialization of heavy goods vehicle driving. In "Human factors in transport research" (Eds. D. J. Oborne and J.A. Levis). (Academic Press, London). vol. I, pp.134-142.
McKenna, F.P. 1982. The human factor in driving accidents. An overview of approaches and problems. Erg. 25,867-877.
Monod, H. 1967. La validitè des mesures de la frequence cardiaque en ergonomie. Erg., 10, 485-537.
Moray, N. 1979. Models and measures of mental workload. In "Mental workload". (Ed. N. Moray). (Plenum Press, New York). pp. 13-22.
Oborne, O.J. and Levis, J.A. 1980. Human factors in transport research. (Academic Press, London).
Parmeggiani, L. et al. (unpublished paper). Stress and absenteeism of italian factory workers.
Rahe, R.H. 1975. Life change and near-future illness reports. In "Emotions their parameters and measurement". (Ed.L.Levi). (Raven Press, New York). pp. 511-530.
Rapaport, D. et al. 1968. Diagnostic psychological testing. (International University Press, New York).
Rose, M.R. and Sachar, E. 1981. Psychoendocrinology. In "Textbook of endocrinology" (Ed. R. Williams). (W.B.Saunders Company, Philadelphia). pp. 646-671.
Rutenfranz, J. 1982. Recent advances in the field of work physiology: heavy work, body positions, shiftwork. (Fondazione Carlo Erba, Milano).
Sanders, A.F. 1979. Some remarks in mental load. In "Mental workload" (Ed. N.Moray). (Plenum Press, New York). pp. 41-77.
Schmale, A.H. and Ader, R. (Eds.) 1980. The challenge of the biopsychologi-

cal model. Psychosom.Med.,42,Supplement.
Selye, H. 1976. Forty years of stress research: principal remaining problems and misconception. Canad.Med.Ass.J., 115, 53-56.
Shagass, C. 1975. Electrophysiological parameters and methods. In "Emotion their parameters and measurement" (Ed. L. Levi). (Raven Press, New York). pp. 279-308.
Swisher, S.N. 1980. The biopsychosocial model: its future for the internist. In "The challenge of the biopsychological model" (Eds. A.H. Schmale and R. Ader), Psychosom.Med., 42, supplement, pp. 113-121.
Tasto, D.L. et al. 1978. Health consequences of shiftwork.(DHEW (NIOSH) Publication No. 78-154).
Torsvall, L. et al. 1981. Age, sleep and irregular workhours. Scand.J.Work Env.Health, 7, 196-203.
Vogt,J. et al. 1973. Motor thermal and sensory factors in heart rate variation. A methodology for indirect estimation of intermittent muscular work and environmental heat loads. Erg., 10, 45-60.
Weber, A. et al. 1980. Psychophysiological effects of repetitive tasks. Erg., 23, 1033-1047.
Welford, A.T. 1974. Stress and performance. In "Man under Stress" (Ed. A. T. Welford). (Taylor and Francis Ltd., London). pp. 1-14.
Yuwiler, A. 1976. Stress anxiety and endocrine function. In "Biological foundations of psychiatry" (Eds. R.G. Genell and S. Gabay). (Raven Press, New York). vol. II, pp. 889-943.